Student Solutions Manual

for

Stewart/Redlin/Watson's

College Algebra

Fourth Edition

John Banks

THOMSON

BROOKS/COLE

Australia • Canada • Mexico • Singapore • Spain • United Kingdom • United States

Printed in Canada

1 2 3 4 5 6 7   07 06 05 04

Printer: Webcom Limited

ISBN: 0-534-40601-7

For more information about our products, contact us at:
Thomson Learning Academic Resource Center
1-800-423-0563

For permission to use material from this text, contact us by:
Phone: 1-800-730-2214
Fax: 1-800-730-2215
Web: http://www.thomsonrights.com

Thomson Brooks/Cole
10 Davis Drive
Belmont, CA 94002-3098
USA

Asia
Thomson Learning
5 Shenton Way #01-01
UIC Building
Singapore 068808

Australia/New Zealand
Thomson Learning
102 Dodds Street
Southbank, Victoria 3006
Australia

Canada
Nelson
1120 Birchmount Road
Toronto, Ontario M1K 5G4
Canada

Europe/Middle East/South Africa
Thomson Learning
High Holborn House
50/51 Bedford Row
London WC1R 4LR
United Kingdom

Latin America
Thomson Learning
Seneca, 53
Colonia Polanco
11560 Mexico D.F.
Mexico

Spain/Portugal
Paraninfo
Calle/Magallanes, 25
28015 Madrid, Spain

Table of Contents

Chapter P

Exercises P.1

1. (a) $M = \dfrac{N}{G} = \dfrac{230}{5.4} = 42.6$ miles/gallon.

 (b) $25 = \dfrac{185}{G} \iff G = \dfrac{185}{25} = 7.4$ gallons.

3. (a) $P = 0.45d + 14.7 = 0.45(180) + 14.7 = 95.7$ lb/in^2.

 (b) $P = 0.45d + 14.7 = 0.45(330) + 14.7 = 163.2$ lb/in^2.

5. (a) $P = 0.06s^3 = 0.06(12^3) = 103.7$ hp.

 (b) $7.5 = 0.06s^3 \iff s^3 = 125$ so $s = 5$ knots.

7. $P = 0.8(1000) - 500 = \$300$.

9. The number N of days in w weeks is $N = 7w$.

11. The average A of two numbers, a and b, is $A = \dfrac{a+b}{2}$.

13. The sum S of two consecutive integers is $S = n + (n+1) = 2n + 1$, where n is the first integer.

15. The sum S of a number n and its square is $S = n + n^2$.

17. The product of two consecutive integers is $P = n(n+1) = n^2 + n$, where n is the smaller integer.

19. The sum S of an integer n and twice the integer $S = n + 2n = 3n$.

21. The time it takes an airplane to travel d miles at r miles per hour is $t = \dfrac{d}{r}$.

23. The area A of a square of side x is $A = x^2$.

25. The length is $4 + x$, so the perimeter P is $P = 2(4 + x) + 2x = 8 + 2x + 2x = 8 + 4x$.

27. The volume V of a cube of side x is $V = x^3$.

29. The surface area A of a cube of side x is the sum of the areas of the 6 sides, each with area x^2; so $A = 6 \cdot x^2 = 6x^2$.

31. The length of one side of the square is $2r$, so its area is $(2r)^2 = 4r^2$. Since the area of the circle is πr^2, the area of what remains is $A = 4r^2 - \pi r^2 = (4 - \pi)r^2$.

33. The race track consists of a rectangle of length x and width $2r$ and two semicircles of radius r. The area of the rectangle is $2rx$ and the area of a circle of radius r is πr^2. So the area enclosed is $A = 2rx + \pi r^2$.

35. This is the volume of the large ball minus the volume of the inside ball. The volume of a ball (sphere) is $\frac{4}{3}\pi(\text{radius})^3$. Thus the volume is $V = \frac{4}{3}\pi R^3 - \frac{4}{3}\pi r^3 = \frac{4}{3}\pi(R^3 - r^3)$.

37. (a) $\$7.95 + 2(\$1.25) = \$7.95 + \$3.75 = \$11.70$.

 (b) The cost C , in dollars, of a pizza with n toppings is $C = 7.95 + 1.25n$.

 (c) Using the model $C = 7.95 + 1.25$ with $C = 14.20$ we get $14.20 = 7.95 + 1.25n$ $\Leftrightarrow$ $1.25n = 6.25$ $\Leftrightarrow$ $n = 5$. So the pizza has 5 toppings.

39. (a) $4\dfrac{\text{ft}}{\text{sec}} \cdot 20 \text{ sec} = 80 \text{ ft}$

 (b) $d = 4\dfrac{\text{ft}}{\text{sec}} \cdot t \text{ sec} = 4t \text{ ft}$

 (c) Since 1 mile $= 5280$ ft,solve $d = 4t = 5280$ for t. We have $4t = 5280$ $\Leftrightarrow$ $t = \frac{5280}{4} = 1320 \text{ sec} = 22$ minutes.

41. (a) If $width = 20$, then $length = 40$, so the $volume = 20 \cdot 20 \cdot 40 = 16{,}000 \text{ in}^3$.

 (b) In terms of width $V = x \cdot x \cdot 2x = 2x^3$.

 (c) Solve $V = 2x^3 = 6750$ for x. We have $2x^3 = 6750$ $\Leftrightarrow$ $x^3 = 3375$ $\Leftrightarrow$ $x = 15$. So $width = 15$ in. and $length = 2(15) = 30$ in. Thus the dimensions are 15 in. $\times$ 15 in. $\times$ 30 in.

43. (a) $\text{GPA} = \dfrac{4a + 3b + 2c + 1d + 0f}{a + b + c + d + f} = \dfrac{4a + 3b + 2c + d}{a + b + c + d + f}$.

 (b) Using $a = 2 \cdot 3 = 6$, $b = 4$, $c = 3 \cdot 3 = 9$, and $d = f = 0$ in the formula from part (a), we obtain $\text{GPA} = \dfrac{4 \cdot 6 + 3 \cdot 4 + 2 \cdot 9}{6 + 4 + 9} = \dfrac{54}{19} = 2.84$.

Exercises P.2

1. (a) natural number 50

   (b) integers $0, -10, 50$

   (c) rational numbers $0, -10, 50, \frac{22}{7}, 0.538, 1.2\overline{3}, -\frac{1}{3}$

   (d) irrational numbers $\sqrt{7}, \sqrt[3]{2}$

3. Commutative Property for addition.   5. Associative Property for addition.

7. Distributive Property.   9. Commutative Property for multiplication.

11. $x + 3 = 3 + x$ 13. $4(A + B) = 4A + 4B$

15. $3(x + y) = 3x + 3y$ 17. $4(2m) = (4 \cdot 2)m = 8m$

19. $-\dfrac{5}{2}(2x - 4y) = -\dfrac{5}{2}(2x) + \dfrac{5}{2}(4y) = -5x + 10y$

21. (a) $\dfrac{3}{10} + \dfrac{4}{15} = \dfrac{9}{30} + \dfrac{8}{30} = \dfrac{17}{30}$ (b) $\dfrac{1}{4} + \dfrac{1}{5} = \dfrac{5}{20} + \dfrac{4}{20} = \dfrac{9}{20}$

23. (a) $\frac{2}{3}(6 - \frac{3}{2}) = \frac{2}{3} \cdot 6 - \frac{2}{3} \cdot \frac{3}{2} = 4 - 1 = 3$

    (b) $0.25(\frac{8}{9} + \frac{1}{2}) = \frac{1}{4}(\frac{16}{18} + \frac{9}{18}) = \frac{1}{4} \cdot \frac{25}{18} = \frac{25}{72}$

25. (a) $\dfrac{2}{\frac{2}{3}} - \dfrac{\frac{2}{3}}{2} = 2 \cdot \frac{3}{2} - \frac{2}{3} \cdot \frac{1}{2} = 3 - \frac{1}{3} = \frac{9}{3} - \frac{1}{3} = \frac{8}{3}.$

    (b) $\dfrac{\frac{1}{12}}{\frac{1}{8} - \frac{1}{9}} = \dfrac{\frac{1}{12}}{\frac{1}{8} - \frac{1}{9}} \cdot \dfrac{72}{72} = \dfrac{6}{9-8} = \dfrac{6}{1} = 6$

27. (a) Since $2 \cdot 3 = 6$ and $2 \cdot \frac{7}{2} = 7$ so $3 < \frac{7}{2}$

    (b) $-6 > -7$

    (c) $3.5 = \frac{7}{2}$

29. (a) False   (b) True

31. (a) False   (b) True

33. (a) $x > 0$ (b) $t < 4$

    (c) $a \geq \pi$ (d) $-5 < x < \frac{1}{3}$

    (e) $|p - 3| \leq 5$

35. (a) $A \cup B = \{1, 2, 3, 4, 5, 6, 7, 8\}$ (b) $A \cap B = \{2, 4, 6\}$

37. (a) $A \cup C = \{1, 2, 3, 4, 5, 6, 7, 8, 9, 10\}$ (b) $A \cap C = \{7\}$

39. (a) $B \cup C = \{x \mid x \le 5\}$ 　　　　　　　(b) $B \cap C = \{x \mid -1 < x < 4\}$

41. $(-3, 0) = \{x \mid -3 < x < 0\}$

43. $[2, 8) = \{x \mid 2 \le x < 8\}$

45. $[2, \infty) = \{x \mid x \ge 2\}$

47. $x \le 1 \quad \Leftrightarrow \quad x \in (-\infty, 1]$

49. $-2 < x \le 1 \quad \Leftrightarrow \quad x \in (-2, 1]$

51. $x > -1 \quad \Leftrightarrow \quad x \in (-1, \infty)$

53. (a) $[-3, 5]$ 　　　　　　　　　　(b) $(-3, 5]$

55. $(-2, 0) \cup (-1, 1) = (-2, 1)$

57. $[-4, 6] \cap [0, 8) = [0, 6]$

59. $(-\infty, -4) \cup (4, \infty)$

61. (a) $|100| = 100$ 　　　　　　　　(b) $|-73| = 73$

63. (a) $\left| |-6| - |-4| \right| = |6 - 4| = |2| = 2$ 　　(b) $\dfrac{-1}{|-1|} = \dfrac{-1}{1} = -1$

65. (a) $|(-2) \cdot 6| = |-12| = 12$ 　　　　(b) $\left| \left(-\frac{1}{3}\right)(-15) \right| = |5| = 5$

67. $|(-2) - 3| = |-5| = 5$

69. (a) $|17 - 2| = 15$ 　　　　　　　(b) $|21 - (-3)| = |21 + 3| = |24| = 24$

　　　(c) $\left| -\frac{3}{10} - \frac{11}{8} \right| = \left| -\frac{12}{40} - \frac{55}{40} \right| = \left| -\frac{67}{40} \right| = \frac{67}{40}$

71. (a) Let $x = 0.777\ldots$. So,

$$10x = 7.7777\ldots$$
$$\underline{x = 0.7777\ldots}$$
$$9x = 7 \qquad \text{Thus } x = \tfrac{7}{9}.$$

　　(b) Let $x = 0.2888\ldots$. So,

$$100x = 28.8888\ldots$$
$$\underline{10x = 2.8888\ldots}$$
$$90x = 26 \qquad \text{Thus } x = \tfrac{26}{90} = \tfrac{13}{45}.$$

　　(c) Let $x = 0.575757\ldots$. So,

$$100x = 57.5757\ldots$$
$$\underline{x = 0.5757\ldots}$$
$$99x = 57 \qquad \text{Thus } x = \tfrac{57}{99} = \tfrac{19}{33}.$$

73. Distributive Property

75. (a) Is $\frac{1}{28}x + \frac{1}{34}y \le 15$ when $x = 165$ and $y = 230$? We have
$\frac{1}{28}(165) + \frac{1}{34}(230) = 5.89 + 6.76 = 12.65 \le 15$, so yes the car can travel 165 city miles and 230 highway miles without running out of gas.

(b) Here we must solve for y when $x = 280$. So $\frac{1}{28}(280) + \frac{1}{34}y = 15$ $\Leftrightarrow$ $10 + \frac{1}{34}y = 15$
$\Leftrightarrow$ $\frac{1}{34}y = 5$ $\Leftrightarrow$ $y = 170$. Thus the car can travel up to 170 highway miles without
running out of gas.

77. (a) Negative since $a > 0$ $\Leftrightarrow$ $-a < 0$.   (b) Positive since $b < 0$ $\Leftrightarrow$ $-b > 0$.

(c) Positive since the product of two negative numbers is positive.

(d) Positive since $a - b = a + (-b)$ is the sum of two positive numbers.

(e) Negative $c - a = c + (-a)$ is the sum of two negative numbers.

(f) Positive since a is positive and bc is positive by part (c).

(g) Negative, since ab is negative (the product of a positive and a negative) and ac is negative for
the same reason, finally the sum of two negative numbers is negative.

(h) Negative, since bc is positive by Part (c) and $a(bc)$ is the product of two positive number,
therefore positive, hence $-abc$ is negative.

79. $\frac{1}{2} + \sqrt{2}$ is irrational. Since the sum of two rational numbers is rational (see Exercise 58), if $\frac{1}{2} + \sqrt{2}$
was rational, then the $\left(\frac{1}{2} + \sqrt{2}\right) + \left(-\frac{1}{2}\right) = \sqrt{2}$ is rational. But this is a contradiction. In general
the sum of a rational number and an irrational number is irrational.
Zero is rational and the product of zero and any real number is zero.
If the rational number is not zero, $\frac{a}{b} \neq 0$, then the product of it and any irrational number, w, is
irrational. Again, if we assume the product is rational, say $\frac{a}{b} \cdot w = \frac{c}{d}$, then since $\frac{a}{b} \neq 0$, we have
$\frac{b}{a} \neq 0$ so $\frac{b}{a}\left(\frac{a}{b} \cdot w\right) = \frac{b}{a} \cdot \frac{c}{d}$ $\Leftrightarrow$ $w = \frac{bc}{ad}$ which is rational. But w is irrational.

81. (a) Construct the number $\sqrt{2}$ on the number
line by transferring the length of the
hypotenuse of a right triangle with legs
of length 1 and 1.

(b) Construct a right triangle with legs of length 1
and 2. By the Pythagorean Theorem, the
length of the hypotenuse is $\sqrt{1^2 + 2^2} = \sqrt{5}$.
Then transfer the length of the hypotenuse to
the number line.

(c) Construct a right triangle with legs of length
$\sqrt{2}$ and 2, (construct $\sqrt{2}$ as in part (a)). By
the Pythagorean Theorem, the length of the
hypotenuse is $\sqrt{(\sqrt{2})^2 + 2^2} = \sqrt{6}$. Then
transfer the length of the hypotenuse to the
number line.

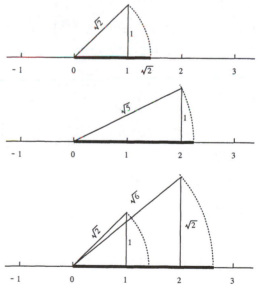

Exercises P.3

1. $5^2 \cdot 5 = 5^3 = 125$

3. $(2^3)^2 = 2^6 = 64$

5. $(-6)^0 = 1$

7. $-3^2 = -9$

9. $(\frac{1}{3})^4 \cdot 3^6 = \frac{1}{3^4} \cdot 3^6 = 3^{-4} \cdot 3^6 = 3^2 = 9$

11. $\dfrac{10^7}{10^4} = 10^{7-4} = 10^3 = 1{,}000$

13. $\dfrac{4^{-3}}{2^{-8}} = \dfrac{(2^2)^{-3}}{2^{-8}} = \dfrac{2^{-6}}{2^{-8}} = 2^{-6-(-8)} = 2^{-6+8} = 2^2 = 4$

15. $(\frac{1}{4})^{-2} = (4)^2 = 16$

17. $(\frac{3}{2})^{-2} \cdot \frac{9}{16} = (\frac{2}{3})^2 \cdot \frac{9}{16} = \frac{2^2}{3^2} \cdot \frac{9}{16} = \frac{4}{9} \cdot \frac{9}{16} = \frac{1}{4}$

19. $(\frac{1}{13})^0 (\frac{2}{3})^6 (\frac{4}{9})^{-3} = 1 \cdot \left(\frac{2^6}{3^6}\right) \left(\frac{9}{4}\right)^3 = \left(\frac{2^6}{3^6}\right) \left(\frac{3^2}{2^2}\right)^3 = \left(\frac{2^6}{3^6}\right) \left(\frac{3^6}{2^6}\right) = 1$

21. $2^{-2} + 2^{-3} = \frac{1}{2^2} + \frac{1}{2^3} = \frac{1}{4} + \frac{1}{8} = \frac{2}{8} + \frac{1}{8} = \frac{3}{8}$

23. $(2x)^4 x^3 = 2^4 x^4 x^3 = 16x^{4+3} = 16x^7$

25. $(-3y)^4 = (-3)^4 y^4 = 3^4 y^4 = 81y^4$

27. $(3z)^2 (6z^2)^{-3} = (3^2 z^2)(6^{-3} z^{2(-3)}) = (3^2 z^2)(6^{-3} z^{-6}) = 3^2 (2 \cdot 3)^{-3} z^{2-6} = 3^2 (2^{-3} \cdot 3^{-3}) z^{2-6}$

     $= 3^{2-3} 2^{-3} z^{-4} = 3^{-1} 2^{-3} z^{-4} = \dfrac{1}{3 \cdot 2^3 z^4} = \dfrac{1}{24z^4}$

29. $\dfrac{5x^2}{25x^5} = \dfrac{5}{25} \cdot \dfrac{x^2}{x^5} = \dfrac{1}{5} \cdot \dfrac{1}{x^{5-2}} = \dfrac{1}{5} \cdot \dfrac{1}{x^3} = \dfrac{1}{5x^3}$

31. $\dfrac{10(x+y)^4}{5(x+y)^3} = \dfrac{10}{5} \cdot \dfrac{(x+y)^4}{(x+y)^3} = 2 \cdot (x+y)^{4-3} = 2(x+y) = 2x + 2y$

33. $a^9 a^{-5} = a^{9-5} = a^4$

35. $(12x^2 y^4)(\frac{1}{2}x^5 y) = (12 \cdot \frac{1}{2}) x^{2+5} y^{4+1} = 6x^7 y^5$

37. $\dfrac{x^9 (2x)^4}{x^3} = 2^4 \cdot x^{9+4-3} = 16x^{10}$

39. $b^4 \left(\dfrac{1}{3} b^2\right) (12b^{-8}) = \dfrac{12}{3} b^{4+2-8} = 4b^{-2} = \dfrac{4}{b^2}$

41. $(rs)^3 (2s)^{-2} (4r)^4 = r^3 s^3 2^{-2} s^{-2} 4^4 r^4 = r^3 s^3 2^{-2} s^{-2} 2^{2 \cdot 4} r^4 = 2^{-2+8} r^{3+4} s^{3-2} = 2^6 r^7 s = 64r^7 s$

43. $\dfrac{(6y^3)^4}{2y^5} = \dfrac{6^4 y^{3 \cdot 4}}{2y^5} = \dfrac{6^4}{2} y^{12-5} = 648y^7$

45. $\dfrac{(x^2y^3)^4(xy^4)^{-3}}{x^2y} = \dfrac{x^8y^{12}x^{-3}y^{-12}}{x^2y} = x^{8-3-2}y^{12-12-1} = x^3y^{-1} = \dfrac{x^3}{y}$

47. $\dfrac{(xy^2z^3)^4}{(x^3y^2z)^3} = \dfrac{x^4y^8z^{12}}{x^9y^6z^3} = x^{4-9}y^{8-6}z^{12-3} = x^{-5}y^2z^9 = \dfrac{y^2z^9}{x^5}$

49. $\left(\dfrac{q^{-1}rs^{-2}}{r^{-5}sq^{-8}}\right)^{-1} = \dfrac{qr^{-1}s^2}{r^5s^{-1}q^8} = q^{1-8}r^{-1-5}s^{2-(-1)} = q^{-7}r^{-6}s^3 = \dfrac{s^3}{q^7r^6}$

51. $69{,}300{,}000 = 6.93 \times 10^7$

53. $0.000028536 = 2.8536 \times 10^{-5}$

55. $129{,}540{,}000 = 1.2954 \times 10^8$

57. $0.0000000014 = 1.4 \times 10^{-9}$

59. $3.19 \times 10^5 = 319{,}000$

61. $2.670 \times 10^{-8} = 0.00000002670$

63. $7.1 \times 10^{14} = 710{,}000{,}000{,}000{,}000$

65. $8.55 \times 10^{-3} = 0.00855$

67. (a) $5{,}900{,}000{,}000{,}000 \text{ mi} = 5.9 \times 10^{12} \text{ mi}$ (b) $0.0000000000004 \text{ cm} = 4 \times 10^{-13} \text{ cm}$

    (c) 33 billion billion molecules $= 33 \times 10^9 \times 10^9 = 3.3 \times 10^{19}$ molecules

69. $(7.2 \times 10^{-9})(1.806 \times 10^{-12}) = 7.2 \times 1.806 \times 10^{-9} \times 10^{-12} \approx 13.0 \times 10^{-21} = 1.3 \times 10^{-20}$

71. $\dfrac{1.295643 \times 10^9}{(3.610 \times 10^{-17})(2.511 \times 10^6)} = \dfrac{1.295643}{3.610 \times 2.511} \times 10^{9+17-6} \approx 0.1429 \times 10^{19} = 1.429 \times 10^{19}$

73. $\dfrac{(0.0000162)(0.01582)}{(594621000)(0.0058)} = \dfrac{(1.62 \times 10^{-5})(1.582 \times 10^{-2})}{(5.94621 \times 10^8)(5.8 \times 10^{-3})} = \dfrac{1.62 \times 1.582}{5.94621 \times 5.8} \times 10^{-5-2-8+3}$

    $0.074 \times 10^{-12} = 7.4 \times 10^{-14}$

75. (a) b^5 is negative since a negative number raised to an odd power is negative.

    (b) b^{10} is positive since a negative number raised to an even power is positive.

    (c) ab^2c^3 we have $(positive)(negative)^2(negative)^3 = (positive)(positive)(negative)$ which is negative.

    (d) Since $b - a$ is negative, $(b - a)^3 = (negative)^3$ which is negative.

    (e) Since $b - a$ is negative, $(b - a)^4 = (negative)^4$ which is positive.

    (f) $\dfrac{a^3c^3}{b^6c^6} = \dfrac{(positive)^3(negative)^3}{(negative)^6(negative)^6} = \dfrac{(positive)(negative)}{(positive)(positive)} = \dfrac{negative}{positive}$ which is negative.

77. Since one light year is 5.9×10^{12} mile, Centauri is about $4.3 \times 5.9 \times 10^{12} \approx 2.54 \times 10^{13}$ miles or $25{,}400{,}000{,}000{,}000$ miles.

79. Volume $= (average\ depth)(area) = (3.7 \times 10^3 \text{ m})(3.6 \times 10^{14} \text{ m}^2)\left(\frac{10^3 \text{ liters}}{\text{m}^3}\right) \approx 1.33 \times 10^{21}$ liters

81. $\begin{aligned} \text{\textit{number of}} \atop \text{\textit{molecules}} &= (\textit{volume}) \cdot \left(\frac{\textit{liters}}{m^3} \right) \cdot \left(\frac{\textit{molecules}}{\text{per } 22.4 \textit{ liters}} \right) \\ &= (5 \cdot 10 \cdot 3) \cdot \left(10^3 \right) \cdot \left(\frac{6.02 \times 10^{23}}{22.4} \right) \approx 4.03 \times 10^{27} \end{aligned}$

83.

Year	Total interest
1	\$152.08
2	308.79
3	470.26
4	636.64
5	808.08

85.   (a) $\dfrac{18^5}{9^5} = \left(\dfrac{18}{9} \right)^5 = 2^5 = 32$

      (b) $20^6 \cdot (0.5)^6 = (20 \cdot 0.5)^6 = 10^6 = 1{,}000{,}000$

Exercises P.4

1. $\dfrac{1}{\sqrt{5}} = 5^{-1/2}$

3. $4^{2/3} = \sqrt[3]{4^2} = \sqrt[3]{16}$

5. $\sqrt[5]{5^3} = 5^{3/5}$

7. $a^{2/5} = \sqrt[5]{a^2}$

9. (a) $\sqrt{16} = \sqrt{4^2} = 4$

    (b) $\sqrt[4]{16} = \sqrt[4]{2^4} = 2$

    (c) $\sqrt[4]{\dfrac{1}{16}} = \sqrt[4]{\left(\dfrac{1}{2}\right)^4} = \dfrac{1}{2}$

11. (a) $\sqrt{\tfrac{4}{9}} = \sqrt{\left(\tfrac{2}{3}\right)^2} = \tfrac{2}{3}$

    (b) $\sqrt[4]{256} = \sqrt[4]{4^4} = 4$

    (c) $\sqrt[6]{\dfrac{1}{64}} = \sqrt[6]{\dfrac{1}{2^6}} = \dfrac{\sqrt[6]{1}}{\sqrt[6]{2^6}} = \dfrac{1}{2}$

13. (a) $\left(\dfrac{4}{9}\right)^{-1/2} = \left(\dfrac{2^2}{3^2}\right)^{-1/2} = \dfrac{2^{-1}}{3^{-1}} = \dfrac{3}{2}$

    (b) $(-32)^{2/5} = \left[(-2)^5\right]^{2/5} = (-2)^2 = 4$

    (c) $(-125)^{-1/3} = \left[(-5)^3\right]^{-1/3} = (-5)^{-1} = \dfrac{1}{-5} = -\dfrac{1}{5}$

15. (a) $\left(\dfrac{1}{32}\right)^{2/5} = \left(\dfrac{1}{2^5}\right)^{2/5} = \left(\dfrac{1}{2}\right)^{5\cdot(2/5)} = \left(\dfrac{1}{2}\right)^2 = \dfrac{1}{4}$

    (b) $(27)^{-4/3} = (3^3)^{-4/3} = (3)^{3\cdot(-4/3)} = 3^{-4} = \dfrac{1}{3^4} = \dfrac{1}{81}$

    (c) $\left(\dfrac{1}{8}\right)^{-2/3} = \left(\dfrac{1}{2^3}\right)^{-2/3} = (2^{-3})^{-2/3} = 2^{(-3)\cdot(-2/3)} = 2^2 = 4$

17. (a) $100^{-1.5} = (10^2)^{-1.5} = 10^{2\cdot(-1.5)} = 10^{-3} = \dfrac{1}{10^3} = \dfrac{1}{1000}$

    (b) $4^{2/3} \cdot 6^{2/3} \cdot 9^{2/3} = (2^2)^{2/3} \cdot (2\cdot 3)^{2/3} \cdot (3^2)^{2/3} = 2^{4/3} \cdot 2^{2/3} \cdot 3^{2/3} \cdot 3^{4/3} = 2^2 \cdot 3^2 = 4\cdot 9 = 36$

    (c) $0.001^{-2/3} = (10^{-3})^{-2/3} = 10^2 = 100$

19. When $x = 3$, $y = 4$, $z = -1$ we have $\sqrt{x^2 + y^2} = \sqrt{3^2 + 4^2} = \sqrt{9 + 16} = \sqrt{25} = 5$.

21. When $x = 3$, $y = 4$, $z = -1$ we have
$$(9x)^{2/3} + (2y)^{2/3} + z^{2/3} = (9\cdot 3)^{2/3} + (2\cdot 4)^{2/3} + (-1)^{2/3} = (3^3)^{2/3} + (2^3)^{2/3} + (1)^{1/3}$$
$$= 3^2 + 2^2 + 1 = 9 + 4 + 1 = 14.$$

23. $\sqrt{32} + \sqrt{18} = \sqrt{16\cdot 2} + \sqrt{9\cdot 2} = \sqrt{4^2\cdot 2} + \sqrt{3^2\cdot 2} = 4\sqrt{2} + 3\sqrt{2} = 7\sqrt{2}$

25. $\sqrt{125} - \sqrt{45} = \sqrt{25 \cdot 5} - \sqrt{9 \cdot 5} = \sqrt{5^2 \cdot 5} - \sqrt{3^2 \cdot 5} = 5\sqrt{5} - 3\sqrt{5} = 2\sqrt{5}$

27. $\sqrt[3]{108} - \sqrt[3]{32} = 3\sqrt[3]{4} - 2\sqrt[3]{4} = \sqrt[3]{4}$ 29. $\sqrt{245} - \sqrt{125} = 7\sqrt{5} - 5\sqrt{5} = 2\sqrt{5}$

31. $\sqrt[5]{96} + \sqrt[5]{3} = \sqrt[5]{32 \cdot 3} + \sqrt[5]{3} = \sqrt[5]{2^5 \cdot 3} + \sqrt[5]{3} = 2\sqrt[5]{3} + \sqrt[5]{3} = 3\sqrt[5]{3}$

33. $\sqrt[4]{x^4} = |x|$ 35. $\sqrt[4]{16x} = \sqrt[4]{2^4 x^8} = 2x^2$

37. $\sqrt[3]{x^3 y} = (x^3)^{1/3} y^{1/3} = x\sqrt[3]{y}$

39. $\sqrt[5]{a^6 b^7} = a^{6/5} b^{7/5} = a \cdot a^{1/5} b \cdot b^{2/5} = ab\sqrt[5]{ab^2}$

41. $\sqrt[3]{\sqrt{64x^6}} = (8|x^3|)^{1/3} = 2|x|$

43. $x^{2/3} x^{1/5} = x^{(10/15 + 3/15)} = x^{13/15}$

45. $(-3a^{1/4})(9a)^{-3/2} = (-1 \cdot 3a^{1/4})(3^2 a)^{-3/2} = (-1 \cdot 3a^{1/4})(3^{-3} a^{-3/2}) = -1 \cdot 3^{1-3} \cdot a^{1/4 - 3/2}$

     $= -1 \cdot 3^{-3} a^{-5/4} = \dfrac{-1}{3^3 \cdot a^{5/4}} = \dfrac{-1}{9a^{5/4}}$

47. $(4b)^{1/2}(8b^{2/5}) = \sqrt{4} \cdot 8b^{1/2} b^{2/5} = 16b^{(5/10 + 4/10)} = 16b^{9/10}$

49. $(c^2 d^3)^{-1/3} = c^{-2/3} d^{-1} = \dfrac{1}{c^{2/3} d}$ 51. $(y^{3/4})^{2/3} = y^{(3/4) \cdot (2/3)} = y^{1/2}$

53. $(2x^4 y^{-4/5})^3 (8y^2)^{2/3} = 2^3 x^{12} y^{-12/5} 8^{2/3} y^{4/3} = 2^{3+2} x^{12} y^{(-12/5 + 4/3)} = \dfrac{32x^{12}}{y^{16/15}}$ (Note that

     $8^{2/3} = (8^{1/3})^2 = 2^2$.)

55. $\left(\dfrac{x^6 y}{y^4}\right)^{5/2} = \dfrac{x^{15} y^{5/2}}{y^{10}} = x^{15} y^{5/2 - 10} = x^{15} y^{-15/2} = \dfrac{x^{15}}{y^{15/2}}$

57. $\left(\dfrac{3a^{-2}}{4b^{-1/3}}\right)^{-1} = \dfrac{3^{-1} a^2}{4^{-1} b^{1/3}} = \dfrac{4a^2}{3b^{1/3}}$

59. $\dfrac{(9st)^{3/2}}{(27s^3 t^{-4})^{2/3}} = \dfrac{27s^{3/2} t^{3/2}}{9s^2 t^{-8/3}} = 3s^{3/2 - 2} t^{3/2 + 8/3} = 3s^{-1/2} t^{25/6} = \dfrac{3t^{25/6}}{s^{1/2}}$

61. (a) $\dfrac{1}{\sqrt{6}} = \dfrac{1}{\sqrt{6}} \cdot \dfrac{\sqrt{6}}{\sqrt{6}} = \dfrac{\sqrt{6}}{6}$ (b) $\dfrac{3}{\sqrt{2}} = \dfrac{3}{\sqrt{2}} \cdot \dfrac{\sqrt{2}}{\sqrt{2}} = \dfrac{3\sqrt{2}}{2}$

     (c) $\dfrac{9}{\sqrt{3}} = \dfrac{9}{\sqrt{3}} \cdot \dfrac{\sqrt{3}}{\sqrt{3}} = \dfrac{9\sqrt{3}}{3} = 3\sqrt{3}$

63. (a) $\dfrac{1}{\sqrt[3]{4}} = \dfrac{1}{\sqrt[3]{2^2}} \cdot \dfrac{\sqrt[3]{2}}{\sqrt[3]{2}} = \dfrac{\sqrt[3]{2}}{2}$

(b) $\dfrac{\cancel{2}^{l}}{\sqrt[4]{3}} = \dfrac{\cancel{2}^{l}}{\sqrt[4]{3}} \cdot \dfrac{\sqrt[4]{3^3}}{\sqrt[4]{3^3}} = \dfrac{\cancel{2}\sqrt[4]{3^3}}{3} = \dfrac{\cancel{2}\sqrt[4]{27}}{3}$

(c) $\dfrac{8}{\sqrt[5]{2}} = \dfrac{8}{\sqrt[5]{2}} \cdot \dfrac{\sqrt[5]{2^4}}{\sqrt[5]{2^4}} = \dfrac{8\sqrt[5]{2^4}}{2} = 4\sqrt[5]{2^4} = 4\sqrt[5]{16}$

65. (a) $\dfrac{1}{\sqrt[3]{x}} = \dfrac{1}{\sqrt[3]{x}} \cdot \dfrac{\sqrt[3]{x^2}}{\sqrt[3]{x^2}} = \dfrac{\sqrt[3]{x^2}}{x}$

(b) $\dfrac{1}{\sqrt[5]{x^2}} = \dfrac{1}{\sqrt[5]{x^2}} \cdot \dfrac{\sqrt[5]{x^3}}{\sqrt[5]{x^3}} = \dfrac{\sqrt[5]{x^3}}{x}$

(c) $\dfrac{1}{\sqrt[7]{x^3}} = \dfrac{1}{\sqrt[7]{x^3}} \cdot \dfrac{\sqrt[7]{x^4}}{\sqrt[7]{x^4}} = \dfrac{\sqrt[7]{x^4}}{x}$

67. First convert 1135 feet to miles. This gives $1135 \text{ ft} = 1135 \cdot \frac{1 \text{ mile}}{5280 \text{ feet}} = 0.215 \text{ mi}$. Thus the distance you can see is given by $D = \sqrt{2rh + h^2} = \sqrt{2(3960)(.215) + (.215)^2} = \sqrt{1702.8} = 41.3$ miles.

69. (a) Substituting we get $0.30(60) + 0.38(3400)^{1/2} - 3(650)^{1/3} \approx 18 + 0.38(58.31) - 3(8.66) \approx 18 + 22.16 - 25.98 \approx 14.18$. Since this value is less than 16, the sail boat qualifies for the race.

(b) Solve for A when $L = 65$ and $V = 600$. Substituting we get:
$0.30(65) + 0.38A^{1/2} - 3(600)^{1/3} \le 16 \quad \Leftrightarrow \quad 19.5 + 0.38A^{1/2} - 25.30 \le 16 \quad \Leftrightarrow$
$0.38A^{1/2} - 5.80 \le 16 \quad \Leftrightarrow \quad 0.38A^{1/2} \le 21.80 \quad \Leftrightarrow \quad A^{1/2} \le 57.38 \quad \Leftrightarrow \quad A \le 3292.0$.
Thus the largest possible sail is 3292 ft^2.

71. Since 1 day = 86,400 sec, 365.25 days = 31,557,600 sec. Substituting we obtain
$$d = \left(\dfrac{6.67 \times 10^{-11} \times 1.99 \times 10^{30}}{4\pi^2} \right)^{1/3} \cdot \left(3.15576 \times 10^7\right)^{2/3} \approx 1.5 \times 10^{11} \text{m} = 1.5 \times 10^8 \text{ km}.$$

73. (a)

n	1	2	5	10	100
$2^{1/n}$	$2^{1/1} = 2$	$2^{1/2} = 1.414$	$2^{1/5} = 1.149$	$2^{1/10} = 1.072$	$2^{1/100} = 1.007$

So when n gets large, $2^{1/n}$ decreases to 1.

(b)

n	1	2	5	10
$\left(\frac{1}{2}\right)^{1/n}$	$\left(\frac{1}{2}\right)^{1/1} = 0.5$	$\left(\frac{1}{2}\right)^{1/2} = 0.707$	$\left(\frac{1}{2}\right)^{1/5} = 0.871$	$\left(\frac{1}{2}\right)^{1/10} = 0.933$

100
$\left(\frac{1}{2}\right)^{1/100} = 0.993$

So when n gets large, $\left(\frac{1}{2}\right)^{1/n}$ increases to 1.

75. Using $v = \dfrac{1}{10}c$ in the formula $m = \dfrac{m_0}{\sqrt{1 - \dfrac{v^2}{c^2}}}$, we get: $m = \dfrac{m_0}{\sqrt{1 - \dfrac{\left(\frac{1}{10}c\right)^2}{c^2}}} = \dfrac{m_0}{\sqrt{1 - \dfrac{1}{100}}}$. Thus

$m = \dfrac{1}{\sqrt{\dfrac{99}{100}}} m_0 = \dfrac{10\sqrt{11}}{33} m_0$. So the rest mass of the spaceship is multiplied by $\dfrac{10\sqrt{11}}{33} \approx 1.005$.

Using $v = \frac{1}{2}c$ in the formula, we get: $m = \dfrac{m_0}{\sqrt{1 - \frac{\left(\frac{1}{2}c\right)^2}{c^2}}} = \dfrac{m_0}{\sqrt{1 - \frac{1}{4}}}$. Thus

$m = \dfrac{1}{\sqrt{\frac{3}{4}}} m_0 = \frac{2\sqrt{3}}{3} m_0$. So the rest mass of the spaceship is multiplied by $\frac{2\sqrt{3}}{3} \approx 1.15$.

Using $v = 0.9c$ in the formula, we get: $m = \dfrac{m_0}{\sqrt{1 - \frac{(.9c)^2}{c^2}}} = \dfrac{m_0}{\sqrt{1 - .81}}$. Thus

$m = \dfrac{1}{\sqrt{0.19}} m_0 \approx 2.29 m_0$. So the rest mass of the spaceship is multiplied by $\dfrac{1}{\sqrt{0.19}} \approx 2.29$. As the

spaceship travels very close to the speed of light, $v \to c$, so the term $\dfrac{v^2}{c^2}$ approaches 1. So $\sqrt{1 - \dfrac{v^2}{c^2}}$

approaches 0, and we obtain $m = \dfrac{m_0}{\text{very small number}}$, which means $m = (\text{very large number})m_0$.

Hence the mass of the space ship becomes arbitrarily large as its speed approaches the speed of light.

The actual value of the speed of light does not affect the calculations.

Exercises P.5

1. Type: trinomial; Terms: x^2, $-3x$, and 7; Degree 2

3. Type: monomial; Terms: -8; Degree 0

5. Type: four-term polynomial; Terms: x, $-x^2$, x^3, and $-x^4$; Degree 4

7. Not a polynomial,

9. Polynomial, degree 3.

11. Not a polynomial.

13. $(12x - 7) - (5x - 12) = 12x - 7 - 5x + 12 = 7x + 5$

15. $(3x^2 + x + 1) + (2x^2 - 3x - 5) = 5x^2 - 2x - 4$

17. $(x^3 + 6x^2 - 4x + 7) - (3x^2 + 2x - 4) = x^3 + 6x^2 - 4x + 7 - 3x^2 - 2x + 4$
    $= x^3 + 3x^2 - 6x + 11$

19. $8(2x + 5) - 7(x - 9) = 16x + 40 - 7x + 63 = 9x + 103$

21. $2(2 - 5t) + t^2(t - 1) - (t^4 - 1) = 4 - 10t + t^3 - t^2 - t^4 + 1 = -t^4 + t^3 - t^2 - 10t + 5$

23. $x^2(2x^2 - x + 1) = x^2(2x^2) - x^2(x) + x^2(1) = 2x^4 - x^3 + x^2$

25. $\sqrt{x}(x - \sqrt{x}) = x^{1/2}(x - x^{1/2}) = x^{1/2}x - x^{1/2}x^{1/2} = x^{3/2} - x$

27. $y^{1/3}(y^2 - 1) = y^{1/3}y^2 - y^{1/3} = y^{7/3} - y^{1/3}$

29. $(3t - 2)(7t - 5) = 21t^2 - 15t - 14t + 10 = 21t^2 - 29t + 10$

31. $(x + 2y)(3x - y) = 3x^2 - xy + 6xy - 2y^2 = 3x^2 + 5xy - 2y^2$

33. $(1 - 2y)^2 = 1 - 4y + 4y^2$

35. $(2x^2 + 3y^2)^2 = (2x^2)^2 + 2(2x^2)(3y^2) + (3y^2)^2 = 4x^4 + 12x^2y^2 + 9y^4$

37. $(2x - 5)(x^2 - x + 1) = 2x^3 - 2x^2 + 2x - 5x^2 + 5x - 5 = 2x^3 - 7x^2 + 7x - 5$

39. $(x^2 - a^2)(x^2 + a^2) = (x^2)^2 - (a^2)^2 = x^4 - a^4$ (the difference of squares)

41. $\left(\sqrt{a} - \dfrac{1}{b}\right)\left(\sqrt{a} + \dfrac{1}{b}\right) = (\sqrt{a})^2 - \left(\dfrac{1}{b}\right)^2 = a - \dfrac{1}{b^2}$ (the difference of squares)

43. $(1 + a^3)^3 = 1 + 3(a^3) + 3(a^3)^2 + (a^3)^3 = 1 + 3a^3 + 3a^6 + a^9$ (perfect cube)

45. $(x^2 + x - 1)(2x^2 - x + 2) = x^2(2x^2 - x + 2) + x(2x^2 - x + 2) - (2x^2 - x + 2)$
     $= 2x^4 - x^3 + 2x^2 + 2x^3 - x^2 + 2x - 2x^2 + x - 2 = 2x^4 + x^3 - x^2 + 3x - 2$

47. $(x^2 + x - 2)(x^3 - x + 1) = x^5 - x^3 + x^2 + x^4 - x^2 + x - 2x^3 + 2x - 2$
     $= x^5 + x^4 - 3x^3 + 3x - 2$

49. $(1 + x^{4/3})(1 - x^{2/3}) = 1 - x^{2/3} + x^{4/3} - x^{6/3} = 1 - x^{2/3} + x^{4/3} - x^2$

51. $(3x^2y + 7xy^2)(x^2y^3 - 2y^2) = 3x^4y^4 - 6x^2y^3 + 7x^3y^5 - 14xy^4$
     $= 3x^4y^4 + 7x^3y^5 - 6x^2y^3 - 14xy^4$ (arranging in decreasing powers of x)

53. $(2x + y - 3)(2x + y + 3) = [(2x + y) - 3][(2x + y) + 3] = (2x + y)^2 - 3^2 = 4x^2 + 4xy + y^2 - 9$

55. $(x + y + z(x - y - z) = [x + (y + z)][x - (y + z)] = x^2 - (y + z)^2 = x^2 - (y^2 + 2yz + z^2)$
     $= x^2 - y^2 - 2yz - z^2$

57. (a) The height of the box is x; the width is $6 - 2x$; and the length is $10 - 2x$. Since
          Volume = height $\times$ width $\times$ length we have $V = x(6 - 2x)(10 - 2x)$.

     (b) $V = x(60 - 32x + 4x^2) = 60x - 32x^2 + 4x^3$; degree 3.

     (c) When $x = 1$ the volume is $V = 60(1) - 32(1^2) + 4(1^3) = 32$ and when $x = 2$ the volume is
          $V = 60(2) - 32(2^2) + 4(2^3) = 24$.

59. (a) $A = 2000(1 + r)^3 = 2000(1 + 3r + 3r^2 + r^3) = 2000 + 6000r + 6000r^2 + 2000r^3$; degree 3.

     (b) Remember that % means divide by 100, so 2% = 0.02.

Interest rate r	2%	3%	4.5%	6%	10%
Amount A	$2122.42	$2185.45	$2282.33	$2382.03	$2662.00

61   (a) When $x = 1$, $(x + 5)^2 = (1 + 5)^2 = 36$ and $x^2 + 25 = 1^2 + 25 = 26$.

     (b) $(x + 5)^2 = x^2 + 10x + 25$.

Exercises P.6

1. $5a - 20 = 5(a - 4)$

3. $-2x^3 + 16x = -2x(x^2 - 8)$

5. $y(y - 6) + 9(y - 6) = (y - 6)(y + 9)$

7. $2x^2y - 6xy^2 + 3xy = xy(2x - 6y + 3)$

9. $x^2 + 2x - 3 = (x - 1)(x + 3)$

11. $6 + y - y^2 = (3 - y)(2 + y)$

13. $8x^2 - 14x - 15 = (2x - 5)(4x + 3)$

15. $(3x + 2)^2 + 8(3x + 2) + 12 = [(3x + 2) + 2][(3x + 2) + 6] = (3x + 4)(3x + 8)$

17. $9a^2 - 16 = (3a)^2 - 4^2 = (3a - 4)(3a + 4)$

19. $27x^3 + y^3 = (3x)^3 + y^3 = (3x + y)[(3x)^2 + 3xy + y^2] = (3x + y)(9x^2 - 3xy + y^2)$

21. $8s^3 - 125t^3 = (2s)^3 - (5t)^3 = (2s - 5t)[(2s)^2 + (2s)(5t) + (5t)^2]$
     $= (2s - 5t)(4s^2 + 10st + 25t^2)$

23. $x^2 + 12x + 36 = x^2 + 2(6x) + 6^2 = (x + 6)^2$

25. $x^3 + 4x^2 + x + 4 = x^2(x + 4) + 1(x + 4) = (x + 4)(x^2 + 1)$

27. $2x^3 + x^2 - 6x - 3 = x^2(2x + 1) - 3(2x + 1) = (2x + 1)(x^2 - 3)$. If irrational coefficients are
     permitted, then this can be further factored as $(2x + 1)(x - \sqrt{3})(x - \sqrt{3})$.

29. $x^3 + x^2 + x + 1 = x^2(x + 1) + 1(x + 1) = (x + 1)(x^2 + 1)$

31. $12x^3 + 18x = 6x(2x^2 + 3)$

33. $6y^4 - 15y^3 = 3y^3(2y - 5)$

35. $x^2 - 2x - 8 = (x - 4)(x + 2)$

37. $y^2 - 8y + 15 = (y - 3)(y - 5)$

39. $2x^2 + 5x + 3 = (2x + 3)(x + 1)$

41. $9x^2 - 36x - 45 = 9(x^2 - 4x - 5) = 9(x - 5)(x + 1)$

43. $6x^2 - 5x - 6 = (3x + 2)(2x - 3)$

45. $4t^2 - 12t + 9 = (2t - 3)^2$

47. $r^2 - 6rs + 9s^2 = (r - 3s)^2$

49. $x^2 - 36 = (x - 6)(x + 6)$

51. $49 - 4y^2 = (7 - 2y)(7 + 2y)$

53. $(a + b)^2 - (a - b)^2 = [(a + b) - (a - b)][(a + b) + (a - b)] = (2b)(2a) = 4ab$

55. $x^2(x^2 - 1) - 9(x^2 - 1) = (x^2 - 1)(x^2 - 9) = (x - 1)(x + 1)(x - 3)(x + 3)$

57. $t^3 + 1 = (t + 1)(t^2 - t + 1)$

59. $8x^3 - 125 = (2x)^3 - 5^3 = (2x - 5)[(2x)^2 + (2x)(5) + 5^2] = (2x - 5)(4x^2 + 10x + 25)$

61. $x^6 - 8y^3 = (x^2)^3 - (2y)^3 = (x^2 - 2y)[(x^2)^2 + (x^2)(2y) + (2y)^2] = (x^2 - 2y)(x^4 + 2x^2y + 4y^2)$

63. $x^3 + 2x^2 + x = x(x^2 + 2x + 1) = x(x + 1)^2$

65. $x^4 + 2x^3 - 3x^2 = x^2(x^2 + 2x - 3) = x^2(x - 1)(x + 3)$

67. $y^3 - 3y^2 - 4y + 12 = (y^3 - 3y^2) + (-4y + 12) = y^2(y - 3) + (-4)(y - 3) = (y - 3)(y^2 - 4)$
     $= (y - 3)(y - 2)(y + 2)$ (factor by grouping)

69. $2x^3 + 4x^2 + x + 2 = (2x^3 + 4x^2) + (x + 2) = 2x^2(x + 2) + (1)(x + 2) = (x + 2)(2x^2 + 1)$
     (factor by grouping)

71. $(x - 1)(x + 2)^2 - (x - 1)^2(x + 2) = (x - 1)(x + 2)\Big[(x + 2) - (x - 1)\Big] = 3(x - 1)(x + 2)$

73. $y^4(y + 2)^3 + y^5(y + 2)^4 = y^4(y + 2)^3\Big[(1) + y(y + 2)\Big] = y^4(y + 2)^3(y^2 + 2y + 1)$
     $= y^4(y + 2)^3(y + 1)^2$

75. Start by factoring $y^2 - 7y + 10$, and then substitute $a^2 + 1$ for y. This gives
     $(a^2 + 1)^2 - 7(a^2 + 1) + 10 = \Big[(a^2 + 1) - 2\Big]\Big[(a^2 + 1) - 5\Big] = (a^2 - 1)(a^2 - 4)$
     $= (a - 1)(a + 1)(a - 2)(a + 2)$.

77. $x^{5/2} - x^{1/2} = x^{1/2}(x^2 - 1) = \sqrt{x}(x - 1)(x + 1)$

79. Start by factoring out the power of x with the smallest exponent, that is, $x^{-3/2}$. So
     $x^{-3/2} + 2x^{-1/2} + x^{1/2} = x^{-3/2}(1 + 2x + x^2) = \dfrac{(1 + x)^2}{x^{3/2}}$.

81. Start by factoring out the power of $(x^2 + 1)$ with the smallest exponent, that is, $(x^2 + 1)^{-1/2}$. So
     $(x^2 + 1)^{1/2} + 2(x^2 + 1)^{-1/2} = (x^2 + 1)^{-1/2}\Big[(x^2 + 1) + 2\Big] = \dfrac{x^2 + 3}{\sqrt{x^2 + 1}}$.

83. $2x^{1/3}(x - 2)^{2/3} - 5x^{4/3}(x - 2)^{-1/3} = x^{1/3}(x - 2)^{-1/3}[2(x - 2) - 5x]$
     $= x^{1/3}(x - 2)^{-1/3}(2x - 4 - 5x) = x^{1/3}(x - 2)^{-1/3}(-3x - 4) = \dfrac{(-3x - 4)\sqrt[3]{x}}{\sqrt[3]{x - 2}}$

85. $3x^2(4x - 12)^2 + x^3(2)(4x - 12)(4) = x^2(4x - 12)[3(4x - 12) + x(2)(4)]$
    $= 4x^2(x - 3)(12x - 36 + 8x) = 4x^2(x - 3)(20x - 36) = 16x^2(x - 3)(5x - 9)$

87. $3(2x - 1)^2(2)(x + 3)^{1/2} + (2x - 1)^3\left(\frac{1}{2}\right)(x + 3)^{-1/2}$
    $= (2x - 1)^2(x + 3)^{-1/2}\left[6(x + 3) + (2x - 1)\left(\frac{1}{2}\right)\right] = (2x - 1)^2(x + 3)^{-1/2}\left(6x + 18 + x - \frac{1}{2}\right)$
    $= (2x - 1)^2(x + 3)^{-1/2}\left(7x + \frac{35}{2}\right)$

89. $(x^2 + 3)^{-1/3} - \frac{2}{3}x^2(x^2 + 3)^{-4/3} = (x^2 + 3)^{-4/3}\left[(x^2 + 3) - \frac{2}{3}x^2\right] = (x^2 + 3)^{-4/3}\left(\frac{1}{3}x^2 + 3\right)$
    $= \dfrac{\frac{1}{3}x^2 + 3}{(x^2 + 3)^{4/3}}$

91. (a) $\frac{1}{2}[(a + b)^2 - (a^2 + b^2)] = \frac{1}{2}[a^2 + 2ab + b^2 - a^2 - b^2] = \frac{1}{2}(2ab) = ab.$

    (b) $(a^2 + b^2)^2 - (a^2 - b^2)^2 = [(a^2 + b^2) - (a^2 - b^2)][(a^2 + b^2) + (a^2 - b^2)]$
        $= (a^2 + b^2 - a^2 + b^2)(a^2 + b^2 + a^2 - b^2) = (2b^2)(2a^2) = 4a^2b^2$

    (c) LHS $= (a^2 + b^2)(c^2 + d^2) = a^2c^2 + a^2d^2 + b^2c^2 + b^2d^2$

        RHS $= (ac + bd)^2 + (ad - bc)^2 = a^2c^2 + 2abcd + b^2d^2 + a^2d^2 - 2abcd + b^2c^2$
        $= a^2c^2 + a^2d^2 + b^2c^2 + b^2d^2$

        So LHS = RHS, that is, $(a^2 + b^2)(c^2 + d^2) = (ac + bd)^2 + (ad - bc)^2.$

    (d) $4a^2c^2 - (c^2 - b^2 + a^2)^2 = (2ac)^2 - (c^2 - b^2 + a^2)$
        $= \left[(2ac) - (c^2 - b^2 + a^2)\right]\left[(2ac) + (c^2 - b^2 + a^2)\right]$ (the difference of squares)
        $= (2ac - c^2 + b^2 - a^2)(2ac + c^2 - b^2 + a^2)$
        $= \left[b^2 - (c^2 - 2ac + a^2)\right]\left[(c^2 + 2ac + a^2) - b^2\right]$ (regrouping)
        $= \left[b^2 - (c - a)^2\right]\left[(c + a)^2 - b^2\right]$ (perfect squares)
        $= [b - (c - a)][b + (c - a)][(c + a) - b][(c + a) + b]$ (each factor is a difference of squares)
        $= (b - c + a)(b + c - a)(c + a - b)(c + a + b)$
        $= (a + b - c)(-a + b + c)(a - b + c)(a + b + c)$

    (e) $x^4 + 3x^2 + 4 = (x^4 + 4x^2 + 4) - x^2 = (x^2 + 2)^2 - x^2 = \left[(x^2 + 2) - x\right]\left[(x^2 + 2) + x\right] =$
        $(x^2 - x + 2)(x^2 + x + 2)$

93. The volume of the shell is the difference between the volumes of the outside cylinder (with radius R) and the inside cylinder (with radius r). Thus
    $$V = \pi R^2 h - \pi r^2 h = \pi(R^2 - r^2)h = \pi(R - r)(R + r)h = 2\pi \cdot \frac{R + r}{2} \cdot h \cdot (R - r).$$
    The average radius is $\dfrac{R + r}{2}$ and $2\pi \cdot \dfrac{R + r}{2}$ is the average circumference (length of the rectangular box), h is the height, and $R - r$ is the thickness of the rectangular box. Thus
    $$V = \pi R^2 h - \pi r^2 h = 2\pi \cdot \frac{R + r}{2} \cdot h \cdot (R - r) = 2\pi \cdot (\textit{average radius}) \cdot (\textit{height}) \cdot (\textit{thickness})$$

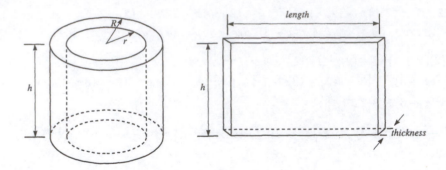

95. (a) $528^2 - 527^2 = (528 - 527)(528 + 527) = 1(1055) = 1055$

     (b) $122^2 - 120^2 = (122 - 120)(122 + 120) = 2(242) = 484$

     (c) $1020^2 - 1010^2 = (1020 - 1010)(1020 + 1010) = 10(2030) = 20{,}300$

     (d) $49 \cdot 51 = (50 - 1)(50 + 1) = 50^2 - 1 = 2500 - 1 = 2499$

     (e) $998 \cdot 1002 = (1000 - 2)(1000 + 2) = 1000^2 - 2^2 = 1{,}000{,}000 - 4 = 999{,}996$

97.
$$
\begin{array}{r}
A + 1 \\
\times \quad A - 1 \\
\hline
-A - 1 \\
A^2 + A \phantom{{}- 1} \\
\hline
A^2 - 1
\end{array}
\qquad
\begin{array}{r}
A^2 + A + 1 \\
\times \quad A - 1 \\
\hline
-A^2 - A - 1 \\
A^3 + A^2 + A \phantom{{}-1} \\
\hline
A^3 - 1
\end{array}
\qquad
\begin{array}{r}
A^3 + A^2 + A + 1 \\
\times \quad A - 1 \\
\hline
-A^3 - A^2 - A - 1 \\
A^4 + A^3 + A^2 + A \phantom{{}-1} \\
\hline
A^4 - 1
\end{array}
$$

Based on the pattern: $A^5 - 1 = (A - 1)(A^4 + A^3 + A^2 + A + 1)$

$$
\begin{array}{r}
A^4 + A^3 + A^2 + A + 1 \\
\times \quad A - 1 \\
\hline
-A^4 - A^3 - A^2 - A - 1 \\
A^5 + A^4 + A^3 + A^2 + A \phantom{{}-1} \\
\hline
A^5 - 1
\end{array}
$$

The generalized pattern is $A^n - 1 = (A - 1)(A^{n-1} + A^{n-2} + \cdots + A^2 + A + 1)$, where n is a positive integer.

Exercises P.7

1. (a)   When $x = 5$ we get $4(5^2) - 10(5) + 3 = 53$.

    (b)   Domain; all real numbers.

3. (a)   When $x = 7$ we get $\dfrac{2(7) + 1}{7 - 4} = \dfrac{15}{3} = 5$.

    (b)   Since $x - 4 \neq 0$ we have $x \neq 4$. Domain: $\{x \mid x \neq 4\}$

5. (a)   When $x = 6$ we get $\sqrt{6 + 3} = \sqrt{9} = 3$.

    (b)   Since $x + 3 \geq 0 \quad \Leftrightarrow \quad x \geq -3$. Domain; $\{x \mid x \geq -3\}$

7. $\dfrac{12x}{6x^2} = \dfrac{6x \cdot 2}{6x \cdot x} = \dfrac{2}{x}$

9. $\dfrac{5y^2}{10y + y^2} = \dfrac{y \cdot 5y}{y \cdot (10 + y)} = \dfrac{5y}{10 + y}$

11. $\dfrac{3(x + 2)(x - 1)}{6(x-1)^2} = \dfrac{3(x - 1) \cdot (x + 2)}{3(x - 1) \cdot 2(x - 1)} = \dfrac{x + 2}{2(x - 1)}$

13. $\dfrac{x - 2}{x^2 - 4} = \dfrac{x - 2}{(x - 2)(x + 2)} = \dfrac{1}{x + 2}$

15. $\dfrac{x^2 + 6x + 8}{x^2 + 5x + 4} = \dfrac{(x + 2)(x + 4)}{(x + 1)(x + 4)} = \dfrac{x + 2}{x + 1}$

7. $\dfrac{y^2 + y}{y^2 - 1} = \dfrac{y(y + 1)}{(y - 1)(y + 1)} = \dfrac{y}{y - 1}$

19. $\dfrac{2x^3 - x^2 - 6x}{2x^2 - 7x + 6} = \dfrac{x(2x^2 - x - 6)}{(2x - 3)(x - 2)} = \dfrac{x(2x + 3)(x - 2)}{(2x - 3)(x - 2)} = \dfrac{x(2x + 3)}{2x - 3}$

21. $\dfrac{4x}{x^2 - 4} \cdot \dfrac{x + 2}{16x} = \dfrac{4x}{(x - 2)(x + 2)} \cdot \dfrac{x + 2}{16x} = \dfrac{1}{4(x - 2)}$

23. $\dfrac{x^2 - x - 12}{x^2 - 9} \cdot \dfrac{3 + x}{4 - x} = \dfrac{(x - 4)(x + 3)}{(x - 3)(x + 3)} \cdot \dfrac{x + 3}{-(x - 4)} = \dfrac{x + 3}{-(x - 3)} = \dfrac{3 + x}{3 - x}$

25. $\dfrac{t - 3}{t^2 + 9} \cdot \dfrac{t + 3}{t^2 - 9} = \dfrac{(t - 3)(t + 3)}{(t^2 + 9)(t - 3)(t + 3)} = \dfrac{1}{t^2 + 9}$

27. $\dfrac{x^2 + 7x + 12}{x^2 + 3x + 2} \cdot \dfrac{x^2 + 5x + 6}{x^2 + 6x + 9} = \dfrac{(x + 3)(x + 4)}{(x + 1)(x + 2)} \cdot \dfrac{(x + 2)(x + 3)}{(x + 3)(x + 3)} = \dfrac{x + 4}{x + 1}$

29. $\dfrac{2x^2 + 3x + 1}{x^2 + 2x - 15} \div \dfrac{x^2 + 6x + 5}{2x^2 - 7x + 3} = \dfrac{2x^2 + 3x + 1}{x^2 + 2x - 15} \cdot \dfrac{2x^2 - 7x + 3}{x^2 + 6x + 5} =$

$$\frac{(2x+1)(x+1)}{(x-3)(x+5)} \cdot \frac{(2x-1)(x-3)}{(x+1)(x+5)} = \frac{(2x+1)(2x-1)}{(x+5)(x+5)} = \frac{(2x+1)(2x-1)}{(x+5)^2}$$

31. $$\frac{\dfrac{x^3}{x+1}}{\dfrac{x}{x^2+2x+1}} = \frac{x^3}{x+1} \cdot \frac{x^2+2x+1}{x} = \frac{x^3(x+1)(x+1)}{(x+1)x} = x^2(x+1)$$

33. $$\frac{x/y}{z} = \frac{x}{y} \cdot \frac{1}{z} = \frac{x}{yz}$$

35. $$2 + \frac{x}{x+3} = \frac{2(x+3)}{x+3} + \frac{x}{x+3} = \frac{2x+6+x}{x+3} = \frac{3x+6}{x+3} = \frac{3(x+2)}{(x+3)}$$

37. $$\frac{1}{x+5} + \frac{2}{x-3} = \frac{x-3}{(x+5)(x-3)} + \frac{2(x+5)}{(x+5)(x-3)} = \frac{x-3+2x+10}{(x+5)(x-3)} = \frac{3x+7}{(x+5)(x-3)}$$

39. $$\frac{1}{x+1} - \frac{1}{x+2} = \frac{x+2}{(x+1)(x+2)} + \frac{-(x+1)}{(x+1)(x+2)} = \frac{x+2-x-1}{(x+1)(x+2)} = \frac{1}{(x+1)(x+2)}$$

41. $$\frac{x}{(x+1)^2} + \frac{2}{x+1} = \frac{x}{(x+1)^2} + \frac{2(x+1)}{(x+1)(x+1)} = \frac{x+2x+2}{(x+1)^2} = \frac{3x+2}{(x+1)^2}$$

43. $$u+1+\frac{u}{u+1} = \frac{(u+1)(u+1)}{u+1} + \frac{u}{u+1} = \frac{u^2+2u+1+u}{u+1} = \frac{u^2+3u+1}{u+1}$$

45. $$\frac{1}{x^2} + \frac{1}{x^2+x} = \frac{1}{x^2} + \frac{1}{x(x+1)} = \frac{x+1}{x^2(x+1)} + \frac{x}{x^2(x+1)} = \frac{2x+1}{x^2(x+1)}$$

47. $$\frac{2}{x+3} - \frac{1}{x^2+7x+12} = \frac{2}{x+3} - \frac{1}{(x+3)(x+4)} = \frac{2(x+4)}{(x+3)(x+4)} + \frac{-1}{(x+3)(x+4)}$$
$$= \frac{2x+8-1}{(x+3)(x+4)} = \frac{2x+7}{(x+3)(x+4)}$$

49. $$\frac{1}{x+3} + \frac{1}{x^2-9} = \frac{1}{x+3} + \frac{1}{(x-3)(x+3)} = \frac{x-3}{(x-3)(x+3)} + \frac{1}{(x-3)(x+3)}$$
$$= \frac{x-2}{(x-3)(x+3)}$$

51. $$\frac{2}{x} + \frac{3}{x-1} - \frac{4}{x^2-x} = \frac{2}{x} + \frac{3}{x-1} - \frac{4}{x(x-1)} = \frac{2(x-1)}{x(x-1)} + \frac{3x}{x(x-1)} + \frac{-4}{x(x-1)}$$
$$= \frac{2x-2+3x-4}{x(x-1)} = \frac{5x-6}{x(x-1)}$$

53. $$\frac{1}{x^2+3x+2} - \frac{1}{x^2-2x-3} = \frac{1}{(x+2)(x+1)} - \frac{1}{(x-3)(x+1)}$$
$$= \frac{x-3}{(x-3)(x+2)(x+1)} + \frac{-(x+2)}{(x-3)(x+2)(x+1)}$$

$$= \frac{x - 3 - x - 2}{(x-3)(x+2)(x+1)} = \frac{-5}{(x-3)(x+2)(x+1)}$$

55.
$$\frac{\dfrac{x}{y} - \dfrac{y}{x}}{\dfrac{1}{x^2} - \dfrac{1}{y^2}} = \frac{\dfrac{x^2 - y^2}{xy}}{\dfrac{y^2 - x^2}{x^2 y^2}} = \frac{x^2 - y^2}{xy} \cdot \frac{x^2 y^2}{y^2 - x^2} = \frac{xy}{-1} = -xy.$$

An alternative method is to multiply the numerator and denominator by the common denominator of both the numerator and denominator, in this case $x^2 y^2$.

$$\frac{\dfrac{x}{y} - \dfrac{y}{x}}{\dfrac{1}{x^2} - \dfrac{1}{y^2}} = \frac{\left(\dfrac{x}{y} - \dfrac{y}{x}\right)}{\left(\dfrac{1}{x^2} - \dfrac{1}{y^2}\right)} \cdot \frac{x^2 y^2}{x^2 y^2} = \frac{x^3 y - x y^3}{y^2 - x^2} = \frac{xy(x^2 - y^2)}{y^2 - x^2} = -xy$$

57.
$$\frac{1 + \dfrac{1}{c-1}}{1 - \dfrac{1}{c-1}} = \frac{\dfrac{c-1}{c-1} + \dfrac{1}{c-1}}{\dfrac{c-1}{c-1} + \dfrac{-1}{c-1}} = \frac{\dfrac{c}{c-1}}{\dfrac{c-2}{c-1}} = \frac{c}{c-1} \cdot \frac{c-1}{c-2} = \frac{c}{c-2}.$$

Using the alternative method we obtain:

$$\frac{1 + \dfrac{1}{c-1}}{1 - \dfrac{1}{c-1}} = \frac{\left(1 + \dfrac{1}{c-1}\right)}{\left(1 - \dfrac{1}{c-1}\right)} \cdot \frac{c-1}{c-1} = \frac{c-1+1}{c-1-1} = \frac{c}{c-2}.$$

59.
$$\frac{\dfrac{5}{x-1} - \dfrac{2}{x+1}}{\dfrac{x}{x-1} + \dfrac{1}{x+1}} = \frac{\dfrac{5(x+1)}{(x-1)(x+1)} + \dfrac{-2(x-1)}{(x-1)(x+1)}}{\dfrac{x(x+1)}{(x-1)(x+1)} + \dfrac{x-1}{(x-1)(x+1)}} = \frac{\dfrac{5x+5-2x+2}{(x-1)(x+1)}}{\dfrac{x^2 + x + x - 1}{(x-1)(x+1)}}$$

$$= \frac{3x+7}{(x-1)(x+1)} \cdot \frac{(x-1)(x+1)}{x^2 + 2x - 1} = \frac{3x+7}{x^2 + 2x - 1}$$

Alternatively,

$$\frac{\dfrac{5}{x-1} - \dfrac{2}{x+1}}{\dfrac{x}{x-1} + \dfrac{1}{x+1}} = \frac{\left(\dfrac{5}{x-1} - \dfrac{2}{x+1}\right)}{\left(\dfrac{x}{x-1} + \dfrac{1}{x+1}\right)} \cdot \frac{(x-1)(x+1)}{(x-1)(x+1)} = \frac{5(x+1) - 2(x-1)}{x(x+1) + (x-1)}$$

$$= \frac{5x+5-2x+2}{x^2 + x + x - 1} = \frac{3x+7}{x^2 + 2x - 1}.$$

61.
$$\frac{x^{-2} - y^{-2}}{x^{-1} + y^{-1}} = \frac{\dfrac{1}{x^2} - \dfrac{1}{y^2}}{\dfrac{1}{x} + \dfrac{1}{y}} = \frac{\dfrac{y^2}{x^2 y^2} - \dfrac{x^2}{x^2 y^2}}{\dfrac{y}{xy} + \dfrac{x}{xy}} = \frac{y^2 - x^2}{x^2 y^2} \cdot \frac{xy}{y+x} = \frac{(y-x)(y+x)xy}{x^2 y^2 (y+x)} = \frac{y-x}{xy}$$

Alternatively,

$$\frac{x^{-2}-y^{-2}}{x^{-1}+y^{-1}} = \frac{\left(\dfrac{1}{x^2}-\dfrac{1}{y^2}\right)}{\left(\dfrac{1}{x}+\dfrac{1}{y}\right)} \cdot \frac{x^2y^2}{x^2y^2} = \frac{y^2-x^2}{xy^2+x^2y} = \frac{(y-x)(y+x)}{xy(y+x)} = \frac{y-x}{xy}.$$

63. $\dfrac{1}{1+a^n} + \dfrac{1}{1+a^{-n}} = \dfrac{1}{1+a^n} + \dfrac{1}{1+a^{-n}} \cdot \dfrac{a^n}{a^n} = \dfrac{1}{1+a^n} + \dfrac{a^n}{a^n+1} = \dfrac{1+a^n}{1+a^n} = 1$

65. $\dfrac{\dfrac{1}{a+h}-\dfrac{1}{a}}{h} = \dfrac{\dfrac{a}{a(a+h)}-\dfrac{a+h}{a(a+h)}}{h} = \dfrac{\dfrac{a-a-h}{a(a+h)}}{h} = \dfrac{-h}{a(a+h)} \cdot \dfrac{1}{h} = \dfrac{-1}{a(a+h)}$

67. $\dfrac{\dfrac{1-(x+h)}{2+(x+h)}-\dfrac{1-x}{2+x}}{h} = \dfrac{\dfrac{(2+x)(1-x-h)}{(2+x)(2+x+h)}-\dfrac{(1-x)(2+x+h)}{(2+x)(2+x+h)}}{h}$

$$= \dfrac{\dfrac{2-x-x^2-2h-xh}{(2+x)(2+x+h)}-\dfrac{2-x-x^2+h-xh}{(2+x)(2+x+h)}}{h} = \dfrac{-3h}{(2+x)(2+x+h)} \cdot \dfrac{1}{h}$$

$$= \dfrac{-3}{(2+x)(2+x+h)}$$

69. $\sqrt{1+\left(\dfrac{x}{\sqrt{1-x^2}}\right)^2} = \sqrt{1+\dfrac{x^2}{1-x^2}} = \sqrt{\dfrac{1-x^2}{1-x^2}+\dfrac{x^2}{1-x^2}} = \sqrt{\dfrac{1}{1-x^2}} = \dfrac{1}{\sqrt{1-x^2}}$

71. $\dfrac{3(x+2)^2(x-3)^2-(x+2)^3(2)(x-3)}{(x-3)^4} = \dfrac{(x+2)^2(x-3)[3(x-3)-(x+2)(2)]}{(x-3)^4}$

$$= \dfrac{(x+2)^2(3x-9-2x-4)}{(x-3)^3} = \dfrac{(x+2)^2(x-13)}{(x-3)^3}$$

73. $\dfrac{2(1+x)^{1/2}-x(1+x)^{-1/2}}{1+x.} = \dfrac{(1+x)^{-1/2}[2(1+x)-x]}{1+x} = \dfrac{x+2}{(1+x)^{3/2}}$

75. $\dfrac{3(1+x)^{1/3}-x(1+x)^{-2/3}}{(1+x)^{2/3}} = \dfrac{(1+x)^{-2/3}[3(1+x)-x]}{(1+x)^{2/3}} = \dfrac{2x+3}{(1+x)^{4/3}}$

77. $\dfrac{1}{2-\sqrt{3}} = \dfrac{1}{2-\sqrt{3}} \cdot \dfrac{2+\sqrt{3}}{2+\sqrt{3}} = \dfrac{2+\sqrt{3}}{4-3} = \dfrac{2+\sqrt{3}}{1} = 2+\sqrt{3}$

79. $\dfrac{2}{\sqrt{2}+\sqrt{7}} = \dfrac{2}{\sqrt{2}+\sqrt{7}} \cdot \dfrac{\sqrt{2}-\sqrt{7}}{\sqrt{2}-\sqrt{7}} = \dfrac{2\left(\sqrt{2}-\sqrt{7}\right)}{2-7} = \dfrac{2\left(\sqrt{2}-\sqrt{7}\right)}{-5} = \dfrac{2\left(\sqrt{7}-\sqrt{2}\right)}{5}$

81. $\dfrac{y}{\sqrt{3}+\sqrt{y}} = \dfrac{y}{\sqrt{3}+\sqrt{y}} \cdot \dfrac{\sqrt{3}-\sqrt{y}}{\sqrt{3}-\sqrt{y}} = \dfrac{y\left(\sqrt{3}-\sqrt{y}\right)}{3-y} = \dfrac{y\sqrt{3}-y\sqrt{y}}{3-y}$

83. $\dfrac{1-\sqrt{5}}{3} = \dfrac{1-\sqrt{5}}{3} \cdot \dfrac{1+\sqrt{5}}{1+\sqrt{5}} = \dfrac{1-5}{3\left(1+\sqrt{5}\right)} = \dfrac{-4}{3\left(1+\sqrt{5}\right)}$

85. $\dfrac{\sqrt{r}+\sqrt{2}}{5} = \dfrac{\sqrt{r}+\sqrt{2}}{5} \cdot \dfrac{\sqrt{r}-\sqrt{2}}{\sqrt{r}-\sqrt{2}} = \dfrac{r-2}{5\left(\sqrt{r}-\sqrt{2}\right)}$

87. $\sqrt{x^2+1} - x = \dfrac{\sqrt{x^2+1}-x}{1} \cdot \dfrac{\sqrt{x^2+1}+x}{\sqrt{x^2+1}+x} = \dfrac{x^2+1-x^2}{\sqrt{x^2+1}+x} = \dfrac{1}{\sqrt{x^2+1}+x}$

89. $\dfrac{16+a}{16} = \dfrac{16}{16} + \dfrac{a}{16} = 1 + \dfrac{a}{16}$, so the statement is true.

91. This statement is false. For example, take $x = 2$, then LHS $= \dfrac{2}{4+x} = \dfrac{2}{4+2} = \dfrac{2}{6} = \dfrac{1}{3}$, while RHS $= \dfrac{1}{2} + \dfrac{2}{x} = \dfrac{1}{2} + \dfrac{2}{2} = \dfrac{3}{2}$, and $\dfrac{1}{3} \neq \dfrac{3}{2}$.

93. This statement is false. For example, take $x = 0$ and $y = 1$. Then substituting into the left side we obtain LHS $= \dfrac{x}{x+y} = \dfrac{0}{0+1} = 0$, while the right side yields RHS $= \dfrac{1}{1+y} = \dfrac{1}{1+1} = \dfrac{1}{2}$, and $0 \neq \frac{1}{2}$.

95. This statement is true: $\dfrac{-a}{b} = (-a)\left(\dfrac{1}{b}\right) = (-1)(a)\left(\dfrac{1}{b}\right) = (-1)\left(\dfrac{a}{b}\right) = -\dfrac{a}{b}$.

97. (a) $R = \dfrac{1}{\dfrac{1}{R_1} + \dfrac{1}{R_2}} = \dfrac{1}{\dfrac{1}{R_1} + \dfrac{1}{R_2}} \cdot \dfrac{R_1 R_2}{R_1 R_2} = \dfrac{R_1 R_2}{R_2 + R_1}$

    (b) Substituting $R_1 = 10$ ohms and $R_2 = 20$ ohms yields $R = \dfrac{(10)(20)}{(20)+(10)} = \dfrac{200}{30} \approx 6.7$ ohms.

99.

x	2.80	2.90	2.95	2.99	2.999	3	3.001	3.01	3.05	3.10	3.20
$\dfrac{x^2-9}{x-3}$	5.80	5.90	5.95	5.99	5.999	?	6.001	6.01	6.05	6.10	6.20

From the table, we see that the expression $\dfrac{x^2-9}{x-3}$ approaches 6 as x approaches 3. We simplify the expression: $\dfrac{x^2-9}{x-3} = \dfrac{(x-3)(x+3)}{x-3} = x+3$ $(x \neq 3)$. Clearly as x approaches $3, x+3$ approaches 6. This explains the result in the table.

101. Answers will vary.

Algebraic Error		Counterexample
$\dfrac{1}{a} + \dfrac{1}{b} \;\cancel{\times}\; \dfrac{1}{a+b}$	$\dfrac{1}{2} + \dfrac{1}{2} \;\cancel{\times}\; \dfrac{1}{2+2}$	LHS $= \dfrac{1}{2} + \dfrac{1}{2} = 1$ RHS $= \dfrac{1}{2+2} = \dfrac{1}{4}$

Algebraic Error	Counterexample	
$(a+b)^2 \neq a^2 + b^2$	$(1+3)^2 \neq 1^2 + 3^2$	$\text{LHS} = (1+3)^2 = 4^2 = 16$ $\text{RHS} = 1^2 + 3^2 = 1 + 9 = 10$
$\sqrt{a^2 + b^2} \neq a + b$	$\sqrt{5^2 + 12^2} \neq 5 + 12$	$\text{LHS} = \sqrt{5^2 + 12^2} = \sqrt{25 + 144}$ $\qquad = \sqrt{169} = 13$ $\text{RHS} = 5 + 12 = 17$
$\dfrac{a+b}{a} \neq b$	$\dfrac{2+6}{2} \neq 6$	$\text{LHS} = \dfrac{2+6}{2} = \dfrac{8}{2} = 4$ $\text{RHS} = 6$
$(a^3 + b^3)^{1/3} \neq a + b$	$(2^3 + 2^3)^{1/3} \neq 2 + 2$	$\text{LHS} = (a^3 + b^3)^{1/3} = (8+8)^{1/3}$ $\qquad = 2\sqrt[3]{2}$ $\text{RHS} = 2 + 2 = 4$
$\dfrac{a^m}{a^n} \neq a^{m/n}$	$\dfrac{3^5}{3^2} \neq 3^{5/2}$	$\text{LHS} = \dfrac{3^5}{3^2} = \dfrac{243}{9} = 27$ $\text{RHS} = 3^{5/2} = 9\sqrt{3}$
$a^{-1/n} \neq \dfrac{1}{a^n}$	$64^{-1/3} \neq \dfrac{1}{64^3}$	$\text{LHS} = 64^{-1/3} = \left(\dfrac{1}{64}\right)^{1/3} = \dfrac{1}{4}$ $\text{RHS} = \dfrac{1}{64^3} = \dfrac{1}{262{,}144}$

Review Exercises for Chapter P

1. Commutative Property for addition.

3. Distributive Property.

5. $[-2, 6) \quad \Leftrightarrow \quad -2 \le x < 6$

7. $(-\infty, 4] \quad \Leftrightarrow \quad x \le 4$

9. $x \ge 5 \quad \Leftrightarrow \quad [5, \infty)$

11. $-1 < x \le 5 \quad \Leftrightarrow \quad (-1, 5]$

13. $|3 - |-9|| = |3 - 9| = |-6| = 6$

15. $2^{-3} - 3^{-2} = \dfrac{1}{8} - \dfrac{1}{9} = \dfrac{9}{72} - \dfrac{8}{72} = \dfrac{1}{72}$

17. $216^{-1/3} = \dfrac{1}{216^{1/3}} = \dfrac{1}{\sqrt[3]{216}} = \dfrac{1}{6}$

19. $\dfrac{\sqrt{242}}{\sqrt{2}} = \sqrt{\dfrac{242}{2}} = \sqrt{121} = 11$

21. $2^{1/2}8^{1/2} = \sqrt{2} \cdot \sqrt{8} = \sqrt{16} = 4$

23. $\dfrac{1}{x^2} = x^{-2}$

25. $x^2 x^m (x^3)^m = x^{2+m+3m} = x^{4m+2}$

27. $x^a x^b x^c = x^{a+b+c}$

29. $x^{c+1}(x^{2c-1})^2 = x^{(c+1)+2(2c-1)} = x^{c+1+4c-2} = x^{5c-1}$

31. $(2x^3 y)^2 (3x^{-1} y^2) = 4x^6 y^2 \cdot 3x^{-1} y^2 = 4 \cdot 3x^{6-1} y^{2+2} = 12x^5 y^4$

33. $\dfrac{x^4 (3x)^2}{x^3} = \dfrac{x^4 \cdot 9x^2}{x^3} = 9x^{4+2-3} = 9x^3$

35. $\sqrt[3]{(x^3 y)^2 y^4} = \sqrt[3]{x^6 y^4 y^2} = \sqrt[3]{x^6 y^6} = x^2 y^2$

37. $\dfrac{x}{2 + \sqrt{x}} = \dfrac{x}{2 + \sqrt{x}} \cdot \dfrac{2 - \sqrt{x}}{2 - \sqrt{x}} = \dfrac{x(2 - \sqrt{x})}{4 - x}$. Here simplify means to rationalize the denominator.

39. $\dfrac{8r^{1/2} s^{-3}}{2r^{-2} s^4} = 4r^{(1/2)-(-2)} s^{-3-4} = 4r^{5/2} s^{-7} = \dfrac{4r^{5/2}}{s^7}$

41. $78{,}250{,}000{,}000 = 7.825 \times 10^{10}$

43. $\dfrac{ab}{c} \approx \dfrac{(0.00000293)(1.582 \times 10^{-14})}{2.8064 \times 10^{12}} = \dfrac{(2.93 \times 10^{-6})(1.582 \times 10^{-14})}{2.8064 \times 10^{12}}$

$= \dfrac{2.93 \cdot 1.582}{2.8064} \times 10^{-6-14-12} \approx 1.65 \times 10^{-32}$

45. $2x^2 y - 6xy^2 = 2xy(x - 3y)$

47. $x^2 - 9x + 18 = (x - 6)(x - 3)$

49. $3x^2 - 2x - 1 = (3x + 1)(x - 1)$

51. $4t^2 - 13t - 12 = (4t + 3)(t - 4)$

53. $25 - 16t^2 = (5 - 4t)(5 + 4t)$

55. $x^6 - 1 = (x^3 - 1)(x^3 + 1) = (x - 1)(x^2 + x + 1)(x + 1)(x^2 - x + 1)$

57. $x^{-1/2} - 2x^{1/2} + x^{3/2} = x^{-1/2}(1 - 2x + x^2) = x^{-1/2}(1 - x)^2$

59. $4x^3 - 8x^2 + 3x - 6 = 4x^2(x - 2) + 3(x - 2) = (4x^2 + 3)(x - 2)$

61. $(x^2 + 2)^{5/2} + 2x(x^2 + 2)^{3/2} + x^2\sqrt{x^2 + 2} = (x^2 + 2)^{1/2}((x^2 + 2)^2 + 2x(x^2 + 2) + x^2)$
$= \sqrt{x^2 + 2}(x^4 + 4x^2 + 4 + 2x^3 + 4x + x^2) = \sqrt{x^2 + 2}(x^4 + 2x^3 + 5x^2 + 4x + 4)$
$= \sqrt{x^2 + 2}(x^2 + x + 2)^2$

63. $a^2y - b^2y = y(a^2 - b^2) = y(a - b)(a + b)$

65. $(x + 1)^2 - 2(x + 1) + 1 = \left[(x + 1) - 1\right]^2 = x^2$. You can also obtain this result by expanding each term and then simplifying.

67. $(2x + 1)(3x - 2) - 5(4x - 1) = 6x^2 - 4x + 3x - 2 - 20x + 5 = 6x^2 - 21x + 3$

69. $(2a^2 - b)^2 = (2a^2)^2 - 2(2a^2)(b) + (b)^2 = 4a^4 - 4a^2b + b^2$

71. $(x - 1)(x - 2)(x - 3) = (x - 1)(x^2 - 5x + 6) = x^3 - 5x^2 + 6x - x^2 + 5x - 6$
$= x^3 - 6x^2 + 11x - 6$

73. $\sqrt{x}(\sqrt{x} + 1)(2\sqrt{x} - 1) = (x + \sqrt{x})(2\sqrt{x} - 1) = 2x\sqrt{x} - x + 2x - \sqrt{x} = 2x^{3/2} + x - x^{1/2}$

75. $x^2(x - 2) + x(x - 2)^2 = x^3 - 2x^2 + x(x^2 - 4x + 4) = x^3 - 2x^2 + x^3 - 4x^2 + 4x$
$= 2x^3 - 6x^2 + 4x$

77. $\dfrac{x^2 - 2x - 3}{2x^2 + 5x + 3} = \dfrac{(x - 3)(x + 1)}{(2x + 3)(x + 1)} = \dfrac{x - 3}{2x + 3}$

79. $\dfrac{x^2 + 2x - 3}{x^2 + 8x + 16} \cdot \dfrac{3x + 12}{x - 1} = \dfrac{(x + 3)(x - 1)}{(x + 4)(x + 4)} \cdot \dfrac{3(x + 4)}{(x - 1)} = \dfrac{3(x + 3)}{x + 4}$

81. $\dfrac{x^2 - 2x - 15}{x^2 - 6x + 5} \div \dfrac{x^2 - x - 12}{x^2 - 1} = \dfrac{(x - 5)(x + 3)}{(x - 5)(x - 1)} \cdot \dfrac{(x - 1)(x + 1)}{(x - 4)(x + 3)} = \dfrac{x + 1}{x - 4}$

83. $\dfrac{1}{x - 1} - \dfrac{x}{x^2 + 1} = \dfrac{x^2 + 1}{(x - 1)(x^2 + 1)} - \dfrac{x(x - 1)}{(x - 1)(x^2 + 1)} = \dfrac{x^2 + 1 - x^2 + x}{(x - 1)(x^2 + 1)} = \dfrac{x + 1}{(x - 1)(x^2 + 1)}$

85. $\dfrac{1}{x - 1} - \dfrac{2}{x^2 - 1} = \dfrac{1}{x - 1} - \dfrac{2}{(x - 1)(x + 1)} = \dfrac{x + 1}{(x - 1)(x + 1)} - \dfrac{2}{(x - 1)(x + 1)}$

$$= \frac{x+1-2}{(x-1)(x+1)} = \frac{x-1}{(x-1)(x+1)} = \frac{1}{x+1}$$

87. $\dfrac{\dfrac{1}{x} - \dfrac{1}{2}}{x-2} = \dfrac{\dfrac{2}{2x} - \dfrac{x}{2x}}{x-2} = \dfrac{2-x}{2x} \cdot \dfrac{1}{x-2} = \dfrac{-1(x-2)}{2x} \cdot \dfrac{1}{x-2} = \dfrac{-1}{2x}$

89. $\dfrac{3(x+h)^2 - 5(x+h) - (3x^2 - 5x)}{h} = \dfrac{3x^2 + 6xh + 3h^2 - 5x - 5h - 3x^2 + 5x}{h}$

$$= \frac{6xh + 3h^2 - 5h}{h} = \frac{h(6x + 3h - 5)}{h} = 6x + 3h - 5$$

91. This statement is false. For example, take $x = 1$ and $y = 1$; then LHS $= (x+y)^3 = (1+1)^3 = 2^3$ $= 8$, while RHS $= x^3 + y^3 = 1^3 + 1^3 = 1 + 1 = 2$, and $8 \neq 2$.

93. This statement is true: $\dfrac{12+y}{y} = \dfrac{12}{y} + \dfrac{y}{y} = \dfrac{12}{y} + 1$.

95. This statement is false. For example, take $a = -1$; then LHS $= \sqrt{a^2} = \sqrt{(-1)^2} = \sqrt{1} = 1$, which does not equal $a = -1$. The true statement is $\sqrt{a^2} = |a|$.

97. This statement is false. For example, take $x = 1$ and $y = 1$, then LHS $= x^3 + y^3 = 1^3 + 1^3$ $= 1 + 1 = 2$, while RHS $= (x+y)(x^2 + xy + y^2) = (1+1)[1^2 + (1)(1) + 1^2] = 2(3) = 6$.

99. Substituting for a and b we obtain $a^2 + b^2 = (2mn)^2 + (m^2 - n^2)^2$ $= 4m^2n^2 + m^4 - 2m^2n^2 + n^4 = m^4 + 2m^2n^2 + n^4 = (m^2 + n^2)^2$. Since this last expression is c^2, we have $a^2 + b^2 = c^2$ for these values.

Chapter P Test

1. (a) $[-2, 5)$

    $(7, \infty)$

    (b) $x \le 3 \quad \Leftrightarrow \quad (-\infty, 3]$ $0 < x \le 8 \quad \Leftrightarrow \quad (0, 8]$

    (c) Distance $= |-12 - 39| = |-51| = 51$

2. (a) $(-3)^4 = 81$ (b) $-3^4 = -81$

    (c) $3^{-4} = \dfrac{1}{3^4} = \dfrac{1}{81}$ (d) $\dfrac{5^{23}}{5^{21}} = 5^{23-21} = 5^2 = 25$

    (e) $\left(\dfrac{2}{3}\right)^{-2} = \dfrac{3^2}{2^2} = \dfrac{9}{4}$ (f) $\dfrac{\sqrt[6]{64}}{\sqrt{32}} = \dfrac{\sqrt[6]{2^6}}{\sqrt{16 \cdot 2}} = \dfrac{2}{4\sqrt{2}} = \dfrac{1}{2\sqrt{2}}$

    (g) $16^{-3/4} = \left(2^4\right)^{-3/4} = 2^{-3} = \frac{1}{8}$

3. (a) $186{,}000{,}000{,}000 = 1.86 \times 10^{11}$ (b) $0.0000003965 = 3.965 \times 10^{-7}$

4. (a) $\sqrt{200} - \sqrt{32} = 10\sqrt{2} - 4\sqrt{2} = 6\sqrt{2}$

    (b) $(3a^2b^3)(4ab^2)^2 = 3a^2b^3 \cdot 4^2a^2b^4 = 48a^4b^7$

    (c) $\sqrt[3]{\dfrac{125}{x^{-9}}} = \sqrt[3]{5^3 \cdot x^9} = 5x^3$

    (d) $\left(\dfrac{3x^{3/2}y^3}{x^2y^{-1/2}}\right)^{-2} = \dfrac{3^{-2}x^{-3}y^{-6}}{x^{-4}y^1} = \frac{1}{9}x^{-3-(-4)}y^{-6-1} = \frac{1}{9}xy^{-7} = \dfrac{x}{9y^7}$

5. (a) $3(x+6) + 4(2x-5) = 3x + 18 + 8x - 20 = 11x - 2$

    (b) $(x+3)(4x-5) = 4x^2 - 5x + 12x - 15 = 4x^2 + 7x - 15$

    (c) $\left(\sqrt{a} + \sqrt{b}\right)\left(\sqrt{a} - \sqrt{b}\right) = \left(\sqrt{a}\right)^2 - \left(\sqrt{b}\right)^2 = a - b$

    (d) $(2x+3)^2 = (2x)^2 + 2(2x)(3) + (3)^2 = 4x^2 + 12x + 9$

    (e) $(x+2)^3 = (x)^3 + 3(x)^2(2) + 3(x)(2)^2 + (2)^3 = x^3 + 6x^2 + 12x + 8$

6. (a) $4x^2 - 25 = (2x - 5)(2x + 5)$

    (b) $2x^2 + 5x - 12 = (2x - 3)(x + 4)$

    (c) $x^3 - 3x^2 - 4x + 12 = x^2(x - 3) - 4(x - 3) = (x - 3)(x^2 - 4) = (x - 3)(x - 2)(x + 2)$

    (d) $x^4 + 27x = x(x^3 + 27) = x(x + 3)(x^2 - 3x + 9)$

    (e) $3x^{3/2} - 9x^{1/2} + 6x^{-1/2} = 3x^{-1/2}(x^2 - 3x + 2) = 3x^{-1/2}(x - 2)(x - 1)$

    (f) $x^3y - 4xy = xy(x^2 - 4) = xy(x - 2)(x + 2)$

7. (a) $\dfrac{x^2 + 3x + 2}{x^2 - x - 2} = \dfrac{(x+1)(x+2)}{(x+1)(x-2)} = \dfrac{x+2}{x-2}$

    (b) $\dfrac{2x^2 - x - 1}{x^2 - 9} \cdot \dfrac{x+3}{2x+1} = \dfrac{(2x+1)(x-1)}{(x-3)(x+3)} \cdot \dfrac{x+3}{2x+1} = \dfrac{x-1}{x-3}$

    (c) $\dfrac{x^2}{x^2 - 4} - \dfrac{x+1}{x+2} = \dfrac{x^2}{(x-2)(x+2)} - \dfrac{x+1}{x+2} = \dfrac{x^2}{(x-2)(x+2)} + \dfrac{-(x+1)(x-2)}{(x-2)(x+2)}$

    $= \dfrac{x^2 - (x^2 - x - 2)}{(x-2)(x+2)} = \dfrac{x+2}{(x-2)(x+2)} = \dfrac{1}{x-2}$

    (d) $\dfrac{\dfrac{y}{x} - \dfrac{x}{y}}{\dfrac{1}{y} - \dfrac{1}{x}} = \dfrac{\dfrac{y}{x} - \dfrac{x}{y}}{\dfrac{1}{y} - \dfrac{1}{x}} \cdot \dfrac{xy}{xy} = \dfrac{y^2 - x^2}{x - y} = \dfrac{(y-x)(y+x)}{x-y} = \dfrac{-(x-y)(y+x)}{x-y} = -(y+x)$

8. (a) $\dfrac{6}{\sqrt[3]{4}} = \dfrac{6}{\sqrt[3]{2^2}} = \dfrac{6}{\sqrt[3]{2^2}} \cdot \dfrac{\sqrt[3]{2}}{\sqrt[3]{2}} = \dfrac{6\sqrt[3]{2}}{2} = 3\sqrt[3]{2}$

    (b) $\dfrac{\sqrt{6}}{2 + \sqrt{3}} = \dfrac{\sqrt{6}}{2 + \sqrt{3}} \cdot \dfrac{2 - \sqrt{3}}{2 - \sqrt{3}} = \dfrac{2\sqrt{6} - \sqrt{18}}{4 - 3} = \dfrac{2\sqrt{6} - \sqrt{9 \cdot 2}}{1} = 2\sqrt{6} - 3\sqrt{2}$

Principles of Problem Solving

1. Let d be the distance traveled to and from work. Let t_1 and t_2 be the times for the trip from home to work and the trip from work to home, respectively. Using $time = \dfrac{distance}{rate}$, we get $t_1 = \dfrac{d}{50}$ and $t_2 = \dfrac{d}{30}$. Since $average\ speed = \dfrac{distance\ traveled}{total\ time}$, we have $average\ speed = \dfrac{2d}{t_1 + t_2} = \dfrac{2d}{\frac{d}{50} + \frac{d}{30}}$

   $\Leftrightarrow\quad average\ speed = \dfrac{150(2d)}{150\left(\frac{d}{50} + \frac{d}{30}\right)} = \dfrac{300d}{3d + 5d} = \dfrac{300}{8} = 37.5$ mi/h.

3. We use the formula $d = rt$ (distance = rate $\times$ time). Since the car and the van each travel at a speed of 40 mi/h, they approach each other at a combined speed of 80 mi/h. (The distance between them decreases at a rate of 80 mi/h.) So the time spent driving till they meet is $t = \dfrac{d}{r} = \dfrac{120}{80} = 1.5$ hours. Thus, the fly flies at a speed of 100 mi/h for 1.5 hours, and therefore travels a distance of $d = rt = (100)(1.5) = 150$ miles.

5. We continue the pattern. Three parallel cuts produce 10 pieces. Thus, each new cut produces an additional 3 pieces. Since the first cut produces 4 pieces, we get the formula $f(n) = 4 + 3(n - 1)$, $n \geq 1$. Since $f(142) = 4 + 3(141) = 427$, we see that 142 parallel cuts produce 427 pieces.

7. George's speed is $\frac{1}{50}$ lap/s, and Sue's is $\frac{1}{30}$ lap/s. So after t seconds, George has run $\frac{t}{50}$ laps, and Sue has run $\frac{t}{30}$ laps. They will next be side by side when Sue has run exactly one more lap than George, that is, when $\frac{t}{30} = \frac{t}{50} + 1$. We solve for t by first multiplying by 150, so $50t = 30t + 150\quad \Leftrightarrow\quad 20t = 150\quad \Leftrightarrow\quad t = 75$. Therefore, they will be even after 75 s.

9. Method 1:
   After the exchanges, the volume of liquid in the pitcher and in the cup is the same as it was to begin with. Thus, any coffee in the pitcher of cream must be replacing an equal amount of cream that has ended up in the coffee cup.
   Method 2:
   Alternatively, look at the drawing of the spoonful of coffee and cream mixture being returned to the pitcher of cream. Suppose it is possible to separate the cream and the coffee, as shown. Then you can see that the coffee going into the cream occupies the same volume as the cream that was left in the coffee

   Method 3 (an algebraic approach):
   Suppose the cup of coffee has y spoonfuls of coffee. When one spoonful of cream is added to the coffee cup, the resulting mixture has the following ratios:

   $$\frac{cream}{mixture} = \frac{1}{y + 1}\qquad \frac{coffee}{mixture} = \frac{y}{y + 1}$$

   So, when we remove a spoonful of the mixture and put it into the pitcher of cream, we are really removing $\dfrac{1}{y + 1}$ of a spoonful of cream and $\dfrac{y}{y + 1}$ spoonful of coffee. Thus the <u>amount of cream</u>

left in the mixture (cream in the coffee) is $1 - \dfrac{1}{y+1} = \dfrac{y}{y+1}$ of a spoonful. This is the same as the amount of coffee we added to the cream.

11. Let r be the radius of the earth in feet. Then the circumference (length of the ribbon) is $2\pi r$. When we increase the radius by 1 foot, the new radius is $r + 1$, so the new circumference is $2\pi(r + 1)$. Thus you need $2\pi(r + 1) - 2\pi r = 2\pi$ extra feet of ribbon.

13. Let $r_1 = 8$.
$r_2 = \frac{1}{2}\left(8 + \frac{72}{8}\right) = \frac{1}{2}(8 + 9) = 8.5$
$r_3 = \frac{1}{2}\left(8.5 + \frac{72}{8.5}\right) = \frac{1}{2}(8.5 + 8.471) = 8.485$
$r_4 = \frac{1}{2}\left(8.485 + \frac{72}{8.485}\right) = \frac{1}{2}(8.485 + 8.486) = 8.485$
Thus $\sqrt{72} \approx 8.49$.

15. The first few powers of 3 are $3^1 = 3$, $3^2 = 9$, $3^3 = 27$, $3^4 = 81$, $3^5 = 243$, $3^6 = 729$. It appears that the final digit cycles in a pattern, namely $3 \to 9 \to 7 \to 1 \to 3 \to 9 \to 7 \to 1$, of length 4. Since $459 = 4 \times 114 + 3$, the final digit is the third in the cycle, namely 7.

17. Let p be an odd prime and label the other integer sides of the triangle as shown
Then $p^2 + b^2 = c^2 \;\Leftrightarrow\; p^2 = c^2 - b^2 \;\Leftrightarrow\; p^2 = (c - b)(c + b)$. Now, $c - b$ and $c + b$ are integer factors of p^2 where p is a prime.
Case 1: $c - b = p$ and $c + b = p$. Subtracting gives $b = 0$ which contradicts that b is the length of a side of a triangle

Case 2: $c - b = 1$ and $c + b = p^2$. Adding gives $2c = 1 + p^2 \;\Leftrightarrow\; c = \dfrac{1 + p^2}{2}$ and $b = \dfrac{p^2 - 1}{2}$.

Since p is odd, p^2 is odd. Thus $\dfrac{1 + p^2}{2}$ and $\dfrac{p^2 - 1}{2}$ both define integers and are the only solutions.
(For example, if we choose $p = 3$, then $c = 5$ and $b = 4$.)

19. The north pole is such a point. And there are others: Consider a point a_1 near the south pole such that the parallel passing through a_1 forms a circle C_1 with circumference exactly one mile. Any point P_1 exactly one mile north of the circle C_1 along a meridian is a point satisfying the conditions in the problem: starting at P_1 she walks one mile south to the point a_1 on the circle C_1, then one mile east along C_1 returning to the point a_1, then north for one mile to P_1. That's not all. If a point a_2 (or $a_3, a_4, a_5, \ldots$) is chosen near the south pole so that the parallel passing through it forms a circle C_2 ($C_3, C_4, C_5, \ldots$) with a circumference of exactly $\frac{1}{2}$ mile ($\frac{1}{3}$ mi, $\frac{1}{4}$ mi, $\frac{1}{5}$ mi, $\ldots$), then the point P_2 ($P_3, P_4, P_5, \ldots$) one mile north of a_2 ($a_3, a_4, a_5, \ldots$) along a meridian satisfies the conditions of the problem: she walks one mile south from P_2 ($P_3, P_4, P_5, \ldots$) arriving at a_2 ($a_3, a_4, a_5, \ldots$) along the circle C_2 ($C_3, C_4, C_5, \ldots$), walks east along the circle for one mile thus traversing the circle twice (three times, four times, five times, $\ldots$) returning to a_2 ($a_3, a_4, a_5, \ldots$), and then walks north one mile to P_2 ($P_3, P_4, P_5, \ldots$).

21.   Let r and R be the radius of the smaller circle and larger circle, respectively. Since the line is tangent to the smaller circle, using the Pythagorean Theorem, we get the following relationship between the radii: $R^2 = 1^2 + r^2$. Thus the area of the shaded region is $\pi R^2 - \pi r^2 = \pi(1 + r^2) - \pi r^2 = \pi$.

23.   Since $\sqrt[3]{1729} \approx 12.0023$, we start with $n = 12$ and find the other perfect cube.

n	$1729 - n^3$
12	$1729 - (12)^3 = 1$
11	$1729 - (11)^3 = 398$
10	$1729 - (10)^3 = 729$

Since 1 and 729 are perfect cubes, the two representations we seek are $1^3 + 12^3 = 1729$ and $9^3 + 10^3 = 1729$.

25.   (a)   Imagine that the rooms are colored alternately black and white (like a checkerboard) with the entrance being a white room. So there are 18 black rooms and 18 white rooms. If a tourist is in a white room, he must next go to a black room, and if he is in a black room, the must next go to a white room.

        (b)   Since the entrance is white, any path that the tourist takes must be of the form W B W B W B .... Since the museum contains an even number of room and the tourist starts in a white room, he must end his tour in a black room. But the exit is in a white room, so the proposed tour is impossible.

Chapter One
Exercises 1.1

1. (a) When $x = -2$, LHS $= 4(-2) + 7 = -8 + 7 = -1$ and RHS $= 9(-2) - 3 = -18 - 3$ $= -21$. Since LHS $\neq$ RHS, $x = -2$ is not a solution.

 (b) When $x = 2$, LHS $= 4(-2) + 7 = 8 + 7 = 15$ and RHS $= 9(2) - 3 = 18 - 3 = 15$. Since LHS $=$ RHS, $x = 2$ is a solution.

3. (a) When $x = 2$, LHS $= 1 - [2 - (3 - (2))] = 1 - [2 - 1] = 1 - 1 = 0$ and RHS $= 4(2) - (6 + (2)) = 8 - 8 = 0$. Since LHS $=$ RHS, $x = 2$ is a solution.

 (b) When $x = 4$ LHS $= 1 - [2 - (3 - (4))] = 1 - [2 - (-1)] = 1 - 3 = -2$ and RHS $= 4(4) - (6 + (4)) = 16 - 10 = 6$. Since LHS $\neq$ RHS, $x = 4$ is not a solution.

5. (a) When $x = -1$, LHS $= 2(-1)^{1/3} - 3 = 2(-1) - 3 = -2 - 3 = -5$. Since LHS $\neq 1$, $x = -1$ is not a solution.

 (b) When $x = 8$ LHS $= 2(8)^{1/3} - 3 = 2(2) - 3 = 4 - 3 = 1 =$ RHS. So $x = 8$ is a solution.

7. (a) When $x = 0$, LHS $= \dfrac{0 - a}{0 - b} = \dfrac{-a}{-b} = \dfrac{a}{b} =$ RHS. So $x = 0$ is a solution.

 (b) When $x = b$, LHS $= \dfrac{b - a}{b - b} = \dfrac{b - a}{0}$ is not defined, so $x = b$ is not a solution.

9. $2x + 7 = 31 \quad \Leftrightarrow \quad 2x = 24 \quad \Leftrightarrow \quad x = 12$

11. $\frac{1}{2}x - 8 = 1 \quad \Leftrightarrow \quad \frac{1}{2}x = 9 \quad \Leftrightarrow \quad x = 18$

13. $x - 3 = 2x + 6 \quad \Leftrightarrow \quad -9 = x$

15. $-7w = 15 - 2w \quad \Leftrightarrow \quad -5w = 15 \quad \Leftrightarrow \quad w = -3$

17. $\frac{1}{2}y - 2 = \frac{1}{3}y \quad \Leftrightarrow \quad 3y - 12 = 2y$ (multiply both sides by the LCD, 6) $\quad \Leftrightarrow \quad y = 12$

19. $2(1 - x) = 3(1 + 2x) + 5 \quad \Leftrightarrow \quad 2 - 2x = 3 + 6x + 5 \quad \Leftrightarrow \quad 2 - 2x = 8 + 6x \quad \Leftrightarrow$ $-6 = 8x \quad \Leftrightarrow \quad x = -\frac{3}{4}$

21. $4\left(y - \frac{1}{2}\right) - y = 6(5 - y) \quad \Leftrightarrow \quad 4y - 2 - y = 30 - 6y \quad \Leftrightarrow \ ' \ 3y - 2 = 30 - 6y \quad \Leftrightarrow$ $9y = 32 \quad \Leftrightarrow \quad y = \frac{32}{9}$

23. $\dfrac{1}{x} = \dfrac{4}{3x} + 1 \quad \Rightarrow \quad 3 = 4 + 3x$ (multiply both sides by the LCD $3x$) $\quad \Leftrightarrow \quad -1 = 3x \quad \Leftrightarrow$ $x = -\frac{1}{3}$

25. $\dfrac{2}{t + 6} = \dfrac{3}{t - 1} \quad \Rightarrow \quad 2(t - 1) = 3(t + 6)$ (multiply both sides by the LCD $(t - 1)(t + 6)$) $\quad \Leftrightarrow$ $2t - 2 = 3t + 18 \quad \Leftrightarrow \quad -20 = t$

27. $r - 2[1 - 3(2r + 4)] = 61 \quad \Leftrightarrow \quad r - 2(1 - 6r - 12) = 61 \quad \Leftrightarrow \quad r - 2(-6r - 11) = 61 \quad \Leftrightarrow$
$r + 12r + 22 = 61 \quad \Leftrightarrow \quad 13r = 39 \quad \Leftrightarrow \quad r = 3$

29. $\sqrt{3}\,x + \sqrt{12} = \dfrac{x + 5}{\sqrt{3}} \quad \Leftrightarrow \quad 3x + 6 = x + 5$ (multiply both sides by $\sqrt{3}$) $\quad \Leftrightarrow \quad 2x = -1$
$\Leftrightarrow \quad x = -\frac{1}{2}$

31. $\dfrac{2}{x} - 5 = \dfrac{6}{x} + 4 \quad \Rightarrow \quad 2 - 5x = 6 + 4x \quad \Leftrightarrow \quad -4 = 9x \quad \Leftrightarrow \quad -\dfrac{4}{9} = x$

33. $\dfrac{3}{x + 1} - \dfrac{1}{2} = \dfrac{1}{3x + 3} \quad \Rightarrow \quad 3(6) - (3x + 3) = 2$ (multiply both sides by the LCD $6(x + 1)$)
$\Leftrightarrow \quad 18 - 3x - 3 = 2 \quad \Leftrightarrow \quad -3x + 15 = 2 \quad \Leftrightarrow \quad -3x = -13 \quad \Leftrightarrow \quad x = \frac{13}{3}$

35. $\dfrac{2x - 7}{2x + 4} = \dfrac{2}{3} \quad \Rightarrow \quad (2x - 7)3 = 2(2x + 4)$ (cross multiply) $\quad \Leftrightarrow \quad 6x - 21 = 4x + 8 \quad \Leftrightarrow$
$2x = 29 \quad \Leftrightarrow \quad x = \frac{29}{2}$

37. $x - \frac{1}{3}x - \frac{1}{2}x - 5 = 0 \quad \Leftrightarrow \quad 6x - 2x - 3x - 30 = 0$ (multiply both sides by the LCD 6) $\quad \Leftrightarrow$
$x = 30$

39. $\dfrac{1}{z} - \dfrac{1}{2z} - \dfrac{1}{5z} = \dfrac{10}{z + 1} \quad \Rightarrow \quad 10(z + 1) - 5(z + 1) - 2(z + 1) = 10(10z)$ (multiply both sides
by the LCD $10z(z + 1)$) $\quad \Leftrightarrow \quad 3(z + 1) = 100z \quad \Leftrightarrow \quad 3z + 3 = 100z \quad \Leftrightarrow \quad 3 = 97z \quad \Leftrightarrow$
$\frac{3}{97} = z$

41. $\dfrac{u}{u - \frac{u+1}{2}} = 4 \quad \Rightarrow \quad u = 4\left(u - \dfrac{u + 1}{2}\right)$ (cross multiply) $\quad \Leftrightarrow \quad u = 4u - 2u - 2 \quad \Leftrightarrow$
$u = 2u - 2 \quad \Leftrightarrow \quad 2 = u$

43. $\dfrac{x}{2x - 4} - 2 = \dfrac{1}{x - 2} \quad \Rightarrow \quad x - 2(2x - 4) = 2$ (multiply both sides by the LCD $2(x - 2)$) $\quad \Leftrightarrow$
$x - 4x + 8 = 2 \quad \Leftrightarrow \quad -3x = -6 \quad \Leftrightarrow \quad x = 2$. But substituting $x = 2$ into the original
equation does not work, since we cannot divide by 0. Thus there is no solution.

45. $\dfrac{3}{x + 4} = \dfrac{1}{x} + \dfrac{6x + 12}{x^2 + 4x} \quad \Rightarrow \quad 3(x) = (x + 4) + 6x + 12$ (multiply both sides by the LCD
$x(x + 4)$) $\quad \Leftrightarrow \quad 3x = 7x + 16 \quad \Leftrightarrow \quad -4x = 16 \quad \Leftrightarrow \quad x = -4$. But substituting $x = -4$ into
the original equation does not work, since we cannot divide by 0. Thus there is no solution.

47. $x^2 = 49 \quad \Rightarrow \quad x = \pm 7$

49. $x^2 - 24 = 0 \quad \Leftrightarrow \quad x^2 = 24 \quad \Rightarrow \quad x = \pm\sqrt{24} = \pm 2\sqrt{6}$

51. $8x^2 - 64 = 0 \quad \Leftrightarrow \quad x^2 - 8 = 0 \quad \Leftrightarrow \quad x^2 = 8 \quad \Rightarrow \quad x = \pm\sqrt{8} = \pm 2\sqrt{2}$

53. $x^2 + 16 = 0 \quad \Leftrightarrow \quad x^2 = -16$ which has no real solution.

55. $(x + 2)^2 = 4 \quad \Leftrightarrow \quad (x + 2)^2 = 4 \quad \Rightarrow \quad x + 2 = \pm 2$. If $x + 2 = 2$, then $x = 0$. If $x + 2 = -2$,
then $x = -4$. The solutions are -4 and 0.

57. $x^3 = 27 \quad \Leftrightarrow \quad x = 27^{1/3} = 3.$

59. $0 = x^4 - 16 = (x^2 + 4)(x^2 - 4) = (x^2 + 4)(x - 2)(x + 2).$ $x^2 + 4 = 0$ has no real solution. If $x - 2 = 0$, then $x = 2$. If $x + 2 = 0$, then $x = -2$. The solutions are ± 2.

61. $x^4 + 64 = 0 \quad \Leftrightarrow \quad x^4 = -64$ which has no real solution.

63. $(x + 2)^4 - 81 = 0 \quad \Leftrightarrow \quad (x + 2)^4 = 81 \quad \Leftrightarrow \quad [(x + 2)^4]^{1/4} = \pm 81^{1/4} \quad \Leftrightarrow \quad x + 2 = \pm 3.$ So $x + 2 = 3$, then $x = 1$. If $x + 2 = -3$, then $x = -5$. The solutions are: $-5, 1$.

65. $3(x - 3)^3 = 375 \quad \Leftrightarrow \quad (x - 3)^3 = 125 \quad \Leftrightarrow \quad (x - 3) = 125^{1/3} = 5 \quad \Leftrightarrow \quad x = 3 + 5 = 8.$

67. $\sqrt[3]{x} = 5 \quad \Leftrightarrow \quad x = 5^3 = 125$

69. $2x^{5/3} + 64 = 0 \quad \Leftrightarrow \quad 2x^{5/3} = -64 \quad \Leftrightarrow \quad x^{5/3} = -32 \quad \Leftrightarrow$
$x = (-32)^{3/5} = (-2^5)^{1/5} = (-2)^3 = -8.$

71. $3.02x + 1.48 = 10.92 \quad \Leftrightarrow \quad 3.02x = 9.44 \quad \Leftrightarrow \quad x = \dfrac{9.44}{3.02} \approx 3.13$

73. $2.15x - 4.63 = x + 1.19 \quad \Leftrightarrow \quad 1.15x = 5.82 \quad \Leftrightarrow \quad x = \dfrac{5.82}{1.19} \approx 5.06$

75. $3.16(x + 4.63) = 4.19(x - 7.24) \quad \Leftrightarrow \quad 3.16x + 14.63 = 4.19x - 30.34 \quad \Leftrightarrow \quad 44.97 = 1.03x$
$\Leftrightarrow \quad x = \dfrac{44.97}{1.03} \approx 43.66$

77. $\dfrac{0.26x - 1.94}{3.03 - 2.44x} = 1.76 \quad \Rightarrow \quad 0.26x - 1.94 = 1.76(3.03 - 2.44x) \quad \Leftrightarrow$
$0.26x - 1.94 = 5.33 - 4.29x \quad \Leftrightarrow \quad 4.55x = 7.27 \quad \Leftrightarrow \quad x = \dfrac{7.27}{4.55} \approx 1.60$

79. $PV = nRT \quad \Leftrightarrow \quad R = \dfrac{PV}{nT}$

81. $\dfrac{1}{R} = \dfrac{1}{R_1} + \dfrac{1}{R_2} \quad \Leftrightarrow \quad R_1 R_2 = R R_2 + R R_1$ (multiply both sides by the LCD, $R R_1 R_2$). Thus
$R_1 R_2 - R R_1 = R R_2 \quad \Leftrightarrow \quad R_1(R_2 - R) = R R_2 \quad \Leftrightarrow \quad R_1 = \dfrac{R R_2}{R_2 - R}$

83. $\dfrac{ax + b}{cx + d} = 2 \quad \Leftrightarrow \quad ax + b = 2(cx + d) \quad \Leftrightarrow \quad ax + b = 2cx + 2d \quad \Leftrightarrow \quad ax - 2cx = 2d - b$
$\Leftrightarrow \quad (a - 2c)x = 2d - b \quad \Leftrightarrow \quad x = \dfrac{2d - b}{a - 2c}$

85. $a^2 x + (a - 1) = (a + 1)x \quad \Leftrightarrow \quad a^2 x - (a + 1)x = -(a - 1) \quad \Leftrightarrow \quad (a^2 - (a + 1))x = -a + 1$
$\Leftrightarrow \quad (a^2 - a - 1)x = -a + 1 \quad \Leftrightarrow \quad x = \dfrac{-a + 1}{a^2 - a - 1}$

87. $V = \dfrac{1}{3}\pi r^2 h \quad \Leftrightarrow \quad r^2 = \dfrac{3V}{\pi h} \quad \Rightarrow \quad r = \pm\sqrt{\dfrac{3V}{\pi h}}$

89. $a^2 + b^2 = c^2 \quad \Leftrightarrow \quad b^2 = c^2 - a^2 \quad \Rightarrow \quad b = \pm\sqrt{c^2 - a^2}$

91. $V = \frac{4}{3}\pi r^3 \quad \Leftrightarrow \quad r^3 = \frac{3V}{4\pi} \quad \Leftrightarrow \quad r = \sqrt[3]{\frac{3V}{4\pi}}$

93. (a) The shrinkage factor when $w = 250$ is $S = \dfrac{0.032(250) - 2.5}{10,000} = \dfrac{8 - 2.5}{10,000} = 0.00055$. So the
     beam shrinks $0.00055 \times 12.025 \approx 0.007$ m, so when it dries it will be
     $12.025 - 0.007 = 12.018$ m long.

     (b) Substituting $S = 0.00050$ we get $0.00050 = \dfrac{0.032w - 2.5}{10,000} \quad \Leftrightarrow \quad 5 = 0.032w - 2.5 \quad \Leftrightarrow$
     $7.5 = 0.032w \quad \Leftrightarrow \quad w = \dfrac{7.5}{0.032} \approx 234.375$. So the water content should be 234.375 kg/m^3.

95. (a) Solving for v when $P = 10,000$ we get $10,000 = 15.6v^3 \quad \Leftrightarrow \quad v^3 \approx 641.02 \quad \Leftrightarrow \quad v \approx 8.6$
     km/h.

     (b) Solving for v when $P = 50,000$ we get $50,000 = 15.6v^3 \quad \Leftrightarrow \quad v^3 \approx 3205.13 \quad \Leftrightarrow$
     $v \approx 14.7$ km/h.

97. (a) $3(0) + k - 5 = k(0) - k + 1 \quad \Leftrightarrow \quad k - 5 = -k + 1 \quad \Leftrightarrow \quad 2k = 6 \quad \Leftrightarrow \quad k = 3$

     (b) $3(1) + k - 5 = k(1) - k + 1 \quad \Leftrightarrow \quad 3 + k - 5 = k - k + 1 \quad \Leftrightarrow \quad k - 2 = 1 \quad \Leftrightarrow$
     $k = 3$

     (c) $3(2) + k - 5 = k(2) - k + 1 \quad \Leftrightarrow \quad 6 + k - 5 = 2k - k + 1 \quad \Leftrightarrow \quad k + 1 = k + 1$. Since
     both sides of this equation are equal, $x = 2$ is a solution for every value of k. That is, $x = 2$ is
     a solution to every member of this family of equations.

99. (a) Using the formulas for volume we have: for the sphere, $V = \frac{4}{3}\pi r^3$; for the cylinder,
     $V = \pi r^2 h_1$; and for the cone, $V = \frac{1}{3}\pi r^2 h_2$. Since these have the same volume we must have
     $\frac{4}{3}\pi r^3 = \pi r^2 h_1$ and $\frac{4}{3}\pi r^3 = \frac{1}{3}\pi r^2 h_2$.

     (b) Setting the volume of the sphere equal to the volume of the cylinder, we have $\frac{4}{3}\pi r^3 = \pi r^2 h_1$
     $\Leftrightarrow \quad \frac{4}{3}r = h_1$. So the height of the cylinder is $h_1 = \frac{4}{3}r$.

     Likewise, setting the volume of the sphere equal to the volume of the cone, we have
     $\frac{4}{3}\pi r^3 = \frac{1}{3}\pi r^2 h_2 \quad \Leftrightarrow \quad 4r = h_2$. So the height of the cone is $h_2 = 4r$.

Exercises 1.2

1. If n is the first integer, then $n + 1$ is the middle integer, and $n + 2$ is the third integer. So the sum of the three consecutive integers is $n + (n + 1) + (n + 2) = 3n + 3$.

3. If s is the third test score, then since the other test scores are 78 and 82, the average of the three test scores is $\dfrac{78 + 82 + s}{3} = \dfrac{160 + s}{3}$.

5. If x dollars are invested at $2\frac{1}{2}\%$ simple interest, then the first year you will receive $0.025x$ dollars in interest.

7. Since w is the width of the rectangle, the length is three times the width, so length $= 3w$. Then the $area = length \times width = 3w \times w = 3w^2$ ft^2.

9. Since $distance = rate \times time$ we have $distance = s \times (45 \text{ min}) \dfrac{1 \text{ hr}}{60 \text{ min}} = \dfrac{3}{4} s$ mi.

11. If x is the gallons of pure water added, the mixture will contain 25 oz of salt and $3 + x$ gallons of water. Thus the concentration is $\dfrac{25}{3 + x}$.

13. Let x be the first integer. Then $x + 1$ and $x + 2$ are the next consecutive integers. So
$x + (x + 1) + (x + 2) = 336 \quad \Leftrightarrow \quad 3x + 3 = 336 \quad \Leftrightarrow \quad 3x = 333 \quad \Leftrightarrow \quad x = 111$. Thus the consecutive integers are 111, 112, and 113.

15. Let m be the amount invested at $4\frac{1}{2}\%$. Then $12{,}000 - m$ is the amount invested at 4%.
Since $total\ interest = (interest\ earned\ at\ 4\frac{1}{2}\%) + (interest\ earned\ at\ 4\%)$, we have
$525 = 0.045m + 0.04(12{,}000 - m) \quad \Leftrightarrow \quad 525 = 0.045m + 480 - 0.04m \quad \Leftrightarrow \quad 45 = 0.005m$
$\Leftrightarrow \quad m = \frac{45}{0.005} = 9{,}000$. Thus \$9,000 is invested at $4\frac{1}{2}\%$, and $\$12{,}000 - 9{,}000 = \$3{,}000$ is invested at 4%.

17. Let r be the annual interest rate. Then $3500r = 262.50 \quad \Leftrightarrow \quad r = \frac{262.50}{3500} = 0.075$. Thus the \$3500 must be invested at 7.5%.

19. Let x be her monthly salary. Since her $annual\ salary = 12 \times (monthly\ salary) + (Christmas\ bonus)$ we have $97{,}300 = 12x + 8{,}500 \quad \Leftrightarrow \quad 88{,}800 = 12x \quad \Leftrightarrow \quad x \approx 7{,}400$. Her monthly salary is \$7,400.

21. Let h be the amount that Craig inherits. So $(x + 22{,}000)$ is the amount that he invests and doubles.
Thus $2(x + 22{,}000) = 134{,}000 \quad \Leftrightarrow \quad 2x + 44{,}000 = 134{,}000 \quad \Leftrightarrow \quad 2x = 90{,}000 \quad \Leftrightarrow$
$x = 45{,}000$. So Craig inherits \$45,000.

23. Let x be the hours the assistant work. Then $2x$ is the hours the plumber worked. Since the
$labor\ charge = plumber's\ labor + assistant's\ labor$, we have
$4025 = 45(2x) + 25x \quad \Leftrightarrow \quad 4025 = 90x + 25x \quad \Leftrightarrow \quad 4025 = 115x \quad \Leftrightarrow \quad x = \frac{4025}{115} = 35$
Thus the assistant works for 35 hours, and the plumber works for $2 \times 35 = 70$ hours.

25. All ages are in terms of the daughter's age 7 years ago. Let y be age of the daughter 7 years ago. Then $11y$ is the age of the movie star 7 years ago. Today, the daughter is $y + 7$, and the movie star is $11y + 7$. But the movie star is also 4 times his daughter's age today. So $4(y + 7) = 11y + 7$ $\Leftrightarrow$ $4y + 28 = 11y + 7$ $\Leftrightarrow$ $21 = 7y$ $\Leftrightarrow$ $y = 3$. Thus the movie star's age today is $11(3) + 7 = 40$ years.

27. Let p be the number of pennies. Then p is the number of nickels and p is the number of dimes. So the *value of the coins in the purse = value of the pennies + value of the nickels + value of the dimes*. Thus $1.44 = 0.01p + 0.05p + 0.10p$ $\Leftrightarrow$ $1.44 = 0.16p$ $\Leftrightarrow$ $p = \frac{1.44}{0.16} = 9$. So the purse contains 9 pennies, 9 nickels, and 9 dimes.

29. Let l be the length of the garden. Since *area = width $\cdot$ length*, we obtain the equation $1125 = 25l$ $\Leftrightarrow$ $l = \frac{1125}{25} = 45$ ft. So the garden is 45 feet long.

31. Let x be the length of a side of the square plot. As shown in the figure to the right,
area of the plot = area of the building +

                        area of the parking lot

So $x^2 = 60(40) + 12,000 = 2,400 + 12,000 = 14,400$ $\Rightarrow$ $x = \pm 120$. So the plot of land is 120 feet by 120 feet.

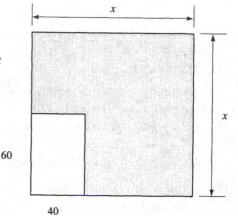

33. The figure is a trapezoid, so the *area* $= \dfrac{base_1 + base_2}{2}(height)$. Putting in the known quantities we have $120 = \dfrac{y + 2y}{2}(y) = \frac{3}{2}y^2$ $\Leftrightarrow$ $y^2 = 80$ $\Rightarrow$ $y = \pm\sqrt{80} = \pm 4\sqrt{5}$. Since length is positive, $y = 4\sqrt{5} \approx 8.94$.

35. Let x be the width of the strip. Then the length of the mat is $20 + 2x$, and the width of the mat is $15 + 2x$. Thus *perimeter* $= 2 \times length + 2 \times width$ $\Leftrightarrow$ $102 = 2(20 + 2x) + 2(15 + 2x)$ $\Leftrightarrow$ $102 = 40 + 4x + 30 + 4x$ $\Leftrightarrow$ $102 = 70 + 8x$ $\Leftrightarrow$ $32 = 8x$ $\Leftrightarrow$ $x = 4$. Thus the strip of mat is 4 inches wide.

37. Let x be the grams of silver added. The weight of the rings is $5 \times 18\,\text{g} = 90\,\text{g}$.

	5 rings	Pure silver	mixture
grams	90	x	$90 + x$
rate (% gold)	0.90	0	0.75
value	$0.90(90)$	$0x$	$0.75(90 + x)$

So $0.90(90) + 0x = 0.75(90 + x)$ $\Leftrightarrow$ $81 = 67.5 + 0.75x$ $\Leftrightarrow$ $0.75x = 13.5$ $\Leftrightarrow$ $x = \frac{13.5}{0.75} = 18$. Thus 18 grams of silver must be added to get the required mixture.

39. Let x be the liters of coolant removed and replaced by water.

	60% antifreeze	60% antifreeze (removed)	water	mixture
liters	3.6	x	x	3.6
rate (% antifreeze)	0.60	0.60	0	0.50
value	0.60(3.6)	$-0.60x$	$0x$	0.50(3.6)

so $0.60(3.6) - 0.60x + 0x = 0.50(3.6)$ $\Leftrightarrow$ $2.16 - 0.6x = 1.8$ $\Leftrightarrow$ $-0.6x = -0.36$ $\Leftrightarrow$ $x = \frac{-0.36}{-0.6} = 0.6$. Thus 0.6 liters must be removed and replaced by water.

41. Let c be the concentration of fruit juice in the cheaper brand. The new mixture that Jill makes will consist of 650 ml of the original fruit punch and 100 ml of the cheaper fruit punch.

	Fruit Punch	Cheaper FP	mixture
ml	650	100	750
concentration	0.50	c	0.48
juice	$0.50 \cdot 650$	$100c$	$0.48 \cdot 750$

So $0.50 \cdot 650 + 100c = 0.48 \cdot 750$ $\Leftrightarrow$ $325 + 100c = 360$ $\Leftrightarrow$ $100c = 35$ $\Leftrightarrow$ $c = 0.35$. Thus the cheaper brand is only 35% fruit juice.

43. Let x be the length of the man's shadow, in meters. Using similar triangles, $\dfrac{10 + x}{6} = \dfrac{x}{2}$ $\Leftrightarrow$ $20 + 2x = 6x$ $\Leftrightarrow$ $4x = 20$ $\Leftrightarrow$ $x = 5$. Thus the man's shadow is 5 meters long.

45. Let x be the distance from the fulcrum to where the mother sits. Then substituting the known values into the formula given, we have $100(8) = 125x$ $\Leftrightarrow$ $800 = 125x$ $\Leftrightarrow$ $x = 6.4$. So the mother should sit 6.4 feet from the fulcrum.

47. Let t be the time in minutes it would take Candy and Tim if they work together. Candy delivers the papers at a rate of $\frac{1}{70}$ of the job per minute, while Tim delivers the paper at a rate of $\frac{1}{80}$ of the job per minute. The sum of the fractions of the job that each can do individually in one minute equals the fraction of the job they can do working together. So we have $\dfrac{1}{t} = \dfrac{1}{70} + \dfrac{1}{80}$ $\Leftrightarrow$ $560 = 8t + 7t$ $\Leftrightarrow$ $560 = 15t$ $\Leftrightarrow$ $t = 37\frac{1}{3}$ minutes. Since $\frac{1}{3}$ of a minute is 20 seconds, it would take them 37 minutes 20 seconds if they worked together.

49. Let t be the time, in hours, it takes Karen to paint a house alone. Then working together, Karen and Betty can paint a house in $\frac{2}{3}t$ hours. The sum of their individual rates equals their rate working together, so $\dfrac{1}{t} + \dfrac{1}{6} = \dfrac{1}{\frac{2}{3}t}$ $\Leftrightarrow$ $\dfrac{1}{t} + \dfrac{1}{6} = \dfrac{3}{2t}$ $\Leftrightarrow$ $6 + t = 9$ $\Leftrightarrow$ $t = 3$. Thus it would take Karen 3 hours to paint a house alone.

51. Let t be the time in hours that Wendy spent on the train. Then $\frac{11}{2} - t$ is the time in hours that Wendy spent on the bus. Using the equation $total\ distance = \left(\begin{smallmatrix}distance\ traveled\\by\ bus\end{smallmatrix}\right) + \left(\begin{smallmatrix}distance\ traveled\\by\ train\end{smallmatrix}\right)$, and the table

	Rate	Time	Distance
By train	40	t	$40t$
By bus	60	$\frac{11}{2} - t$	$60\left(\frac{11}{2} - t\right)$

we get the equation $300 = 40t + 60\left(\frac{11}{2} - t\right)$ $\Leftrightarrow$ $300 = 40t + 330 - 60t$ $\Leftrightarrow$ $-30 = -20t$ $\Leftrightarrow$ $t = \frac{-30}{-20} = 1.5$ hours. So the time spent on the train is $5.5 - 1.5 = 4$ hours.

53. Let r be the speed of the plane from Montreal to Los Angeles. Then $r + 0.20r = 1.20r$ is the speed of the plane from Los Angeles to Montreal.

	Rate	Time	Distance
Montreal to L.A.	r	$\dfrac{2500}{r}$	2500
L.A. to Montreal	$1.2r$	$\dfrac{2500}{1.2r}$	2500

Using the equation $total\ time = \left(\begin{smallmatrix} time\ traveling \\ Montreal\ to\ LA \end{smallmatrix}\right) + \left(\begin{smallmatrix} time\ traveling \\ LA\ to\ Montreal \end{smallmatrix}\right)$, we have

$9\frac{1}{6} = \dfrac{2500}{r} + \dfrac{2500}{1.2r}$ $\Leftrightarrow$ $\dfrac{55}{6} = \dfrac{2500}{r} + \dfrac{2500}{1.2r}$ $\Leftrightarrow$ $55 \cdot 1.2r = 2500 \cdot 6 \cdot 1.2 + 2500 \cdot 6$ $\Leftrightarrow$ $66r = 18000 + 15000$ $\Leftrightarrow$ $66r = 33000$ $\Leftrightarrow$ $r = \dfrac{33000}{66} = 500$. Thus the plane flew at a speed of 500 mph on the trip from Montreal to Los Angeles.

55. Let l be the length of the lot in feet. Then the length of the diagonal is $l + 10$. We apply the Pythagorean Theorem with the hypotenuse as the diagonal. So $l^2 + 50^2 = (l + 10)^2$ $\Leftrightarrow$ $l^2 + 2500 = l^2 + 20l + 100$ $\Leftrightarrow$ $20l = 2400$ $\Leftrightarrow$ $l = 120$. Thus the length of the lot is 120 feet.

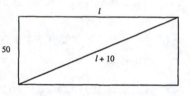

57. Let h be the height in feet of the structure. The structure is composed of a right cylinder with radius 10 and height $\frac{2}{3}h$ and a cone with base radius 10 and height $\frac{1}{3}h$. Using the formulas for the volume of a cylinder and that of a cone, we obtain the equation $1400\pi = \pi(10)^2\left(\frac{2}{3}h\right) + \frac{1}{3}\pi(10)^2\left(\frac{1}{3}h\right)$ $\Leftrightarrow$ $1400\pi = \frac{200\pi}{3}h + \frac{100\pi}{9}h$ $\Leftrightarrow$ $126 = 6h + h$ (multiply both sides by $\frac{9}{100\pi}$) $\Leftrightarrow$ $126 = 7h$ $\Leftrightarrow$ $h = 18$. Thus the height of the structure is 18 feet.

Exercises 1.3

1. $x^2 + x - 12 = 0 \iff (x-3)(x+4) = 0 \iff x - 3 = 0$ or $x + 4 = 0$. Thus $x = 3$ or $x = -4$.

3. $x^2 - 7x + 12 = 0 \iff (x-4)(x-3) = 0 \iff x - 4 = 0$ or $x - 3 = 0$. Thus $x = 4$ or $x = 3$.

5. $3x^2 - 5x - 2 = 0 \iff (3x+1)(x-2) = 0 \iff 3x + 1 = 0$ or $x - 2 = 0$. Thus $x = -\frac{1}{3}$ or $x = 2$.

7. $2y^2 + 7y + 3 = 0 \iff (2y+1)(y+3) = 0 \iff 2y + 1 = 0$ or $y + 3 = 0$. Thus $y = -\frac{1}{2}$ or $y = -3$.

9. $6x^2 + 5x = 4 \iff 6x^2 + 5x - 4 = 0 \iff (2x-1)(3x+4) = 0 \iff 2x - 1 = 0$ or $3x + 4 = 0$. If $2x - 1 = 0$, then $x = \frac{1}{2}$; if $3x + 4 = 0$, then $x = -\frac{4}{3}$.

11. $x^2 = 5(x + 100) \iff x^2 = 5x + 500 \iff x^2 - 5x - 500 = 0 \iff (x-25)(x+20) = 0 \iff x - 25 = 0$ or $x + 20 = 0$. Thus $x = 25$ or $x = -20$.

13. $x^2 + 2x - 5 = 0 \iff x^2 + 2x = 5 \iff x^2 + 2x + 1 = 5 + 1 \iff (x+1)^2 = 6 \implies x + 1 = \pm\sqrt{6} \iff x = -1 \pm \sqrt{6}$.

15. $x^2 - 6x - 11 = 0 \iff x^2 - 6x = 11 \iff x^2 - 6x + 9 = 11 + 9 \iff (x-3)^2 = 20 \implies x - 3 = \pm 2\sqrt{5} \iff x = 3 \pm 2\sqrt{5}$.

17. $x^2 + x - \frac{3}{4} = 0 \iff x^2 + x = \frac{3}{4} \iff x^2 + x + \frac{1}{4} = \frac{3}{4} + \frac{1}{4} \iff (x + \frac{1}{2})^2 = 1 \implies x + \frac{1}{2} = \pm 1 \iff x = -\frac{1}{2} \pm 1$. So $x = -\frac{1}{2} - 1 = \frac{-3}{2}$ and $x = -\frac{1}{2} + 1 = \frac{1}{2}$.

19. $x^2 + 22x + 21 = 0 \iff x^2 + 22x = -21 \iff x^2 + 22x + 11^2 = -21 + 11^2 = -21 + 121 \iff (x+11)^2 = 100 \implies x + 11 = \pm 10 \iff x = -11 \pm 10$. Thus $x = -1$ or $x = -21$.

21. $2x^2 + 8x + 1 = 0 \iff x^2 + 4x + \frac{1}{2} = 0 \iff x^2 + 4x = -\frac{1}{2} \iff x^2 + 4x + 4 = -\frac{1}{2} + 4 \iff (x+2)^2 = \frac{7}{2} \implies x + 2 = \pm\sqrt{\frac{7}{2}} \iff x = -2 \pm \frac{\sqrt{14}}{2}$.

23. $4x^2 - x = 0 \iff x^2 - \frac{1}{4}x = 0 \iff x^2 - \frac{1}{4}x + \frac{1}{64} = \frac{1}{64} \iff (x - \frac{1}{8})^2 = \frac{1}{64} \implies x - \frac{1}{8} = \pm\frac{1}{8} \iff x = \frac{1}{8} \pm \frac{1}{8}$, so $x = \frac{1}{8} - \frac{1}{8} = 0$ or $x = \frac{1}{8} + \frac{1}{8} = \frac{1}{4}$.

25. $x^2 - 2x - 15 = 0 \iff (x+3)(x-5) = 0 \iff x + 3 = 0$ or $x - 5 = 0$. Thus $x = -3$ or $x = 5$.

27. $x^2 - 7x + 10 = 0 \iff (x-5)(x-2) = 0 \iff x - 5 = 0$ or $x - 2 = 0$. Thus $x = 5$ or $x = 2$.

29. $2x^2 + x - 3 = 0 \iff (x-1)(2x+3) = 0 \iff x - 1 = 0$ or $2x + 3 = 0$. If $x - 1 = 0$, then $x = 1$; if $2x + 3 = 0$, then $x = -\frac{3}{2}$.

31. $x^2 + 3x + 1 = 0 \quad \Leftrightarrow \quad x^2 + 3x = -1 \quad \Leftrightarrow \quad x^2 + 3x + \frac{9}{4} = -1 + \frac{9}{4} = \frac{5}{4} \quad \Leftrightarrow \quad (x + \frac{3}{2})^2 = \frac{5}{4}$

$\Rightarrow \quad x + \frac{3}{2} = \pm\sqrt{\frac{5}{4}} = \pm\frac{\sqrt{5}}{2} \quad \Leftrightarrow \quad x = -\frac{3}{2} \pm \frac{\sqrt{5}}{2}.$

33. $x^2 + 12x - 27 = 0 \quad \Leftrightarrow \quad x^2 + 12x = 27 \quad \Leftrightarrow \quad x^2 + 12x + 36 = 27 + 36 \quad \Leftrightarrow$

$(x + 6)^2 = 63 \quad \Rightarrow \quad x + 6 = \pm 3\sqrt{7} \quad \Leftrightarrow \quad x = -6 \pm 3\sqrt{7}.$

35. $3x^2 + 6x - 5 = 0 \quad \Leftrightarrow \quad x^2 + 2x - \frac{5}{3} = 0 \quad \Leftrightarrow \quad x^2 + 2x = \frac{5}{3} \quad \Leftrightarrow \quad x^2 + 2x + 1 = \frac{5}{3} + 1$

$\Leftrightarrow \quad (x + 1)^2 = \frac{8}{3} \quad \Rightarrow \quad x + 1 = \pm\sqrt{\frac{8}{3}} \quad \Leftrightarrow \quad x = -1 \pm \frac{2\sqrt{6}}{3}.$

37. $2y^2 - y - \frac{1}{2} = 0 \quad \Rightarrow \quad y = \dfrac{-b \pm \sqrt{b^2 - 4ac}}{2a} = \dfrac{-(-1) \pm \sqrt{(-1)^2 - 4(2)(-\frac{1}{2})}}{2(2)} = \dfrac{1 \pm \sqrt{1 + 4}}{4}$

$= \dfrac{1 \pm \sqrt{5}}{4}.$

39. $4x^2 + 16x - 9 = 0 \quad \Leftrightarrow \quad (2x - 1)(2x + 9) = 0 \quad \Leftrightarrow \quad 2x - 1 = 0 \text{ or } 2x + 9 = 0. \text{ If}$

$2x - 1 = 0$, then $x = \frac{1}{2}$; if $2x + 9 = 0$, then $x = -\frac{9}{2}$.

41. $3 + 5z + z^2 = 0 \quad \Rightarrow \quad z = \dfrac{-b \pm \sqrt{b^2 - 4ac}}{2a} = \dfrac{-(5) \pm \sqrt{(5)^2 - 4(1)(3)}}{2(1)} = \dfrac{-5 \pm \sqrt{25 - 12}}{2}$

$= \dfrac{-5 \pm \sqrt{13}}{2}.$

43. $x^2 - \sqrt{5}x + 1 = 0 \quad \Rightarrow \quad x = \dfrac{-b \pm \sqrt{b^2 - 4ac}}{2a} = \dfrac{-\left(-\sqrt{5}\right) \pm \sqrt{\left(-\sqrt{5}\right)^2 - 4(1)(1)}}{2(1)} =$

$\dfrac{\sqrt{5} \pm \sqrt{5 - 4}}{2} = \dfrac{\sqrt{5} \pm 1}{2}.$

45. $10y^2 - 16y + 5 = 0 \quad \Rightarrow \quad x = \dfrac{-b \pm \sqrt{b^2 - 4ac}}{2a} = \dfrac{-(-16) \pm \sqrt{(-16)^2 - 4(10)(5)}}{2(10)} =$

$\dfrac{16 \pm \sqrt{256 - 200}}{20} = \dfrac{16 \pm \sqrt{56}}{20} = \dfrac{16 \pm 2\sqrt{14}}{20} = \dfrac{8 \pm \sqrt{14}}{10}.$

47. $3x^2 + 2x + 2 = 0 \quad \Rightarrow \quad x = \dfrac{-b \pm \sqrt{b^2 - 4ac}}{2a} = \dfrac{-(2) \pm \sqrt{(2)^2 - 4(3)(2)}}{2(3)}$

$= \dfrac{-2 \pm \sqrt{4 - 24}}{6} = \; = \dfrac{-2 \pm \sqrt{-20}}{6}.$ Since the discriminant is less than 0, the equation has no

real solutions.

49. $x^2 - 0.011x - 0.064 = 0 \quad \Rightarrow \quad x = \dfrac{-(-0.011) \pm \sqrt{(-0.011)^2 - 4(1)(-0.064)}}{2(1)}$

$= \dfrac{0.011 \pm \sqrt{0.000121 + 0.256}}{2} = \dfrac{0.011 \pm \sqrt{0.256121}}{2} \approx \dfrac{0.011 \pm 0.506}{2}$

Thus $x \approx \dfrac{0.011 + 0.506}{2} = 0.259$ or $x \approx \dfrac{0.011 - 0.506}{2} = -0.248$.

51. $x^2 - 2.450x + 1.501 = 0 \quad \Rightarrow \quad x = \dfrac{-(-2.450) \pm \sqrt{(-2.450)^2 - 4(1)(1.501)}}{2(1)}$

$= \dfrac{2.450 \pm \sqrt{6.0025 - 6.004}}{2} = \dfrac{2.450 \pm \sqrt{-0.0015}}{2}$. Thus there are no real solutions.

53. $2.232x^2 - 4.112x = 6.219 \quad \Leftrightarrow \quad 2.232x^2 - 4.112x - 6.219 = 0 \quad \Rightarrow$

$x = \dfrac{-(-4.112) \pm \sqrt{(-4.112)^2 - 4(2.232)(-6.219)}}{2(2.232)} = \dfrac{4.112 \pm \sqrt{16.9085 + 55.5232}}{4.464} =$

$\dfrac{4.112 \pm \sqrt{72.4317}}{4.464} = \dfrac{4.112 \pm 8.511}{4.464}$. Thus $x = \dfrac{4.112 - 8.511}{4.464} = -0.985$ or

$x = \dfrac{4.112 + 8.511}{4.464} = 2.828$.

55. $h = \frac{1}{2}gt^2 + v_0 t \quad \Leftrightarrow \quad \frac{1}{2}gt^2 + v_0 t - h = 0$. Using the quadratic formula,

$t = \dfrac{-(v_0) \pm \sqrt{(v_0)^2 - 4\left(\frac{1}{2}g\right)(-h)}}{2\left(\frac{1}{2}g\right)} = \dfrac{-v_0 \pm \sqrt{v_0^2 + 2gh}}{g}$.

57. $A = 2x^2 + 4xh \quad \Leftrightarrow \quad 2x^2 + 4xh - A = 0$. Using the quadratic formula,

$x = \dfrac{-(4h) \pm \sqrt{(4h)^2 - 4(2)(-A)}}{2(2)} = \dfrac{-4h \pm \sqrt{16h^2 + 8A}}{4} = \dfrac{-4h \pm \sqrt{4(4h^2 + 2A)}}{4} =$

$\dfrac{-4h \pm 2\sqrt{4h^2 + 2A}}{4} = \dfrac{2(-2h \pm \sqrt{4h^2 + 2A})}{4} = \dfrac{-2h \pm \sqrt{4h^2 + 2A}}{2}$.

59. $\dfrac{1}{s+a} + \dfrac{1}{s+b} = \dfrac{1}{c} \quad \Leftrightarrow \quad c(s+b) + c(s+a) = (s+a)(s+b) \quad \Leftrightarrow$

$cs + bc + cs + ac = s^2 + as + bs + ab \quad \Leftrightarrow \quad s^2 + (a+b-2c)s + (ab - ac - bc) = 0$. Using

the quadratic formula, $s = \dfrac{-(a+b-2c) \pm \sqrt{(a+b-2c)^2 - 4(1)(ab - ac - bc)}}{2(1)}$

$= \dfrac{-(a+b-2c) \pm \sqrt{a^2 + b^2 + 4c^2 + 2ab - 4ac - 4bc - 4ab + 4ac + 4bc}}{2}$

$= \dfrac{-(a+b-2c) \pm \sqrt{a^2 + b^2 + 4c^2 - 2ab}}{2}$.

61. $D = b^2 - 4ac = (-6)^2 - 4(1)(1) = 32$. Since D is positive, this equation has two real solutions.

63. $D = b^2 - 4ac = (2.20)^2 - 4(1)(1.21) = 4.84 - 4.84 = 0$. Since $D = 0$, this equation has one
real solution.

65. $D = b^2 - 4ac = (5)^2 - 4(4)(\frac{13}{8}) = 25 - 26 = -1$. Since D is negative, this equation has no real
solutions.

67. $D = b^2 - 4ac = (r)^2 - 4(1)(-s) = r^2 + 4s$. Since D is positive, this equation has two real
solutions.

69. $a^2x^2 + 2ax + 1 = 0$ $\Leftrightarrow$ $(ax+1)^2 = 0$ $\Leftrightarrow$ $ax+1 = 0$. So $ax+1 = 0$ then $ax = -1$ $\Leftrightarrow$ $x = -\dfrac{1}{a}$.

71. $ax^2 - (2a+1)x + (a+1) = 0$ $\Leftrightarrow$ $[ax - (a+1)](x-1) = 0$ $\Leftrightarrow$ $ax - (a+1) = 0$ or $x - 1 = 0$. If $ax - (a+1) = 0$, then $x = \dfrac{a+1}{a}$; if $x - 1 = 0$, then $x = 1$.

73. We want to find the values of k that make the discriminant 0. Thus $k^2 - 4(4)(25) = 0$ $\Leftrightarrow$ $k^2 = 400$ $\Leftrightarrow$ $k = \pm 20$.

75. Let n be one number. Then the other number must be $55 - n$, since $n + (55 - n) = 55$. Because the product is 684, we have $(n)(55 - n) = 684$ $\Leftrightarrow$ $55n - n^2 = 684$ $\Leftrightarrow$ $n^2 - 55n + 684 = 0$ $\Rightarrow$

$$n = \frac{-(-55) \pm \sqrt{(-55)^2 - 4(1)(684)}}{2(1)} = \frac{55 \pm \sqrt{3025 - 2736}}{2} = \frac{55 \pm \sqrt{289}}{2} = \frac{55 \pm 17}{2}. \text{ So}$$

$n = \frac{55+17}{2} = \frac{72}{2} = 36$ or $n = \frac{55-17}{2} = \frac{38}{2} = 19$. In either case, the two numbers are 19 and 36.

77. Let w be the width of the garden in feet. Then the length is $w + 10$. Thus $875 = w(w+10)$ $\Leftrightarrow$ $w^2 + 10w - 875 = 0$ $\Leftrightarrow$ $(w+35)(w-25) = 0$. So $w + 35 = 0$ in which case $w = -35$, which is not possible, or $w - 25 = 0$ and so $w = 25$. Thus the width is 25 feet and the length is 35 feet.

79. Let w be the width of the garden in feet. We use the perimeter to express the length of the garden in terms of width. Since the *perimeter* $= 2 \times width + 2 \times length$, we have $200 = 2w + 2length$ $\Leftrightarrow$ $2length = 200 - 2w$ $\Leftrightarrow$ $length = 100 - w$. Using the formula for area, we have $2400 = w(100 - w) = 100w - w^2$ $\Leftrightarrow$ $w^2 - 100w + 2400 = 0$ $\Leftrightarrow$ $(w-40)(w-60) = 0$. So $w - 40 = 0$ $\Leftrightarrow$ $w = 40$, or $w - 60 = 0$ $\Leftrightarrow$ $w = 60$. If $w = 40$, then $length = 100 - 40 = 60$. And if $w = 60$, then $length = 100 - 60 = 40$. So the length is 60 feet and the width is 40 feet.

81. This shape is a trapezoid whose area is given by the formula $area = \dfrac{base_1 + base_2}{2} \times height$. Substituting 1 for the top base, $x + 1$ for the bottom base, and x for height, we have

$$1200 = \frac{1 + (1 + x)}{2}x = \frac{1}{2}(x+2)x \quad \Leftrightarrow \quad 2400 = x^2 + 2x \quad \Leftrightarrow \quad x^2 + 2x - 2400 = 0 \quad \Leftrightarrow$$

$(x+50)(x-48) = 0$. So $x = -50$ or $x = 48$. Since x represents a length, x is positive, so $x = 48$ cm.

83. Let x be the length of one side of the cardboard, so we start with a piece of cardboard x by x. When 4 inches are removed from each side, the base of the box is $(x-8)$ by $(x-8)$. Since the volume is 100 in^3, we get $4(x-8)^2 = 100$ $\Leftrightarrow$ $x^2 - 16x + 64 = 25$ $\Leftrightarrow$ $x^2 - 16x + 39 = 0$ $\Leftrightarrow$ $(x-3)(x-13) = 0$. So $x = 3$ or $x = 13$. But $x = 3$ is not possible, since then the length of the base would be $3 - 8 = -5$, and all lengths must be positive. Thus $x = 13$, and the piece of cardboard is 13 inches by 13 inches.

85. Let w be the width of the lot in feet. Then the length is $w + 6$. Using the Pythagorean Theorem, we have $w^2 + (w+6)^2 = (174)^2$ $\Leftrightarrow$ $w^2 + w^2 + 12w + 36 = 30276$ $\Leftrightarrow$

$2w^2 + 12w - 30240 = 0 \quad \Leftrightarrow \quad w^2 + 6w - 15120 = 0 \quad \Leftrightarrow \quad (w + 126)(w - 120) = 0$. So either $w + 126 = 0$ in which case $w = -126$, which is not possible, or $w - 120 = 0$ in which case $w = 120$. Thus the width is 120 feet and the length is 126 feet.

87. Using $h_0 = 288$, we solve $0 = -16t^2 + 288$, for $t \geq 0$. So $0 = -16t^2 + 288 \quad \Leftrightarrow \quad 16t^2 = 288$ $\Leftrightarrow \quad t^2 = 18 \quad \Rightarrow \quad t = \pm\sqrt{18} = \pm3\sqrt{2}$. Thus it takes $3\sqrt{2} \approx 4.24$ seconds for the ball the hit the ground.

89. We are given $v_o = 40$ ft/s

   (a) Setting $h = 24$, we have $24 = -16t^2 + 40t \quad \Leftrightarrow \quad 16t^2 - 40t + 24 = 0 \quad \Leftrightarrow$ $8(2t^2 - 5t + 3) = 0 \quad \Leftrightarrow \quad 8(2t - 3)(t - 1) = 0 \quad \Leftrightarrow \quad t = 1$ or $t = 1\frac{1}{2}$. Therefore, the ball reaches 24 feet in 1 second (on the ascent) and again after $1\frac{1}{2}$ seconds (on its descent).

   (b) Setting $h = 48$, we have $48 = -16t^2 + 40t \quad \Leftrightarrow \quad 16t^2 - 40t + 48 = 0 \quad \Leftrightarrow$ $2t^2 - 5t + 6 = 0 \quad \Leftrightarrow \quad t = \frac{5 \pm \sqrt{25-48}}{4} = \frac{5 \pm \sqrt{-23}}{4}$. However, since the discriminant $D < 0$, there are no real solutions, and hence the ball never reaches a height of 48 feet.

   (c) The greatest height h is reached only once. So $h = -16t^2 + 40t \quad \Leftrightarrow \quad 16t^2 - 40t + h = 0$ has only one solution. Thus $D = (-40)^2 - 4(16)(h) = 0 \quad \Leftrightarrow \quad 1600 - 64h = 0 \quad \Leftrightarrow$ $h = 25$. So the greatest height reached by the ball is 25 feet.

   (d) Setting $h = 25$, we have $25 = -16t^2 + 40t \quad \Leftrightarrow \quad 16t^2 - 40t + 25 = 0 \quad \Leftrightarrow$ $(4t - 5)^2 = 0 \quad \Leftrightarrow \quad t = 1\frac{1}{4}$. Thus the ball reaches the highest point of its path after $1\frac{1}{4}$ seconds.

   (e) Setting $h = 0$ (ground level), we have $0 = -16t^2 + 40t \quad \Leftrightarrow \quad 2t^2 - 5t = 0 \quad \Leftrightarrow$ $t(2t - 5) = 0 \quad \Leftrightarrow \quad t = 0$ (start) or $t = 2\frac{1}{2}$. So the ball hits the ground in $2\frac{1}{2}$ seconds.

91. (a) The fish population on January 1, 2002 corresponds to $t = 0$, so $F = 1000(30 + 17(0) - (0)^2) = 30,000$. To find when the population will again reach this value, we set $F = 30,000$, giving $30000 = 1000(30 + 17t - t^2) = 30000 + 17000t - 1000t^2$ $\Leftrightarrow \quad 0 = 17000t - 1000t^2 = 1000t(17 - t) \quad \Leftrightarrow \quad t = 0$ or $t = 17$. Thus the fish population will again be the same 17 years later, that is, on January 1, 2019.

   (b) Setting $F = 0$, we have $0 = 1000(30 + 17t - t^2) \quad \Leftrightarrow \quad t^2 - 17t - 30 = 0 \quad \Leftrightarrow$ $t = \frac{17 \pm \sqrt{289+120}}{-2} = \frac{17 \pm \sqrt{409}}{-2} = \frac{17 \pm 20.22}{2}$. Thus $t \approx -1.612$ or $t \approx 18.612$. Since $t < 0$ is inadmissible, it follows that the fish in the lake will have died out 18.612 years after January 1, 2002, that is on Aug. 12, 2020.

93. Let x be the rate, in mi/h, at which the salesman drove between Ajax and Barrington.

Cities	Distance	Rate	Time
Ajax $\rightarrow$ Barrington	120	x	$\dfrac{120}{x}$
Barrington $\rightarrow$ Collins	150	$x + 10$	$\dfrac{150}{x + 10}$

We have used the equation $Time = \frac{Distance}{Rate}$ to fill in the "Time" column of the table. Since the second part of the trip took 6 min. (or $1/10$ hours) more than the first, we can use the $Time$ column to get the equation $\dfrac{120}{x} + \dfrac{1}{10} = \dfrac{150}{x + 10} \quad \Rightarrow \quad 120(10)(x + 10) + x(x + 10) = 150(10x)$

$\Leftrightarrow \quad 1200x + 12000 + x^2 + 10x = 1500x \quad \Leftrightarrow \quad x^2 - 290x + 12000 = 0 \quad \Leftrightarrow$

$x = \dfrac{-(-290) \pm \sqrt{(-290)^2 - 4(1)(12000)}}{2} = \dfrac{290 \pm \sqrt{84100 - 48000}}{2} = \dfrac{290 \pm \sqrt{36100}}{2} =$

$\frac{290 \pm 190}{2} = 145 \pm 95$. Hence, the salesman drove either 50 mi/h or 240 mi/h between Ajax and Barrington. (The first choice seems more likely!)

95. Let r be the rowing rate in km/h of the crew in still water. Then their rate upstream was $r - 3$ km/h, and their rate downstream was $r + 3$ km/h.

	Distance	Rate	Time
Upstream	6	$r - 3$	$\dfrac{6}{r - 3}$
Downstream	6	$r + 3$	$\dfrac{6}{r + 3}$

Since the time to row upstream plus the time to row downstream was 2 hours 40 minutes $= \frac{8}{3}$ hour,

we get the equation $\dfrac{6}{r - 3} + \dfrac{6}{r + 3} = \dfrac{8}{3} \quad \Leftrightarrow \quad 6(3)(r + 3) + 6(3)(r - 3) = 8(r - 3)(r + 3)$

$\Leftrightarrow \quad 18r + 54 + 18r - 54 = 8r^2 - 72 \quad \Leftrightarrow \quad 0 = 8r^2 - 36r - 72 = 4(2r^2 - 9r - 18)$

$= 4(2r + 3)(r - 6)$. Since $2r + 3 = 0 \quad \Leftrightarrow \quad r = -\frac{3}{2}$ is impossible, the solution is $r - 6 = 0$

$\Leftrightarrow \quad r = 6$. So the rate of the rowing crew in still water is 6 km/h.

97. Let w be the uniform width of the lawn. With w cut off each end, the area of the factory is $(240 - 2w)(180 - 2w)$. Since the lawn and the factory are equal in size this area, is $\frac{1}{2} \cdot 240 \cdot 180$. So $21600 = 43200 - 480w - 360w + 4w^2 \quad \Leftrightarrow \quad 0 = 4w^2 - 840w + 21600$ $= 4(w^2 - 210w + 5400) = 4(w - 30)(w - 180) \quad \Rightarrow \quad w = 30$ or $w = 180$. Since 180 is too wide, the width of the lawn is 30 feet, and the factory is 120 by 180.

99. Let t be the time, in hours it takes Irene to wash all the windows. Then it takes Henry $t + \frac{3}{2}$ hours to wash all the windows, and the sum of the fraction of the job per hour they can do individually equals the fraction of the job they can do together. Since 1 hour 48 minutes $= 1 + \frac{48}{60} = 1 + \frac{4}{5} = \frac{9}{5}$, we

have $\dfrac{1}{t} + \dfrac{1}{t + \frac{3}{2}} = \dfrac{1}{\frac{9}{5}} \quad \Leftrightarrow \quad \dfrac{1}{t} + \dfrac{2}{2t + 3} = \dfrac{5}{9} \quad \Rightarrow \quad 9(2t + 3) + 2(9t) = 5t(2t + 3) \quad \Leftrightarrow$

$18t + 27 + 18t = 10t^2 + 15t \quad \Leftrightarrow \quad 10t^2 - 21t - 27 = 0 \quad \Leftrightarrow$

$t = \dfrac{-(-21) \pm \sqrt{(-21)^2 - 4(10)(-27)}}{2(10)} = \dfrac{21 \pm \sqrt{441 + 1080}}{20} = \dfrac{21 \pm 39}{20}$. So $t = \dfrac{21 - 39}{20} =$

$-\frac{9}{10}$ or $t = \frac{21 + 39}{20} = 3$. Since $t < 0$ is impossible, all the windows are washed by Irene alone in 3 hours and by Henry alone in $3 + \frac{3}{2} = 4\frac{1}{2}$ hours.

101. Let x be the distance from the center of the earth to the dead spot (in thousand of mile). Now setting

$F = 0$, we have $0 = \dfrac{-K}{x^2} + \dfrac{0.012K}{(239 - x)^2} \quad \Leftrightarrow \quad \dfrac{K}{x^2} = \dfrac{0.012K}{(239 - x)^2} \quad \Leftrightarrow$

$K(239 - x)^2 = 0.012Kx^2 \quad \Leftrightarrow \quad 57121 - 478x + x^2 = 0.012x^2 \quad \Leftrightarrow$

$0.988x^2 - 478x + 57121 = 0$. Using the quadratic formula, we obtain

$x = \dfrac{-(-478) \pm \sqrt{(-478)^2 - 4(0.988)(57121)}}{2(0.988)} = \dfrac{478 \pm \sqrt{228484 - 225742.192}}{1.976} =$

$\dfrac{478 \pm \sqrt{2741.808}}{1.976} \approx \dfrac{478 \pm 52.362}{1.976} \approx 241.903 \pm 26.499$. So either $x \approx 241.903 + 26.499 \approx$

268 or $x \approx 241.903 - 26.499 \approx 215$. Since 268 is greater than the distance from the earth to the moon, we reject it; thus $x \approx 215$ thousand miles $= 215{,}000$ miles.

103. Let x equal the original length of the reed in cubits. Then $x - 1$ is the piece that fits 60 times along the length of the field, that is, the length is $60(x - 1)$. The width is $30x$. Then converting cubits to ninda, we have $375 = 60(x - 1) \cdot 30x \cdot \frac{1}{12^2} = \frac{25}{2}x(x - 1)$ $\Leftrightarrow$ $30 = x^2 - x$ $\Leftrightarrow$ $x^2 - x - 30 = 0$ $\Leftrightarrow$ $(x - 6)(x + 5) = 0$. So $x = 6$ or $x = -5$. Since x must be positive, the original length of the reed is 6 cubits.

Exercises 1.4

1. $5 - 7i$: real part 5, imaginary part -7.

3. $\dfrac{-2 - 5i}{3} = -\dfrac{2}{3} - \dfrac{5}{3}i$: real part $-\dfrac{2}{3}$, imaginary part $-\dfrac{5}{3}$.

5. 3: real part 3, imaginary part 0.

7. $-\dfrac{2}{3}i$: real part 0, imaginary part $-\dfrac{2}{3}$.

9. $\sqrt{3} + \sqrt{-4} = \sqrt{3} + 2i$: real part $\sqrt{3}$, imaginary part 2.

11. $(2 - 5i) + (3 + 4i) = (2 + 3) + (-5 + 4)i = 5 - i$.

13. $(-6 + 6i) + (9 - i) = (-6 + 9) + (6 - 1)i = 3 + 5i$.

15. $(7 - \frac{1}{2}i) - (5 + \frac{3}{2}i) = (7 - 5) + (-\frac{1}{2} - \frac{3}{2})i = 2 - 2i$.

17. $(-12 + 8i) - (7 + 4i) = -12 + 8i - 7 - 4i = (-12 - 7) + (8 - 4)i = -19 + 4i$.

19. $4(-1 + 2i) = -4 + 8i$.

21. $(7 - i)(4 + 2i) = 28 + 14i - 4i - 2i^2 = (28 + 2) + (14 - 4)i = 30 + 10i$.

23. $(3 - 4i)(5 - 12i) = 15 - 36i - 20i + 48i^2 = (15 - 48) + (-36 - 20)i = -33 - 56i$.

25. $(6 + 5i)(2 - 3i) = 12 - 18i + 10i - 15i^2 = (12 + 15) + (-18 + 10)i = 27 - 8i$.

27. $\dfrac{1}{i} = \dfrac{1}{i} \cdot \dfrac{i}{i} = \dfrac{i}{i^2} = \dfrac{i}{-1} = -i$.

29. $\dfrac{2 - 3i}{1 - 2i} = \dfrac{2 - 3i}{1 - 2i} \cdot \dfrac{1 + 2i}{1 + 2i} = \dfrac{2 + 4i - 3i - 6i^2}{1 - 4i^2} = \dfrac{(2 + 6) + (4 - 3)i}{1 + 4} = \dfrac{8 + i}{5}$ or $\dfrac{8}{5} + \dfrac{1}{5}i$.

31. $\dfrac{26 + 39i}{2 - 3i} = \dfrac{26 + 39i}{2 - 3i} \cdot \dfrac{2 + 3i}{2 + 3i} = \dfrac{52 + 78i + 78i + 117i^2}{4 - 9i^2} = \dfrac{(52 - 117) + (78 + 78)i}{4 + 9} =$

    $\dfrac{-65 + 156i}{13} = \dfrac{13(-5 + 12i)}{13} = -5 + 12i$.

33. $\dfrac{10i}{1 - 2i} = \dfrac{10i}{1 - 2i} \cdot \dfrac{1 + 2i}{1 + 2i} = \dfrac{10i + 20i^2}{1 - 4i^2} = \dfrac{-20 + 10i}{1 + 4} = \dfrac{5(-4 + 2i)}{5} = -4 + 2i$.

35. $\dfrac{4 + 6i}{3i} = \dfrac{4 + 6i}{3i} \cdot \dfrac{i}{i} = \dfrac{4i + 6i^2}{3i^2} = \dfrac{-6 + 4i}{-3} = \dfrac{-6}{-3} + \dfrac{4}{-3}i = 2 - \dfrac{4}{3}i$.

37. $\dfrac{1}{1 + i} - \dfrac{1}{1 - i} = \dfrac{1}{1 + i} \cdot \dfrac{1 - i}{1 - i} - \dfrac{1}{1 - i} \cdot \dfrac{1 + i}{1 + i} = \dfrac{1 - i}{1 - i^2} - \dfrac{1 + i}{1 - i^2} = \dfrac{1 - i}{2} + \dfrac{-1 - i}{2} = -i$.

39. $i^3 = i^2 \cdot i = -1 \cdot i = -i$.

41. $i^{100} = (i^4)^{25} = (1)^{25} = 1$.

43. $\sqrt{-25} = 5i$.

45. $\sqrt{-3}\,\sqrt{-12} = i\sqrt{3} \cdot 2i\sqrt{3} = 6i^2 = -6$.

47. $\left(3 - \sqrt{-5}\right)\left(1 + \sqrt{-1}\right) = \left(3 - i\sqrt{5}\right)(1 + i) = 3 + 3i - i\sqrt{5} - i^2\sqrt{5} = \left(3 + \sqrt{5}\right) + \left(3 - \sqrt{5}\right)i.$

49. $\dfrac{2 + \sqrt{-8}}{1 + \sqrt{-2}} = \dfrac{2 + 2i\sqrt{2}}{1 + i\sqrt{2}} = \dfrac{2 + 2i\sqrt{2}}{1 + i\sqrt{2}} \cdot \dfrac{1 - i\sqrt{2}}{1 - i\sqrt{2}} = \dfrac{2 - 2i\sqrt{2} + 2i\sqrt{2} - 4i^2}{1 - 2i^2} =$

$\dfrac{(2 + 4) + (-2\sqrt{2} + 2\sqrt{2})i}{1 + 2} = \dfrac{6}{3} = 2.$

51. $\dfrac{\sqrt{-36}}{\sqrt{-2}\sqrt{-9}} = \dfrac{6i}{i\sqrt{2} \cdot 3i} = \dfrac{2}{i\sqrt{2}} \cdot \dfrac{i\sqrt{2}}{i\sqrt{2}} = \dfrac{2i\sqrt{2}}{2i^2} = \dfrac{i\sqrt{2}}{-1} = -i\sqrt{2}.$

53. $x^2 + 9 = 0 \;\Leftrightarrow\; x^2 = -9 \;\Rightarrow\; x = \pm 3i.$

55. $x^2 - 4x + 5 = 0 \;\Rightarrow\; x = \dfrac{-(-4) \pm \sqrt{(-4)^2 - 4(1)(5)}}{2(1)} = \dfrac{4 \pm \sqrt{16 - 20}}{2} = \dfrac{4 \pm \sqrt{-4}}{2}$

$= \dfrac{4 \pm 2i}{2} = 2 \pm i.$

57. $x^2 + x + 1 = 0 \;\Rightarrow\; x = \dfrac{-(1) \pm \sqrt{(1)^2 - 4(1)(1)}}{2(1)} = \dfrac{-1 \pm \sqrt{1 - 4}}{2} = \dfrac{-1 \pm \sqrt{-3}}{2} =$

$\dfrac{-1 \pm i\sqrt{3}}{2} = -\dfrac{1}{2} \pm \dfrac{i\sqrt{3}}{2}.$

59. $2x^2 - 2x + 1 = 0 \;\Rightarrow\; x = \dfrac{-(-2) \pm \sqrt{(-2)^2 - 4(2)(1)}}{2(2)} = \dfrac{2 \pm \sqrt{4 - 8}}{4} = \dfrac{2 \pm \sqrt{-4}}{4} =$

$\dfrac{2 \pm 2i}{4} = \dfrac{1}{2} \pm \dfrac{1}{2}i.$

61. $t + 3 + \dfrac{3}{t} = 0 \;\Leftrightarrow\; t^2 + 3t + 3 = 0 \;\Rightarrow\; t = \dfrac{-(3) \pm \sqrt{(3)^2 - 4(1)(3)}}{2(1)} = \dfrac{-3 \pm \sqrt{9 - 12}}{2}$

$= \dfrac{-3 \pm \sqrt{-3}}{2} = \dfrac{-3 \pm i\sqrt{3}}{2} = -\dfrac{3}{2} \pm \dfrac{i\sqrt{3}}{2}.$

63. $6x^2 + 12x + 7 = 0 \;\Rightarrow\; x = \dfrac{-(12) \pm \sqrt{(12)^2 - 4(6)(7)}}{2(6)} = \dfrac{-12 \pm \sqrt{144 - 168}}{12} =$

$\dfrac{-12 \pm \sqrt{-24}}{12} = \dfrac{-12 \pm 2i\sqrt{6}}{12} = \dfrac{-12}{12} \pm \dfrac{2i\sqrt{6}}{12} = -1 \pm \dfrac{i\sqrt{6}}{6}.$

65. $\frac{1}{2}x^2 - x + 5 = 0 \;\Rightarrow\; x = \dfrac{-(-1) \pm \sqrt{(-1)^2 - 4(\frac{1}{2})(5)}}{2(\frac{1}{2})} = \dfrac{1 \pm \sqrt{1 - 10}}{1} = 1 \pm \sqrt{-9} =$

$1 \pm 3i.$

67. LHS $= \overline{z} + \overline{w} = \overline{(a + bi)} + \overline{(c + di)} = a - bi + c - di = (a + c) + (-b - d)i = (a + c) - (b + d)i.$

RHS $= \overline{z + w} = \overline{(a + bi) + (c + di)} = \overline{(a + c) + (b + d)i} = (a + c) - (b + d)i.$

Since LHS $=$ RHS, this proves the statement.

69. LHS $= (\overline{z})^2 = \left(\overline{(a + b\,i)}\right)^2 = (a - bi)^2 = a^2 - 2ab\,i + b^2\,i^2 = (a^2 - b^2) - 2ab\,i$.

RHS $= \overline{z^2} = \overline{(a + b\,i)^2} = \overline{a^2 + 2ab\,i + b^2\,i^2} = \overline{(a^2 - b^2) + 2ab\,i} = (a^2 - b^2) - 2ab\,i$.

Since LHS $=$ RHS, this proves the statement.

71. $z + \overline{z} = (a + b\,i) + \overline{(a + b\,i)} = a + b\,i + a - b\,i = 2a$, which is a real number.

73. $z \cdot \overline{z} = (a + b\,i) \cdot \overline{(a + b\,i)} = (a + b\,i) \cdot (a - b\,i) = a^2 - b^2 i^2 = a^2 + b^2$, which is a real number.

75. Using the quadratic formula, the solutions to the equation are $x = \dfrac{-b \pm \sqrt{b^2 - 4ac}}{2a}$. Since both solutions are imaginary, we have $b^2 - 4ac < 0 \quad \Leftrightarrow \quad 4ac - b^2 > 0$, so the solutions are $x = \dfrac{-b}{2a} \pm \dfrac{\sqrt{4ac - b^2}}{2a}\, i$, where $\sqrt{4ac - b^2}$ is a real number. Thus the solutions are complex conjugates of each other.

Exercises 1.5

1. $x^3 = 16x$ $\Leftrightarrow$ $0 = x^3 - 16x = x(x^2 - 16) = x(x - 4)(x + 4)$. So $x = 0$, $x - 4 = 0$ $\Leftrightarrow$ $x = 4$, or $x + 4 = 0$ $\Leftrightarrow$ $x = -4$. The solutions are 0 and ± 4.

3. $0 = x^6 - 81x^2 = x^2(x^4 - 81) = x^2(x^2 - 9)(x^2 + 9) = x^2(x - 3)(x + 3)(x^2 + 9)$. So $x^2 = 0$ $\Leftrightarrow$ $x = 0$, or $x - 3 = 0$ $\Leftrightarrow$ $x = 3$, or $x + 3 = 0$ $\Leftrightarrow$ $x = -3$. $x^2 + 9 = 0$ $\Leftrightarrow$ $x^2 = -9$ which has no real solutions. The solutions are 0 and ± 3.

5. $0 = x^5 + 8x^2 = x^2(x^3 + 8) = x^2(x + 2)(x^2 - 2x + 4)$ $\Leftrightarrow$ $x^2 = 0$, $x + 2 = 0$, or $x^2 - 2x + 4 = 0$. If $x^2 = 0$, then $x = 0$; if $x + 2 = 0$, then $x = -2$. And $x^2 - 2x + 4 = 0$ has no real solutions. Thus the solutions are $x = 0$ and $x = -2$.

7. $0 = x^3 - 5x^2 + 6x = x(x^2 - 5x + 6) = x(x - 2)(x - 3)$ $\Leftrightarrow$ $x = 0$, $x - 2 = 0$, or $x - 3 = 0$. Thus $x = 0$, or $x = 2$, or $x = 3$. The solutions are $x = 0$, $x = 2$, and $x = 3$.

9. $0 = x^4 + 4x^3 + 2x^2 = x^2(x^2 + 4x + 2)$. So either $x^2 = 0$ $\Leftrightarrow$ $x = 0$, or using the quadratic formula on $x^2 + 4x + 2 = 0$, we have $x = \dfrac{-4 \pm \sqrt{4^2 - 4(1)(2)}}{2(1)} = \dfrac{-4 \pm \sqrt{16 - 8}}{2} = \dfrac{-4 \pm \sqrt{8}}{2}$
   $= \dfrac{-4 \pm 2\sqrt{2}}{2} = -2 \pm \sqrt{2}$. The solutions are 0, $-2 - \sqrt{2}$, and $-2 + \sqrt{2}$.

11. $0 = x^3 - 5x^2 - 2x + 10 = x^2(x - 5) - 2(x - 5) = (x - 5)(x^2 - 2)$. If $x - 5 = 0$, then $x = 5$. If $x^2 - 2 = 0$, then $x^2 = 2$ $\Leftrightarrow$ $x = \pm\sqrt{2}$. The solutions are 5 and $\pm\sqrt{2}$.

13. $x^3 - x^2 + x - 1 = x^2 + 1$ $\Leftrightarrow$
    $0 = x^3 - 2x^2 + x - 2 = x^2(x - 2) + (x - 2) = (x - 2)(x^2 + 1)$. Since $x^2 + 1 = 0$ has no real solution, the only solution comes from $x - 2 = 0$ $\Leftrightarrow$ $x = 2$.

15. $\dfrac{1}{x - 1} + \dfrac{1}{x + 2} = \dfrac{5}{4}$ $\Leftrightarrow$ $4(x - 1)(x + 2)\left(\dfrac{1}{x - 1} + \dfrac{1}{x + 2}\right) = 4(x - 1)(x + 2)\left(\dfrac{5}{4}\right)$ $\Leftrightarrow$
    $4(x + 2) + 4(x - 1) = 5(x - 1)(x + 2)$ $\Leftrightarrow$ $4x + 8 + 4x - 4 = 5x^2 + 5x - 10$ $\Leftrightarrow$
    $5x^2 - 3x - 14 = 0$ $\Leftrightarrow$ $(5x + 7)(x - 2) = 0$. If $5x + 7 = 0$, then $x = -\frac{7}{5}$; if $x - 2 = 0$, then $x = 2$. The solutions are $-\frac{7}{5}$ and 2.

17. $\dfrac{x^2}{x + 100} = 50$ $\Rightarrow$ $x^2 = 50(x + 100) = 50x + 5000$ $\Leftrightarrow$ $x^2 - 50x - 5000 = 0$ $\Leftrightarrow$
    $(x - 100)(x + 50) = 0$ $\Leftrightarrow$ $x - 100 = 0$ or $x + 50 = 0$. Thus $x = 100$ or $x = -50$. The solutions are 100 and -50.

19. $\dfrac{x + 5}{x - 2} = \dfrac{5}{x + 2} + \dfrac{28}{x^2 - 4}$ $\Rightarrow$ $(x + 2)(x + 5) = 5(x - 2) + 28$ $\Leftrightarrow$
    $x^2 + 7x + 10 = 5x - 10 + 28$ $\Leftrightarrow$ $x^2 + 2x - 8 = 0$ $\Leftrightarrow$ $(x - 2)(x + 4) = 0$ $\Leftrightarrow$
    $x - 2 = 0$ or $x + 4 = 0$ $\Leftrightarrow$ $x = 2$ or $x = -4$. However, $x = 2$ is inadmissible since we can't divide by 0 in the original equation, so the only solution is -4.

21. $\dfrac{1}{x-1} - \dfrac{2}{x^2} = 0 \quad \Leftrightarrow \quad x^2 - 2(x-1) = 0 \quad \Leftrightarrow \quad x^2 - 2x + 2 = 0 \quad \Rightarrow$

    $x = \dfrac{-(-2) \pm \sqrt{(-2)^2 - 4(1)(2)}}{2(1)} = \dfrac{2 \pm \sqrt{4-8}}{2} = \dfrac{2 \pm \sqrt{-4}}{2}$. Since the radicand is negative,

    there are no real solutions.

23. $0 = (x+5)^2 - 3(x+5) - 10 = [(x+5) - 5][(x+5) + 2] = x(x+7) \quad \Leftrightarrow \quad x = 0 \text{ or } x = -7$.
    The solutions are 0 and -7.

25. Let $w = \dfrac{1}{x+1}$. Then $\left(\dfrac{1}{x+1}\right)^2 - 2\left(\dfrac{1}{x+1}\right) - 8 = 0$ becomes $w^2 - 2w - 8 = 0 \quad \Leftrightarrow$

    $(w-4)(w+2) = 0$. So $w - 4 = 0 \quad \Leftrightarrow \quad w = 4$, and $w + 2 = 0 \quad \Leftrightarrow \quad w = -2$. When $w = 4$,

    we have $\dfrac{1}{x+1} = 4 \quad \Leftrightarrow \quad 1 = 4x + 4 \quad \Leftrightarrow \quad -3 = 4x \quad \Leftrightarrow \quad x = -\tfrac{3}{4}$. When $w = -2$, we

    have $\dfrac{1}{x+1} = -2 \quad \Leftrightarrow \quad 1 = -2x - 2 \quad \Leftrightarrow \quad 3 = -2x \quad \Leftrightarrow \quad x = -\tfrac{3}{2}$. Solutions are $-\tfrac{3}{4}$ and

    $-\tfrac{3}{2}$.

27. Let $w = x^2$. Then $x^4 - 13x^2 + 40 = (x^2)^2 - 13x^2 + 40 = 0$ becomes $w^2 - 13w + 40 = 0 \quad \Leftrightarrow$
    $(w-5)(w-8) = 0$. So $w - 5 = 0 \quad \Leftrightarrow \quad w = 5$, and $w - 8 = 0 \quad \Leftrightarrow \quad w = 8$. When $w = 5$,
    we have $x^2 = 5 \quad \Rightarrow \quad x = \pm\sqrt{5}$. When $w = 8$, we have $x^2 = 8 \quad \Rightarrow \quad x = \pm\sqrt{8} = \pm 2\sqrt{2}$.
    The solutions are $\pm\sqrt{5}$ and $\pm 2\sqrt{2}$.

29. $2x^4 + 4x^2 + 1 = 0$. The LHS is the sum of two nonnegative numbers and a positive number, so
    $2x^4 + 4x^2 + 1 \geq 1 \neq 0$. This equation has no real solutions.

31. $0 = x^6 - 26x^3 - 27 = (x^3 - 27)(x^3 + 1)$. If $x^3 - 27 = 0 \quad \Leftrightarrow \quad x^3 = 27$, so $x = 3$. If
    $x^3 + 1 = 0 \quad \Leftrightarrow \quad x^3 = -1$, so $x = -1$. The solutions are 3 and -1.

33. Let $u = x^{2/3}$. Then $0 = x^{4/3} - 5x^{2/3} + 6$ becomes $u^2 - 5u + 6 = 0 \quad \Leftrightarrow \quad (u-3)(u-2) = 0$
    $\Leftrightarrow \quad u - 3 = 0 \text{ or } u - 2 = 0$. If $u - 3 = 0$, then $x^{2/3} - 3 = 0 \quad \Leftrightarrow \quad x^{2/3} = 3 \quad \Rightarrow$
    $x = \pm 3^{3/2} = \pm 3\sqrt{3}$. If $u - 2 = 0$, then $x^{2/3} - 2 = 0 \quad \Leftrightarrow \quad x^{2/3} = 2 \quad \Rightarrow$
    $x = \pm 2^{3/2} = 2\sqrt{2}$. The solutions are $\pm 3\sqrt{3}$ and $\pm 2\sqrt{2}$.

35. $4(x+1)^{1/2} - 5(x+1)^{3/2} + (x+1)^{5/2} = 0 \quad \Leftrightarrow \quad \sqrt{x+1}[4 - 5(x+1) + (x+1)^2] = 0 \quad \Leftrightarrow$
    $\sqrt{x+1}(4 - 5x - 5 + x^2 + 2x + 1) = 0 \quad \Leftrightarrow \quad \sqrt{x+1}(x^2 - 3x) = 0 \quad \Leftrightarrow$
    $\sqrt{x+1} \cdot x(x-3) = 0 \quad \Leftrightarrow \quad x = -1 \text{ or } x = 0 \text{ or } x = 3$. The solutions are $-1, 0,$ and 3.

37. $0 = x^{3/2} + 8x^{1/2} + 16x^{-1/2} = x^{-1/2}(x^2 + 8x + 16) = x^{-1/2}(x+4)^2$. Now $x^{-1/2} = \dfrac{1}{\sqrt{x}}$ so

    $x \neq 0$. So $x + 4 = 0 \quad \Leftrightarrow \quad x = -4$. But $\sqrt{-4}$ is not a real number, so this equation has no real
    solutions. Alternatively, we see that this is the sum of three positive real numbers (remember
    $x \neq 0$), so it never equals zero.

39. Let $u = x^{1/6}$. (We choose the exponent $\tfrac{1}{6}$ because the LCD of 2, 3, and 6 is 6.) Then
    $x^{1/2} - 3x^{1/3} = 3x^{1/6} - 9 \quad \Leftrightarrow \quad x^{3/6} - 3x^{2/6} = 3x^{1/6} - 9 \quad \Leftrightarrow \quad u^3 - 3u^2 = 3u - 9 \quad \Leftrightarrow$
    $0 = u^3 - 3u^2 - 3u + 9 = u^2(u-3) - 3(u-3) = (u-3)(u^2 - 3)$. So $u - 3 = 0$ or $u^2 - 3 = 0$.

If $u - 3 = 0$, then $x^{1/6} - 3 = 0$ $\Leftrightarrow$ $x^{1/6} = 3$ $\Leftrightarrow$ $x = 3^6 = 729$. If $u^2 - 3 = 0$, then $x^{1/3} - 3 = 0$ $\Leftrightarrow$ $x^{1/3} = 3$ $\Leftrightarrow$ $x = 3^3 = 27$. The solutions are 729 and 27.

41. $\dfrac{1}{x^3} + \dfrac{4}{x^2} + \dfrac{4}{x} = 0$ $\Rightarrow$ $1 + 4x + 4x^2 = 0$ $\Leftrightarrow$ $(1 + 2x)^2 = 0$ $\Leftrightarrow$ $1 + 2x = 0$ $\Leftrightarrow$ $2x = -1$ $\Leftrightarrow$ $x = -\frac{1}{2}$. The solution is $-\frac{1}{2}$.

43. $\sqrt{2x+1} + 1 = x$ $\Leftrightarrow$ $\sqrt{2x+1} = x - 1$ $\Rightarrow$ $2x + 1 = (x-1)^2$ $\Leftrightarrow$ $2x + 1 = x^2 - 2x + 1$ $\Leftrightarrow$ $0 = x^2 - 4x = x(x-4)$. Potential solutions are $x = 0$ and $x - 4$ $\Leftrightarrow$ $x = 4$. These are only potential solutions since *squaring* is not a reversible operation. We must check each potential solution in the original equation.

Checking $x = 0$: $\sqrt{2(0)+1} + 1 \stackrel{?}{=} (0)$, $\sqrt{1} + 1 \stackrel{?}{=} 0$, NO!

Checking $x = 4$: $\sqrt{2(4)+1} + 1 \stackrel{?}{=} (4)$, $\sqrt{9} + 1 \stackrel{?}{=} 4$, $3 + 1 \stackrel{?}{=} 4$, Yes. The only solution is $x = 4$.

45. $\sqrt{5-x} + 1 = x - 2$ $\Leftrightarrow$ $\sqrt{5-x} = x - 3$ $\Rightarrow$ $5 - x = (x-3)^2$ $\Leftrightarrow$ $5 - x = x^2 - 6x + 9$ $\Leftrightarrow$ $0 = x^2 - 5x + 4 = (x-4)(x-1)$. Potential solutions are $x = 4$ and $x = 1$. We must check each potential solution in the original equation.

Checking $x = 4$: $\sqrt{5-(4)} + 1 \stackrel{?}{=} (4) - 2$, $\sqrt{1} + 1 \stackrel{?}{=} 4 - 2$, $1 + 1 \stackrel{?}{=} 2$, Yes.

Checking $x = 1$: $\sqrt{5-(1)} + 1 \stackrel{?}{=} (1) - 2$, $\sqrt{4} + 1 \stackrel{?}{=} -1$, $2 + 1 \stackrel{?}{=} -1$, NO! The only solution is $x = 4$.

47. $x - \sqrt{x+3} = \dfrac{x}{2}$ $\Leftrightarrow$ $\dfrac{x}{2} = \sqrt{x+3}$ $\Rightarrow$ $\left(\dfrac{x}{2}\right)^2 = x + 3$ $\Leftrightarrow$ $\dfrac{x^2}{4} = x + 3$ $\Leftrightarrow$ $x^2 = 4x + 12$ $\Leftrightarrow$ $0 = x^2 - 4x - 12 = (x-6)(x+2)$. Potential solutions are $x = 6$ and $x = -2$. We must check each potential solution in the original equation.

Checking $x = 6$: $6 - \sqrt{6+3} \stackrel{?}{=} \dfrac{6}{2}$, $6 - \sqrt{6+3} = 6 - \sqrt{9} = 6 - 3$ and $\dfrac{6}{2} = 3$, hence this is a solution.

Checking $x = -2$: $-2 - \sqrt{-2+3} \stackrel{?}{=} \dfrac{-2}{2}$, $-2 - \sqrt{-2+3} = -2 - \sqrt{1} = -2 - 1 = -3$ and $\dfrac{-2}{2} = -1$, hence this not a solution. The only solution is $x = 6$.

49. $\sqrt{\sqrt{x-5}+x} = 5$. Squaring both sides, we get $\sqrt{x-5} + x = 25$ $\Leftrightarrow$ $\sqrt{x-5} = 25 - x$. Squaring both sides again, we get $x - 5 = (25-x)^2$ $\Leftrightarrow$ $x - 5 = 625 - 50x + x^2$ $\Leftrightarrow$ $0 = x^2 - 51x + 630 = (x-30)(x-21)$. Potential solutions are $x = 30$ and $x = 21$. We must check each potential solution in the original equation.

Checking $x = 30$: $\sqrt{\sqrt{(30)-5}+(30)} \stackrel{?}{=} 5$, $\sqrt{\sqrt{(30)-5}+(30)} = \sqrt{\sqrt{25}+30} = \sqrt{35} > 5$, hence not a solution.

Checking $x = 21$: $\sqrt{\sqrt{(21)-5}+21} \stackrel{?}{=} 5$, $\sqrt{\sqrt{(21)-5}+21} = \sqrt{\sqrt{16}+21} = \sqrt{25} = 5$, hence a solution. The only solution is $x = 21$.

51. $x^2\sqrt{x+3} = (x+3)^{3/2}$ $\Leftrightarrow$ $0 = x^2\sqrt{x+3} - (x+3)^{3/2}$ $\Leftrightarrow$ $0 = \sqrt{x+3}[(x^2) - (x+3)]$
$\Leftrightarrow$ $0 = \sqrt{x+3}(x^2 - x - 3)$. If $(x+3)^{1/2} = 0$, then $x+3 = 0$ $\Leftrightarrow$ $x = -3$. If
$x^2 - x - 3 = 0$, then using the quadratic formula $x = \frac{1\pm\sqrt{13}}{2}$. The solutions are -3 and $\frac{1\pm\sqrt{13}}{2}$.

53. $\sqrt{x+\sqrt{x+2}} = 2$. Squaring both sides, we get $x + \sqrt{x+2} = 4$ $\Leftrightarrow$ $\sqrt{x+2} = 4 - x$.
Squaring both sides again, we get $x+2 = (4-x)^2 = 16 - 8x + x^2$ $\Leftrightarrow$ $0 = x^2 - 9x + 14$
$\Leftrightarrow$ $0 = (x-7)(x-2)$. If $x - 7 = 0$, then $x = 7$. If $x - 2 = 0$, then $x = 2$. So $x = 2$ is a
solution but $x = 7$ is not, since it does not satisfy the original equation.

55. $x^3 = 1$ $\Leftrightarrow$ $x^3 - 1 = 0$ $\Leftrightarrow$ $(x-1)(x^2 + x + 1) = 0$ $\Leftrightarrow$ $x - 1 = 0$ or $x^2 + x + 1 = 0$.
If $x - 1 = 0$, then $x = 1$. If $x^2 + x + 1 = 0$, then using the quadratic formula $x = \frac{-1\pm i\sqrt{3}}{2}$. The
solutions are 1 and $\frac{-1\pm i\sqrt{3}}{2}$.

57. $x^3 + x^2 + x = 0$ $\Leftrightarrow$ $x(x^2 + x + 1) = 0$ $\Leftrightarrow$ $x = 0$ or $x = \frac{-1\pm i\sqrt{3}}{2}$. The solutions are 0 and
$\frac{-1\pm i\sqrt{3}}{2}$.

59. $x^4 - 6x^2 + 8 = 0$ $\Leftrightarrow$ $(x^2 - 4)(x^2 - 2) = 0$ $\Leftrightarrow$ $x = \pm 2$ or $x = \pm\sqrt{2}$. The solutions are
± 2 and $\pm\sqrt{2}$.

61. $x^6 - 9x^3 + 8 = 0$ $\Leftrightarrow$ $(x^3 - 8)(x^3 - 1) = 0$ $\Leftrightarrow$
$(x-2)(x^2 + 2x + 4)(x-1)(x^2 + x + 1) = 0$ $\Leftrightarrow$ $x = 2$ or $x = \frac{-2\pm 2i\sqrt{3}}{2} = -1 \pm i\sqrt{3}$ or
$x = 1$ or $x = \frac{-1\pm i\sqrt{3}}{2}$. The solutions are $2, -1 \pm i\sqrt{3}, 1$, and $\frac{-1\pm i\sqrt{3}}{2}$.

63. $\sqrt{x^2+1} + \dfrac{8}{\sqrt{x^2+1}} = \sqrt{x^2+9}$. Squaring both sides, we have $(x^2 + 1) + 16 + \dfrac{64}{x^2+1} = x^2 + 9$
$\Leftrightarrow$ $\dfrac{64}{x^2+1} = -8$ $\Leftrightarrow$ $\dfrac{8}{x^2+1} = -1$ $\Leftrightarrow$ $x^2 + 1 = -8$ $\Leftrightarrow$ $x^2 = -9$ $\Leftrightarrow$ $x = \pm 3$
i. We must check each potential solution in the original equation.

Checking $x = \pm 3\,i$: $\sqrt{(\pm 3\,i)^2 + 1} + \dfrac{8}{\sqrt{(\pm 3\,i)^2 + 1}} \overset{?}{=} \sqrt{(\pm 3\,i)^2 + 9}$,

LHS $= \sqrt{-8} + \dfrac{8}{\sqrt{-8}} = 2\sqrt{2}\,i + \dfrac{8}{2\sqrt{2}\,i} = 2\sqrt{2}\,i - 2\sqrt{2}\,i = 0$.

RHS $= \sqrt{(\pm 3\,i)^2 + 9} = \sqrt{-9 + 9} = 0$. Since LHS $=$ RHS, $-3\,i$ and $3\,i$ are solutions.

65. $0 = x^4 + 5ax^2 + 4a^2 = (x^2 + a)(x^2 + 4a)$. Since a is positive, $x^2 + a = 0$ $\Leftrightarrow$ $x^2 = -a$ $\Leftrightarrow$
$x = \pm i\sqrt{a}$. Again, since a is positive, $x^2 + 4a = 0$ $\Leftrightarrow$ $x^2 = -4a$ $\Leftrightarrow$ $x = \pm 2i\sqrt{a}$. Thus
the four solutions are: $\pm i\sqrt{a}, \pm 2i\sqrt{a}$.

67. $\sqrt{x+a} + \sqrt{x-a} = \sqrt{2}\sqrt{x+6}$. Squaring both sides, we have
$x + a + 2(\sqrt{x+a})(\sqrt{x-a}) + x - a = 2(x+6)$ $\Leftrightarrow$
$2x + 2(\sqrt{x+a})(\sqrt{x-a}) = 2x + 12$ $\Leftrightarrow$ $2(\sqrt{x+a})(\sqrt{x-a}) = 12$ $\Leftrightarrow$
$(\sqrt{x+a})(\sqrt{x-a}) = 6$. Squaring both sides again we have: $(x+a)(x-a) = 36$ $\Leftrightarrow$
$x^2 - a^2 = 36$ $\Leftrightarrow$ $x^2 = a^2 + 36$ $\Leftrightarrow$ $x = \pm\sqrt{a^2 + 36}$. Checking these answers, we see that

$x = -\sqrt{a^2 + 36}$ is not a solution (for example, try substituting $a = 8$), but $x = \sqrt{a^2 + 36}$ is a solution.

69. Let x be the number of people originally intended to take the trip. Then originally, the cost of the trip is $\dfrac{900}{x}$. After 5 people cancel, there are now $x - 5$ people, each paying $\dfrac{900}{x} + 2$. Thus

$$900 = (x - 5)\left(\frac{900}{x} + 2\right) \quad \Leftrightarrow \quad 900 = 900 + 2x - \frac{4500}{x} - 10 \quad \Leftrightarrow \quad 0 = 2x - 10 - \frac{4500}{x}$$

$\Leftrightarrow \quad 0 = 2x^2 - 10x - 4500 = (2x - 100)(x + 45)$. Thus either $2x - 100 = 0$, so $x = 50$, or $x + 45 = 0$, $x = -45$. Since the number of people on the trip must be positive, originally 50 people intended to take the trip.

71. We want to solve for t when $P = 500$. Letting $u = \sqrt{t}$ and substituting, we have

$500 = 3t + 10\sqrt{t} + 140 \quad \Leftrightarrow \quad 500 = 3u^2 + 10u + 140 \quad \Leftrightarrow \quad 0 = 3u^2 + 10u - 360 \quad \Rightarrow$

$u = \dfrac{-5 \pm \sqrt{1105}}{3}$. Since $u = \sqrt{t}$, we must have $u \geq 0$. So $\sqrt{t} = u = \dfrac{-5 + \sqrt{1105}}{3} \approx 9.414 \quad \Rightarrow$

$t = \approx 88.62$. So it will take 89 days for the fish population to reach 500.

73. We have that the volume is 180 ft^3, so $x(x - 4)(x + 9) = 180 \quad \Leftrightarrow \quad x^3 + 5x^2 - 36x = 180$
$\Leftrightarrow \quad x^3 + 5x^2 - 36x - 180 = 0 \quad \Leftrightarrow \quad x^2(x + 5) - 36(x + 5) = 0 \quad \Leftrightarrow$
$(x + 5)(x^2 - 36) = 0 \quad \Leftrightarrow \quad (x + 5)(x + 6)(x - 6) = 0 \quad \Rightarrow \quad x = 6$ is the only positive solution. So the box is 2 feet by 6 feet by 15 feet.

75. Let x be the length, in miles, of the abandoned road to be used. Then the length of the abandoned road not used is $40 - x$, and the length of the new road is $\sqrt{10^2 + (40 - x)^2}$ miles, by the Pythagorean Theorem. Since the cost of the road is *cost per mile* × *number of miles*, we have $100{,}000x + 200{,}000\sqrt{x^2 - 80x + 1700} = 6{,}800{,}000 \quad \Leftrightarrow \quad 2\sqrt{x^2 - 80x + 1700} = 68 - x$. Squaring both sides, we get $4x^2 - 320x + 6800 = 4624 - 136x + x^2 \quad \Leftrightarrow$ $3x^2 - 184x + 2176 = 0 \quad \Leftrightarrow \quad x = \dfrac{184 \pm \sqrt{33856 - 26112}}{6} = \dfrac{184 \pm 88}{6} \quad \Leftrightarrow \quad x = \dfrac{136}{3}$ or $x = 16$. Since $45\frac{1}{3}$ is longer than the existing road, 16 miles of the abandoned road should be used. A completely new road would have length $\sqrt{10^2 + 40^2}$, (let $x = 0$), and would cost $\sqrt{1700} \times 200{,}000 \approx 8.3$ million dollars. So no, it would not be cheaper.

77. Let x be the height of the pile in feet. Then the diameter is $3x$ and the radius is $\dfrac{3x}{2}$ feet. Since the volume of the cone is 1000 ft^3, we have $\dfrac{\pi}{3}\left(\dfrac{3x}{2}\right)^2 x = 1000 \quad \Leftrightarrow \quad \dfrac{3\pi x^3}{4} = 1000 \quad \Leftrightarrow$

$x^3 = \dfrac{4000}{3\pi} \quad \Leftrightarrow \quad x = \sqrt[3]{\dfrac{4000}{3\pi}} \approx 7.52$ feet.

79. Let x be the length of the hypotenuse of the triangle, in feet. Then one of the other sides has length $x - 7$ feet, and since the perimeter is 392 feet, the remaining side must have length

$392 - x - (x - 7) = 399 - 2x$. From the Pythagorean Theorem, we get
$(x - 7)^2 + (399 - 2x)^2 = x^2 \quad \Leftrightarrow \quad 4x^2 - 1610x + 159250 = 0$.
Using the quadratic formula, we get

$$x = \frac{1610 \pm \sqrt{1610^2 - 4(4)(159250)}}{2(4)} = \frac{1610 \pm \sqrt{44100}}{8} = \frac{1610 \pm 210}{8},$$

and so $x = 227.5$ or $x = 175$. But if $x = 227.5$, then the side of length $x - 7$
combined with the hypotenuse already exceeds the perimeter of 392 feet, and so
we must have $x = 175$. Thus the other sides have length $175 - 7 = 168$ and $399 - 2(175) = 49$.
The lot has sides of length 49 feet, 168 feet, and 175 feet.

81. Since the total time is 3 s, we have $3 = \dfrac{\sqrt{d}}{4} + \dfrac{d}{1090}$. Letting $w = \sqrt{d}$, we have $3 = \frac{1}{4}w + \frac{1}{1090}w^2$

$\Leftrightarrow \quad \frac{1}{1090}w^2 + \frac{1}{4}w - 3 = 0 \quad \Leftrightarrow \quad 2w^2 + 545w - 6540 = 0 \quad \Rightarrow \quad w = \dfrac{-545 \pm 591.054}{4}$.

Since $w \geq 0$, we have $\sqrt{d} = w \approx 11.51$, so $d = 132.56$. The well is 132.6 ft deep.

Exercises 1.6

1. $x = -2$: $-2 - 3 \overset{?}{>} 0$. No, $-5 \not> 0$.

   $x = 0$: $0 - 3 \overset{?}{>} 0$. No, $-3 \not> 0$.

   $x = 1$: $1 - 3 \overset{?}{>} 0$. No, $-2 \not> 0$.

   $x = 2$: $2 - 3 \overset{?}{>} 0$. No, $-1 \not> 0$

   Only 4 satisfies the inequality.

   $x = -1$: $-1 - 3 \overset{?}{>} 0$. No, $-4 \not> 0$.

   $x = \frac{1}{2}$: $\frac{1}{2} - 3 \overset{?}{>} 0$. No, $-\frac{5}{2} \not> 0$.

   $x = \sqrt{2}$: $\sqrt{2} - 3 \overset{?}{>} 0$. No, $\sqrt{2} - 3 \not> 0$.

   $x = 4$: $4 - 3 \overset{?}{>} 0$. Yes, $1 > 0$.

3. $x = -2$: $3 - 2(-2) \overset{?}{\leq} \frac{1}{2}$. No, $7 \not\leq \frac{1}{2}$.

   $x = 0$: $3 - 2(0) \overset{?}{\leq} \frac{1}{2}$. No, $3 \not\leq \frac{1}{2}$.

   $x = 1$: $3 - 2(1) \overset{?}{\leq} \frac{1}{2}$. No, $1 \not\leq \frac{1}{2}$.

   $x = 2$: $3 - 2(2) \overset{?}{\leq} \frac{1}{2}$. Yes, $-1 \leq \frac{1}{2}$.

   $x = -1$: $3 - 2(-1) \overset{?}{\leq} \frac{1}{2}$. No, $6 \not\leq \frac{1}{2}$.

   $x = \frac{1}{2}$: $3 - 2(\frac{1}{2}) \overset{?}{\leq} \frac{1}{2}$. No, $2 \not\leq \frac{1}{2}$.

   $x = \sqrt{2}$: $3 - 2(\sqrt{2}) \overset{?}{\leq} \frac{1}{2}$. Yes, $3 - 2\sqrt{2} \leq \frac{1}{2}$.

   $x = 4$: $3 - 2(4) \overset{?}{\leq} \frac{1}{2}$. Yes, $5 \leq \frac{1}{2}$.

   The elements $\sqrt{2}$, 2, and 4 all satisfies the inequality.

5. $x = -2$: $1 \overset{?}{<} 2(-2) - 4 \overset{?}{\leq} 7$. No, since $2(-2) - 4 = -8$ and $1 \not< -8$.

   $x = -1$: $1 \overset{?}{<} 2(-1) - 4 \overset{?}{\leq} 7$. No, since $2(-1) - 4 = -6$ and $1 \not< -6$.

   $x = 0$: $1 \overset{?}{<} 2(0) - 4 \overset{?}{\leq} 7$. No, since $2(0) - 4 = -4$ and $1 \not< -4$.

   $x = \frac{1}{2}$: $1 \overset{?}{<} 2(\frac{1}{2}) - 4 \overset{?}{\leq} 7$. No, since $2(\frac{1}{2}) - 4 = -3$ and $1 \not< -3$.

   $x = 1$: $1 \overset{?}{<} 2(1) - 4 \overset{?}{\leq} 7$. No, since $2(1) - 4 = -2$ and $1 \not< -2$.

   $x = \sqrt{2}$: $1 \overset{?}{<} 2(\sqrt{2}) - 4 \overset{?}{\leq} 7$. No, since $2\sqrt{2} - 4 < 0$ and $1 \not< 0$.

   $x = 2$: $1 \overset{?}{<} 2(2) - 4 \overset{?}{\leq} 7$. No, since $2(1) - 4 = -2$ and $1 \not< -2$.

   $x = 4$: $1 \overset{?}{<} 2(4) - 4 \overset{?}{\leq} 7$. Yes, $2(4) - 4 = 4$ and $1 < 4 \leq 7$.

   Only 4 satisfies the inequality.

7. $x = -2$: $\frac{1}{(-2)} \overset{?}{\leq} \frac{1}{2}$. Yes, $-\frac{1}{2} \leq \frac{1}{2}$.

   $x = 0$: $\frac{1}{0} \overset{?}{\leq} \frac{1}{2}$. No, $\frac{1}{0}$ is not defined.

   $x = 1$: $\frac{1}{1} \overset{?}{\leq} \frac{1}{2}$. No, $1 \not\leq \frac{1}{2}$.

   $x = 2$: $\frac{1}{2} \overset{?}{\leq} \frac{1}{2}$. Yes, $\frac{1}{2} \leq \frac{1}{2}$.

   $x = -1$: $\frac{1}{(-1)} \overset{?}{\leq} \frac{1}{2}$. Yes, $-1 \leq \frac{1}{2}$.

   $x = \frac{1}{2}$: $\frac{1}{1/2} \overset{?}{\leq} \frac{1}{2}$. No, $2 \not\leq \frac{1}{2}$.

   $x = \sqrt{2}$: $\frac{1}{\sqrt{2}} \overset{?}{\leq} \frac{1}{2}$. No, $2 \not\leq \sqrt{2}$.

   $x = 4$: $\frac{1}{4} \overset{?}{\leq} \frac{1}{2}$. Yes.

   The elements -2, -1, 2, and 4 all satisfies the inequality.

9. $2x \leq 7$ $\Leftrightarrow$ $x \leq \frac{7}{2}$.   Interval: $(-\infty, \frac{7}{2}]$.   Graph:

11. $2x - 5 > 3$ $\Leftrightarrow$ $2x > 8$ $\Leftrightarrow$ $x > 4$.   Interval: $(4, \infty)$.

      Graph:

13. $7 - x \geq 5$ $\Leftrightarrow$ $-x \geq -2$ $\Leftrightarrow$ $x \leq 2$.   Interval: $(-\infty, 2]$.

      Graph:

15. $2x + 1 < 0$ $\Leftrightarrow$ $2x < -1$ $\Leftrightarrow$ $x < -\frac{1}{2}$.   Interval: $\left(-\infty, -\frac{1}{2}\right)$.

      Graph:

17. $3x + 11 \leq 6x + 8$ $\Leftrightarrow$ $3 \leq 3x$ $\Leftrightarrow$ $1 \leq x$.   Interval: $[1, \infty)$.

      Graph:

19. $\frac{1}{2}x - \frac{2}{3} > 2$ $\Leftrightarrow$ $\frac{1}{2}x > \frac{8}{3}$ $\Leftrightarrow$ $x > \frac{16}{3}$.   Interval: $\left(\frac{16}{3}, \infty\right)$.

      Graph:

21. $\frac{1}{3}x + 2 < \frac{1}{6}x - 1$ $\Leftrightarrow$ $\frac{1}{6}x < -3$ $\Leftrightarrow$ $x < -18$ Interval: $(-\infty, -18)$

      Graph:

23. $4 - 3x \leq -(1 + 8x)$ $\Leftrightarrow$ $4 - 3x \leq -1 - 8x$ $\Leftrightarrow$ $5x \leq -5$ $\Leftrightarrow$ $x \leq -1$.

      Interval: $(-\infty, -1]$.   Graph:

25. $2 \leq x + 5 < 4$ $\Leftrightarrow$ $-3 \leq x < -1$.   Interval: $[-3, -1)$.

      Graph:

27. $-1 < 2x - 5 < 7$ $\Leftrightarrow$ (add 5 to each expression) $4 < 2x < 12$ $\Leftrightarrow$ (divide each expression

      by 2) $2 < x < 6$. Interval: $(2, 6)$.   Graph:

29. $-2 < 8 - 2x \leq -1$ $\Leftrightarrow$ $-10 < -2x \leq -9$ $\Leftrightarrow$ $5 > x \geq \frac{9}{2}$ $\Leftrightarrow$ $\frac{9}{2} \leq x < 5$.

      Interval: $\left[\frac{9}{2}, 5\right)$.   Graph:

31. $\frac{2}{3} \geq \frac{2x - 3}{12} > \frac{1}{6}$ $\Leftrightarrow$ (multiply each expression by 12) $8 \geq 2x - 3 > 2$ $\Leftrightarrow$ $11 \geq 2x > 5$

      $\Leftrightarrow$ $\frac{11}{2} \geq x > \frac{5}{2}$ $\Leftrightarrow$ (expressing in standard form) $\frac{5}{2} < x \leq \frac{11}{2}$.   Interval: $\left(\frac{5}{2}, \frac{11}{2}\right]$.

      Graph:

33. $(x + 2)(x - 3) < 0$. The expression on the left of the inequality changes sign where $x = -2$ and where $x = 3$. Thus we must check the intervals in the following table.

Interval	$(-\infty, -2)$	$(-2, 3)$	$(3, \infty)$
Sign of $x + 2$	$-$	$+$	$+$
Sign of $x - 3$	$-$	$-$	$+$
Sign of $(x + 2)(x - 3)$	$+$	$-$	$+$

From the table, the solution set is $\{x \mid -2 < x < 3\}$.
Interval: $(-2, 3)$.   Graph:

35. $x(2x + 7) \geq 0$. The expression on the left of the inequality changes sign where $x = 0$ and where $x = -\frac{7}{2}$. Thus we must check the intervals in the following table.

Interval	$\left(-\infty, -\frac{7}{2}\right)$	$\left(-\frac{7}{2}, 0\right)$	$(0, \infty)$
Sign of x	$-$	$-$	$+$
Sign of $2x + 7$	$-$	$+$	$+$
Sign of $x(2x + 7)$	$+$	$-$	$+$

From the table, the solution set is $\left\{x \mid x \leq -\frac{7}{2} \text{ or } 0 \leq x\right\}$.
Interval: $\left(-\infty, -\frac{7}{2}\right] \cup [0, \infty)$.   Graph:

37. $x^2 - 3x - 18 \leq 0 \quad \Leftrightarrow \quad (x + 3)(x - 6) \leq 0$. The expression on the left of the inequality changes sign where $x = 6$ and where $x = -3$. Thus we must check the intervals in the following table.

Interval	$(-\infty, -3)$	$(-3, 6)$	$(6, \infty)$
Sign of $x + 3$	$-$	$+$	$+$
Sign of $x - 6$	$-$	$-$	$+$
Sign of $(x + 3)(x - 6)$	$+$	$-$	$+$

From the table, the solution set is $\{x \mid -3 \leq x \leq 6\}$.
Interval: $[-3, 6]$.   Graph:

39. $2x^2 + x \geq 1 \quad \Leftrightarrow \quad 2x^2 + x - 1 \geq 0 \quad \Leftrightarrow \quad (x + 1)(2x - 1) \geq 0$. The expression on the left of the inequality changes sign where $x = -1$ and where $x = \frac{1}{2}$. Thus we must check the intervals in the following table.

Interval	$(-\infty, -1)$	$\left(-1, \frac{1}{2}\right)$	$\left(\frac{1}{2}, \infty\right)$
Sign of $x + 1$	$-$	$+$	$+$
Sign of $2x - 1$	$-$	$-$	$+$
Sign of $(x + 1)(2x - 1)$	$+$	$-$	$+$

From the table, the solution set is $\left\{x \mid x \leq -1 \text{ or } \frac{1}{2} \leq x\right\}$.
Interval: $\left(-\infty, -1\right] \cup \left[\frac{1}{2}, \infty\right)$.   Graph:

41. $3x^2 - 3x < 2x^2 + 4 \quad \Leftrightarrow \quad x^2 - 3x - 4 < 0 \quad \Leftrightarrow \quad (x + 1)(x - 4) < 0$. The expression on the left of the inequality changes sign where $x = -1$ and where $x = 4$. Thus we must check the intervals in the following table.

Interval	$(-\infty, -1)$	$(-1, 4)$	$(4, \infty)$
Sign of $x + 1$	$-$	$+$	$+$
Sign of $x - 4$	$-$	$-$	$+$
Sign of $(x + 1)(x - 4)$	$+$	$-$	$+$

From the table, the solution set is $\{x \mid -1 < x < 4\}$.

Interval: $(-1, 4)$.   Graph:

43. $x^2 > 3(x + 6)$ $\Leftrightarrow$ $x^2 - 3x - 18 > 0$ $\Leftrightarrow$ $(x + 3)(x - 6) > 0$. The expression on the left of the inequality changes sign where $x = 6$ and where $x = -3$. Thus we must check the intervals in the following table.

Interval	$(-\infty, -3)$	$(-3, 6)$	$(6, \infty)$
Sign of $x + 3$	$-$	$+$	$+$
Sign of $x - 6$	$-$	$-$	$+$
Sign of $(x + 3)(x - 6)$	$+$	$-$	$+$

From the table, the solution set is $\{x \mid x < -3 \text{ or } 6 < x\}$.
Interval: $(-\infty, -3) \cup (6, \infty)$.   Graph:

45. $x^2 < 4$ $\Leftrightarrow$ $x^2 - 4 < 0$ $\Leftrightarrow$ $(x + 2)(x - 2) < 0$. The expression on the left of the inequality changes sign where $x = -2$ and where $x = 2$. Thus we must check the intervals in the following table.

Interval	$(-\infty, -2)$	$(-2, 2)$	$(2, \infty)$
Sign of $x + 2$	$-$	$+$	$+$
Sign of $x - 2$	$-$	$-$	$+$
Sign of $(x + 2)(x - 2)$	$+$	$-$	$+$

From the table, the solution set is $\{x \mid -2 < x < 2\}$.
Interval: $(-2, 2)$.   Graph:

47. $-2x^2 \leq 4$ $\Leftrightarrow$ $-2x^2 - 4 \leq 0$ $\Leftrightarrow$ $-2(x^2 + 1) \leq 0$. Since $x^2 + 1 > 0$, $-2(x^2 + 1) \leq 0$ for all x.   Interval: $(-\infty, \infty)$ Graph:

49. $x^3 - 4x > 0$ $\Leftrightarrow$ $x(x^2 - 4) > 0$ $\Leftrightarrow$ $x(x + 2)(x - 2) > 0$. The expression on the left of the inequality changes sign where $x = 0$, $x = -2$ and where $x = 4$. Thus we must check the intervals in the following table.

Interval	$(-\infty, -2)$	$(-2, 0)$	$(0, 2)$	$(2, \infty)$
Sign of x	$-$	$-$	$+$	$+$
Sign of $x + 2$	$-$	$+$	$+$	$+$
Sign of $x - 2$	$-$	$-$	$-$	$+$
Sign of $x(x + 2)(x - 2)$	$-$	$+$	$-$	$+$

From the table, the solution set is $\{x \mid -2 < x < 0 \text{ or } x > 2\}$.
Interval: $(-2, 0) \cup (2, \infty)$.   Graph:

51. $\dfrac{x - 3}{x + 1} \geq 0$. The expression on the left of the inequality changes sign where $x = -1$ and where $x = 3$. Thus we must check the intervals in the following table.

Interval	$(-\infty, -1)$	$(-1, 3)$	$(3, \infty)$
Sign of $x + 1$	$-$	$+$	$+$
Sign of $x - 3$	$-$	$-$	$+$
Sign of $\dfrac{x - 3}{x + 1}$	$+$	$-$	$+$

From the table, the solution set is $\{x \mid x < -1 \text{ or } x \leq 3\}$. Since the denominator cannot equal 0 we must have $x \neq -1$.

Interval: $(-\infty, -1) \cup [3, \infty)$.   Graph:

53. $\dfrac{4x}{2x+3} > 2 \quad \Leftrightarrow \quad \dfrac{4x}{2x+3} - 2 > 0 \quad \Leftrightarrow \quad \dfrac{4x}{2x+3} - \dfrac{2(2x+3)}{2x+3} > 0 \quad \Leftrightarrow \quad \dfrac{-6}{2x+3} > 0$. The

expression on the left of the inequality changes sign where $x = -\dfrac{3}{2}$. Thus we must check the

intervals in the following table.

Interval	$\left(-\infty, -\frac{3}{2}\right)$	$\left(-\frac{3}{2}, \infty\right)$
Sign of -6	$-$	$-$
Sign of $2x+3$	$-$	$+$
Sign of $\dfrac{-6}{2x+3}$	$+$	$-$

From the table, the solution set is $\left\{x \mid x < -\frac{3}{2}\right\}$.

Interval: $\left(-\infty, -\frac{3}{2}\right)$.   Graph:

55. $\dfrac{2x+1}{x-5} \le 3 \quad \Leftrightarrow \quad \dfrac{2x+1}{x-5} - 3 \le 0 \quad \Leftrightarrow \quad \dfrac{2x+1}{x-5} - \dfrac{3(x-5)}{x-5} \le 0 \quad \Leftrightarrow \quad \dfrac{-x+16}{x-5} \le 0$. The

expression on the left of the inequality changes sign where $x = 16$ and where $x = 5$. Thus we must

check the intervals in the following table.

Interval	$(-\infty, 5)$	$(5, 16)$	$(16, \infty)$
Sign of $-x+16$	$+$	$+$	$-$
Sign of $x-5$	$-$	$+$	$+$
Sign of $\dfrac{-x+16}{x-5}$	$-$	$+$	$-$

From the table, the solution set is $\{x \mid x < 5 \text{ or } x \ge 16\}$. Since the denominator cannot equal 0, we

must have $x \ne 5$.

Interval: $(-\infty, 5) \cup [16, \infty)$.   Graph:

57. $\dfrac{4}{x} < x \quad \Leftrightarrow \quad \dfrac{4}{x} - x < 0 \quad \Leftrightarrow \quad \dfrac{4}{x} - \dfrac{x \cdot x}{x} < 0 \quad \Leftrightarrow \quad \dfrac{4-x^2}{x} < 0 \quad \Leftrightarrow$

$\dfrac{(2-x)(2+x)}{x} \le 0$. The expression on the left of the inequality changes sign where $x = 0$, where

$x = -2$, and where $x = 2$. Thus we must check the intervals in the following table.

Interval	$(-\infty, -2)$	$(-2, 0)$	$(0, 2)$	$(2, \infty)$
Sign of $2+x$	$-$	$+$	$+$	$+$
Sign of x	$-$	$-$	$+$	$+$
Sign of $2-x$	$+$	$+$	$+$	$-$
Sign of $\dfrac{(2-x)(2+x)}{x}$	$+$	$-$	$+$	$-$

From the table, the solution set is $\{x \mid -2 < x < 0 \text{ or } 2 < x\}$.

Interval: $(-2, 0) \cup (2, \infty)$.   Graph:

59. $1 + \dfrac{2}{x+1} \le \dfrac{2}{x} \quad \Leftrightarrow \quad 1 + \dfrac{2}{x+1} - \dfrac{2}{x} \le 0 \quad \Leftrightarrow \quad \dfrac{x(x+1)}{x(x+1)} + \dfrac{2x}{x(x+1)} - \dfrac{2(x+1)}{x(x+1)} \le 0 \quad \Leftrightarrow$

$\dfrac{x^2 + x + 2x - 2x - 2}{x(x+1)} \le 0 \quad \Leftrightarrow \quad \dfrac{x^2 + x - 2}{x(x+1)} \le 0 \quad \Leftrightarrow \quad \dfrac{(x+2)(x-1)}{x(x+1)} \le 0$. The expression on the left of the inequality changes sign where $x = -2$, where $x = -1$, where $x = 0$, and where $x = 1$. Thus we must check the intervals in the following table.

Interval	$(-\infty, -2)$	$(-2, -1)$	$(-1, 0)$	$(0, 1)$	$(1, \infty)$
Sign of $x + 2$	$-$	$+$	$+$	$+$	$+$
Sign of $x - 1$	$-$	$-$	$-$	$-$	$+$
Sign of x	$-$	$-$	$-$	$+$	$+$
Sign of $x + 1$	$-$	$-$	$+$	$+$	$+$
Sign of $\dfrac{(x+2)(x-1)}{x(x+1)}$	$+$	$-$	$+$	$-$	$+$

Since $x = -1$ and $x = 0$ yield undefined expressions, we cannot include them in the solution. From the table, the solution set is $\{x \mid -2 \le x < -1 \text{ or } 0 < x \le 1\}$.

Interval: $[-2, -1) \cup (0, 1]$.   Graph:

61. $\dfrac{6}{x-1} - \dfrac{6}{x} \ge 1 \quad \Leftrightarrow \quad \dfrac{6}{x-1} - \dfrac{6}{x} - 1 \ge 0 \quad \Leftrightarrow \quad \dfrac{6x}{x(x-1)} - \dfrac{6(x-1)}{x(x-1)} - \dfrac{x(x-1)}{x(x-1)} \ge 0 \quad \Leftrightarrow$

$\dfrac{6x - 6x + 6 - x^2 + x}{x(x-1)} \ge 0 \quad \Leftrightarrow \quad \dfrac{-x^2 + x + 6}{x(x-1)} \ge 0 \quad \Leftrightarrow \quad \dfrac{(-x+3)(x+2)}{x(x-1)} \ge 0.$ The

expression on the left of the inequality changes sign where $x = 3$, where $x = -2$, where $x = 0$, and where $x = 1$. Thus we must check the intervals in the following table.

Interval	$(-\infty, -2)$	$(-2, 0)$	$(0, 1)$	$(1, 3)$	$(3, \infty)$
Sign of $-x + 3$	$+$	$+$	$+$	$+$	$-$
Sign of $x + 2$	$-$	$+$	$+$	$+$	$+$
Sign of x	$-$	$-$	$+$	$+$	$+$
Sign of $x - 1$	$-$	$-$	$-$	$+$	$+$
Sign of $\dfrac{(-x+3)(x+2)}{x(x-1)}$	$-$	$+$	$-$	$+$	$-$

From the table, the solution set is $\{x \mid -2 \le x < 0 \text{ or } 1 < x \le 3\}$. The points $x = 0$ and $x = 1$ are excluded from the solution set because they make the denominator zero.

Interval: $[-2, 0) \cup (1, 3]$ Graph:

63. $\dfrac{x+2}{x+3} < \dfrac{x-1}{x-2} \quad \Leftrightarrow \quad \dfrac{x+2}{x+3} - \dfrac{x-1}{x-2} < 0 \quad \Leftrightarrow \quad \dfrac{(x+2)(x-2)}{(x+3)(x-2)} - \dfrac{(x-1)(x+3)}{(x-2)(x+3)} < 0 \quad \Leftrightarrow$

$\dfrac{x^2 - 4 - x^2 - 2x + 3}{(x+3)(x-2)} < 0 \quad \Leftrightarrow \quad \dfrac{-2x - 1}{(x+3)(x-2)} < 0.$ The expression on the left of the inequality

changes sign where $x = -\frac{1}{2}$, where $x = -3$, and where $x = 2$. Thus we must check the intervals in the following table.

Interval	$(-\infty, -3)$	$(-3, -\frac{1}{2})$	$(-\frac{1}{2}, 2)$	$(2, \infty)$
Sign of $-2x - 1$	$+$	$+$	$-$	$-$
Sign of $x + 3$	$-$	$+$	$+$	$+$
Sign of $x - 2$	$-$	$-$	$-$	$+$
Sign of $\dfrac{-2x - 1}{(x+3)(x-2)}$	$+$	$-$	$+$	$-$

From the table, the solution set is $\{x \mid -3 < x < -\frac{1}{2} \text{ or } 2 < x\}$.

Interval: $\left(-3, -\frac{1}{2}\right) \cup (2, \infty)$.   Graph:

65. $x^4 > x^2$ $\Leftrightarrow$ $x^4 - x^2 > 0$ $\Leftrightarrow$ $x^2(x^2 - 1) > 0$ $\Leftrightarrow$ $x^2(x-1)(x+1) > 0$. The expression on the left of the inequality changes sign where $x = 0$, where $x = 1$, and where $x = -1$. Thus we must check the intervals in the following table.

Interval	$(-\infty, -1)$	$(-1, 0)$	$(0, 1)$	$(1, \infty)$
Sign of x^2	$+$	$+$	$+$	$+$
Sign of $x - 1$	$-$	$-$	$-$	$+$
Sign of $x + 1$	$-$	$+$	$+$	$+$
Sign of $x^2(x-1)(x+1)$	$+$	$-$	$-$	$+$

From the table, the solution set is $\{x \mid x < -1 \text{ or } 1 < x\}$
Interval: $(-\infty, -1) \cup (1, \infty)$.   Graph:

67. For $\sqrt{16 - 9x^2}$ to be defined as a real number we must have $16 - 9x^2 \geq 0$ $\Leftrightarrow$ $(4 - 3x)(4 + 3x) \geq 0$. The expression in the inequality changes sign at $x = \frac{4}{3}$ and $x = -\frac{4}{3}$.

Interval	$\left(-\infty, -\frac{4}{3}\right)$	$\left(-\frac{4}{3}, \frac{4}{3}\right)$	$\left(\frac{4}{3}, \infty\right)$
Sign of $4 - 3x$	$+$	$+$	$-$
Sign of $4 + 3x$	$-$	$+$	$+$
Sign of $(4 - 3x)(4 + 3x)$	$-$	$+$	$-$

Thus $-\frac{4}{3} \leq x \leq \frac{4}{3}$.

69. For $\left(\dfrac{1}{x^2 - 5x - 14}\right)^{1/2}$ to be defined as a real number we must have $x^2 - 5x - 14 > 0$ $\Leftrightarrow$ $(x - 7)(x + 2) > 0$. The expression in the inequality changes sign at $x = 7$ and $x = -2$.

Interval	$(-\infty, -2)$	$(-2, 7)$	$(7, \infty)$
Sign of $x - 7$	$-$	$-$	$+$
Sign of $x + 2$	$-$	$+$	$+$
Sign of $(x - 7)(x + 2)$	$+$	$-$	$+$

Thus $x < -2$ or $7 < x$, and the solution set is $(-\infty, -2) \cup (7, \infty)$.

71. (a) $a(bx - c) \geq bc$ (where $a, b, c > 0$) $\Leftrightarrow$ $bx - c \geq \dfrac{bc}{a}$ $\Leftrightarrow$ $bx \geq \dfrac{bc}{a} + c$ $\Leftrightarrow$

$x \geq \dfrac{1}{b}\left(\dfrac{bc}{a} + c\right) = \dfrac{c}{a} + \dfrac{c}{b}$ $\Leftrightarrow$ $x \geq \dfrac{c}{a} + \dfrac{c}{b}$.

(b) We have $a \leq bx + c < 2a$, where $a, b, c > 0$ $\Leftrightarrow$ $a - c \leq bx < 2a - c$ $\Leftrightarrow$

$\dfrac{a - c}{b} \leq x < \dfrac{2a - c}{b}$.

73. Inserting the relationship $C = \frac{5}{9}(F - 32)$, we have $20 \leq C \leq 30$ $\Leftrightarrow$ $20 \leq \frac{5}{9}(F - 32) \leq 30$ $\Leftrightarrow$ $36 \leq F - 32 \leq 54$ $\Leftrightarrow$ $68 \leq F \leq 86$.

75. Let x be the *average miles driven a day*. Each day the cost of Plan A is $30 + 0.10x$, and the cost of Plan B is 50. Plan B saves money when $50 < 30 + 0.10x$ $\Leftrightarrow$ $20 < 0.1x$ $\Leftrightarrow$ $200 < x$. So Plan B saves money when you average more than 200 miles a day.

77. We need to solve $6400 \leq 0.35m + 2200 \leq 7100$ for m. So $6400 \leq 0.35m + 2200 \leq 7100$ ⇔ $4200 \leq 0.35m \leq 4900$ ⇔ $12000 \leq m \leq 14000$. She plans on driving between 12,000 and 14,000 miles.

79. (a) Let x be the number of \$3 increases. Then the number of seats sold is $120 - x$. So
$P = 200 + 3x$ ⇔ $3x = P - 200$ ⇔ $x = \frac{1}{3}(P - 200)$. Substituting for x we have
that the number of seats sold is $120 - x = 120 - \frac{1}{3}(P - 200) = -\frac{1}{3}P + \frac{560}{3}$.

   (b) $90 \leq -\frac{1}{3}P + \frac{560}{3} \leq 115$ ⇔ $270 \leq 360 - P + 200 \leq 345$ ⇔
$270 \leq -P + 560 \leq 345$ ⇔ $-290 \leq -P \leq -215$ ⇔ $290 \geq P \geq 215$. Putting this
into standard order, we have $215 \leq P \leq 290$. So the ticket prices are between \$215 and \$290.

81. $0.0004 \leq \dfrac{4,000,000}{d^2} \leq 0.01$. Since $d^2 \geq 0$ and $d \neq 0$, we can multiply each expression by d^2 to
obtain $0.0004d^2 \leq 4,000,000 \leq 0.01d^2$. Solving each pair, we have $0.0004d^2 \leq 4,000,000$ ⇔
$d^2 \leq 10,000,000,000$ ⇒ $d \leq 100,000$ (recall that d represents distance, so it is always
nonnegative). Solving $4,000,000 \leq 0.01d^2$ ⇔ $400,000,000 \leq d^2 \Rightarrow 20,000 \leq d$. Putting
these together, we have $20,000 \leq d \leq 100,000$.

83. $128 + 16t - 16t^2 \geq 32$ ⇔ $-16t^2 + 16t + 96 \geq 0$ ⇔ $-16(t^2 - t - 6) \geq 0$ ⇔
$-16(t - 3)(t + 2) \geq 0$. The expression on the left of the inequality changes sign at $x = -2$, at
$t = 3$, and at $t = -2$. However, $t \geq 0$, so the only endpoint is $t = 3$. Thus we check the intervals in
the following table.

Interval	$(0, 3)$	$(3, \infty)$
Sign of -16	$-$	$-$
Sign of $t - 3$	$-$	$+$
Sign of $t + 2$	$+$	$+$
Sign of $-16(t - 3)(t + 2)$	$+$	$-$

So $0 \leq t \leq 3$.

85. $240 > v + \dfrac{v^2}{20}$ ⇔ $\frac{1}{20}v^2 + v - 240 < 0$ ⇔ $\left(\frac{1}{20}v - 3\right)(v + 80) < 0$. The expression in the
inequality changes sign at $v = 60$ and $v = -80$. However, since v represents the speed, we must
have $v \geq 0$.

Interval	$(0, 60)$	$(60, \infty)$
Sign of $\frac{1}{20}v - 3$	$-$	$+$
Sign of $v + 80$	$+$	$+$
Sign of $\left(\frac{1}{20}v - 3\right)(v + 80)$	$-$	$+$

So Kerry must drive less than 60 mph.

87. Let n be the number of people in the group. Then the bus fare is $\dfrac{360}{n}$, and the cost of the theater
tickets is $30 - 0.25n$. We want the total cost to be less than \$39 per person, that is,
$\dfrac{360}{n} + (30 - 0.25n) < 39$. If we multiply this inequality by n, we will not change the direction of
the inequality; n is positive since it represents the number of people. So we get
$360 + n(30 - 0.25n) < 39n$ ⇔ $-0.25n^2 + 30n + 360 < 39n$ ⇔

$-0.25n^2 - 9n + 360 < 0 \quad \Leftrightarrow \quad (0.25n + 15)(-n + 24)$. The expression on the left of the inequality changes sign when $n = -60$ and $n = 24$. Since $n > 0$, we check the intervals in the following table.

Interval	(0, 24)	(24, ∞)
Sign of $0.25n + 15$	+	+
Sign of $-n + 24$	+	−
Sign of $(0.25n + 15)(-n + 24)$	+	−

So the group must have more than 24 people in order that the cost of the theater tour is less than $39.

89. **Case 1:** $\underline{a < b < 0.}$ We have $a \cdot a > a \cdot b$, since $a < 0$, and $b \cdot a > b \cdot b$, since $b < 0$. So $a^2 > a \cdot b > b^2$, that is $a < b < 0 \quad \Rightarrow \quad a^2 > b^2$. Continuing, we have $a \cdot a^2 < a \cdot b^2$, since $a < 0$ and $b^2 \cdot a < b^2 \cdot b$, since $b^2 > 0$. So $a^3 < ab^2 < b^3$. Thus $a < b < 0 \quad \Rightarrow \quad a^3 > b^3$. So $a < b < 0 \quad \Rightarrow \quad a^n > b^n$, if n is even, and $a^n < b$, if n is odd.

Case 2: $\underline{0 < a < b.}$ We have $a \cdot a < a \cdot b$, since $a > 0$, and $b \cdot a < b \cdot b$, since $b < 0$. So $a^2 < a \cdot b < b^2$. Thus $0 < a < b \quad \Rightarrow \quad a^2 < b^2$. Likewise, $a^2 \cdot a < a^2 \cdot b$ and $b \cdot a^2 < b \cdot b^2$, thus $a^3 < b^3$. So $0 < a < b \quad \Rightarrow \quad a^n < b^n$, for all positive integers n.

Case 3: $\underline{a < 0 < b.}$ If n is odd, then $a^n < b^n$, because a^n is negative and b^n is positive. If n is even, then we could have either $a^n < b^n$ or $a^n > b^n$. For example, $-1 < 2$ and $(-1)^2 < 2^2$, but $-3 < 2$ and $(-3)^2 > 2^2$.

Exercises 1.7

1. $|4x| = 24 \iff 4x = \pm 24 \iff x = \pm 6.$

3. $5|x| + 3 = 28 \iff 5|x| = 25 \iff |x| = 5 \iff x = \pm 5.$

5. $|x - 3| = 2$ is equivalent to $x - 3 = \pm 2 \iff x = 3 \pm 2 \iff x = 1$ or $x = 5.$

7. $|x + 4| = 0.5$ is equivalent to $x + 4 = \pm 0.5 \iff x = -4 \pm 0.5 \iff x = -4.5$ or $x = -3.5.$

9. $|4x + 7| = 9$ is equivalent to either $4x + 7 = 9 \iff 4x = 2 \iff x = \frac{1}{2};$ or $4x + 7 = -9$ $\iff 4x = -16 \iff x = -4.$ The two solutions are $x = \frac{1}{2}$ and $x = -4.$

11. $4 - |3x + 6| = 1 \iff -|3x + 6| = -3 \iff |3x + 6| = 3,$ which is equivalent to either $3x + 6 = 3 \iff 3x = -3 \iff x = -1;$ or $3x + 6 = -3 \iff 3x = -9 \iff x = -3.$ The two solutions are $x = -1$ and $x = -3.$

13. $3|x + 5| + 6 = 15 \iff 3|x + 5| = 9 \iff |x + 5| = 3,$ which is equivalent to either $x + 5 = 3 \iff x = -2;$ or $x + 5 = -3 \iff x = -8.$ The two solutions are $x = -2$ and $x = -8.$

15. $8 + 5\left|\frac{1}{3}x - \frac{5}{6}\right| = 33 \iff 5\left|\frac{1}{3}x - \frac{5}{6}\right| = 25 \iff \left|\frac{1}{3}x - \frac{5}{6}\right| = 5,$ which is equivalent to either $\frac{1}{3}x - \frac{5}{6} = 5 \iff \frac{1}{3}x = \frac{35}{6} \iff x = \frac{35}{2};$ or $\frac{1}{3}x - \frac{5}{6} = -5 \iff \frac{1}{3}x = -\frac{25}{6} \iff$ $x = -\frac{25}{2}.$ The two solutions are $x = -\frac{25}{2}$ and $x = \frac{35}{2}.$

17. $|x - 1| = |3x + 2|,$ which is equivalent to either $x - 1 = 3x + 2 \iff -2x = 3 \iff x = -\frac{3}{2};$ or $x - 1 = -(3x + 2) \iff x - 1 = -3x - 2 \iff 4x = -1 \iff x = -\frac{1}{4}.$ The two solutions are $x = -\frac{3}{2}$ and $x = -\frac{1}{4}.$

19. $|x| \le 4 \iff -4 \le x \le 4.$ Interval: $[-4, 4].$

21. $|2x| > 7$ is equivalent to $2x > 7 \iff x > \frac{7}{2};$ or $2x < 7 \iff x < -\frac{7}{2}.$ Interval: $\left(-\infty, -\frac{7}{2}\right) \cup \left(\frac{7}{2}, \infty\right).$

23. $|x - 5| \le 3 \iff -3 \le x - 5 \le 3 \iff 2 \le x \le 8.$ Interval: $[2, 8].$

25. $|x + 1| \ge 1$ is equivalent to $x + 1 \ge 1 \iff x \ge 0;$ or $x + 1 \le -1 \iff x \le -2.$ Interval: $(-\infty, -2] \cup [0, \infty).$

27. $|x + 5| \ge 2$ is equivalent to $x + 5 \ge 2 \iff x \ge -3;$ or $x + 5 \le -2 \iff x \le -7.$ Interval: $(-\infty, -7] \cup [-3, \infty).$

29. $|2x - 3| \le 0.4 \iff -0.4 \le 2x - 3 \le 0.4 \iff 2.6 \le 2x \le 3.4 \iff 1.3 \le x \le 1.7.$ Interval: $[1.3, 1.7].$

31. $\left|\dfrac{x-2}{3}\right| < 2 \quad \Leftrightarrow \quad -2 < \dfrac{x-2}{3} < 2 \quad \Leftrightarrow \quad -6 < x - 2 < 6 \quad \Leftrightarrow \quad -4 < x < 8.$
   Interval: $(-4,\, 8)$.

33. $|x+6| < 0.001 \quad \Leftrightarrow \quad -0.001 < x + 6 < 0.001 \quad \Leftrightarrow \quad -6.001 < x < -5.999.$
   Interval: $(-6.001,\, -5.999)$.

35. $4|x+2| - 3 < 13 \quad \Leftrightarrow \quad 4|x+2| < 16 \quad \Leftrightarrow \quad |x+2| < 4 \quad \Leftrightarrow \quad -4 < x + 2 < 4 \quad \Leftrightarrow$
   $-6 < x < 2.$ Interval: $(-6,\, 2)$.

37. $8 - |2x-1| \geq 6 \quad \Leftrightarrow \quad -|2x-1| \geq -2 \quad \Leftrightarrow \quad |2x-1| \leq 2 \quad \Leftrightarrow \quad -2 \leq 2x - 1 \leq 2 \quad \Leftrightarrow$
   $-1 \leq 2x \leq 3 \quad \Leftrightarrow \quad -\frac{1}{2} \leq x \leq \frac{3}{2}.$ Interval: $\left[-\frac{1}{2},\, \frac{3}{2}\right]$.

39. $\frac{1}{2}\left|4x + \frac{1}{3}\right| > \frac{5}{6} \quad \Leftrightarrow \quad \left|4x + \frac{1}{3}\right| > \frac{5}{3}$, which is equivalent to $4x + \frac{1}{3} > \frac{5}{3} \quad \Leftrightarrow \quad 4x > \frac{4}{3} \quad \Leftrightarrow$
   $x > \frac{1}{3}$, or $4x + \frac{1}{3} < -\frac{5}{3} \quad \Leftrightarrow \quad 4x < -2 \quad \Leftrightarrow \quad x < -\frac{1}{2}.$ Interval: $\left(-\infty,\, -\frac{1}{2}\right) \cup \left(\frac{1}{3},\, \infty\right)$.

41. $1 \leq |x| \leq 4.$ If $x \geq 0$, then this is equivalent to $1 \leq x \leq 4.$ If $x < 0$, then this is equivalent to
   $1 \leq -x \leq 4 \quad \Leftrightarrow \quad -1 \geq x \geq -4 \quad \Leftrightarrow \quad -4 \leq x \leq -1.$ Interval: $[-4,\, -1] \cup [1,\, 4]$.

43. $\dfrac{1}{|x+7|} > 2 \quad \Leftrightarrow \quad 1 > 2|x+7| \ (x \neq -7) \quad \Leftrightarrow \quad |x+7| < \frac{1}{2} \quad \Leftrightarrow \quad -\frac{1}{2} < x + 7 < \frac{1}{2} \quad \Leftrightarrow$
   $-\frac{15}{2} < x < -\frac{13}{2}$ and $x \neq -7.$ Interval: $\left(-\frac{15}{2},\, -7\right) \cup \left(-7,\, -\frac{13}{2}\right)$.

45. $|x| < 3$

47. $|x - 7| \geq 5$

49. $|x| \leq 2$

53. (a) Let x be the thickness of the laminate. Then $|x - 0.020| \leq 0.003$.

    (b) $|x - 0.020| \leq 0.003 \quad \Leftrightarrow \quad -0.003 \leq x - 0.020 \leq 0.003 \quad \Leftrightarrow \quad 0.017 \leq x \leq 0.023$.

55. $|x - 1|$ is the distance between x and 1;
   $|x - 3|$ is the distance between x and 3.
   So $|x - 1| < |x - 3|$ represents those points closer to 1 than to 3, and the solution is $x < 2$, since 2
   is the point half way between 1 and 3. If $a < b$, then the solution to $|x - a| < |x - b|$ is $x < \frac{a+b}{2}$.

Review Exercises for Chapter 1

1. $5x + 11 = 36 \quad \Leftrightarrow \quad 5x = 25 \quad \Leftrightarrow \quad x = 5.$

3. $3x + 12 = 24 \quad \Leftrightarrow \quad 3x = 12 \quad \Leftrightarrow \quad x = 4.$

5. $7x - 6 = 4x + 9 \quad \Leftrightarrow \quad 3x = 15 \quad \Leftrightarrow \quad x = 5.$

7. $\frac{1}{3}x - \frac{1}{2} = 2 \quad \Leftrightarrow \quad 2x - 3 = 12 \quad \Leftrightarrow \quad 2x = 15 \quad \Leftrightarrow \quad x = \frac{15}{2}.$

9. $2(x + 3) - 4(x - 5) = 8 - 5x \quad \Leftrightarrow \quad 2x + 6 - 4x + 20 = 8 - 5x \quad \Leftrightarrow \quad -2x + 26 = 8 - 5x$
   $\Leftrightarrow \quad 3x = -18 \quad \Leftrightarrow \quad x = -6.$

11. $\dfrac{x + 1}{x - 1} = \dfrac{2x - 1}{2x + 1} \quad \Leftrightarrow \quad (x + 1)(2x + 1) = (2x - 1)(x - 1) \quad \Leftrightarrow \quad 2x^2 + 3x + 1 = 2x^2 - 3x + 1$
    $\Leftrightarrow \quad 6x = 0 \quad \Leftrightarrow \quad x = 0.$

13. $x^2 = 144 \quad \Rightarrow \quad x = \pm 12.$

15. $5x^4 - 16 = 0 \quad \Leftrightarrow \quad 5x^4 = 16 \quad \Leftrightarrow \quad x^4 = \frac{16}{5} \quad \Rightarrow \quad x^2 = \pm \frac{4}{\sqrt{5}}.$ Now $x^2 = -\frac{4}{\sqrt{5}}$ has no real

    solutions, so we solve $x^2 = \frac{4}{\sqrt{5}} \quad \Rightarrow \quad x = \pm\frac{2}{\sqrt[4]{5}} = \pm\frac{2\sqrt[4]{5^3}}{5}$

17. $5x^3 - 15 = 0 \quad \Leftrightarrow \quad 5x^3 = 15 \quad \Leftrightarrow \quad x^3 = 3 \quad \Leftrightarrow \quad x = \sqrt[3]{3}.$

19. $(x + 1)^3 = -64 \quad \Leftrightarrow \quad x + 1 = -4 \quad \Leftrightarrow \quad x = -1 - 4 = -5.$

21. $\sqrt[3]{x} = -3 \quad \Leftrightarrow \quad x = (-3)^3 = -27.$

23. $4x^{3/4} - 500 = 0 \quad \Leftrightarrow \quad 4x^{3/4} = 500 \quad \Leftrightarrow \quad x^{3/4} = 125 \quad \Leftrightarrow \quad x = 125^{4/3} = 5^4 = 625.$

25. $\dfrac{x + 1}{x - 1} = \dfrac{3x}{3x - 6} = \dfrac{3x}{3(x - 2)} = \dfrac{x}{x - 2} \quad \Leftrightarrow \quad (x + 1)(x - 2) = x(x - 1) \quad \Leftrightarrow$
    $x^2 - x - 2 = x^2 - x \quad \Leftrightarrow \quad -2 = 0.$ Since this last equation is never true, there are no real
    solutions to the original equation.

27. $x^2 - 9x + 14 = 0 \quad \Leftrightarrow \quad (x - 7)(x - 2) = 0 \quad \Leftrightarrow \quad x = 7 \text{ or } x = 2.$

29. $2x^2 + x = 1 \quad \Leftrightarrow \quad 2x^2 + x - 1 = 0 \quad \Leftrightarrow \quad (2x - 1)(x + 1) = 0.$ So either $2x - 1 = 0 \quad \Leftrightarrow$
    $2x = 1 \quad \Leftrightarrow \quad x = \frac{1}{2};$ or $x + 1 = 0 \quad \Leftrightarrow \quad x = -1.$

31. $0 = 4x^3 - 25x = x(4x^2 - 25) = x(2x - 5)(2x + 5) = 0.$ So either $x = 0;$ or $2x - 5 = 0 \quad \Leftrightarrow$
    $2x = 5 \quad \Leftrightarrow \quad x = \frac{5}{2};$ or $2x + 5 = 0 \quad \Leftrightarrow \quad 2x = -5 \quad \Leftrightarrow \quad x = -\frac{5}{2}.$

33. $3x^2 + 4x - 1 = 0 \quad \Rightarrow \quad x = \dfrac{-b \pm \sqrt{b^2 - 4ac}}{2a} = \dfrac{-(4) \pm \sqrt{(4)^2 - 4(3)(-1)}}{2(3)}$
    $= \dfrac{-4 \pm \sqrt{16 + 12}}{6} = \dfrac{-4 \pm \sqrt{28}}{6} = \dfrac{-4 \pm 2\sqrt{7}}{6} = \dfrac{2(-2 \pm \sqrt{7})}{6} = \dfrac{-2 \pm \sqrt{7}}{3}.$

35. $\dfrac{1}{x} + \dfrac{2}{x-1} = 3 \quad \Leftrightarrow \quad (x-1) + 2(x) = 3(x)(x-1) \quad \Leftrightarrow \quad x - 1 + 2x = 3x^2 - 3x \quad \Leftrightarrow$

$0 = 3x^2 - 6x + 1 \quad \Rightarrow \quad x = \dfrac{-b \pm \sqrt{b^2 - 4ac}}{2a} = \dfrac{-(-6) \pm \sqrt{(-6)^2 - 4(3)(1)}}{2(3)}$

$= \dfrac{6 \pm \sqrt{36 - 12}}{6} = \dfrac{6 \pm \sqrt{24}}{6} = \dfrac{6 \pm 2\sqrt{6}}{6} = \dfrac{2(3 \pm \sqrt{6})}{6} = \dfrac{3 \pm \sqrt{6}}{3}.$

37. $x^4 - 8x^2 - 9 = 0 \quad \Leftrightarrow \quad (x^2 - 9)(x^2 + 1) = 0 \quad \Leftrightarrow \quad (x-3)(x+3)(x^2+1) = 0 \quad \Rightarrow$
$x - 3 = 0 \quad \Leftrightarrow \quad x = 3,$ or $x + 3 = 0 \quad \Leftrightarrow \quad x = -3,$ however $x^2 + 1 = 0$ has no real
solution. The solutions are $x = \pm 3$.

39. $x^{-1/2} - 2x^{1/2} + x^{3/2} = 0 \quad \Leftrightarrow \quad x^{-1/2}(1 - 2x + x^2) = 0 \quad \Leftrightarrow \quad x^{1/2}(1-x)^2 = 0.$ Since
$x^{-1/2} \neq 0$, the only solution comes from $(1-x)^2 = 0 \quad \Leftrightarrow \quad 1 - x = 0 \quad \Leftrightarrow \quad x = 1.$

41. $|x - 7| = 4 \quad \Leftrightarrow \quad x - 7 = \pm 4 \quad \Leftrightarrow \quad x = 7 \pm 4,$ so $x = 11$ or $x = 3$.

43. $|2x - 5| = 9$ is equivalent to $2x - 5 = \pm 9 \quad \Leftrightarrow \quad 2x = 5 \pm 9 \quad \Leftrightarrow \quad x = \frac{5 \pm 9}{2}.$ So $x = -2$ or
$x = 7$.

45. Let x be the number of pounds of raisins. Then the number of pounds of nuts is $50 - x$.

	raisins	nuts	mixture
pounds	x	$50 - x$	50
rate (cost per pound)	3.20	2.40	2.72
value	$3.20x$	$2.40(50 - x)$	$2.72(50)$

So $3.20x + 2.40(50 - x) = 2.72(50) \quad \Leftrightarrow \quad 3.20x + 120 - 2.40x = 136 \quad \Leftrightarrow \quad 0.8x = 16$
$\Leftrightarrow \quad x = 20.$ Thus the mixture uses 20 pounds raisins and $50 - 20 = 30$ pounds of nuts.

47. Let r be the rate the woman runs in mi/h. Then she cycles at $r + 8$ mi/h.

	Rate	Time	Distance
Cycle	$r + 8$	$\dfrac{4}{r+8}$	4
Run	r	$\dfrac{2.5}{r}$	2.5

Since the total time of the workout is 1 hour, we have $\dfrac{4}{r+8} + \dfrac{2.5}{r} = 1.$ Multiplying by $2r(r+8)$,
we get $4(2r) + 2.5(2)(r+8) = 2r(r+8) \quad \Leftrightarrow \quad 8r + 5r + 40 = 2r^2 + 16r \quad \Leftrightarrow$
$0 = 2r^2 + 3r - 40 \quad \Rightarrow \quad r = \dfrac{-3 \pm \sqrt{(3)^2 - 4(2)(-40)}}{2(2)} = \dfrac{-3 \pm \sqrt{9 + 320}}{4} = \dfrac{-3 \pm \sqrt{329}}{4}.$
Since $r \geq 0$, we reject the negative value. She runs at $r = \dfrac{-3 + \sqrt{329}}{4} \approx 3.78$ mi/h.

49. Let x be the length of one side in cm. Then $28 - x$ is the length of the other side. Using the
Pythagorean theorem, we have $x^2 + (28 - x)^2 = 20^2 \quad \Leftrightarrow \quad x^2 + 784 - 56x + x^2 = 400 \quad \Leftrightarrow$
$2x^2 - 56x + 384 = 0 \quad \Leftrightarrow \quad 2(x^2 - 28x + 192) = 0 \quad \Leftrightarrow \quad 2(x - 12)(x - 16) = 0.$ So $x = 12$
or $x = 16$. If $x = 12$, then the other side is $28 - 12 = 16$. Similarly, if $x = 16$, then the other side
is 12. The sides are 12 cm and 16 cm.

51. Let w be width of the pool. Then the length of the pool is $2w$. Thus the volume of the pool is
    $8(w)(2w) = 8464 \quad \Leftrightarrow \quad 16w^2 = 8464 \quad \Leftrightarrow \quad w^2 = 529 \quad \Rightarrow \quad w = \pm\, 23$. Since $w > 0$, we
    reject the negative value. The pool is 23 feet wide, $2(23) = 46$ feet long, and 8 feet deep.

53. $(3 - 5\,i) - (6 + 4\,i) = 3 - 5\,i - 6 - 4\,i = -3 - 9\,i$.

55. $(2 + 7\,i)(6 - i) = 12 - 2\,i + 42\,i - 7\,i^2 = 12 + 40\,i + 7 = 19 + 40\,i$.

57. $\dfrac{2 - 3\,i}{2 + 3\,i} = \dfrac{2 - 3\,i}{2 + 3\,i} \cdot \dfrac{2 - 3\,i}{2 - 3\,i} = \dfrac{4 - 12\,i + 9\,i^2}{4 - 9\,i^2} = \dfrac{4 - 12i - 9}{4 + 9} = \dfrac{-5 - 12\,i}{13} = -\dfrac{5}{13} - \dfrac{12}{13}\,i$.

59. $i^{45} = i^{44}\,i = (i^4)^{11}\,i = (1)^{11}\,i = i$.

61. $(1 - \sqrt{-3}\,)(2 + \sqrt{-4}\,) = (1 - \sqrt{3}\,i)(2 + 2\,i) = 2 + 2i - 2\sqrt{3}\,i - 2\sqrt{3}\,i^2$

    $= 2 + (2 - 2\sqrt{3}\,)i + 2\sqrt{3} = (2 + 2\sqrt{3}\,) + (2 - 2\sqrt{3}\,)\,i$.

63. $x^2 + 16 = 0 \quad \Leftrightarrow \quad x^2 = -16 \quad \Rightarrow \quad x = \pm\,\sqrt{-16} = \pm 4\,i$.

65. $x^2 + 6x + 10 = 0 \quad \Rightarrow \quad x = \dfrac{-6 \pm \sqrt{6^2 - 4(1)(10)}}{2(1)} = \dfrac{-6 \pm \sqrt{36 - 40}}{2} = \dfrac{-6 \pm \sqrt{-4}}{2}$

    $= \dfrac{-6 \pm 2\,i}{2} = -3 \pm i$.

67. $x^4 - 256 = 0 \quad \Leftrightarrow \quad (x^2 + 16)(x^2 - 16) = 0$. Thus either $x^2 + 16 = 0 \quad \Leftrightarrow \quad x^2 = -16 \quad \Rightarrow$
    $x = \pm\,\sqrt{-16} \quad \Leftrightarrow \quad x = \pm 4i$; or $x^2 - 16 = 0 \quad \Leftrightarrow \quad (x - 4)(x + 4) = 0 \quad \Rightarrow \quad x - 4 = 0$
    $\Leftrightarrow \quad x = 4$; or $x + 4 = 0 \quad \Leftrightarrow \quad x = -4$. The solutions are $\pm 4i$ and ± 4.

69. $x^2 + 4x = (2x + 1)^2 \quad \Leftrightarrow \quad x^2 + 4x = 4x^2 + 4x + 1 \quad \Leftrightarrow \quad 0 = 3x^2 + 1 \quad \Leftrightarrow \quad 3x^2 = -1$
    $\Leftrightarrow \quad x^2 = -\frac{1}{3} \quad \Rightarrow \quad x = \pm\sqrt{-\frac{1}{3}} = \pm\frac{\sqrt{3}}{3}\,i$.

71. $3x - 2 > -11 \quad \Leftrightarrow \quad 3x > -9 \quad \Leftrightarrow \quad x > -3$.   Interval: $(-3, \infty)$.
    Graph:

73. $-1 < 2x + 5 \le 3 \quad \Leftrightarrow \quad -6 < 2x \le -2 \quad \Leftrightarrow \quad -3 < x \le -1$ Interval: $(-3, -1]$.
    Graph:

75. $x^2 + 4x - 12 > 0 \quad \Leftrightarrow \quad (x - 2)(x + 6) > 0$. The expression on the left of the inequality
    changes sign where $x = 2$ and where $x = -6$. Thus we must check the intervals in the following
    table.

Interval	$(-\infty, -6)$	$(-6, 2)$	$(2, \infty)$
Sign of $x - 2$	$-$	$-$	$+$
Sign of $x + 6$	$-$	$+$	$+$
Sign of $(x - 2)(x + 6)$	$+$	$-$	$+$

Interval: $(-\infty, -6) \cup (2, \infty)$.   Graph:

77. $\dfrac{2x+5}{x+1} \le 1 \quad \Leftrightarrow \quad \dfrac{2x+5}{x+1} - 1 \le 0 \quad \Leftrightarrow \quad \dfrac{2x+5}{x+1} - \dfrac{x+1}{x+1} \le 0 \quad \Leftrightarrow \quad \dfrac{x+4}{x+1} \le 0.$ The expression on the left of the inequality changes sign where $x = -1$ and where $x = -4$. Thus we must check the intervals in the following table.

Interval	$(-\infty, -4)$	$(-4, -1)$	$(-1, \infty)$
Sign of $x + 4$	$-$	$+$	$+$
Sign of $x + 1$	$-$	$-$	$+$
Sign of $\dfrac{x+4}{x+1}$	$+$	$-$	$+$

We exclude $x = -1$, since the expression is not defined at this value. Thus the solution is $[-4, -1)$.

Graph:

79. $\dfrac{x-4}{x^2-4} \le 0 \quad \Leftrightarrow \quad \dfrac{x-4}{(x-2)(x+2)} \le 0.$ The expression on the left of the inequality changes sign where $x = -2$, where $x = 2$, and where $x = 4$. Thus we must check the intervals in the following table.

Interval	$(-\infty, -2)$	$(-2, 2)$	$(2, 4)$	$(4, \infty)$
Sign of $x - 4$	$-$	$-$	$-$	$+$
Sign of $x - 2$	$-$	$-$	$+$	$+$
Sign of $x + 2$	$-$	$+$	$+$	$+$
Sign of $\dfrac{x-4}{(x-2)(x+2)}$	$-$	$+$	$-$	$+$

Since the expression is not defined when $x = \pm 2$, we exclude these values and the solution is $(-\infty, -2) \cup (2, 4]$.   Graph:

81. $|x - 5| \le 3 \quad \Leftrightarrow \quad -3 \le x - 5 \le 3 \quad \Leftrightarrow \quad 2 \le x \le 8.$ Interval: $[2, 8]$.

Graph:

83. $|2x + 1| \ge 1$ is equivalent to $2x + 1 \ge 1$ or $2x + 1 \le -1$.

Case 1: $2x + 1 \ge 1 \quad \Leftrightarrow \quad 2x \ge 0 \quad \Leftrightarrow \quad x \ge 0.$

Case 2: $2x + 1 \le -1 \quad \Leftrightarrow \quad 2x \le -2 \quad \Leftrightarrow \quad x \le -1.$

Interval: $(-\infty, -1] \cup [0, \infty)$.   Graph:

85. (a) For $\sqrt{24 - x - 3x^2}$ to define a real number, we must have $24 - x - 3x^2 \ge 0 \quad \Leftrightarrow \quad (8 - 3x)(3 + x) \ge 0.$ The expression on the left of the inequality changes sign where $8 - 3x = 0 \quad \Leftrightarrow \quad -3x = -8 \quad \Leftrightarrow \quad x = \frac{8}{3}$; or where $x = -3$. Thus we must check the intervals in the following table.

Interval	$(-\infty, -3)$	$\left(-3, \frac{8}{3}\right)$	$\left(\frac{8}{3}, \infty\right)$
Sign of $8 - 3x$	$+$	$+$	$-$
Sign of $3 + x$	$-$	$+$	$+$
Sign of $(8 - 3x)(3 + x)$	$-$	$+$	$-$

Interval: $\left[-3, \frac{8}{3}\right]$.   Graph:

(b)   For $\dfrac{1}{\sqrt[4]{x-x^4}}$ to define a real number we must have $x - x^4 > 0 \quad \Leftrightarrow \quad x(1 - x^3) > 0 \quad \Leftrightarrow$

$x(1-x)(1+x+x^2) > 0$. The expression on the left of the inequality changes sign where

$x = 0$; or where $x = 1$; or where $1 + x + x^2 = 0 \quad \Rightarrow \quad x = \dfrac{-1 \pm \sqrt{1^2 - 4(1)(1)}}{2(1)}$

$= \frac{1 \pm \sqrt{1-4}}{2}$ which has no real solution. Thus we must check the intervals in the following
table.

Interval	$(-\infty, 0)$	$(0, 1)$	$(1, \infty)$
Sign of x	$-$	$+$	$+$
Sign of $1 - x$	$+$	$+$	$-$
Sign of $1 + x + x^2$	$+$	$+$	$+$
Sign of $x(1-x)(1+x+x^2)$	$-$	$+$	$-$

Interval: $(0, 1)$.   Graph:

Chapter 1 Test

1. (a) $3x - 7 = \frac{1}{2}x \quad \Leftrightarrow \quad -7 = -\frac{5}{2}x \quad \Leftrightarrow \quad x = \frac{14}{5}$.

   (b) $x^{2/3} - 9 = 0 \quad \Leftrightarrow \quad x^{2/3} = 9 \quad \Leftrightarrow \quad x^{1/3} = \pm 3 \quad \Leftrightarrow \quad x = \pm 27$.

   (c) $\dfrac{2x}{2x - 5} = \dfrac{x + 2}{x - 1} \quad \Leftrightarrow \quad 2x(x - 1) = (2x - 5)(x + 2) \quad \Leftrightarrow \quad 2x^2 - 2x = 2x^2 - x - 10$
   $\Leftrightarrow \quad -x = -10 \quad \Leftrightarrow \quad x = 10$.

2. Let $d =$ the distance between Ajax and Bixby. Then we have the following table.

	Rate	Time	Distance
Ajax to Bixby	50	$\dfrac{d}{50}$	d
Bixby to Ajax	60	$\dfrac{d}{60}$	d

   We use the fact that the total time is $4\frac{2}{5}$ hours to get the equation $\dfrac{d}{50} + \dfrac{d}{60} = \dfrac{22}{5} \quad \Leftrightarrow$
   $6d + 5d = 1320 \quad \Leftrightarrow \quad 11d = 1320 \quad \Leftrightarrow \quad d = 120$. Thus Ajax and Bixby are 120 miles apart.

3. (a) $(6 - 2i) - (7 - \frac{1}{2}i) = 6 - 2i - 7 + \frac{1}{2}i = -1 - \frac{3}{2}i$.

   (b) $(1 + i)(3 - 2i) = 3 - 2i + 3i - 2i^2 = 3 + i + 2 = 5 + i$.

   (c) $\dfrac{5 + 10i}{3 - 4i} = \dfrac{5 + 10i}{3 - 4i} \cdot \dfrac{3 + 4i}{3 + 4i} = \dfrac{15 + 20i + 30i + 40i^2}{9 - 16i^2} = \dfrac{15 + 50i - 40}{9 + 16} = \dfrac{-25 + 50i}{25}$
   $= -1 + 2i$.

   (d) $i^{50} = i^{48} \cdot i^2 = (i^4)^{12} \cdot i^2 = (1)^{12} \cdot (-1) = -1$.

   (e) $(2 - \sqrt{-2})(\sqrt{8} + \sqrt{-4}) = (2 - \sqrt{2}\,i)(2\sqrt{2} + 2i) = 4\sqrt{2} + 4i - 4i - 2\sqrt{2}\,i^2 =$
   $4\sqrt{2} + 2\sqrt{2} = 6\sqrt{2}$.

4. (a) $x^2 - x - 12 = 0 \quad \Leftrightarrow \quad (x - 4)(x + 3) = 0$. So $x = 4$ or $x = -3$.

   (b) $2x^2 + 4x + 3 = 0 \quad \Rightarrow \quad x = \dfrac{-4 \pm \sqrt{4^2 - 4(2)(3)}}{2(2)} = \dfrac{-4 \pm \sqrt{16 - 24}}{4} = \dfrac{-4 \pm \sqrt{-8}}{4}$
   $= \dfrac{-4 \pm 2\sqrt{2}\,i}{4} = \dfrac{-2 \pm \sqrt{2}\,i}{2}$.

   (c) $\sqrt{3 - \sqrt{x + 5}} = 2 \quad \Rightarrow \quad 3 - \sqrt{x + 5} = 4 \quad \Leftrightarrow \quad -1 = \sqrt{x + 5}$. Squaring both sides
   again, we get $1 = x + 5 \quad \Leftrightarrow \quad x = -4$. But this does not satisfy the original equation, so
   there is no solution. (You must always check your final answers if you have squared both sides
   when solving an equation, since extraneous answers may be introduced.)
   *Some students might stop at the statement $-1 = \sqrt{x + 5}$ and recognize that this is
   impossible, so there can be no solution.

   (d) $x^{1/2} - 3x^{1/4} + 2 = 0$. Let $u = x^{1/4}$, then we have $u^2 - 3u + 2 = 0 \quad \Leftrightarrow$
   $(u - 2)(u - 1) = 0$. So $u - 2 = x^{1/4} - 2 = 0 \quad \Leftrightarrow \quad x^{1/4} = 2 \quad \Rightarrow \quad x = 2^4 = 16$; or

$u - 1 = x^{1/4} - 1 = 0 \quad \Leftrightarrow \quad x^{1/4} = 1 \quad \Rightarrow \quad x = 1^4 = 1$. Thus the solutions are $x = 16$ and $x = 1$.

(e) $x^4 - 16x^2 = 0 \quad \Leftrightarrow \quad x^2(x^2 - 16) = 0 \quad \Leftrightarrow \quad x^2(x - 4)(x + 4) = 0$. So $x = 0$, or $x = -4$, or $x = 4$.

(f) $3|x - 4| - 10 = 0 \quad \Leftrightarrow \quad 3|x - 4| = 10 \quad \Leftrightarrow \quad |x - 4| = \frac{10}{3} \quad \Leftrightarrow \quad x - 4 = \pm\frac{10}{3} \quad \Leftrightarrow$
$x = 4 \pm \frac{10}{3}$. So $x = 4 - \frac{10}{3} = \frac{2}{3}$ or $x = 4 + \frac{10}{3} = \frac{22}{3}$. Thus the solutions are $x = \frac{2}{3}$ and $x = \frac{22}{3}$.

5. Let w be the width of the building lot. Then the length of the lot is $w + 70$. So
$w^2 + (w + 70)^2 = 130^2 \quad \Leftrightarrow \quad w^2 + w^2 + 140w + 4900 = 16900 \quad \Leftrightarrow$
$2w^2 + 140w - 12000 = 0 \quad \Leftrightarrow \quad 2(w^2 + 70w - 6000) = 0 \quad \Leftrightarrow \quad 2(w + 120)(w - 50) = 0$. So
$w = -120$ (which we reject because $w > 0$) or $w = 50$. Thus the lot is 50 by 120.

6. (a) $-1 \le 5 - 2x < 10 \quad \Leftrightarrow \quad -6 \le -2x < 5 \quad \Leftrightarrow \quad 3 \ge x > -\frac{5}{2}$. Expressing in standard form we have: $-\frac{5}{2} < x \le 3$. Interval: $\left(-\frac{5}{2}, 3\right]$ Graph:

(b) $x(x - 1)(x - 2) > 0$. The expression on the left of the inequality changes sign when $x = 0$, $x = 1$, and $x = 2$. Thus we must check the intervals in the following table.

Interval	$(-\infty, 0)$	$(0, 1)$	$(1, 2)$	$(2, \infty)$
Sign of x	$-$	$+$	$+$	$+$
Sign of $x - 1$	$-$	$-$	$+$	$+$
Sign of $x - 2$	$-$	$-$	$-$	$+$
Sign of $x(x - 1)(x - 2)$	$-$	$+$	$-$	$+$

From the table, the solution set is $\{x \mid 0 < x < 1 \text{ or } 2 < x\}$
Interval: $(0, 1) \cup (2, \infty)$.   Graph:

(c) $|x - 3| < 2$ is equivalent to $-2 < x - 3 < 2 \quad \Leftrightarrow \quad 1 < x < 5$.
Interval: $(1, 5)$.   Graph:

(d) $\dfrac{2x + 5}{x + 1} \le 1 \quad \Leftrightarrow \quad \dfrac{2x + 5}{x + 1} - 1 \le 0 \quad \Leftrightarrow \quad \dfrac{2x + 5}{x + 1} - \dfrac{x + 1}{x + 1} \le 0 \quad \Leftrightarrow \quad \dfrac{x + 4}{x + 1} \le 0$. The expression on the left of the inequality changes sign where $x = -4$ and where $x = -1$. Thus we must check the intervals in the following table.

Interval	$(-\infty, -4)$	$(-4, -1)$	$(-1, \infty)$
Sign of $x + 4$	$-$	$+$	$+$
Sign of $x + 1$	$-$	$-$	$+$
Sign of $\dfrac{x + 4}{x + 1}$	$+$	$-$	$+$

Since $x = -1$ makes the expression in the inequality undefined, we exclude this value.
Interval: $[-4, -1)$ Graph:

7. $5 \le \frac{5}{9}(F - 32) \le 10 \quad \Leftrightarrow \quad 9 \le F - 32 \le 18 \quad \Leftrightarrow \quad 41 \le F \le 50$. Thus the medicine is to be stored at a temperature between 41°F to 50°F.

8. For $\sqrt{4x - x^2}$ to be defined as a real number $4x - x^2 \geq 0 \quad \Leftrightarrow \quad x(4 - x) \geq 0$. The expression on the left of the inequality changes sign when $x = 0$ and $x = 4$. Thus we must check the intervals in the following table.

Interval	$(-\infty, 0)$	$(0, 4)$	$(4, \infty)$
Sign of x	$-$	$+$	$+$
Sign of $4 - x$	$+$	$+$	$-$
Sign of $x(4 - x)$	$-$	$+$	$-$

From the table, we see that $\sqrt{4x - x^2}$ is defined when $0 \leq x < 4$.

Focus on Modeling

1. (a) $Cost = \left(\dfrac{cost\,of}{copier}\right) + \left(\dfrac{maintenance}{cost}\right) \times months + \left(\dfrac{copy}{cost}\right) \times months$. Each month the copy
    cost is $8000 \cdot 0.03 = 240$. Thus we get $C_1(n) = 5800 + 25n + 240n = 5800 + 265n$.

    (b) $Cost = \left(\dfrac{rental}{cost}\right) \times months + \left(\dfrac{copy}{cost}\right) \times months$. Each month the copy cost is
    $8000 \cdot 0.06 = 480$. Thus we get $C_2(n) = 95n + 480n = 575n$.

    (c)

Years	Purchase	Rental
1 ($n = 12$)	8,980	6,900
2 ($n = 24$)	12,160	13,800
3 ($n = 36$)	15,340	20,700
4 ($n = 48$)	18,520	27,600
5 ($n = 60$)	21,700	34,500
6 ($n = 72$)	24,880	41,400

    (d) The cost will be the same when $C_1(n) = C_2(n)$. So $5800 + 265n = 575n$ $\Leftrightarrow$
    $5800 = 310n$ $\Leftrightarrow$ $n \approx 18.71$ months.

3. (a) $Cost = \left(\dfrac{set\,up}{cost}\right) + \left(\dfrac{cost\,per}{tire}\right) \times tires$. So $C(x) = 8000 + 22x$.

    (b) $Revenue = \left(\dfrac{price\,per}{tire}\right) \times tires$. So $R(x) = 49x$.

    (c) $Profit = Revenue - Cost$. So $P(x) = R(x) - C(x) = 49x - (8000 + 22x) = 27x - 8000$.

    (d) Break even is when $Profit = 0$. Thus $27x - 8000 = 0$ $\Leftrightarrow$ $27x = 8000$ $\Leftrightarrow$
    $x \approx 296.3$. So they need to sell at least 297 tires to break even.

5. (a) Design 1 is a square and the perimeter of a square is $4 \times side$, we have $24 = 4x$ so each side, x,
    is 6 feet. Thus the area is $6^2 = 36$ ft^2.
    Design 2 is a circle and the perimeter of a circle is $2\pi r$ and the area is πr^2. Thus we must solve
    $2\pi r = 24$ for r. But $r = \dfrac{12}{\pi}$. Thus the $area = \pi \left(\dfrac{12}{\pi}\right)^2 = \dfrac{144}{\pi} \approx 45.8$ ft^2.
    Thus design 2 would give the greatest area.

    (b) In design 1, the cost is the $\$3 \times$ perimeter, p, so $120 = 3p$ the perimeter is 40 feet. By part (a),
    each side is then $\frac{40}{4} = 10$ feet. So the area is $10^2 = 100$ ft^2.
    In design 2, the cost is the $\$4 \times$ perimeter, since the perimeter is $2\pi r$, we get $120 = 4(2\pi r)$ so
    $r = \dfrac{120}{8\pi} = \dfrac{15}{\pi}$. The area is $\pi r^2 = \pi \left(\dfrac{15}{\pi}\right)^2 = \dfrac{225}{\pi} \approx 71.6$ ft^2.
    Design 1 would yield the largest greatest area.

7. (a)

	Plan A $39, 400 minutes	Plan B $49, 400 minutes	Plan B $69, 400 minutes
500	$39 + 100(0.20) = \$59$	$49	$69
600	$39 + 200(0.20) = \$79$	$49	$69
700	$39 + 300(0.20) = \$99$	$49 + 100(0.15) = \$64$	$69
800	$39 + 400(0.20) = \$119$	$49 + 200(0.15) = \$79$	$69
900	$39 + 500(0.20) = \$139$	$49 + 300(0.15) = \$94$	$69
1000	$39 + 600(0.20) = \$159$	$49 + 400(0.15) = \$109$	$69
1100	$39 + 700(0.20) = \$179$	$49 + 500(0.15) = \$124$	$69 + 100(0.15) = \$84$

(b) If Jane uses 500 minute, Plan B is the cheapest. If Jane 700 minutes Plan B is still cheaper. And if Jane talks for 1100 minutes then Plan C is the cheapest.

Chapter Two

Exercises 2.1

1.

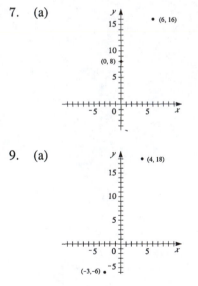

3. The two points are $(0, 2)$ and $(3, 0)$.

(a) $d = \sqrt{(3 - 0)^2 + (0 - (-2))^2} = \sqrt{3^2 + 2^2} = \sqrt{9 + 4} = \sqrt{13}$

(b) midpoint: $\left(\frac{3+0}{2}, \frac{0+2}{2}\right) = \left(\frac{3}{2}, 1\right)$

5. The two points are $(-3, 3)$ and $(5, -3)$.

(a) $d = \sqrt{(-3 - 5)^2 + (3 - (-3))^2} = \sqrt{(-8)^2 + 6^2} = \sqrt{64 + 36} = \sqrt{100} = 10$

(b) midpoint: $\left(\frac{-3+5}{2}, \frac{3+(-3)}{2}\right) = (1, 0)$

7. (a)

(b) $d = \sqrt{(0 - 6)^2 + (8 - 16)^2}$
$= \sqrt{(-6)^2 + (-8)^2} = \sqrt{100} = 10$

(c) midpoint: $\left(\frac{0+6}{2}, \frac{8+16}{2}\right) = (3, 12)$

9. (a)

(b) $d = \sqrt{(-3 - 4)^2 + (-6 - 18)^2}$
$= \sqrt{(-7)^2 + (-24)^2} = \sqrt{49 + 576}$
$= \sqrt{625} = 25$

(c) midpoint: $\left(\frac{-3+4}{2}, \frac{-6+18}{2}\right) = \left(\frac{1}{2}, 6\right)$

11. (a)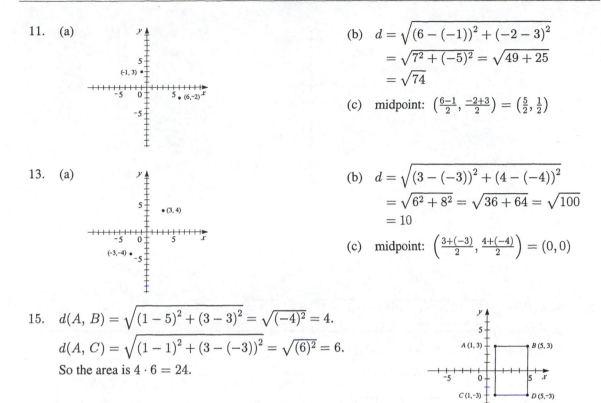

(b) $d = \sqrt{(6 - (-1))^2 + (-2 - 3)^2}$

$= \sqrt{7^2 + (-5)^2} = \sqrt{49 + 25}$

$= \sqrt{74}$

(c) midpoint: $\left(\frac{6-1}{2}, \frac{-2+3}{2}\right) = \left(\frac{5}{2}, \frac{1}{2}\right)$

13. (a)

(b) $d = \sqrt{(3 - (-3))^2 + (4 - (-4))^2}$

$= \sqrt{6^2 + 8^2} = \sqrt{36 + 64} = \sqrt{100}$

$= 10$

(c) midpoint: $\left(\frac{3+(-3)}{2}, \frac{4+(-4)}{2}\right) = (0, 0)$

15. $d(A, B) = \sqrt{(1 - 5)^2 + (3 - 3)^2} = \sqrt{(-4)^2} = 4$.

$d(A, C) = \sqrt{(1 - 1)^2 + (3 - (-3))^2} = \sqrt{(6)^2} = 6$.

So the area is $4 \cdot 6 = 24$.

17. From the graph, the quadrilateral $ABCD$ has a pair of parallel sides,

so $ABCD$ is a trapezoid. The area is $\left(\dfrac{b_1 + b_2}{2}\right) h$. From the graph

we see that $b_1 = d(A, B) = \sqrt{(1 - 5)^2 + (0 - 0)^2} = \sqrt{4^2} = 4$;

$b_2 = d(C, D) = \sqrt{(4 - 2)^2 + (3 - 3)^2} = \sqrt{2^2} = 2$; and h is

the difference in y-coordinates $= |3 - 0| = 3$. Thus the area of the

trapezoid is $\left(\dfrac{4 + 2}{2}\right) 3 = 9$.

19.   21.   23.

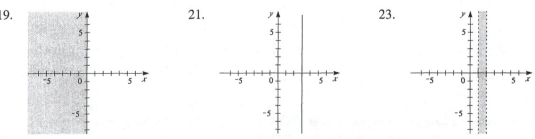

25.   27.   29.

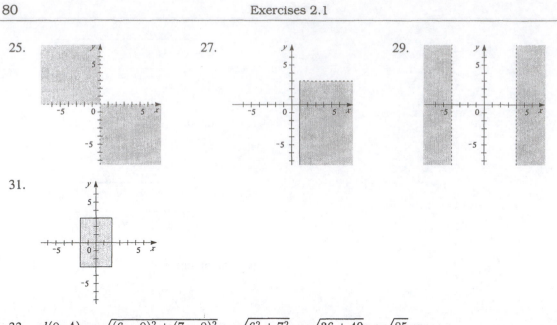

31.

33. $d(0, A) = \sqrt{(6 - 0)^2 + (7 - 0)^2} = \sqrt{6^2 + 7^2} = \sqrt{36 + 49} = \sqrt{85}.$

$d(0, B) = \sqrt{(-5 - 0)^2 + (8 - 0)^2} = \sqrt{(-5)^2 + 8^2} = \sqrt{25 + 64} = \sqrt{89}.$

Thus point $A(6, 7)$ is closer to the origin.

35. $d(P, R) = \sqrt{(-1 - 3)^2 + (-1 - 1)^2} = \sqrt{(-4)^2 + (-2)^2} = \sqrt{16 + 4} = \sqrt{20} = 2\sqrt{5}.$

$d(Q, R) = \sqrt{(-1 - (-1))^2 + (-1 - 3)^2} = \sqrt{0 + (-4)^2} = \sqrt{16} = 4.$ Thus point $Q(-1, 3)$ is closer to point R.

37. Since we do not know which pair are isosceles, we find the length of all three sides.

$d(A, B) = \sqrt{(-3 - 0)^2 + (-1 - 2)^2} = \sqrt{(-3)^2 + (-3)^2} = \sqrt{9 + 9} = \sqrt{18} = 3\sqrt{2}.$

$d(C, B) = \sqrt{(-3 - (-4))^2 + (-1 - 3)^2} = \sqrt{1^2 + (-4)^2} = \sqrt{1 + 16} = \sqrt{17}.$

$d(A, C) = \sqrt{(0 - (-4))^2 + (2 - 3)^2} = \sqrt{4^2 + (-1)^2} = \sqrt{16 + 1} = \sqrt{17}.$ So sides AC and CB have the same length.

39. (a) Here we have $A = (2, 2)$, $B = (3, -1)$, and $C = (-3, -3)$. So

$d(A, B) = \sqrt{(3 - 2)^2 + (-1 - 2)^2} = \sqrt{1^2 + (-3)^2} = \sqrt{1 + 9} = \sqrt{10};$

$d(C, B) = \sqrt{(3 - (-3))^2 + (-1 - (-3))^2} = \sqrt{6^2 + 2^2} = \sqrt{36 + 4} = \sqrt{40} = 2\sqrt{10};$

$d(A, C) = \sqrt{(-3 - 2)^2 + (-3 - 2)^2} = \sqrt{(-5)^2 + (-5)^2} = \sqrt{25 + 25} = \sqrt{50} = 5\sqrt{2}.$

Since $[d(A, B)]^2 + [d(C, B)]^2 = [d(A, C)]^2$, we conclude that the triangle is a right triangle.

(b) The area of the triangle is $\frac{1}{2} \cdot d(C, B) \cdot d(A, B) = \frac{1}{2} \cdot \sqrt{10} \cdot 2\sqrt{10} = 10.$

41. We show that all sides are the same length (its a rhombus) and then show that the diagonals are equal. Here we have $A = (-2, 9)$, $B = (4, 6)$, $C = (1, 0)$, and $D = (-5, 3)$. So

$d(A, B) = \sqrt{(4 - (-2))^2 + (6 - 9)^2} = \sqrt{6^2 + (-3)^2} = \sqrt{36 + 9} = \sqrt{45};$

$d(B,C) = \sqrt{(1-4)^2 + (0-6)^2} = \sqrt{(-3)^2 + (-6)^2} = \sqrt{9+36} = \sqrt{45};$

$d(C,D) = \sqrt{(-5-1)^2 + (3-0)^2} = \sqrt{(-6)^2 + (-3)^2} = \sqrt{36+9} = \sqrt{45};$

$d(D,A) = \sqrt{(-2-(-5))^2 + (9-3)^2} = \sqrt{3^2 + 6^2} = \sqrt{9+36} = \sqrt{45}.$ So the points form a rhombus. Also

$d(A,C) = \sqrt{(1-(-2))^2 + (0-9)^2} = \sqrt{3^2 + (-9)^2} = \sqrt{9+81} = \sqrt{90} = 3\sqrt{10},$

and $d(B,D) = \sqrt{(-5-4)^2 + (3-6)^2} = \sqrt{(-9)^2 + (-3)^2} = \sqrt{81+9} = \sqrt{90} = 3\sqrt{10}.$ Since the diagonals are equal, the rhombus is a square.

43. Let $P = (0, y)$ be such a point. Setting the distances equal we get

$\sqrt{(0-5)^2 + (y-(-5))^2} = \sqrt{(0-1)^2 + (y-1)^2} \quad \Leftrightarrow$

$\sqrt{25 + y^2 + 10y + 25} = \sqrt{1 + y^2 - 2y + 1} \quad \Rightarrow \quad y^2 + 10y + 50 = y^2 - 2y + 2 \quad \Leftrightarrow$

$12y = -48 \quad \Leftrightarrow \quad y = -4.$ Thus, the point is $P = (0, -4).$ Check:

$\sqrt{(0-5)^2 + (-4-(-5))^2} = \sqrt{(-5)^2 + 1^2} = \sqrt{25+1} = \sqrt{26};$

$\sqrt{(0-1)^2 + (-4-1)^2} = \sqrt{(-1)^2 + (-5)^2} = \sqrt{25+1} = \sqrt{26}.$

45. We find the midpoint, M, of PQ and then the midpoint of PM. So $M = \left(\frac{-1+7}{2}, \frac{3+5}{2}\right) = (3,4),$ and the midpoint of PM is $\left(\frac{-1+3}{2}, \frac{3+4}{2}\right) = \left(1, \frac{7}{2}\right).$

47. As indicated by Example 5, we must find a point $S(x_1, y_1)$ such that he midpoints of PR and of QS are the same. Thus

$\left(\frac{4+(-1)}{2}, \frac{2+(-4)}{2}\right) = \left(\frac{x_1+1}{2}, \frac{y_1+1}{2}\right).$ Setting the

x-coordinates equal, we get $\frac{4+(-1)}{2} = \frac{x_1+1}{2} \quad \Leftrightarrow$

$4 - 1 = x_1 + 1 \quad \Leftrightarrow \quad x_1 = 2.$ Setting the y-coordinates equal, we

get $\frac{2+(-4)}{2} = \frac{y_1+1}{2} \quad \Leftrightarrow \quad 2 - 4 = y_1 + 1 \quad \Leftrightarrow$

$y_1 = -3.$ Thus $S = (2, -3).$

49. (a)

(b) The midpoint of AC is $\left(\frac{-2+7}{2}, \frac{-1+7}{2}\right) = \left(\frac{5}{2}, 3\right),$ the midpoint of BD is $\left(\frac{4+1}{2}, \frac{2+4}{2}\right) = \left(\frac{5}{2}, 3\right).$

(c) Since the they have the same midpoint, we conclude the diagonals bisect each other.

51. (a) $d(A,B) = \sqrt{3^2 + 4^2} = \sqrt{25} = 5.$

(b) We want the distances from $C = (4,2)$ to $D = (11, 26).$
The walking distance is $|4 - 11| + |2 - 26| = 7 + 24 = 31$ blocks.
Straight-line distance is $\sqrt{(4-11)^2 + (2-26)^2} = \sqrt{7^2 + 24^2} = \sqrt{625} = 25.$

(c) The two points are on the same avenue or the same street.

53. The midpoint of the line segment is $(66, 45)$. The pressure experienced by an ocean diver at a depth of 66 feet is 45 lb/in^2.

55. (a) The point $(3, 7)$ is reflected to the point $(-3, 7)$.

(b) The point (a, b) is reflected to the point $(-a, b)$.

(c) Since the point $(-a, b)$ is the reflection of (a, b), the point $(-4, -1)$ is the reflection of $(4, -1)$.

(d) $A = (3, 3)$ so $A' = (-3, 3)$;
$B = (6, 1)$ so $B' = (-6, 1)$;
$C = (1, -4)$ so $C' = (-1, -4)$.

57. We need to find a point $S(x_1, y_1)$ such that $PQRS$ is a parallelogram. As indicated by Example 5, this will be the case if the diagonals PR and QS bisect each other. So the midpoints of PR and QS are the same. Thus $\left(\dfrac{(-1) + 4}{2}, \dfrac{(-4) + 2}{2} \right) = \left(\dfrac{x_1 + 1}{2}, \dfrac{y_1 + 1}{2} \right)$. Setting the x-coordinates equal, we get $\dfrac{4 + (-1)}{2} = \dfrac{x_1 + 1}{2} \quad \Leftrightarrow \quad 4 - 1 = x_1 + 1 \quad \Leftrightarrow \quad x_1 = 2$.

Setting the y-coordinates equal, we get $\dfrac{2 + (-4)}{2} = \dfrac{y_1 + 1}{2} \quad \Leftrightarrow 2 - 4 = y_1 + 1 \quad \Leftrightarrow \quad y_1 = -3$.
Thus $S = (2, -3)$.

Exercises 2.2

1. $(0,2)$: $2 \overset{?}{=} 3(0) - 2$ $\Leftrightarrow$ $2 \overset{?}{=} -2$ No.
    $(\frac{1}{3}, 1)$: $1 \overset{?}{=} 3(\frac{1}{3}) - 2$ $\Leftrightarrow$ $1 \overset{?}{=} 1 - 2$ No.
    $(1,1)$: $1 \overset{?}{=} 3(1) - 2$ $\Leftrightarrow$ $1 \overset{?}{=} 3 - 2$ Yes.
    So $(1,1)$ is on the graph of this equation.

3. $(0,0)$: $0 - 2(0) - 1 \overset{?}{=} 0$ $\Leftrightarrow$ $-1 \overset{?}{=} 0$ No.
    $(1,0)$: $1 - 2(0) - 1 \overset{?}{=} 0$ $\Leftrightarrow$ $-1 + 1 \overset{?}{=} 0$ Yes.
    $(-1,-1)$: $(-1) - 2(-1) - 1 \overset{?}{=} 0$ $\Leftrightarrow$ $-1 + 2 - 1 \overset{?}{=} 0$ Yes.
    So $(1,0)$ and $(-1,-1)$ are points on the graph of this equation.

5. $(0,-2)$: $(0)^2 + (0)(-2) + (-2)^2 \overset{?}{=} 4$ $\Leftrightarrow$ $0 + 0 + 4 \overset{?}{=} 4$ Yes.
    $(1,-2)$: $(1)^2 + (1)(-2) + (-2)^2 \overset{?}{=} 4$ $\Leftrightarrow$ $1 - 2 + 4 \overset{?}{=} 4$ No.
    $(2,-2)$: $(2)^2 + (2)(-2) + (-2)^2 \overset{?}{=} 4$ $\Leftrightarrow$ $4 - 4 + 4 \overset{?}{=} 4$ Yes.
    So $(0,-2)$ and $(2,-2)$ are points on the graph of this equation.

7. To find x-intercepts, set $y = 0$. This gives $0 = 4x - x^2$ $\Leftrightarrow$ $0 = x(4 - x)$ $\Leftrightarrow$ $0 = x$ or $x = 4$, so the x-intercept are 0 and 4.
    To find y-intercepts, set $x = 0$. This gives $y = 4(0) - 0^2$ $\Leftrightarrow$ $y = 0$, so the y-intercept is 0.

9. To find x-intercepts, set $y = 0$. This gives $x^4 + 0^2 - x(0) = 16$ $\Leftrightarrow$ $x^4 = 16$ $\Leftrightarrow$ $x = \pm 2$. So the x-intercept are -2 and 2.
    To find y-intercepts, set $x = 0$. This gives $0^4 + y^2 - (0)y = 16$ $\Leftrightarrow$ $y^2 = 16$ $\Leftrightarrow$ $y = \pm 4$. So the y-intercept are -4 and 4.

11. To find x-intercepts, set $y = 0$. This gives $0 = x - 3$ $\Leftrightarrow$ $x = 3$, so the x-intercept is 3.
    To find y-intercepts, set $x = 0$. This gives $y = 0 - 3$ $\Leftrightarrow$ $y = -3$, so the y-intercept is -3.

13. To find x-intercepts, set $y = 0$. This gives $0 = x^2 - 9$ $\Leftrightarrow$ $x^2 = 9$ $\Rightarrow$ $x = \pm 3$, so the x-intercept are ± 3. To find y-intercepts, set $x = 0$. This gives $y = (0)^2 - 9$ $\Leftrightarrow$ $y = -9$, so the y-intercept is -9.

15. To find x-intercepts, set $y = 0$. This gives $x^2 + (0)^2 = 4$ $\Leftrightarrow$ $x^2 = 4$ $\Rightarrow$ $x = \pm 2$, so the x-intercept are ± 2. To find y-intercepts, set $x = 0$. This gives $(0)^2 + y^2 = 4$ $\Leftrightarrow$ $y^2 = 4$ $\Rightarrow$ $y = \pm 2$, so the y-intercept are ± 2.

17. To find x-intercepts, set $y = 0$. This gives $x(0) = 5$ $\Leftrightarrow$ $0 = 5$, which is impossible, so there is no x-intercept. Likewise, to find y-intercepts, set $x = 0$. This gives $(0)y = 5$ $\Leftrightarrow$ $0 = 5$, which is again impossible, so there is no y-intercept.

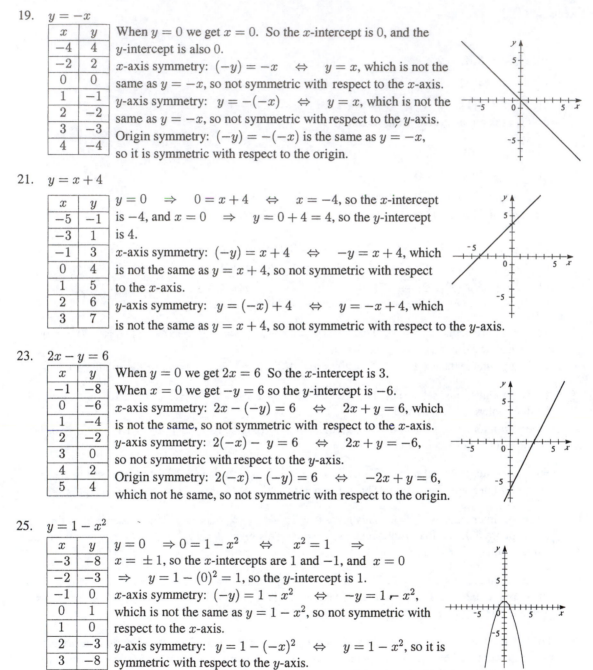

19. $y = -x$

x	y
-4	4
-2	2
0	0
1	-1
2	-2
3	-3
4	-4

When $y = 0$ we get $x = 0$. So the x-intercept is 0, and the y-intercept is also 0.

x-axis symmetry: $(-y) = -x$ $\Leftrightarrow$ $y = x$, which is not the same as $y = -x$, so not symmetric with respect to the x-axis.

y-axis symmetry: $y = -(-x)$ $\Leftrightarrow$ $y = x$, which is not the same as $y = -x$, so not symmetric with respect to the y-axis.

Origin symmetry: $(-y) = -(-x)$ is the same as $y = -x$, so it is symmetric with respect to the origin.

21. $y = x + 4$

x	y
-5	-1
-3	1
-1	3
0	4
1	5
2	6
3	7

$y = 0$ $\Rightarrow$ $0 = x + 4$ $\Leftrightarrow$ $x = -4$, so the x-intercept is -4, and $x = 0$ $\Rightarrow$ $y = 0 + 4 = 4$, so the y-intercept is 4.

x-axis symmetry: $(-y) = x + 4$ $\Leftrightarrow$ $-y = x + 4$, which is not the same as $y = x + 4$, so not symmetric with respect to the x-axis.

y-axis symmetry: $y = (-x) + 4$ $\Leftrightarrow$ $y = -x + 4$, which is not the same as $y = x + 4$, so not symmetric with respect to the y-axis.

23. $2x - y = 6$

x	y
-1	-8
0	-6
1	-4
2	-2
3	0
4	2
5	4

When $y = 0$ we get $2x = 6$ So the x-intercept is 3. When $x = 0$ we get $-y = 6$ so the y-intercept is -6.

x-axis symmetry: $2x - (-y) = 6$ $\Leftrightarrow$ $2x + y = 6$, which is not the same, so not symmetric with respect to the x-axis.

y-axis symmetry: $2(-x) - y = 6$ $\Leftrightarrow$ $2x + y = -6$, so not symmetric with respect to the y-axis.

Origin symmetry: $2(-x) - (-y) = 6$ $\Leftrightarrow$ $-2x + y = 6$, which not he same, so not symmetric with respect to the origin.

25. $y = 1 - x^2$

x	y
-3	-8
-2	-3
-1	0
0	1
1	0
2	-3
3	-8

$y = 0$ $\Rightarrow 0 = 1 - x^2$ $\Leftrightarrow$ $x^2 = 1$ $\Rightarrow$ $x = \pm 1$, so the x-intercepts are 1 and -1, and $x = 0$ $\Rightarrow$ $y = 1 - (0)^2 = 1$, so the y-intercept is 1.

x-axis symmetry: $(-y) = 1 - x^2$ $\Leftrightarrow$ $-y = 1 - x^2$, which is not the same as $y = 1 - x^2$, so not symmetric with respect to the x-axis.

y-axis symmetry: $y = 1 - (-x)^2$ $\Leftrightarrow$ $y = 1 - x^2$, so it is symmetric with respect to the y-axis.

Origin symmetry: $(-y) = 1 - (-x)^2$ $\Leftrightarrow$ $-y = 1 - x^2$ which is not the same as $y = 1 - x^2$. Not symmetric with respect to the origin.

27. $4y = x^2$ $\Leftrightarrow$ $y = \frac{1}{4}x^2$

x	y
-6	9
-4	4
-2	1
0	0
2	1
4	4
3	2

$y = 0$ $\Rightarrow$ $0 = \frac{1}{4}x^2$ $\Leftrightarrow$ $x^2 = 0$ $\Rightarrow$ $x = 0$, so the x-intercept is 0, and $x = 0$ $\Rightarrow$ $y = \frac{1}{4}(0)^2 = 0$, so the y-intercept is 0.

x-axis symmetry: $(-y) = \frac{1}{4}x^2$, which is not the same as $y = \frac{1}{4}x^2$, so not symmetric with respect to the x-axis.

y-axis symmetry: $y = \frac{1}{4}(-x)^2$ $\Leftrightarrow$ $y = \frac{1}{4}x^2$, so it is symmetric with respect to the y-axis.

Origin symmetry: $(-y) = \frac{1}{4}(-x)^2$ $\Leftrightarrow$ $-y = \frac{1}{4}x^2$, which is not the same as $y = \frac{1}{4}x^2$, so not symmetric with respect to the origin.

29. $y = x^2 - 9$

x	y
-4	7
-3	0
-2	-5
-1	-8
0	-9
1	-8
2	-5
3	0
4	7

$y = 0$ $\Rightarrow 0 = x^2 - 9$ $\Leftrightarrow$ $x^2 = 9$ $\Rightarrow$ $x = \pm 3$, so the x-intercepts are 3 and -3, and $x = 0$ $\Rightarrow$ $y = (0)^2 - 9 = -9$, so the y-intercept is -9.

x-axis symmetry: $(-y) = x^2 - 9$, which is not the same as $y = x^2 - 9$, so not symmetric with respect to the x-axis.

y-axis symmetry: $y = (-x)^2 - 9$ $\Leftrightarrow$ $y = x^2 - 9$, so it is symmetric with respect to the y-axis.

Origin symmetry: $(-y) = (-x)^2 - 9$ $\Leftrightarrow$ $-y = x^2 - 9$, which is not the same as $y = x^2 - 9$, so not symmetric with respect to the origin.

31. $xy = 2$ $\Leftrightarrow$ $y = \dfrac{2}{x}$

x	y
-4	$-\frac{1}{2}$
-2	-1
-1	-2
$-\frac{1}{2}$	-4
$-\frac{1}{4}$	-8

x	y
$\frac{1}{4}$	8
$\frac{1}{2}$	4
1	2
2	1
4	$\frac{1}{2}$

$y = 0$ or $x = 0$ $\Rightarrow$ $0 = 2$, which is impossible, so this equation has no x-intercept and no y-intercept.

x-axis symmetry: $x(-y) = 2$ $\Leftrightarrow$ $-xy = 2$, which is not the same as $xy = 2$, so not symmetric with respect to the x-axis.

y-axis symmetry: $(-x)y = 2$ $\Leftrightarrow$ $-xy = 2$,

which is not the same as $xy = 2$, so not symmetric with respect to the y-axis.

Origin symmetry: $(-x)(-y) = 2$ $\Leftrightarrow$ $xy = 2$, so it is symmetric with respect to the origin.

33. $y = \sqrt{x}$

x	y
0	0
$\frac{1}{4}$	$\frac{1}{2}$
1	1
2	$\sqrt{2}$
4	2
9	3
16	4

$y = 0$ $\Rightarrow$ $0 = \sqrt{x}$ $\Leftrightarrow$ $x = 0$. so the x-intercept is 0, and $x = 0$ $\Rightarrow$ $y = \sqrt{0} = 0$, so the y-intercept is 0. Since we are graphing real numbers and $\sqrt{x}$ is defined to be a nonnegative number, the equation is not symmetric with respect to the x-axis nor with respect to the y-axis. Also, the equation is not symmetric with respect to the origin.

35. $y = \sqrt{4 - x^2}$. Since the radicand (the inside of the square root) cannot be negative, we must have
$4 - x^2 \geq 0 \;\Leftrightarrow\; x^2 \leq 4 \;\Leftrightarrow\; |x| \leq 2$.

x	y
-2	0
-1	$\sqrt{3}$
0	4
1	$\sqrt{3}$
2	0

$y = 0 \;\Rightarrow\; 0 = \sqrt{4 - x^2} \;\Leftrightarrow\; 4 - x^2 = 0 \;\Leftrightarrow$
$x^2 = 4 \;\Rightarrow\; x = \pm 2$, so the x-intercept are -2 and 2,
and $x = 0 \;\Rightarrow\; y = \sqrt{4 - (0)^2} = \sqrt{4} = 2$, so the
y-intercept is 2.
Since $y \geq 0$, the graph is not symmetric with respect to the
x-axis.

y-axis symmetry: $y = \sqrt{4 - (-x)^2} = \sqrt{4 - x^2}$, so the graph is symmetric with respect to the
y-axis. Also, since $y \geq 0$ the graph is not symmetric with respect to the origin.

37. $y = |x|$

x	y
-3	3
-2	2
-1	1
0	0
1	1
2	2
3	3

$y = 0 \;\Rightarrow\; 0 = |x| \;\Leftrightarrow\; x = 0$, so the x-intercept is
0, and $x = 0 \;\Rightarrow\; y = |0| = 0$, so the y-intercept is 0.
Since $y \geq 0$, the graph is not symmetric with respect to the
x-axis.
y-axis symmetry: $y = |-x| = |x|$, so the graph is
symmetric with respect to the y-axis.
Since $y \geq 0$, the graph is not symmetric with respect to the
origin.

39. $y = 4 - |x|$

x	y
-6	-2
-4	0
-2	2
0	4
2	2
4	0
6	-2

$y = 0 \;\Rightarrow\; 0 = 4 - |x| \;\Leftrightarrow\; |x| = 4 \;\Rightarrow\; x = \pm 4$, so
the x-intercepts are -4 and 4, and $x = 0 \;\Rightarrow$
$y = 4 - |0| = 4$, so the y-intercept is 4.
x-axis symmetry: $(-y) = 4 - |x| \;\Leftrightarrow\; y = -4 + |x|$,
which is not the same as $y = 4 - |x|$, so not symmetric with
respect to the x-axis.
y-axis symmetry: $y = 4 - |-x| = 4 - |x|$, so it is symmetric
with respect to the y-axis.

Origin symmetry: $(-y) = 4 - |-x| \;\Leftrightarrow\; y = -4 + |x|$, which is not the same as $y = 4 - |x|$, so
not symmetric with respect to the origin.

41. Since $x = y^3$ is solved for x in terms of y, we insert values for y and find the corresponding values
of x.

x	y
-27	-3
-8	-2
-1	-1
0	0
1	1
8	2
27	3

$y = 0 \;\Rightarrow\; x = (0)^3 = 0$, so the x-intercept is 0, and
$x = 0 \;\Rightarrow\; 0 = y^3 \;\Rightarrow\; y = 0$, so the y-intercept is 0.
x-axis symmetry: $x = (-y)^3 = -y^3$, which is not the same
as $x = y^3$, so not symmetric with respect to the x-axis.
y-axis symmetry: $(-x) = y^3 \;\Leftrightarrow\; x = -y^3$, which is not
the same as $x = y^3$, so not symmetric with respect to the
y-axis.

Origin symmetry: $(-x) = (-y)^3$ $\Leftrightarrow$ $-x = -y^3$ $\Leftrightarrow$ $x = y^3$, so it is symmetric with respect to the origin.

43. $y = x^4$

x	y
-3	81
-2	16
-1	1
0	0
1	1
2	16
3	81

$y = 0$ $\Rightarrow$ $0 = x^4$ $\Rightarrow$ $x = 0$, so the x-intercept is 0, and $x = 0$ $\Rightarrow$ $y = 0^4 = 0$, so the y-intercept is 0.

x-axis symmetry: $(-y) = x^4$ $\Leftrightarrow$ $y = -x^4$, which is not the same as $y = x^4$, so not symmetric with respect to the x-axis.

y-axis symmetry: $y = (-x)^4 = x^4$, so it is symmetric with respect to the y-axis.

Origin symmetry: $(-y) = (-x)^4$ $\Leftrightarrow$ $-y = x^4$, which is not the same as $y = x^4$, so not symmetric with respect to the origin.

45. x-axis symmetry: $(-y) = x^4 + x^2$ $\Leftrightarrow$ $y = -x^4 - x^2$, which is not the same as $y = x^4 + x^2$, so not symmetric with respect to the x-axis.

y-axis symmetry: $y = (-x)^4 + (-x)^2 = x^4 + x^2$, so it is symmetric with respect to the y-axis.

Origin symmetry: $(-y) = (-x)^4 + (-x)^2$ $\Leftrightarrow$ $-y = x^4 + x^2$, which is not the same as $y = x^4 + x^2$, so not symmetric with respect to the origin.

47. x-axis symmetry: $x^2(-y)^2 + x(-y) = 1$ $\Leftrightarrow$ $x^2y^2 - xy = 1$, which is not the same as $x^2y^2 + xy = 1$, so not symmetric with respect to the x-axis.

y-axis symmetry: $(-x)^2y^2 + (-x)y = 1$ $\Leftrightarrow$ $x^2y^2 - xy = 1$, which is not the same as $x^2y^2 + xy = 1$, so not symmetric with respect to the y-axis.

Origin symmetry: $(-x)^2(-y)^2 + (-x)(-y) = 1$ $\Leftrightarrow$ $x^2y^2 + xy = 1$, so it is symmetric with respect to the origin.

49. x-axis symmetry: $(-y) = x^3 + 10x$ $\Leftrightarrow$ $y = -x^3 - 10x$, which is not the same as $y = x^3 + 10x$, so not symmetric with respect to the x-axis.

y-axis symmetry: $y = (-x)^3 + 10(-x)$ $\Leftrightarrow$ $y = -x^3 - 10x$, which is not the same as $y = x^3 + 10x$, so not symmetric with respect to the y-axis.

Origin symmetry: $(-y) = (-x)^3 + 10(-x)$ $\Leftrightarrow$ $-y = -x^3 - 10x$ $\Leftrightarrow$ $y = x^3 + 10x$, so it is symmetric with respect to the origin.

51. Symmetric with respect to the y-axis.

53. Symmetric with respect to the origin.

55. Using $h = 2$, $k = -1$, and $r = 3$, we get $(x - 2)^2 + (y - (-1))^2 = 3^2$ $\Leftrightarrow$ $(x - 2)^2 + (y + 1)^2 = 9$.

57. The equation of a circle centered at the origin is $x^2 + y^2 = r^2$. Using the point $(4, 7)$ we solve for r^2. This gives $(4)^2 + (7)^2 = r^2$ $\Leftrightarrow$ $16 + 49 = 65 = r^2$. Thus, the equation of the circle is $x^2 + y^2 = 65$.

59. The center is at the midpoint of the line segment, which is $\left(\frac{-1+5}{2}, \frac{1+5}{2}\right) = (2, 3)$. The radius is one half the diameter, so $r = \frac{1}{2}\sqrt{(-1-5)^2 + (1-5)^2} = \frac{1}{2}\sqrt{36 + 16} = \frac{1}{2}\sqrt{52} = \sqrt{13}$. Thus, the equation of the circle is $(x - 2)^2 + (y - 3)^2 = \left(\sqrt{13}\right)^2$ or $(x - 2)^2 + (y - 3)^2 = 13$.

61. Since the circle is tangent to the x-axis, it must contain the point $(7, 0)$, so the radius is the change in the y-coordinates. That is, $r = |-3 - 0| = 3$. So the equation of the circle is $(x - 7)^2 + (y - (-3))^2 = 3^2$, which is $(x - 7)^2 + (y + 3)^2 = 9$.

63. From the figure, the center of the circle is at $(-2, 2)$. The radius is the change in the y-coordinates, so $r = |2 - 0| = 2$. Thus the equation of the circle is $(x - (-2))^2 + (y - 2)^2 = 2^2$, which is $(x + 2)^2 + (y - 2)^2 = 4$.

65. Completing the square gives $x^2 + y^2 - 2x + 4y + 1 = 0$ $\Leftrightarrow$
$x^2 - 2x + \underline{} + y^2 + 4y + \underline{} = -1$ $\Leftrightarrow$
$x^2 - 2x + \left(\frac{-2}{2}\right)^2 + y^2 + 4y + \left(\frac{4}{2}\right)^2 = -1 + \left(\frac{-2}{2}\right)^2 + \left(\frac{4}{2}\right)^2$ $\Leftrightarrow$
$x^2 - 2x + 1 + y^2 + 4y + 4 = -1 + 1 + 4$ $\Leftrightarrow$ $(x - 1)^2 + (y + 2)^2 = 4$.
Thus, the center is $(1, -2)$, and the radius is 2.

67. Completing the square gives $x^2 + y^2 - 4x + 10y + 13 = 0$ $\Leftrightarrow$
$x^2 - 4x + \underline{} + y^2 + 10y + \underline{} = -13$ $\Leftrightarrow$
$x^2 - 4x + \left(\frac{-4}{2}\right)^2 + y^2 + 10y + \left(\frac{10}{2}\right)^2 = -13 + \left(\frac{4}{2}\right)^2 + \left(\frac{10}{2}\right)^2$ $\Leftrightarrow$
$x^2 - 4x + 4 + y^2 + 10y + 25 = -13 + 4 + 25$ $\Leftrightarrow$ $(x - 2)^2 + (y + 5)^2 = 16$.
Thus, the center is $(2, -5)$, and the radius is 4.

69. Completing the square gives $x^2 + y^2 + x = 0$ $\Leftrightarrow$ $x^2 + x + \underline{} + y^2 = 0$ $\Leftrightarrow$
$x^2 + x + \left(\frac{1}{2}\right)^2 + y^2 = \left(\frac{1}{2}\right)^2$ $\Leftrightarrow$ $x^2 + x + \frac{1}{4} + y^2 = \frac{1}{4}$ $\Leftrightarrow$ $\left(x + \frac{1}{2}\right)^2 + y^2 = \frac{1}{4}$. Thus, the center: $\left(-\frac{1}{2}, 0\right)$, and the radius is $\frac{1}{2}$.

71. Completing the square gives $x^2 + y^2 - \frac{1}{2}x + \frac{1}{2}y = \frac{1}{8}$ $\Leftrightarrow$ $x^2 - \frac{1}{2}x + \underline{} + y^2 + \frac{1}{2}y + \underline{} = \frac{1}{8}$
$\Leftrightarrow$ $x^2 - \frac{1}{2}x + \left(\frac{-1/2}{2}\right)^2 + y^2 + \frac{1}{2}y + \left(\frac{1/2}{2}\right)^2 = \frac{1}{8} + \left(\frac{-1/2}{2}\right)^2 + \left(\frac{1/2}{2}\right)^2$ $\Leftrightarrow$
$x^2 - \frac{1}{2}x + \frac{1}{16} + y^2 + \frac{1}{2}y + \frac{1}{16} = \frac{1}{8} + \frac{1}{16} + \frac{1}{16} = \frac{2}{8} = \frac{1}{4}$ $\Leftrightarrow$ $\left(x - \frac{1}{4}\right)^2 + \left(y + \frac{1}{4}\right)^2 = \frac{1}{4}$. Thus, the center is $\left(\frac{1}{4}, -\frac{1}{4}\right)$, and the radius is $\frac{1}{2}$.

73. Completing the square gives $x^2 + y^2 + 4x - 10y = 21$ $\Leftrightarrow$
$x^2 + 4x + \underline{} + y^2 - 10y + \underline{} = 21$ $\Leftrightarrow$
$x^2 + 4x + \left(\frac{4}{2}\right)^2 + y^2 - 10y + \left(\frac{-10}{2}\right)^2 = 21 + \left(\frac{4}{2}\right)^2 + \left(\frac{-10}{2}\right)^2$ $\Leftrightarrow$
$(x + 2)^2 + (y - 5)^2 = 21 + 4 + 25 = 50$.
Thus, the center is $(-2, 5)$, and the radius is $\sqrt{50} = 5\sqrt{2}$.

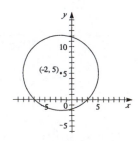

75. Completing the square gives $x^2 + y^2 + 6x - 12y + 45 = 0$ $\Leftrightarrow$
$x^2 + 6x + \underline{} + y^2 - 12y + \underline{} = -45$ $\Leftrightarrow$
$x^2 + 6x + \left(\frac{6}{2}\right)^2 + y^2 - 12y + \left(\frac{-12}{2}\right)^2 = -45 + \left(\frac{6}{2}\right)^2 + \left(\frac{-12}{2}\right)^2$ $\Leftrightarrow$
$(x + 3)^2 + (y - 6)^2 = -45 + 9 + 36 = 0$. Thus, the center is $(-3, 6)$,
and the radius is 0. This is a degenerate circle whose graph consists only
of the point $(-3, 6)$.

77. $\{(x, y)|\ x^2 + y^2 \le 1\}$. This is the set of points inside (and on) the circle
$x^2 + y^2 = 1$.

79. $\{(x, y)|\ 1 \le x^2 + y^2 < 9\}$. This is the set of points outside the circle
$x^2 + y^2 = 1$ and inside (but not on) the circle $x^2 + y^2 = 9$.

81. Completing the square gives $x^2 + y^2 - 4y - 12 = 0$ $\Leftrightarrow$ $x^2 + y^2 - 4y + \underline{} = 12$ $\Leftrightarrow$
$x^2 + y^2 - 4y + \left(\frac{-4}{2}\right)^2 = 12 + \left(\frac{-4}{2}\right)^2$ $\Leftrightarrow$ $x^2 + (y - 2)^2 = 16$. Thus, the center is $(0, 2)$, and the
radius is 4. So the circle $x^2 + y^2 = 4$, with center $(0, 0)$ and radius 2, sits completely inside the
larger circle. Thus, the area is $\pi 4^2 - \pi 2^2 = 16\pi - 4\pi = 12\pi$.

83. (a)   In 1980: 14%; in 199: 6%; in 1999: 2%.

   (b)   It exceeded 6% from 1975 to 1976 and from 1978 to 1982.

   (c)   Between 1980 and 1985 the inflation rate generally decreased. And between 1987 and 1992
         the inflation rate generally increased.

   (d)   Highest was about 14% in 1980. Lowest was about 1% in 2002.

85. Completing the square gives $x^2 + y^2 + ax + by + c = 0$ $\Leftrightarrow$
$x^2 + ax + \underline{} + y^2 + by + \underline{} = -c$ $\Leftrightarrow$
$x^2 + ax + \left(\frac{a}{2}\right)^2 + y^2 + by + \left(\frac{b}{2}\right)^2 = -c + \left(\frac{a}{2}\right)^2 + \left(\frac{b}{2}\right)^2$ $\Leftrightarrow$
$\left(x + \frac{a}{2}\right)^2 + \left(y + \frac{b}{2}\right)^2 = -c + \frac{a^2 + b^2}{4}$.

This equation represents a circle only when $-c + \dfrac{a^2 + b^2}{4} > 0$. This equation represents a point

when $-c + \dfrac{a^2 + b^2}{4} = 0$, and this equation represents the empty set when $-c + \dfrac{a^2 + b^2}{4} < 0$.

When the equation represents a circle, the center is $\left(-\dfrac{a}{2}, -\dfrac{b}{2}\right)$, and the radius is

$$\sqrt{-c + \dfrac{a^2 + b^2}{4}} = \dfrac{1}{2}\sqrt{a^2 + b^2 - 4ac}.$$

87. (a) Symmetric about the x-axis.   (b) Symmetric about the y-axis.

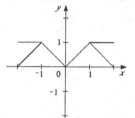

(c) Symmetric about the origin.

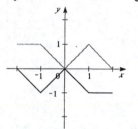

Exercises 2.3

1. $y = x^4 + 2$
    (a) $[-2, 2]$ by $[-2, 2]$ (b) $[0, 4]$ by $[0, 4]$

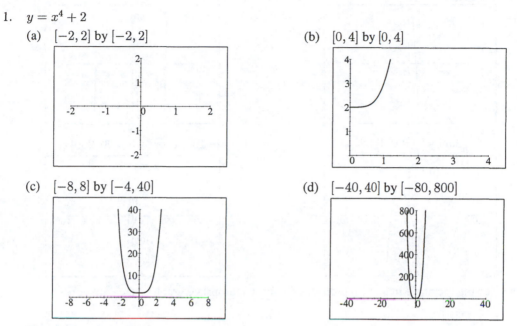

    (c) $[-8, 8]$ by $[-4, 40]$ (d) $[-40, 40]$ by $[-80, 800]$

The viewing rectangle in part (c) produces the most appropriate graph of the equation.

3. $y = 100 - x^2$
    (a) $[-4, 4]$ by $[-4, 4]$ (b) $[-10, 10]$ by $[-10, 10]$

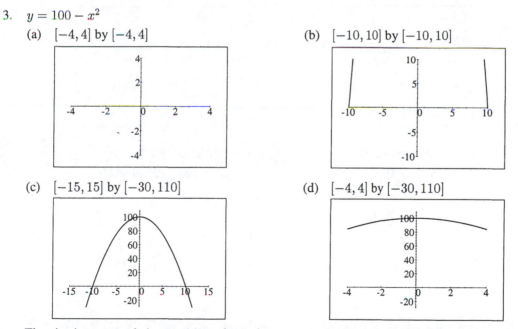

    (c) $[-15, 15]$ by $[-30, 110]$ (d) $[-4, 4]$ by $[-30, 110]$

The viewing rectangle in part (c) produces the most appropriate graph of the equation.

5. $y = 10 + 25x - x^3$

    (a) $[-4, 4]$ by $[-4, 4]$ (b) $[-10, 10]$ by $[-10, 10]$

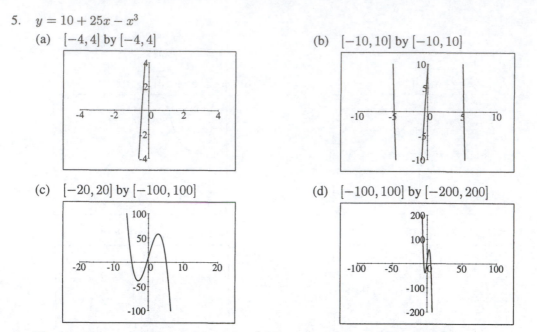

    (c) $[-20, 20]$ by $[-100, 100]$ (d) $[-100, 100]$ by $[-200, 200]$

    The viewing rectangle in part (c) produces the most appropriate graph of the equation.

7. $y = 100x^2$
    $[-2, 2]$ by $[-10, 400]$

9. $y = 4 + 6x - x^2$
    $[-4, 10]$ by $[-10, 20]$

11. $y = \sqrt[4]{256 - x^2}$. We require that
    $256 - x^2 \geq 0 \quad \Rightarrow \quad -16 \leq x \leq 16$,
    so we graph $y = \sqrt[4]{256 - x^2}$ in the
    viewing rectangle $[-20, 20]$ by $[-1, 5]$.

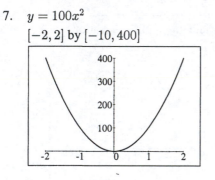

13. $y = 0.01x^3 - x^2 + 5$
    $[-50, 150]$ by $[-2000, 2000]$

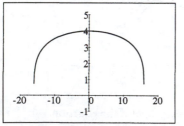

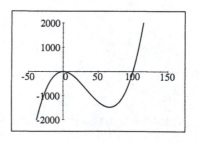

15. $y = x^4 - 4x^3$
$[-4, 6]$ by $[-50, 100]$

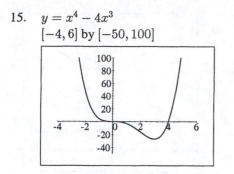

17. $y = 1 + |x - 1|$; $[-3, 5]$ by $[-1, 5]$

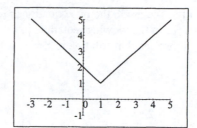

19. $x^2 + y^2 = 9 \iff y^2 = 9 - x^2 \implies$
$y = \pm\sqrt{9 - x^2}$. So we graph the functions
$y_1 = \sqrt{9 - x^2}$ and $y_2 = -\sqrt{9 - x^2}$ in the
viewing rectangle $[-6, 6]$ by $[-4, 4]$.

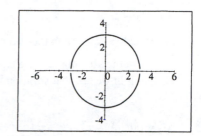

21. $4x^2 + 2y^2 = 1 \iff 2y^2 = 1 - 4x^2 \iff$
$y^2 = \dfrac{1 - 4x^2}{2} \implies y = \pm\sqrt{\dfrac{1 - 4x^2}{2}}$. So we graph
the functions $y_1 = \sqrt{\dfrac{1 - 4x^2}{2}}$ and $y_2 = -\sqrt{\dfrac{1 - 4x^2}{2}}$ in
the viewing rectangle $[-1.2, 1.2]$ by $[-0.8, 0.8]$.

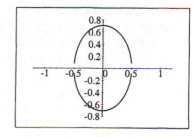

23. Although the graphs of $y = -3x^2 + 6x - \frac{1}{2}$ and
$y = \sqrt{7 - 7x^2/12}$ appear to intersect in the viewing
rectangle $[-4, 4]$ by $[-1, 3]$, there are no points of
intersection. You can verify that this is not an intersection
by zooming in.

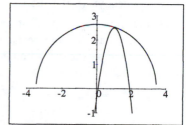

25. The graphs of $y = 6 - 4x - x^2$ and $y = 3x + 18$
appear to have two points of intersection in the
viewing rectangle $[-6, 2]$ by $[-5, 20]$. You can verify
that $x = -4$ and $x = -3$ are exact solutions.

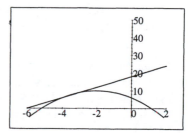

27. Algebraically: $x - 4 = 5x + 12$ $\Leftrightarrow$ $-16 = 4x$ $\Leftrightarrow$
     $x = -4$.
     Graphically: We graph the two equations $y_1 = x - 4$ and
     $y_2 = 5x + 12$ in the viewing rectangle $[-6, 4]$ by $[-10, 2]$.
     Zooming in we see that solution is $x = -4$.

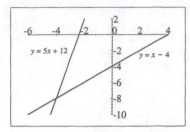

29. Algebraically: $\dfrac{2}{x} + \dfrac{1}{2x} = 7$ $\Leftrightarrow$ $2x\left(\dfrac{2}{x} + \dfrac{1}{2x}\right) = 2x(7)$

     $\Leftrightarrow$ $4 + 1 = 14x$ $\Leftrightarrow$ $x = \frac{5}{14}$.

     Graphically: We graph the two equations $y_1 = \dfrac{2}{x} + \dfrac{1}{2x}$ and
     $y_2 = 7$ in the viewing rectangle $[-2, 2]$ by $[-2, 28]$.
     Zooming in we see that solution is $x \approx 0.36$.

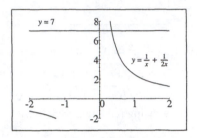

31. Algebraically: $x^2 - 32 = 0$ $\Leftrightarrow$ $x^2 = 32$ $\Rightarrow$ $x = \pm\sqrt{32} = \pm 4\sqrt{2}$.
     Graphically: We graph the equation $y_1 = x^2 - 32$ and
     determine where this curve intersects the x-axis. We use
     the viewing rectangle $[-10, 10]$ by $[-5, 5]$.
     Zooming in, we see that solutions are $x \approx 5.66$
     and $x \approx -5.66$.

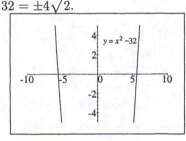

33. Algebraically: $16x^4 = 625$ $\Leftrightarrow$ $x^4 = \frac{625}{16}$ $\Rightarrow$ $x = \pm\frac{5}{2} = \pm 2.5$.
     Graphically: We graph the two equations $y_1 = 16x^4$ and
     $y_2 = 625$ in the viewing rectangle $[-5, 5]$ by $[610, 640]$.
     Zooming in we see that solutions are $x = \pm 2.5$.

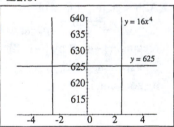

35. Algebraically: $(x - 5)^4 - 80 = 0$ $\Leftrightarrow$ $(x - 5)^4 = 80$ $\Rightarrow$ $x - 5 = \pm\sqrt[4]{80} = \pm 2\sqrt[4]{5}$ $\Leftrightarrow$
     $x = 5 \pm 2\sqrt[4]{5}$.
     Graphically: We graph the equation $y_1 = (x - 5)^4 - 80$ and
     determine where this curve intersects the x-axis. We use
     the viewing rectangle $[-1, 9]$ by $[-5, 5]$.
     Zooming in, we see that solutions are $x \approx 2.01$ and $x \approx 7.99$.

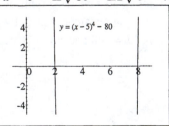

37. We graph $y = x^2 - 7x + 12$ in the viewing rectangle
     $[0, 6]$ by $[-0.1, 0.1]$. The solutions are $x = 3.00$ and 4.00.

39. We graph $y = x^3 - 6x^2 + 11x - 6$ in the viewing
     rectangle $[-1, 4]$ by $[-0.1, 0.1]$. The solutions are
     $x = 1.00$, $x = 2.00$, and $x = 3.00$.

41. We first graph
     $y = x - \sqrt{x+1}$ in the viewing rectangle $[-1, 5]$ by
     $[-0.1, 0.1]$ and find that the solution is near 1.6. Zooming in,
     we see that solutions is $x \approx 1.62$.

43. We graph $y = x^{1/3} - x$ in the viewing rectangle $[-3, 3]$ by
     $[-1, 1]$. The solutions are $x = -1.00$, $x = 0.00$, and
     $x = 1.00$.

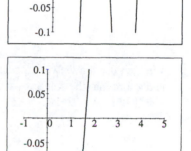

45. $x^3 - 2x^2 - x - 1 = 0$, so we start by graphing the function
     $y = x^3 - 2x^2 - x - 1$ in the viewing rectangle $[-10, 10]$ by
     $[-100, 100]$. There appear to be two solutions, one near
     $x = 0$ and another one between $x = 2$ and $x = 3$.

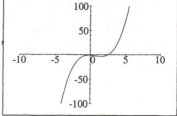

We then use the viewing rectangle $[-1, 5]$ by $[-1, 1]$
and zoom into the only solution at $x \approx 2.55$.

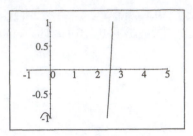

47. $x(x-1)(x+2) = \frac{1}{6}x \quad \Leftrightarrow \quad x(x-1)(x+2) - \frac{1}{6}x = 0.$
      We start by graphing the function $y = x(x-1)(x+2) - \frac{1}{6}x$
      in the viewing rectangle $[-5, 5]$ by $[-10, 10]$. There
      appear to be three solutions.

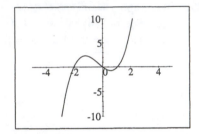

We then use the viewing rectangle $[-2.5, 2.5]$ by $[-1, 1]$
and zoom into the solutions at $x \approx -2.05$, $x = 0.00$, and
$x \approx 1.05$.

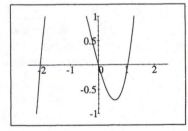

49.   We graph $y = x^2 - 3x - 10$ in the viewing rectangle
      $[-5, 6]$ by $[-12, 2]$. Thus the solution to the inequality
      is: $[-2.0, 5.0]$.

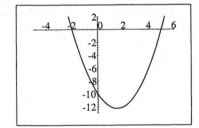

51.   Since $x^3 + 11x \le 6x^2 + 6 \quad \Leftrightarrow \quad x^3 - 6x^2 + 11x - 6 \le 0$,
      we graph $y = x^3 - 6x^2 + 11x - 6$ in the viewing
      rectangle $[0, 5]$ by $[-5, 5]$. The solution set is
      $(-\infty, 1.0] \cup [2.0, 3.0]$.

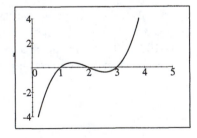

53. Since $x^{1/3} \le x \iff x^{1/3} - x < 0$, we graph
$y = x^{1/3} - x$ in the viewing rectangle $[-3, 3]$ by $[-1, 1]$.
From this, we find that the solution set is $(-1.0, 0) \cup (1.0, \infty)$.

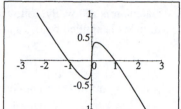

55. Since $(x + 1)^2 < (x - 1)^2 \iff (x + 1)^2 - (x - 1)^2 < 0$,
we graph $y = (x + 1)^2 - (x - 1)^2$ in the viewing rectangle
$[-2, 2]$ by $[-5, 5]$. The solution set is $(-\infty, 0)$.

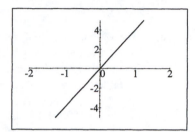

57. As in Example 4, we graph the equation
$y = x^3 - 6x^2 + 9x - \sqrt{x}$ in the viewing rectangle $[0, 10]$ by
$[-2, 15]$. We see the two solutions found in Example 4
and what appears to be an additional solution near $x = 0$.

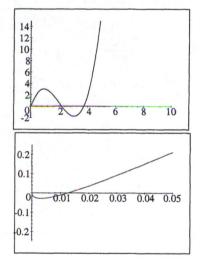

In the viewing rectangle $[0, 0.05]$ by $[-0.25, 0.25]$, we
find two more solutions at $x \approx 0.01$ and $x = 0$. We can
verify that $x = 0$ is an exact solution by substitution.

59. (a)

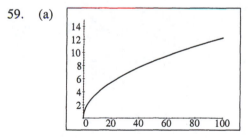

(b) Using the ZOOM and TRACE keys, we find that $y \ge 10$ for $x \ge 66.7$. We could estimate this
since if $x < 100$, then $\left(\dfrac{x}{5280}\right)^2 \le 0.00036$. So for $x < 100$ we have

$\sqrt{1.5x + \left(\dfrac{x}{5280}\right)^2} \approx \sqrt{1.5x}$. Solving $\sqrt{1.5x} > 10$ we get $1.5 > 100$ or $x > \dfrac{100}{1.5} = 66.7$ mi.

61. Calculators follow the following order of operations: exponents are applied before division and division is applied before addition. Therefore, $Y_1 = x\char`\^1/3$ is interpreted as

    $y = (x\char`\^1)/3 = \dfrac{x\char`\^1}{3} = \frac{1}{3}x$, which is the equation of a line. Likewise, $Y_2 = x/x + 4$ is interpreted as

    $y = (x/x) + 4 = 1 + 4 = 5$, again, the equation of a line. The student should have entered the following information into his calculator:

    $Y_1 = x\char`\^(1/3)$

    $Y_2 = x/(x + 4)$.

63. (a) We graph $y_1 = x^3 - 3x$ and $y_2 = k$ for $k = -4, -2, 0,$ 2, and 4 in the viewing rectangle $[-5, 5]$ by $[-10, 10]$. The number of solutions and the solutions are shown in the table below.

    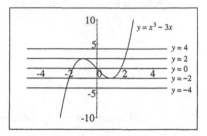

k	number of solutions	Solutions
-4	1	$x \approx -2.20$
-2	2	$x = -2, x = 1$
0	3	$x \approx \pm 1.73, x = 0$
2	2	$x = -1, x = 2$
4	1	$x \approx 2.20$

    (b) The equation $x^3 - 3x = k$ will have one solution for all $k < -2$ or $k > 2$, it will have exactly two solutions when $k = \pm 2$, and it will have three solutions for $-2 < k < 2$.

Exercises 2.4

1. $m = \dfrac{y_2 - y_1}{x_2 - x_1} = \dfrac{4 - 0}{2 - 0} = \dfrac{4}{2} = 2.$

3. $m = \dfrac{y_2 - y_1}{x_2 - x_1} = \dfrac{0 - 2}{10 - 2} = \dfrac{-2}{8} = -\dfrac{1}{4}.$

5. $m = \dfrac{y_2 - y_1}{x_2 - x_1} = \dfrac{4 - 12}{2 - 4} = \dfrac{-8}{-2} = 4.$

7. $m = \dfrac{y_2 - y_1}{x_2 - x_1} = \dfrac{6 - (-3)}{-1 - 1} = \dfrac{9}{-2} = -\dfrac{9}{2}.$

9. For ℓ_1, we find two points, $(-1, 2)$ and $(0, 0)$ that lie on the line. Thus the slope of ℓ_1 is $m = \dfrac{y_2 - y_1}{x_2 - x_1} = \dfrac{2 - 0}{-1 - 0} = -2.$ For ℓ_2, we find two points $(0, 2)$ and $(2, 3)$. Thus, the slope of ℓ_2 is $m = \dfrac{y_2 - y_1}{x_2 - x_1} = \dfrac{3 - 2}{2 - 0} = \dfrac{1}{2}.$ For ℓ_3 we find the points $(2, -2)$ and $(3, 1)$. Thus, the slope of ℓ_3 is $m = \dfrac{y_2 - y_1}{x_2 - x_1} = \dfrac{1 - (-2)}{3 - 2} = 3.$ For ℓ_4, we find the points $(-2, -1)$ and $(2, -2)$. Thus, the slope of ℓ_4 is $m = \dfrac{y_2 - y_1}{x_2 - x_1} = \dfrac{-2 - (-1)}{2 - (-2)} = \dfrac{-1}{4} = -\dfrac{1}{4}.$

11. First we find two points, $(0, 4)$ and $(4, 0)$ that lie on the line. So the slope is $m = \frac{0-4}{4-0} = -1.$ Since the y-intercept is 4, the equation of the line is $y = mx + b = -1x + 4.$ So $y = -x + 4$, or $x + y - 4 = 0.$

13. We choose the two intercepts as points, $(0, -3)$ and $(2, 0)$. So the slope is $m = \frac{0-(-3)}{2-0} = \frac{3}{2}.$ Since the y-intercept is -3, the equation of the line is $y = mx + b = \frac{3}{2}x - 3$, or $3x - 2y - 6 = 0.$

15. Using the equation $y - y_1 = m(x - x_1)$, we get $y - 3 = 1(x - 2)$ $\Leftrightarrow$ $-x + y = 1$ $\Leftrightarrow$ $x - y + 1 = 0.$

17. Using the equation $y - y_1 = m(x - x_1)$, we get $y - 7 = \frac{2}{3}(x - 1)$ $\Leftrightarrow$ $3y - 21 = 2x - 2$ $\Leftrightarrow$ $-2x + 3y = 19$ $\Leftrightarrow$ $2x - 3y + 19 = 0.$

19. First we find the slope, which is $m = \dfrac{y_2 - y_1}{x_2 - x_1} = \dfrac{6 - 1}{1 - 2} = \dfrac{5}{-1} = -5.$ Substituting into $y - y_1 = m(x - x_1)$, we get $y - 6 = -5(x - 1)$ $\Leftrightarrow$ $y - 6 = -5x + 5$ $\Leftrightarrow$ $5x + y - 11 = 0.$

21. Using $y = mx + b$, we have $y = 3x + (-2)$ or $3x - y - 2 = 0.$

23. We are given two points, $(1, 0)$ and $(0, -3)$. Thus, the slope is $m = \dfrac{y_2 - y_1}{x_2 - x_1} = \dfrac{-3 - 0}{0 - 1} = \dfrac{-3}{-1} = 3.$ Using the y-intercept, we have $y = 3x + (-3)$ or $y = 3x - 3$ or $3x - y - 3 = 0.$

25. Since the equation of a horizontal line passing through (a, b) is $y = b$, the equation of the horizontal line passing through $(4, 5)$ is $y = 5$.

27. Since $x + 2y = 6$ $\Leftrightarrow$ $2y = -x + 6$ $\Leftrightarrow$ $y = -\frac{1}{2}x + 3$, the slope of this line is $-\frac{1}{2}$. Thus, the line we seek is given by $y - (-6) = -\frac{1}{2}(x - 1)$ $\Leftrightarrow$ $2y + 12 = -x + 1$ $\Leftrightarrow$ $x + 2y + 11 = 0$.

29. Any line parallel to $x = 5$ will have undefined slope and be of the form $x = a$. Thus the equation of the line is $x = -1$.

31. First find the slope of $2x + 5y + 8 = 0$. This gives $2x + 5y + 8 = 0$ $\Leftrightarrow$ $5y = -2x - 8$ $\Leftrightarrow$ $y = -\frac{2}{5}x - \frac{8}{5}$. So the slope of the line that is perpendicular to $2x + 5y + 8 = 0$ is $m = -\frac{1}{-2/5} = \frac{5}{2}$. The equation of the line we seek is $y - (-2) = \frac{5}{2}(x - (-1))$ $\Leftrightarrow$ $2y + 4 = 5x + 5$ $\Leftrightarrow$ $5x - 2y + 1 = 0$.

33. First find the slope of the line passing through $(2, 5)$ and $(-2, 1)$. This gives $m = \frac{1-5}{-2-2} = \frac{-4}{-4} = 1$, and so the equation of the line we seek is $y - 7 = 1(x - 1)$ $\Leftrightarrow$ $x - y + 6 = 0$.

35. (a)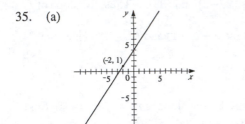

(b) $y - 1 = \frac{3}{2}(x - (-2))$ $\Leftrightarrow$ $2y - 2 = 3(x + 2)$ $\Leftrightarrow$ $2y - 2 = 3x + 6$ $\Leftrightarrow$ $3x - 2y + 8 = 0$.

37.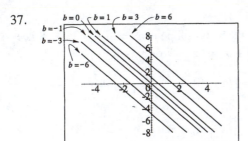

$y = -2x + b$, $b = 0, \pm 1, \pm 3, \pm 6$.
They have the same slope, so they are parallel.

39.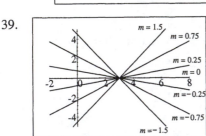

$y = m(x - 3)$, $m = 0, \pm 0.25, \pm 0.75, \pm 1.5$.
Each of the lines contains the point $(3, 0)$ because the point $(3, 0)$ satisfies each equation $y = m(x - 3)$. Since $(3, 0)$ is on the x-axis, we could also say that they all have the same x-intercept.

41. $x + y = 3 \quad \Leftrightarrow \quad y = -x + 3$. So the slope
is -1, and the y-intercept is 3.

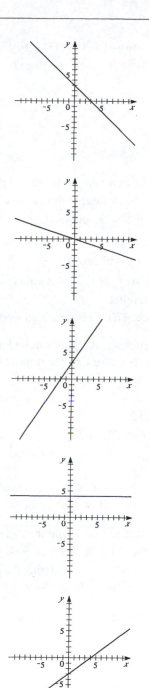

43. $x + 3y = 0 \quad \Leftrightarrow \quad 3y = -x \quad \Leftrightarrow$
$y = -\frac{1}{3}x$. So the slope is $-\frac{1}{3}$, and the
y-intercept is 0.

45. $\frac{1}{2}x - \frac{1}{3}y + 1 = 0 \quad \Leftrightarrow \quad -\frac{1}{3}y = -\frac{1}{2}x - 1$
$\Leftrightarrow \quad y = \frac{3}{2}x + 3$. So the slope is $\frac{3}{2}$, and the
y-intercept is 3.

47. $y = 4$ can also be expressed as $y = 0x + 4$.
So the slope is 0, and the y-intercept is 4.

49. $3x - 4y = 12 \quad \Leftrightarrow \quad -4y = -3x + 12 \quad \Leftrightarrow$
$y = \frac{3}{4}x - 3$. So the slope is $\frac{3}{4}$, and the y-intercept is -3.

51. $3x + 4y - 1 = 0 \quad \Leftrightarrow \quad 4y = -3x + 1 \quad \Leftrightarrow$

     $y = -\frac{3}{4}x + \frac{1}{4}$. So the slope is $-\frac{3}{4}$, and the y-intercept is $\frac{1}{4}$.

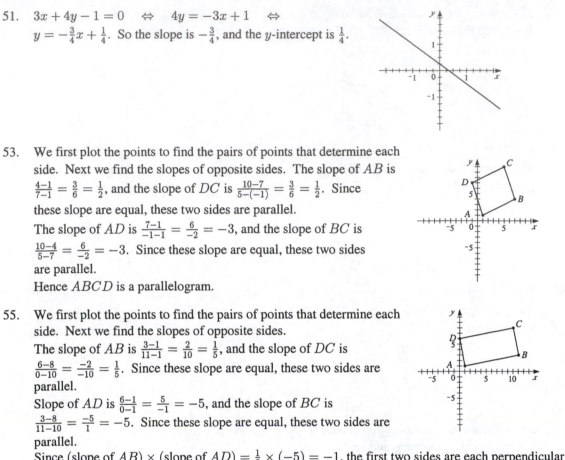

53. We first plot the points to find the pairs of points that determine each
     side. Next we find the slopes of opposite sides. The slope of AB is
     $\frac{4-1}{7-1} = \frac{3}{6} = \frac{1}{2}$, and the slope of DC is $\frac{10-7}{5-(-1)} = \frac{3}{6} = \frac{1}{2}$. Since
     these slope are equal, these two sides are parallel.
     The slope of AD is $\frac{7-1}{-1-1} = \frac{6}{-2} = -3$, and the slope of BC is
     $\frac{10-4}{5-7} = \frac{6}{-2} = -3$. Since these slope are equal, these two sides
     are parallel.
     Hence $ABCD$ is a parallelogram.

55. We first plot the points to find the pairs of points that determine each
     side. Next we find the slopes of opposite sides.
     The slope of AB is $\frac{3-1}{11-1} = \frac{2}{10} = \frac{1}{5}$, and the slope of DC is
     $\frac{6-8}{0-10} = \frac{-2}{-10} = \frac{1}{5}$. Since these slope are equal, these two sides are
     parallel.
     Slope of AD is $\frac{6-1}{0-1} = \frac{5}{-1} = -5$, and the slope of BC is
     $\frac{3-8}{11-10} = \frac{-5}{1} = -5$. Since these slope are equal, these two sides are
     parallel.
     Since (slope of AB) $\times$ (slope of AD) $= \frac{1}{5} \times (-5) = -1$, the first two sides are each perpendicular
     to the second two sides. So the sides form a rectangle.

57. We need the slope and the midpoint of the line AB. The midpoint of AB is $\left(\frac{1+7}{2}, \frac{4-2}{2}\right) = (4, 1)$,
     and the slope of AB is $m = \frac{-2-4}{7-1} = \frac{-6}{6} = -1$. The slope of the perpendicular bisector will have
     slope $\frac{-1}{m} = \frac{-1}{-1} = 1$. Using the point-slope form, the equation of the perpendicular bisector is
     $y - 1 = 1(x - 4)$ or $x - y - 3 = 0$.

59. (a) We start with the two points $(a, 0)$ and $(0, b)$. The slope of the line that contains them is
          $\frac{b-0}{0-a} = -\frac{b}{a}$. So the equation of the line containing them is $y = -\frac{b}{a}x + b$ (using the slope-
          intercept form). Dividing by b (since $b \neq 0$) gives $\frac{y}{b} = -\frac{x}{a} + 1 \quad \Leftrightarrow \quad \frac{x}{a} + \frac{y}{b} = 1$.

     (b) Setting $a = 6$ and $b = -8$, we get $\frac{x}{6} + \frac{y}{-8} = 1 \quad \Leftrightarrow \quad 4x - 3y = 24 \quad \Leftrightarrow$
          $4x - 3y - 24 = 0$.

61. Let h be the change in your horizontal distance, in feet. Then $-\frac{6}{100} = \frac{-1000}{h} \quad \Leftrightarrow$
     $h = \frac{100,000}{6} \approx 16{,}667$. So the change in your horizontal distance is about $16{,}667$ feet.

63. (a) The slope is $0.0417D = 0.0417(200) = 8.34$. It represents the increase in dosage for each one-year increase in the child's age.

     (b) When $a = 0$, $c = 8.34(0 + 1) = 8.34$ mg.

65. (a)

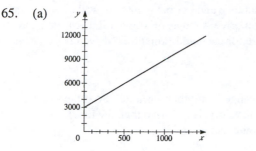

     (b) The slope is the cost per toaster oven, $6. The y-intercept, $3000, is the monthly fixed cost, the cost that is incurred no matter how many toaster ovens are produced.

67. (a) Using n in place of x and t in place of y, we find that the slope is $\dfrac{t_2 - t_1}{n_2 - n_1} = \dfrac{80 - 70}{168 - 120} = \dfrac{10}{48}$ $= \frac{5}{24}$. So the linear equation is $t - 80 = \frac{5}{24}(n - 168)$ $\Leftrightarrow$ $t - 80 = \frac{5}{24}n - 35$ $\Leftrightarrow$ $t = \frac{5}{24}n + 45$.

     (b) When $n = 150$, the temperature is approximately given by $t = \frac{5}{24}(150) + 45 = 76.25°F$ $\approx 76°F$.

69. (a) We are given $\dfrac{\text{change in pressure}}{10 \text{ feet change in depth}} = \dfrac{4.34}{10} = 0.434$. Using P for pressure and d for depth, and using the point $P = 15$ when $d = 0$, we have $P - 15 = 0.434(d - 0)$ $\Leftrightarrow$ $P = 0.434d + 15$.

     (b)

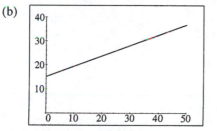

     (c) The slope represents the increase in pressure per foot of descent. The y-intercept represents the pressure at the surface.

     (d) When $P = 100$, then $100 = 0.434d + 15$ $\Leftrightarrow$ $0.434d = 85$ $\Leftrightarrow$ $d = 195.9$ feet. Thus the pressure is 100 lb/in^3 at a depth of approximately 196 feet.

71. (a) Using d in place of x and C in place of y, we find the slope $= \dfrac{C_2 - C_1}{d_2 - d_1} = \dfrac{460 - 380}{800 - 480} = \dfrac{80}{320}$ $= \frac{1}{4}$. So the linear equations is $C - 460 = \frac{1}{4}(d - 800)$ $\Leftrightarrow$ $C - 460 = \frac{1}{4}d - 200$ $\Leftrightarrow$ $C = \frac{1}{4}d + 260$.

     (b) Substituting $d = 1500$ we get $C = \frac{1}{4}(1500) + 260 = 635$. Thus, the cost of driving 1500 miles is $635.

(c)

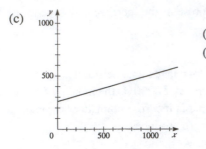

The slope of the line represents the cost per mile, $0.25.

(d)   The y-intercept represents the fixed cost, $260.

(e)   It is a suitable model because you have fixed monthly costs such as insurance and car payments, as well as costs that occur as you drive, such as gasoline, oil, tires, etc., and the cost of these for each additional mile driven is a constant.

73.   Slope is the rate of change of one variable per unit change in another variable. So if the slope is positive, then the temperature is rising. Likewise, if the slope is negative then the temperature is decreasing. If the slope is 0, then the temperature is not changing.

Exercises 2.5

1. $T = kx$, where k is constant.

3. $v = \dfrac{k}{z}$, where k is constant.

5. $y = \dfrac{ks}{t}$, where k is constant.

7. $z = k\sqrt{y}$, where k is constant.

9. $V = klwh$, where k is constant.

11. $R = \dfrac{ki}{Pt}$, where k is constant.

13. Since y is directly proportional to x, $y = kx$. Since $y = 42$ when $x = 6$, we have $42 = k(6)$ $\Leftrightarrow$ $k = 7$. So $y = 7x$.

15. Since M varies directly as x and inversely as y, $M = \dfrac{kx}{y}$. Since $M = 5$ when $x = 2$ and $y = 6$, we have $5 = \dfrac{k(2)}{6}$ $\Leftrightarrow$ $k = 15$. Therefore $M = \dfrac{15x}{y}$.

17. Since W is inversely proportional to the square of r, $W = \dfrac{k}{r^2}$. Since $W = 10$ when $r = 6$, we have $10 = \dfrac{k}{(6)^2}$ $\Leftrightarrow$ $k = 360$. So $W = \dfrac{360}{r^2}$.

19. Since C is jointly proportional to l, w, and h, we have $C = klwh$. Since $C = 128$ when $l = w = h = 2$, we have $128 = k(2)(2)(2)$ $\Leftrightarrow$ $128 = 8k$ $\Leftrightarrow$ $k = 16$. Therefore, $C = 16lwh$.

21. Since s is inversely proportional to the square root of t, we have $s = \dfrac{k}{\sqrt{t}}$. Since $s = 100$ when $t = 25$, we have $100 = \dfrac{k}{\sqrt{25}}$ $\Leftrightarrow$ $100 = \dfrac{k}{5}$ $\Leftrightarrow$ $k = 500$. So $s = \dfrac{500}{\sqrt{t}}$.

23. (a) The force F needed is $F = kx$.

    (b) Since $F = 40$ when $x = 5$, we have $40 = k(5)$ $\Leftrightarrow$ $k = 8$.

    (c) From part (b), we have $F = 8x$. Substituting $x = 4$ into $F = 8x$ gives $F = 8(4) = 32$ N.

25. (a) $C = kpm$.

    (b) Since $C = 60{,}000$ when $p = 120$ and $m = 4000$, we get $60{,}000 = k(120)(4000)$ $\Leftrightarrow$ $k = \frac{1}{8}$. So $C = \frac{1}{8}pm$.

    (c) Substituting $p = 92$ and $m = 5{,}000$, we get $C = \frac{1}{8}(92)(5{,}000) = \$57{,}500$ and $k = \frac{1}{8}$.

27. (a) $P = ks^3$.

    (b) Since $P = 96$ when $s = 20$, we get $96 = k \cdot 20^3$ $\Leftrightarrow$ $k = 0.012$. So $P = 0.012s^3$.

    (c) Substituting $x = 30$, we get $P = 0.012 \cdot 30^3 = 324$ watts.

29. $L = \dfrac{k}{d^2}$. Since $L = 70$ when $d = 10$, we have $70 = \dfrac{k}{10^2}$ so $k = 7{,}000$. Thus $L = \dfrac{7{,}000}{d^2}$. When
     $d = 100$ we get $L = \dfrac{7{,}000}{100^2} = 0.7$ dB.

31. $P = kAv^3$. If $A = \frac{1}{2}A_0$ and $v = 2v_0$, then $P = k\left(\frac{1}{2}A_0\right)(2v_0)^3 = \frac{1}{2}kA_0(8v_0^3) = 4kA_0v_0^3$. The power
     is increased by a factor of 4.

33. $F = kAs^2$. Since $F = 220$ when $A = 40$ and $s = 5$. Solving for k we have $220 = k(40)(5)^2$ $\Leftrightarrow$
     $220 = 1000k$ $\Leftrightarrow$ $k = 0.22$. Now when $A = 28$ and $F = 175$ we get $175 = 0.220(28)s^2$ $\Leftrightarrow$
     $28.4090 = s^2$ so $s = \sqrt{28.4090} = 5.33$ mi/h.

35. (a) $R = \dfrac{kL}{d^2}$.

     (b)   Since $R = 140$ when $L = 1.2$ and $d = 0.005$, we get $140 = \dfrac{k(1.2)}{(0.005)^2}$ $\Leftrightarrow$
           $k = \frac{7}{2400} = 0.00291\overline{6}$.

     (c)   Substituting $L = 3$ and $d = 0.008$, we have $R = \dfrac{7}{2400} \cdot \dfrac{3}{(0.008)^2} = \dfrac{4375}{32} \approx 137\ \Omega$.

37. $E = kT^4$.
     (a)   For the sun, $E_S = k6000^4$ and for earth $E_E = k300^4$. Thus
           $\dfrac{E_S}{E_E} = \dfrac{k6000^4}{k300^4} = \left(\dfrac{6000}{300}\right)^4 = 20^4 = 160{,}000$. So the sun produces 160,000 times the radiation
           energy per unit area than the Earth.

     (b)   The surface area of the sun is $4\pi(435{,}000)^2$ and the surface area of the Earth is $4\pi(3{,}960)^2$. So
           the sun has $\dfrac{4\pi(435{,}000)^2}{4\pi(3{,}960)^2} = \left(\dfrac{435{,}000}{3{,}960}\right)^2$ times the surface area of the Earth. Thus the total
           radiation emitted by the sun is $160{,}000 \times \left(\frac{435{,}000}{3{,}960}\right)^2 = 1{,}930{,}670{,}340$ time the total radiation
           emitted by the Earth.

39. Let S be the final size of the cabbage, in pounds, let N be the amount of nutrients it receives, in
     ounces, and let c be the number of other cabbages around it. Then $S = k\dfrac{N}{c}$. When $N = 20$ and
     $c = 12$, we have $S = 30$, so substituting, we have $30 = k\frac{20}{12}$ $\Leftrightarrow$ $k = 18$. Thus $S = 18\dfrac{N}{c}$.
     When $N = 10$ and $c = 5$, the final size is $S = 18\left(\frac{10}{5}\right) = 36$ lb.

41. (a)   Since f is inversely proportional to L, we have $f = \dfrac{k}{L}$, where k is a positive constant.

     (b)   If we replace L by $2L$ we have $\dfrac{k}{(2L)} = \frac{1}{2} \cdot \dfrac{k}{L} = \frac{1}{2} \cdot f$. So the frequency of the vibration is cut
           in half.

43. Examples include radioactive decay and exponential growth in biology.

Review Exercises for Chapter 2

1. (a)

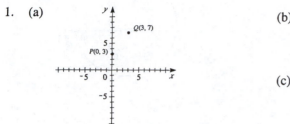

(b) The distance from P to Q is
$$d(P,Q) = \sqrt{(0-3)^2 + (3-7)^2}$$
$$= \sqrt{9+16} = \sqrt{25} = 5.$$

(c) The midpoint is $\left(\frac{0+3}{2}, \frac{3+7}{2}\right) = \left(\frac{1}{2}, 5\right)$.

(d) The line has slope $m = \frac{3-7}{0-3} = \frac{4}{3}$, and has equation $y - 3 = \frac{4}{3}(x - 0)$ $\Leftrightarrow$
$$y - 3 = \frac{4}{3}x \quad \Leftrightarrow \quad y = \frac{4}{3}x + 3.$$

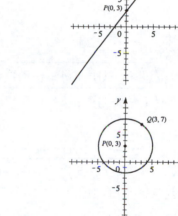

(e) The radius of this circle was found in part (b). It is $r = d(P,Q) = 5$. So the equation is
$$(x - 0)^2 + (y - 3)^2 = (5)^2 \quad \Leftrightarrow$$
$$x^2 + (y - 3)^2 = 25.$$

3. (a)

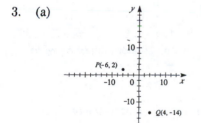

(b) The distance from P to Q is
$$d(P,Q) = \sqrt{(-6-4)^2 + [2-(-14)]^2}$$
$$= \sqrt{100 + 256} = \sqrt{356} = 2\sqrt{89}.$$

(c) The midpoint is $\left(\frac{-6+4}{2}, \frac{2+(-14)}{2}\right) = (-1, -6)$.

(d) The line has slope $m = \frac{2-(-14)}{-6-4} = \frac{16}{-10} = -\frac{8}{5}$.
The equation is then $y - 2 = -\frac{8}{5}(x + 6)$ $\Leftrightarrow$
$$y - 2 = -\frac{8}{5}x - \frac{48}{5} \quad \Leftrightarrow \quad y = -\frac{8}{5}x - \frac{38}{5}.$$

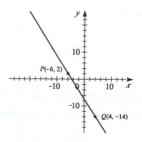

(e) The radius of this circle was found in part (b).
It is $r = d(P, Q) = 2\sqrt{89}$. So the equation is
$$[x - (-6)]^2 + (y - 2)^2 = \left(2\sqrt{89}\right)^2 \quad \Leftrightarrow$$
$$(x + 6)^2 + (y - 2)^2 = 356.$$

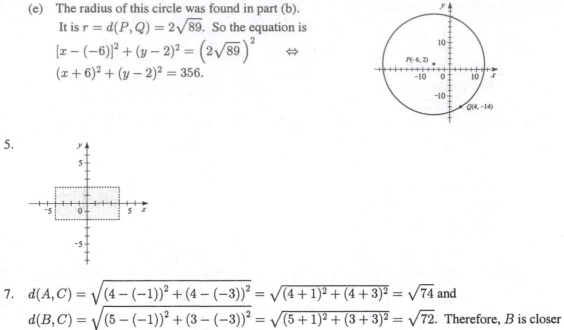

5.

7. $d(A, C) = \sqrt{(4 - (-1))^2 + (4 - (-3))^2} = \sqrt{(4 + 1)^2 + (4 + 3)^2} = \sqrt{74}$ and
$d(B, C) = \sqrt{(5 - (-1))^2 + (3 - (-3))^2} = \sqrt{(5 + 1)^2 + (3 + 3)^2} = \sqrt{72}$. Therefore, B is closer
to C.

9. The center is $C = (-5, -1)$, and the point $P = (0, 0)$ is on the circle. The radius of the circle is
$r = d(P, C) = \sqrt{(0 - (-5))^2 + (0 - (-1))^2} = \sqrt{(0 + 5)^2 + (0 + 1)^2} = \sqrt{26}$. Thus, the equation
of the circle is $(x + 5)^2 + (y + 1)^2 = 26$.

11. $x^2 + y^2 + 2x - 6y + 9 = 0 \quad \Leftrightarrow \quad (x^2 + 2x) + (y^2 - 6y) = -9 \quad \Leftrightarrow$
$(x^2 + 2x + 1) + (y^2 - 6y + 9) = -9 + 1 + 9 \quad \Leftrightarrow \quad (x + 1)^2 + (y - 3)^2 = 1$. This equation
represents a circle with center at $(-1, 3)$ and radius 1.

13. $x^2 + y^2 + 72 = 12x \quad \Leftrightarrow \quad (x^2 - 12x) + y^2 = -72 \quad \Leftrightarrow \quad (x^2 - 12x + 36) + y^2 = -72 + 36$
$\Leftrightarrow \quad (x - 6)^2 + y^2 = -36$. Since the left side of this equation must be greater than or equal to
zero, this equation has no graph.

15. $y = 2x - 3$
x-axis symmetry: $(-y) = 2 - 3x \quad \Leftrightarrow \quad y = -2 + 3x$, which is not the same as the original
equation, so not symmetric with respect to the x-axis.

y-axis symmetry: $y = 2 - 3(-x) \quad \Leftrightarrow \quad y = 2 + 3x$, which is not the
same as the original equation, so not symmetric with respect to the y-axis.

Origin symmetry: $(-y) = 2 - 3(-x) \quad \Leftrightarrow$
$-y = 2 + 3x \quad \Leftrightarrow \quad y = -2 - 3x$, which is not the
same as the original equation, so not symmetric with
respect to the origin.
Hence the graph has no symmetry.

x	y
-2	8
0	2
$\frac{2}{3}$	0

17. $x + 3y = 21$ $\Leftrightarrow$ $y = -\frac{1}{3}x + 7$

x-axis symmetry: $x + 3(-y) = 21$ $\Leftrightarrow$ $x - 3y = 21$, which is not the same as the original equation, so not symmetric with respect to the x-axis.

y-axis symmetry: $(-x) + 3y = 21$ $\Leftrightarrow$ $x - 3y = -21$, which is not the same as the original equation, so not symmetric with respect to the y-axis.

Origin symmetry: $(-x) + 3(-y) = 21$ $\Leftrightarrow$ $x + 3y = -21$, which is not the same as the original equation, so not symmetric with respect to the origin.

Hence the graph has no symmetry.

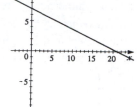

x	y
-3	8
0	7
21	0

19. $\dfrac{x}{2} - \dfrac{y}{7} = 1$ $\Leftrightarrow$ $y = \dfrac{7}{2}x - 7$

x-axis symmetry: $\dfrac{x}{2} - \dfrac{(-y)}{7} = 1$ $\Leftrightarrow$ $\dfrac{x}{2} + \dfrac{y}{7} = 1$, which is not the same as the original equation, so not symmetric with respect to the x-axis.

y-axis symmetry: $\dfrac{(-x)}{2} - \dfrac{y}{7} = 1$ $\Leftrightarrow$ $\dfrac{x}{2} + \dfrac{y}{7} = -1$, which is not the same as the original equation, so not symmetric with respect to the y-axis.

Origin symmetry: $\dfrac{(-x)}{2} - \dfrac{(-y)}{7} = 1$ $\Leftrightarrow$ $\dfrac{x}{2} - \dfrac{y}{7} = -1$, which is not the same as the original equation, so not symmetric with respect to the origin.

Hence, the graph has no symmetry.

x	y
-2	-14
0	-7
2	0

21. $y = 16 - x^2$

x-axis symmetry: $(-y) = 16 - x^2$ $\Leftrightarrow$ $y = -16 + x^2$, which is not the same as the original equation, so not symmetric with respect to the x-axis.

y-axis symmetry: $y = 16 - (-x)^2$ $\Leftrightarrow$ $y = 16 - x^2$, which is the same as the original equation, so its is symmetric with respect to the y-axis.

Origin symmetry: $(-y) = 16 - (-x)^2$ $\Leftrightarrow$ $y = -16 + x^2$, which is not the same as the original equation, so not symmetric with respect to the origin.

Hence, the graph is symmetric with respect to the y-axis.

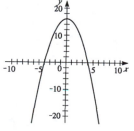

x	y
-3	7
-1	15
0	16
1	15
3	7

23. $x = \sqrt{y}$

x-axis symmetry: $x = \sqrt{-y}$, which is not the same as the original equation, so not symmetric with respect to the x-axis.

y-axis symmetry: $(-x) = \sqrt{y} \iff x = -\sqrt{y}$, which is not the same as the original equation, so not symmetric with respect to the y-axis.

Origin symmetry: $(-x) = \sqrt{-y}$, which is not the same as the original equation, so not symmetric with respect to the origin.

Hence, the graph has no symmetry.

x	y
0	0
1	1
2	4
3	9

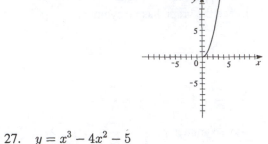

25. $y = x^2 - 6x$
Viewing rectangle $[-10, 10]$ by $[-10, 10]$.

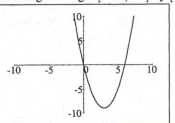

27. $y = x^3 - 4x^2 - 5$
Viewing rectangle $[-4, 10]$ by $[-30, 20]$.

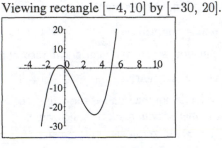

29. $x^2 - 4x = 2x + 7$. We graph the equations $y_1 = x^2 - 4x$ and $y_2 = 2x + 7$ in the viewing rectangle rectangle $[-10, 10]$ by $[-5, 25]$. Using Zoom and/or Trace, we get the solutions $x = -1$ and $x = 7$.

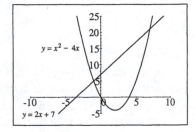

31. $x^4 - 9x^2 = x - 9$. We graph the equations $y_1 = x^4 - 9x^2$ and $y_2 = x - 9$ in the viewing rectangle $[-5, 5]$ by $[-25, 10]$. Using Zoom and/or Trace, we get the solutions $x \approx -2.72$, $x \approx -1.15$, $x = 1.00$, and $x \approx 2.87$.

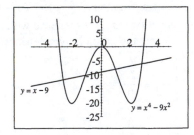

33. $4x - 3 \geq x^2$. We graph the equations $y_1 = 4x - 3$ and $y_2 = x^2$ in the viewing rectangle $[-5, 5]$ by $[0, 15]$. Using Zoom and/or Trace, we find the points of intersection are at $x = 1$ and $x = 3$. Since we want $4x - 3 \geq x^2$, the solution is the interval $[1, 3]$.

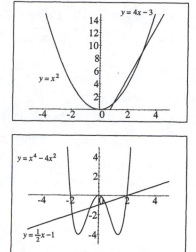

35. $x^4 - 4x^2 < \frac{1}{2}x - 1$. We graph the equations $y_1 = x^4 - 4x^2$ and $y_2 = \frac{1}{2}x - 1$ in the viewing rectangle $[-5, 5]$ by $[-5, 5]$. We find the points of intersection are at $x \approx -1.85$, $x \approx -0.60$, $x \approx 0.45$, and $x = 2.00$. Since we want $x^4 - 4x^2 < \frac{1}{2}x - 1$, the solution is $(-1.85, -0.60) \cup (0.45, 2.00)$.

37. The line has slope $m = \frac{-4+6}{2+1} = \frac{2}{3}$, and so, by the point-slope formula, the equation is
$$y + 4 = \frac{2}{3}(x - 2) \quad \Leftrightarrow \quad y = \frac{2}{3}x - \frac{16}{3} \quad \Leftrightarrow \quad 2x - 3y - 16 = 0.$$

39. The x-intercept is 4, and the y-intercept is 12, so the slope is $m = \frac{12-0}{0-4} = -3$. Therefore, by the slope-intercept formula, the equation of the line is $y = -3x + 12 \quad \Leftrightarrow \quad 3x + y - 12 = 0$.

41. We first find the slope of the line $3x + 15y = 22$. This gives $3x + 15y = 22 \quad \Leftrightarrow \quad 15y = -3x + 22 \quad \Leftrightarrow \quad y = -\frac{1}{5}x + \frac{22}{15}$. So this line has slope $m = -\frac{1}{5}$, as does any line parallel to it. Then the parallel line passing through the origin has equation $y - 0 = -\frac{1}{5}(x - 0) \quad \Leftrightarrow \quad x + 5y = 0$.

43. (a) The slope, 0.3, represents the increase in length of the spring for each unit increase in weight w. The S-intercept is the resting or natural length of the spring.

     (b) When $w = 5$, $S = 0.3(5) + 2.5 = 1.5 + 2.5 = 4.0$ inches.

45. Since M varies directly as z we have $M = kz$. Substituting $M = 120$ when $z = 15$, we find $120 = k(15) \quad \Leftrightarrow \quad k = 8$. Therefore, $M = 8z$.

47. (a) The intensity I varies inversely as the square of the distance d, so $I = \dfrac{k}{d^2}$.

     (b) Substituting $I = 1000$ when $d = 8$, we get $1000 = \frac{k}{(8)^2} \quad \Leftrightarrow \quad k = 64{,}000$.

     (c) From parts (a) and (b), we have $I = \dfrac{64{,}000}{d^2}$. Substituting $d = 20$, we get $I = \dfrac{64{,}000}{(20)^2} = 160$ candles.

49. Let v be the terminal velocity of the parachutist in mi/h and w be his weight in pounds. Since the terminal velocity is directly proportional to the square root of the weight, we have $v = k\sqrt{w}$. Substituting $v = 9$ when $w = 160$, we solve for k. This gives $9 = k\sqrt{160} \quad \Leftrightarrow$

$k = \frac{9}{\sqrt{160}} \approx 0.712$. Thus $v = 0.712\sqrt{w}$. When $w = 240$, the terminal velocity is

$v = 0.712\sqrt{240} \approx 11$ mi/h.

51. Here the center is at $(0,0)$, and the circle passes through the point $(-5, 12)$, so the radius is
$r = \sqrt{(-5-0)^2 + (12-0)^2} = \sqrt{25 + 144} = \sqrt{169} = 13$. The equation of the circle is
$x^2 + y^2 = 13^2 \quad \Leftrightarrow \quad x^2 + y^2 = 169$. The line shown is the tangent that passes through the point
$(-5, 12)$, so it is perpendicular to the line through the points $(0,0)$ and $(-5, 12)$. This line has slope
$m_1 = \frac{12-0}{-5-0} = -\frac{12}{5}$. The slope of the line we seek is $m_2 = -\dfrac{1}{m_1} = -\dfrac{1}{-12/5} = \frac{5}{12}$. Thus, the
equation of the tangent line is $y - 12 = \frac{5}{12}(x+5) \quad \Leftrightarrow \quad y - 12 = \frac{5}{12}x + \frac{25}{12} \quad \Leftrightarrow$
$y = \frac{5}{12}x + \frac{169}{12} \quad \Leftrightarrow \quad 5x - 12y + 169 = 0$.

Chapter 2 Test

1. (a)

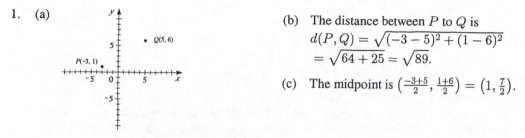

   (b) The distance between P to Q is
$$d(P,Q) = \sqrt{(-3-5)^2 + (1-6)^2}$$
$$= \sqrt{64 + 25} = \sqrt{89}.$$

   (c) The midpoint is $\left(\frac{-3+5}{2}, \frac{1+6}{2}\right) = \left(1, \frac{7}{2}\right)$.

   (d) The slope of the line is $\frac{1-6}{-3-5} = \frac{-5}{-8} = \frac{5}{8}$.

   (e) The perpendicular bisector of PQ contains the midpoint, $(1, \frac{7}{2})$, and it slope is the negative reciprocal of $\frac{5}{8}$. Thus the slope is $-\frac{1}{5/8} = -\frac{8}{5}$. Hence the equation is $y - \frac{7}{2} = -\frac{8}{5}(x-1)$ $\Leftrightarrow$ $y = -\frac{8}{5}x + \frac{8}{5} + \frac{7}{2} = -\frac{8}{5}x + \frac{51}{10}$. That is, $y = -\frac{8}{5}x + \frac{51}{10}$.

   (f) The center of the circle is the midpoint, $(1, \frac{7}{2})$, and the length of the radius is $\frac{1}{2}\sqrt{89}$. Thus the equation of the circle whose diameter is PQ is $(x-1)^2 + (y - \frac{7}{2})^2 = \left(\frac{1}{2}\sqrt{89}\right)^2$ $\Leftrightarrow$ $(x-1)^2 + (y - \frac{7}{2})^2 = \frac{89}{4}$.

2. (a) $x^2 + y^2 = 25 = 5^2$
   Center: $(0,0)$
   Radius: 5

   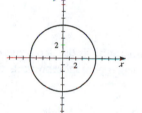

   (b) $(x-2)^2 + (y+1)^2 = 9 = 3^2$
   Center: $(2,-1)$
   Radius: 3

   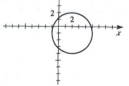

   (c) $x^2 + 6x + y^2 - 2y + 6 = 0$
   $x^2 + 6x + \underline{} + y^2 - 2y + \underline{} = -6$
   $x^2 + 6x + 9 + y^2 - 2y + 1 = -6 + 9 + 1$
   $(x+3)^2 + (y-1)^2 = 4 = 2^2$
   Center: $(-3,1)$
   Radius: 2

3. $2x - 3y = 15$ $\Leftrightarrow$ $-3y = -2x + 15$ $\Leftrightarrow$

    $y = \frac{2}{3}x - 5$.

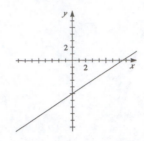

4. (a) $3x + y - 10 = 0$ $\Leftrightarrow$ $y = -3x + 10$, so the slope of the line we seek is -3. Using the
        point-slope, $y - (-6) = -3(x - 3)$ $\Leftrightarrow$ $y + 6 = -3x + 9$ $\Leftrightarrow$ $3x + y - 3 = 0$.

    (b) Using the intercept form we get $\dfrac{x}{6} + \dfrac{y}{4} = 1$ $\Leftrightarrow$ $2x + 3y = 12$ $\Leftrightarrow$ $2x + 3y - 12 = 0$.

5. (a) When $x = 100$ we have $T = 0.08(100) - 4 = 8 - 4 = 4$, so the temperature at one meter is
        4°C.

    (b)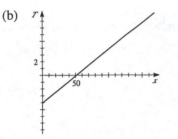

    (c) The slope represents the raise in temperature as the depth increase. The T-intercept is the
        surface temperature of the soil and the x-intercept represents the depth of the "frost line", where
        the soil below is not frozen.

6. (a) We graphed $y = x^3 - 9x - 1$ in the viewing rectangle
        $[-10, 10]$ by $[-10, 10]$. From this we can see there are
        three solutions. Using ZOOM and/or TRACE, we find
        the solutions (to two decimal places) are $x = -2.94$,
        $x = -0.11$, and $x = 3.05$.

    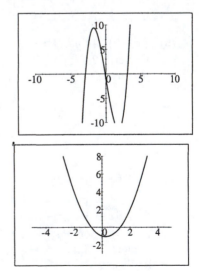

    (b) $\frac{1}{2}x + 2 \geq x^2 + 1$ $\Leftrightarrow$ $0 \geq x^2 - \frac{1}{2}x - 1$. Graphing
        $y = x^2 - \frac{1}{2}x - 1$ and using ZOOM and/or TRACE we
        get the solution is $[-0.78, 1.28]$.

7. (a) $M = k\dfrac{wh^2}{L}$.

(b) Substituting $w = 4$, $h = 6$, $L = 12$, and $M = 4800$, we have $4800 = k\frac{(4)(6^2)}{12}$ $\Leftrightarrow$ $k = 400$. Thus $M = 400\dfrac{wh^2}{L}$.

(c) Now if $L = 10$, $w = 3$, and $h = 10$, then $M = 400\frac{(3)(10^2)}{10} = 12,000$. So the beam can support $12,000$ pounds.

Principles of Modeling

1. (a)

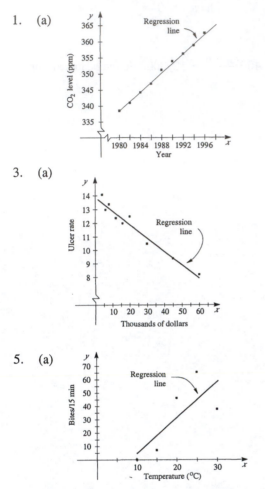

(b) Using a graphing calculator, we obtain the regression line $y = 1.5125x - 2656.4$.

(c) Using $x = 1998$ in the equation $y = 1.5125x - 2656.4$, we get $y \approx 365.6$ ppm CO_2. This value is close to the observed value of 366.7 ppm CO_2.

3. (a)

(b) Using a graphing calculator, we obtain the regression line $y = -0.0995x + 13.9$, where x is measured in thousands of dollars.

(c) Using the regression line equation, $y = -0.0995x + 13.9$, we get an estimated $y = 11.4$ ulcers per 100 population when $x = \$25,000$.

(d) Again using $y = -0.0995x + 13.9$, we get an estimated $y = 5.8$ ulcers per 100 population when $x = \$80,000$.

5. (a)

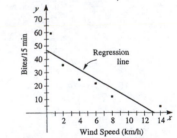

(b) Using a graphing calculator, we obtain the regression line for the temperature-bite data $y = 2.6768x - 22.19$, and for the wind speed-bite data $y = -3.5336x + 46.73$.

(c) The correlation coefficient for the temperature-bite data is $r = 0.77$, so a linear model is not appropriate. The correlation coefficient for the wind-bite data is $r = -0.89$, so a linear model is more suitable in this case.

7. (a)

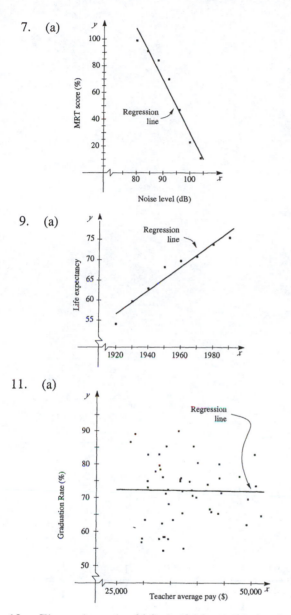

(b) Using a graphing calculator, we obtain
$y = -3.9018x + 419.7$.

(c) The correlation coefficient is $r = -0.98$,
so linear model is appropriate for x
between 80 dB and 104 dB.

(d) Substituting $x = 94$ into the regression
equation, we get
$y = -3.9018(94) + 419.7 \approx 53$. So
the intelligibility is about 53%.

9. (a)

(b) Using a graphing calculator, we obtain
$y = 0.2708x - 462.9$.

(c) We use $x = 2000$ in the equation
$y = 0.2708x - 462.9$, to get $y = 78.2$,
that is, a life expectancy of 78.2 years.

11. (a)

(b) Using a graphing calculator, we obtain
$y = -0.000012x + 72.0$

(c) The correlation coefficient is
$r = -0.00812$, so there is very little
correlation. It appears that teachers' salaries
do not affect graduation rates.

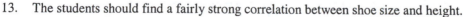

13. The students should find a fairly strong correlation between shoe size and height.

Chapter Three

Exercises 3.1

1. $f(x) = 7x + 2$

3. $f(x) = (x - 4)^2$

5. Divide by 2, then add 7.

7. Square, multiply by 3, then subtract 2.

9. Machine diagram for $f(x) = \sqrt{x}$.

11. $f(x) = 2x^2 + 1$

x	$f(x)$
-1	$2(-1)^2 + 1 = 3$
0	$2(0)^2 + 1 = 1$
1	$2(1)^2 + 1 = 3$
2	$2(2)^2 + 1 = 9$
3	$2(3)^2 + 1 = 19$

13. $f(1) = 2(1) + 1 = 3$; $f(-2) = 2(-2) + 1 = -3$; $f(\frac{1}{2}) = 2(\frac{1}{2}) + 1 = 2$;
    $f(a) = 2(a) + 1 = 2a + 1$; $f(-a) = 2(-a) + 1 = -2a + 1$;
    $f(a + b) = 2(a + b) + 1 = 2a + 2b + 1$.

15. $g(2) = \dfrac{1 - (2)}{1 + (2)} = \dfrac{-1}{3} = -\dfrac{1}{3}$; $g(-2) = \dfrac{1 - (-2)}{1 + (-2)} = \dfrac{3}{-1} = -3$; $g(\frac{1}{2}) = \dfrac{1 - (\frac{1}{2})}{1 + (\frac{1}{2})} = \dfrac{\frac{1}{2}}{\frac{3}{2}} = \dfrac{1}{3}$;
    $g(a) = \dfrac{1 - (a)}{1 + (a)} = \dfrac{1 - a}{1 + a}$; $g(a - 1) = \dfrac{1 - (a - 1)}{1 + (a - 1)} = \dfrac{1 - a + 1}{1 + a - 1} = \dfrac{2 - a}{a}$;
    $g(-1) = \dfrac{1 - (-1)}{1 + (-1)} = \dfrac{2}{0}$, so $g(-1)$ is not defined.

17. $f(0) = 2(0)^2 + 3(0) - 4 = -4$; $f(2) = 2(2)^2 + 3(2) - 4 = 8 + 6 - 4 = 10$;
    $f(-2) = 2(-2)^2 + 3(-2) - 4 = 8 - 6 - 4 = -2$;
    $f(\sqrt{2}) = 2(\sqrt{2})^2 + 3(\sqrt{2}) - 4 = 4 + 3\sqrt{2} - 4 = 3\sqrt{2}$;

$f(x+1) = 2(x+1)^2 + 3(x+1) - 4 = 2x^2 + 4x + 2 + 3x + 3 - 4 = 2x^2 + 7x + 1$;

$f(-x) = 2(-x)^2 + 3(-x) - 4 = 2x^2 - 3x - 4$.

19. $f(-2) = 2|-2 - 1| = 2(3) = 6$; $\quad f(0) = 2|0 - 1| = 2(1) = 2$; $\quad f(\frac{1}{2}) = 2|\frac{1}{2} - 1| = 2(\frac{1}{2}) = 1$;

$f(2) = 2|2 - 1| = 2(1) = 2$; $\quad f(x+1) = 2|(x+1) - 1| = 2|x|$;

$f(x^2 + 2) = 2|(x^2 + 2) - 1| = 2|x^2 + 1| = 2x^2 + 2$ (since $x^2 + 1 > 0$).

21. Since $-2 < 0$, we have $f(-2) = (-2)^2 = 4$. Since $-1 < 0$, we have $f(-1) = (-1)^2 = 1$. Since $0 \geq 0$, we have $f(0) = 0 + 1 = 1$. Since $1 \geq 0$, we have $f(1) = 1 + 1 = 2$. Since $2 \geq 0$, we have $f(2) = 2 + 1 = 3$.

23. Since $-4 \leq -1$, we have $f(-4) = (-4)^2 + 2(-4) = 16 - 8 = 8$. Since $-\frac{3}{2} \leq -1$, we have $f\left(-\frac{3}{2}\right) = \left(-\frac{3}{2}\right)^2 + 2\left(-\frac{3}{2}\right) = \frac{9}{4} - 3 = -\frac{3}{4}$. Since $-1 \leq -1$, we have $f(-1) = (-1)^2 + 2(-1) = 1 - 2 = -1$. Since $0 > -1$, we have $f(0) = 0$. Since $1 > -1$, we have $f(1) = 1$.

25. $f(x+2) = (x+2)^2 + 1 = x^2 + 4x + 4 + 1 = x^2 + 4x + 5$;

$f(x) + f(2) = x^2 + 1 + (2)^2 + 1 = x^2 + 1 + 4 + 1 = x^2 + 6$.

27. $f(x^2) = x^2 + 4$; $\quad [f(x)]^2 = [x + 4]^2 = x^2 + 8x + 16$.

29. $f(a) = 3(a) + 2 = 3a + 2$; $\quad f(a+h) = 3(a+h) + 2 = 3a + 3h + 2$;

$\dfrac{f(a+h) - f(a)}{h} = \dfrac{(3a + 3h + 2) - (3a + 2)}{h} = \dfrac{3a + 3h + 2 - 3a - 2}{h} = \dfrac{3h}{h} = 3$.

31. $f(a) = 5$; $\quad f(a+h) = 5$; $\quad \dfrac{f(a+h) - f(a)}{h} = \dfrac{5 - 5}{h} = 0$.

33. $f(a) = 3 - 5a + 4a^2$; $\quad f(a+h) = 3 - 5(a+h) + 4(a+h)^2$

$= 3 - 5a - 5h + 4(a^2 + 2ah + h^2) = 3 - 5a - 5h + 4a^2 + 8ah + 4h^2$;

$\dfrac{f(a+h) - f(a)}{h} = \dfrac{(3 - 5a - 5h + 4a^2 + 8ah + 4h^2) - (3 - 5a + 4a^2)}{h}$

$= \dfrac{3 - 5a - 5h + 4a^2 + 8ah + 4h^2 - 3 + 5a - 4a^2}{h} = \dfrac{-5h +'8ah + 4h^2}{h} = \dfrac{h(-5 + 8a + 4h)}{h}$

$= -5 + 8a + 4h$.

35. $f(x) = 2x$. Since there is no restrictions, the domain is the set of real numbers, $(-\infty, \infty)$.

37. $f(x) = 2x$. The domain is restricted by the exercise to $[-1, 5]$.

39. $f(x) = \dfrac{1}{x - 3}$. Since the denominator cannot equal 0 we have $x - 3 \neq 0 \quad \Leftrightarrow \quad x \neq 3$. Thus the domain is $\{x| x \neq 3\}$. In interval notation, the domain is $(-\infty, 3) \cup (3, \infty)$.

41. $f(x) = \dfrac{x+2}{x^2-1}$. Since the denominator cannot equal 0 we have $x^2 - 1 \neq 0 \quad \Leftrightarrow \quad x^2 \neq 1 \quad \Rightarrow$ $x \neq \pm 1$. Thus the domain is $\{x \mid x \neq \pm 1\}$. In interval notation, the domain is $(-\infty, -1) \cup (-1, 1) \cup (1, \infty)$.

43. $f(x) = \sqrt{x-5}$. We require $x - 5 \geq 0 \quad \Leftrightarrow \quad x \geq 5$. Thus the domain is $\{x \mid x \geq 5\}$. The domain can also be expressed in interval notation as $[5, \infty)$.

45. $f(t) = \sqrt[3]{t-1}$. Since the odd root is defined for all real numbers, the domain is the set of real numbers, $(-\infty, \infty)$.

47. $h(x) = \sqrt{2x-5}$. Since the square root is defined as a real number only for nonnegative numbers, we require that $2x - 5 \geq 0 \quad \Leftrightarrow \quad 2x \geq 5 \quad \Leftrightarrow \quad x \geq \frac{5}{2}$. So the domain is $\{x \mid x \geq \frac{5}{2}\}$. In interval notation, the domain is $[\frac{5}{2}, \infty)$.

49. $g(x) = \dfrac{\sqrt{2+x}}{3-x}$. We require $2 + x \geq 0$, and the denominator cannot equal 0. Now $2 + x \geq 0$ $\Leftrightarrow \quad x \geq -2$, and $3 - x \neq 0 \quad \Leftrightarrow \quad x \neq 3$. Thus the domain is $\{x \mid x \geq -2 \text{ and } x \neq 3\}$, which can be expressed in interval notation as $[-2, 3) \cup (3, \infty)$.

51. $g(x) = \sqrt[4]{x^2 - 6x}$. Since the input to an even root must be nonnegative, we have $x^2 - 6x \geq 0$ $\Leftrightarrow \quad x(x - 6) \geq 0$. Using the method from Chapter 1, we have

	$(-\infty, 0)$	$(0, 6)$	$(6, \infty)$
Sign of x	$-$	$+$	$+$
Sign of $x - 6$	$-$	$-$	$+$
Sign of $x(x-6)$	$+$	$-$	$+$

Thus the domain is $(-\infty, 0] \cup [6, \infty)$.

53. $f(x) = \dfrac{3}{\sqrt{x-4}}$. Since the input to an even root must be nonnegative and the denominator cannot equal 0, we have $x - 4 > 0 \quad \Leftrightarrow \quad x > 4$. Thus the domain is $(4, \infty)$.

55. $f(x) = \dfrac{(x+1)^2}{\sqrt{2x-1}}$. Since the input to an even root must be nonnegative and the denominator cannot equal 0, we have $2x - 1 > 0 \quad \Leftrightarrow \quad x > \frac{1}{2}$. Thus the domain is $(\frac{1}{2}, \infty)$.

57. (a) $C(10) = 1500 + 3(10) + 0.02(10)^2 + 0.0001(10)^3 = 1500 + 30 + 2 + 0.1 = 1532.1$
       $C(100) = 1500 + 3(100) + 0.02(100)^2 + 0.0001(100)^3 = 1500 + 300 + 200 + 100 = 2100$

    (b) $C(10)$ represents the cost of producing 10 yards of fabric and $C(100)$ represents the cost of producing 100 yards of fabric.

    (c) $C(0) = 1500 + 3(0) + 0.02(0)^2 + 0.0001(0)^3 = 1500$

59. (a) $D(0.1) = \sqrt{2(3960)(0.1) + (0.1)^2} = \sqrt{792.01} \approx 28.1$ miles
       $D(0.2) = \sqrt{2(3960)(0.2) + (0.2)^2} = \sqrt{1584.04} \approx 39.8$ miles

(b) 1135 feet $= \frac{1135}{5280}$ miles ≈ 0.215 miles.

   $D(0.215) = \sqrt{2(3960)(0.215) + (0.215)^2} = \sqrt{1702.846} \approx 41.3$ miles

(c) $D(7) = \sqrt{2(3960)(7) + (7)^2} = \sqrt{55489} \approx 235.6$ miles

61. (a) $v(0.1) = 18500(0.25 - 0.1^2) = 4440$, $v(0.4) = 18500(0.25 - 0.4^2) = 1665$.

(b) That it flows much fast (about 2.75 times faster) 0.1 cm from the center than 0.1 cm from the edge.

(c)

r	$v(r)$
0	4625
0.1	4440
0.2	3885
0.3	2960
0.4	1665
0.5	0

63. (a) $L(0.5c) = 10\sqrt{1 - \dfrac{(0.5c)^2}{c^2}} \approx 8.66$ m, $L(0.75c) = 10\sqrt{1 - \dfrac{(0.75c)^2}{c^2}} \approx 6.61$ m, and

   $L(0.9c) = 10\sqrt{1 - \dfrac{(0.9c)^2}{c^2}} \approx 4.36$ m.

(b) It will appear to get shorter.

65. (a) $C(75) = 75 + 15 = \$90$; $C(90) = 90 + 15 = \$105$; $C(100) = \$100$; and $C(105) = \$105$.

(b) The total price of the books purchased, including shipping.

67. (a) $F(x) = \begin{cases} 15(40 - x) & \text{if } 0 < x < 40 \\ 0 & \text{if } 40 \le x \le 65. \\ 15(x - 65) & \text{if } x > 65 \end{cases}$

(b) $F(30) = 15(40 - 10) = 15 \cdot 10 = \$150$; $F(50) = \$0$; and

   $F(75) = 15(75 - 65)15 \cdot 10 = \$150$.

(c) The fines for violating the speed limits on the freeway.

69. We assume the grass grows linearly.

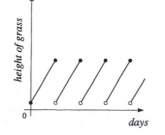

71.

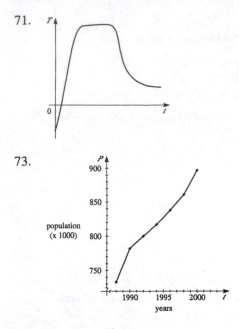

73.

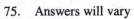

75.   Answers will vary

Exercises 3.2

1.

x	$f(x) = 2$
-9	2
-6	2
-3	2
0	2
3	2
6	2

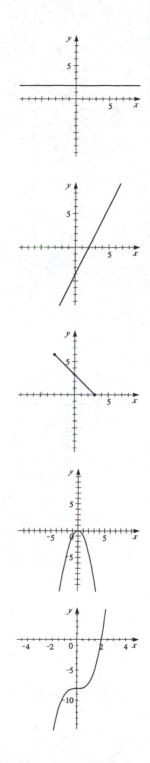

3.

x	$f(x) = 2x - 4$
-1	-6
0	-4
1	-2
2	0
3	2
4	4
5	6

5.

x	$f(x) = -x + 3,\ -3 \le x \le 3$
-3	6
-2	5
0	3
1	2
2	1
3	0

7.

x	$f(x) = -x^2$
± 4	-16
± 3	-9
± 2	-4
± 1	-1
0	0

9.

x	$g(x) = x^3 - 8$
-2	-16
-1	-9
0	-8
1	-7
2	0
3	19

11.

x	$g(x) = \sqrt{x+4}$
-4	0
-3	1
-2	1.414
-1	1.732
0	2
1	2.236

13.

x	$F(x) = \frac{1}{x}$
-5	-0.2
-1	-1
-0.1	-10
0.1	10
1	1
5	0.2

15.

x	$H(x) = |2x|$
± 5	10
± 4	8
± 3	6
± 2	4
± 1	2
0	0

17.

x	$G(x) = |x| + x$
-5	0
-2	0
0	0
1	2
2	4
5	10

19.

x	$f(x) = |2x - 2|$
-5	12
-2	8
0	2
1	0
2	2
5	8

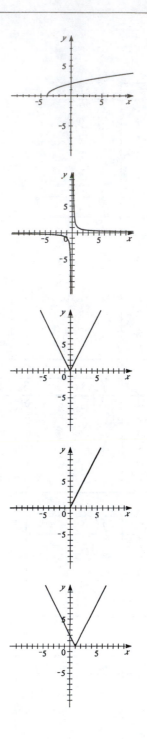

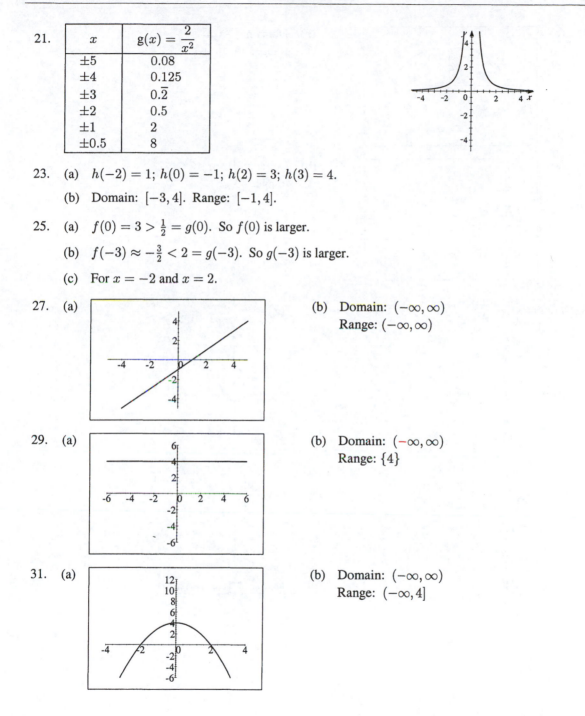

21.

x	$g(x) = \dfrac{2}{x^2}$
± 5	0.08
± 4	0.125
± 3	$0.\overline{2}$
± 2	0.5
± 1	2
± 0.5	8

23. (a) $h(-2) = 1$; $h(0) = -1$; $h(2) = 3$; $h(3) = 4$.

   (b) Domain: $[-3, 4]$. Range: $[-1, 4]$.

25. (a) $f(0) = 3 > \frac{1}{2} = g(0)$. So $f(0)$ is larger.

   (b) $f(-3) \approx -\frac{3}{2} < 2 = g(-3)$. So $g(-3)$ is larger.

   (c) For $x = -2$ and $x = 2$.

27. (a)

   (b) Domain: $(-\infty, \infty)$
       Range: $(-\infty, \infty)$

29. (a)

   (b) Domain: $(-\infty, \infty)$
       Range: $\{4\}$

31. (a)

   (b) Domain: $(-\infty, \infty)$
       Range: $(-\infty, 4]$

33. (a)
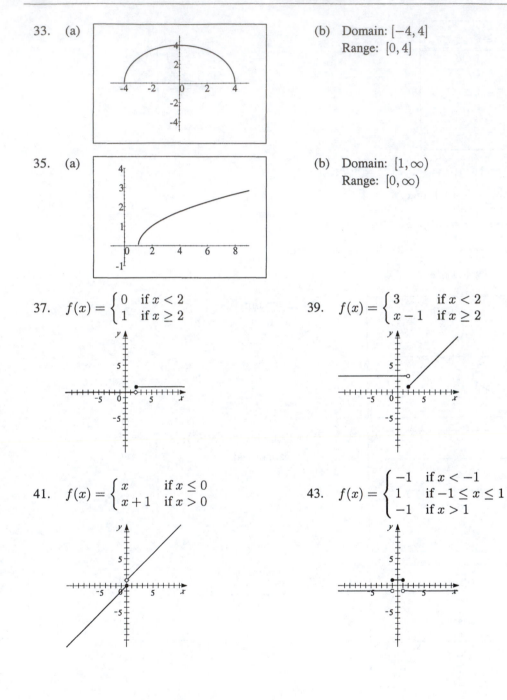

(b) Domain: $[-4, 4]$
Range: $[0, 4]$

35. (a)

(b) Domain: $[1, \infty)$
Range: $[0, \infty)$

37. $f(x) = \begin{cases} 0 & \text{if } x < 2 \\ 1 & \text{if } x \geq 2 \end{cases}$

39. $f(x) = \begin{cases} 3 & \text{if } x < 2 \\ x - 1 & \text{if } x \geq 2 \end{cases}$

41. $f(x) = \begin{cases} x & \text{if } x \leq 0 \\ x + 1 & \text{if } x > 0 \end{cases}$

43. $f(x) = \begin{cases} -1 & \text{if } x < -1 \\ 1 & \text{if } -1 \leq x \leq 1 \\ -1 & \text{if } x > 1 \end{cases}$

45. $f(x) = \begin{cases} 2 & \text{if } x \leq -1 \\ x^2 & \text{if } x > -1 \end{cases}$

47. $f(x) = \begin{cases} 0 & \text{if } |x| \leq 2 \\ 3 & \text{if } |x| > 2 \end{cases}$

49. $f(x) = \begin{cases} 4 & \text{if } x < -2 \\ x^2 & \text{if } -2 \leq x \leq 2 \\ -x+6 & \text{if } x > 2 \end{cases}$

51. $f(x) = \begin{cases} x+2 & \text{if } x \leq -1 \\ x^2 & \text{if } x > -1 \end{cases}$

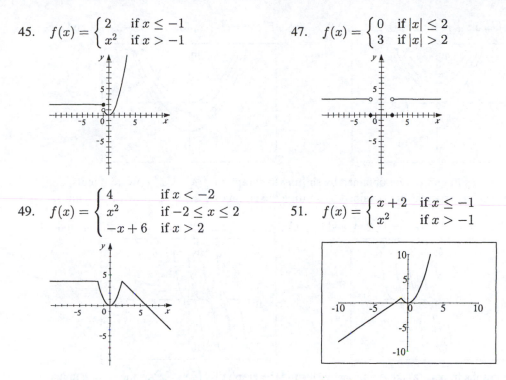

53. The curves in parts (a) and (c) are graphs of a function of x, by the vertical line test.

55. The given curve is the graph of a function of x. Domain: $[-3, 2]$. Range: $[-2, 2]$.

57. No, the given curve is not the graph of a function of x, by the vertical line test.

59. Solving for y in terms of x gives: $x^2 + 2y = 4 \quad \Leftrightarrow \quad 2y = 4 - x^2 \quad \Leftrightarrow \quad y = 2 - \frac{1}{2}x^2$. This defines y as a function of x.

61. Solving for y in terms of x gives: $x = y^2 \quad \Leftrightarrow \quad y = \pm\sqrt{x}$. The last equation gives two values of y for a given value of x. Thus, this equation does not define y as a function of x.

63. Solving for y in terms of x gives: $x + y^2 = 9 \quad \Leftrightarrow \quad y^2 = 9 - x \quad \Leftrightarrow \quad y = \pm\sqrt{9 - x}$. The last equation gives two values of y for a given value of x. Thus, this equation does not define y as a function of x.

65. Solving for y in terms of x gives: $x^2y + y = 1 \quad \Leftrightarrow \quad y(x^2 + 1) = 1 \quad \Leftrightarrow \quad y = \dfrac{1}{x^2 + 1}$. This defines y as a function of x.

67. Solving for y in terms of x gives: $2|x| + y = 0 \quad \Leftrightarrow \quad y = -2|x|$. This defines y as a function of x.

69. Solving for y in terms of x gives: $x = y^3 \quad \Leftrightarrow \quad y = \sqrt[3]{x}$. This defines y as a function of x.

71. (a) $f(x) = x^2 + c$, for $c = 0, 2, 4$, and 6.   (b) $f(x) = x^2 + c$, for $c = 0, -2, -4$, and -6.

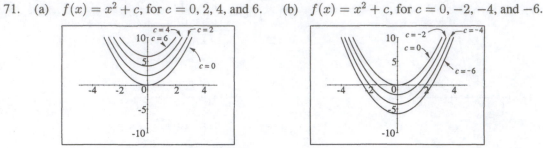

(c) The graphs in part (a) are obtained by shifting the graph of $f(x) = x^2$ upward c units, $c > 0$.
     The graphs in part (b) are obtained by shifting the graph of $f(x) = x^2$ downward c units.

73. (a) $f(x) = (x - c)^3$, for $c = 0, 2, 4$, and 6. (b) $f(x) = (x - c)^3$, for $c = 0, -2, -4$, and -6.

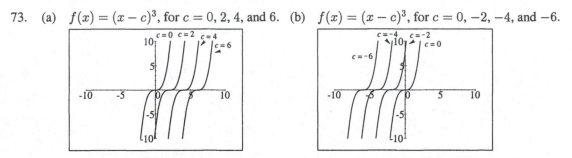

(c) The graphs in part (a) are obtained by shifting the graph of $f(x) = x^3$ to the right c units,
     $c > 0$. The graphs in part (b) are obtained by shifting the graph of $f(x) = x^3$ to the left $|c|$
     units, $c < 0$.

75. (a) $f(x) = x^c$, for $c = \frac{1}{2}, \frac{1}{4}$, and $\frac{1}{6}$.   (b) $f(x) = x^c$, for $c = 1, \frac{1}{3}$, and $\frac{1}{5}$.

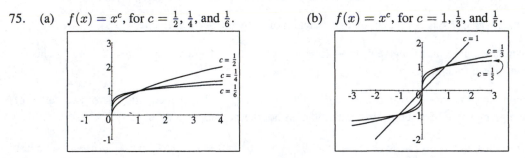

(c) Graphs of even roots are similar to $y = \sqrt{x}$, graphs of odd roots are similar to $y = \sqrt[3]{x}$. As c
     increases, the graph of $y = \sqrt[c]{x}$ becomes steeper near $x = 0$ and flatter when $x > 1$.

77. The slope of the line segment joining the points $(-2, 1)$ and $(4, -6)$ is $m = \frac{-6-1}{4-(-2)} = -\frac{7}{6}$. Using
     the point-slope form, we have $y - 1 = -\frac{7}{6}(x + 2)$ $\Leftrightarrow$ $y = -\frac{7}{6}x - \frac{7}{3} + 1$ $\Leftrightarrow$ $y = -\frac{7}{6}x - \frac{4}{3}$.
     Thus the function is $f(x) = -\frac{7}{6}x - \frac{4}{3}$ for $-2 \le x \le 4$.

79. First solve the circle for y: $x^2 + y^2 = 9$ $\Leftrightarrow$ $y^2 = 9 - x^2$ $\Rightarrow$ $y = \pm\sqrt{9 - x^2}$. Since we
     seek the top half of the circle, we choose $y = \sqrt{9 - x^2}$. So the function is $f(x) = \sqrt{9 - x^2}$,
     $-3 < x < 3$.

81. This person appears to gain weight steadily until the age of 21 when this person's weight gain slows
     down. At age 30, this person experiences a sudden weight loss, but recovers, and by about age 40,
     this person's weight seems to level off at around 200 pounds. It then appears that the person dies at
     about age 68. The sudden weight loss could be due to a number of reasons, among them major
     illness, weight loss program, etc.

83. A won the race. All runners finished the race. Runner B fell, but got up and finished the race.

85. (a) The first noticeable movements occurred at time $t = 5$ seconds.

     (b) It seemed to end at time $t = 30$ seconds.

     (c) Maximum intensity was reached at $t = 17$ seconds.

87. $C(x) = \begin{cases} 2.00 & \text{if } 0 < x \le 1 \\ 2.20 & \text{if } 1 < x \le 1.1 \\ 2.40 & \text{if } 1.1 < x \le 1.2 \\ \vdots \\ 4.00 & \text{if } 1.9 < x < 2 \end{cases}$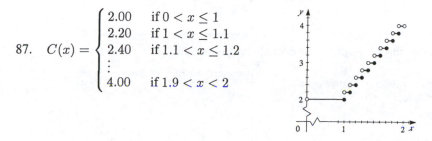

89. The graph of $x = y^2$ is not the graph of a function because both $(1, 1)$ and $(-1, 1)$ satisfy the
     equation $x = y^2$. The graph of $x = y^3$ is the graph of a function because $x = y^3 \Leftrightarrow x^{1/3} = y$.
     If n is even, then both $(1, 1)$ and $(-1, 1)$ satisfies the equation $x = y^n$, so the graph of $x = y^n$ is not
     the graph of a function. When n is odd, $y = x^{1/n}$ is defined for all real numbers, and since $y = x^{1/n}$
     $\Leftrightarrow x = y^n$, the graph of $x = y^n$ is the graph of a function.

91. $f(x) = [\![x]\!]$ $g(x) = [\![2x]\!]$ $h(x) = [\![3x]\!]$

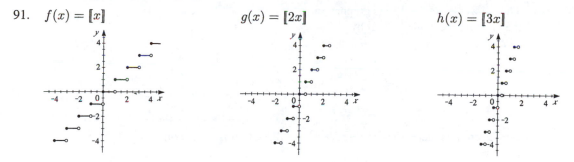

The graph of $k(x) = [\![nx]\!]$ is a step function whose steps are each $\dfrac{1}{n}$ wide.

Exercises 3.3

1. The function is increasing on $[-1, 1]$ and $[2, 4]$. It is decreasing on $[1, 2]$.

3. The function is increasing on $[-2, -1]$ and $[1, 2]$. It is decreasing on $[-3, -2]$, $[-1, 1]$, and $[2, 3]$.

5. (a) $f(x) = x^{2/5}$ is graphed in the viewing rectangle $[-10, 10]$ by $[-5, 5]$.

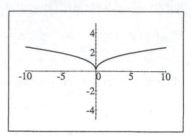

    (b) The function is increasing on $[0, \infty)$. It is decreasing on $(-\infty, 0]$.

7. (a) $f(x) = x^2 - 5x$ is graphed in the viewing rectangle $[-2, 7]$ by $[-10, 10]$.

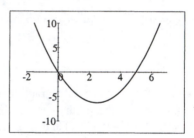

    (b) The function is increasing on $[2.5, \infty)$. It is decreasing on $(-\infty, 2.5]$.

9. (a) $f(x) = 2x^3 - 3x^2 - 12x$ is graphed in the viewing rectangle $[-3, 5]$ by $[-25, 20]$.

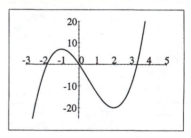

    (b) The function is increasing on $(-\infty, -1]$ and $[2, \infty)$. It is decreasing on $[-1, 2]$.

11. (a) $f(x) = x^3 + 2x^2 - x - 2$ is graphed in the viewing rectangle $[-5, 5]$ by $[-3, 3]$.

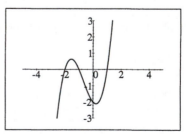

    (b) The function is increasing on $(-\infty, -1.55]$ and $[0.22, \infty)$. It is decreasing on $[-1.55, 0.22]$.

13. We use the points $(1, 3)$ and $(4, 5)$, so the average rate of change $= \frac{5-3}{4-1} = \frac{2}{3}$.

15. We use the points $(0, 6)$ and $(5, 2)$, so the average rate of change $= \frac{2-6}{5-0} = \frac{-4}{5}$.

17. Average rate of change $= \dfrac{f(3) - f(2)}{3 - 2} = \dfrac{[3(3) - 2] - [3(2) - 2]}{1} = 7 - 4 = 3$.

19. Average rate of change $= \dfrac{h(4) - h(-1)}{4 - (-1)} = \dfrac{[4^2 + 2(4)] - [(-1)^2 + 2(-1)]}{5} = \dfrac{24 - (-1)}{5} = 5$.

21. Average rate of change $= \dfrac{f(10) - f(0)}{10 - 0} = \dfrac{[10^3 - 4(10^2)] - [0^3 - 4(0^2)]}{10 - 0} = \dfrac{600 - 0}{10} = 60$.

23. Average rate of change $= \dfrac{f(2 + h) - f(2)}{(2 + h) - 2} = \dfrac{[3(2 + h)^2] - [3(2^2)]}{h} = \dfrac{12 + 12h + 3h^2 - 12}{h}$

$= \dfrac{12h + 3h^2}{h} = \dfrac{h(12 + 3h)}{h} = 12 + 3h$.

25. Average rate of change $= \dfrac{g(1) - g(a)}{1 - a} = \dfrac{\frac{1}{1} - \frac{1}{a}}{1 - a} \cdot \dfrac{a}{a} = \dfrac{a - 1}{a(1 - a)} = \dfrac{-1(1 - a)}{a(1 - a)} = \dfrac{-1}{a}$.

27. Average rate of change $= \dfrac{f(a + h) - f(a)}{(a + h) - a} = \dfrac{\frac{2}{a + h} - \frac{2}{a}}{h} \cdot \dfrac{a(a + h)}{a(a + h)} = \dfrac{2a - 2(a + h)}{ah(a + h)}$

$= \dfrac{-2h}{ah(a + h)} = \dfrac{-2}{a(a + h)}$.

29. (a) Average rate of change $= \dfrac{f(a + h) - f(a)}{(a + h) - a} = \dfrac{\left[\frac{1}{2}(a + h) + 3\right] - \left[\frac{1}{2}a + 3\right]}{h}$

$= \dfrac{\frac{1}{2}a + \frac{1}{2}h + 3 - \frac{1}{2}a - 3}{h} = \dfrac{\frac{1}{2}h}{h} = \dfrac{1}{2}$.

(b) The slope of the line $f(x) = \frac{1}{2}x + 3$ is $\frac{1}{2}$, which is also the average rate of change.

31. (a) The function P is increasing on $[0, 150]$ and $[300, 365]$ and decreasing on $[150, 300]$.

(b) Average rate of change $= \dfrac{W(200) - W(100)}{200 - 100} = \dfrac{50 - 75}{200 - 100} = \dfrac{-25}{100} = -\dfrac{1}{4}$ ft/day.

33. (a) Average rate of change of population $= \frac{1{,}591 - 856}{1994 - 1991} = \frac{735}{3} = 245$ persons/yr.

(b) Average rate of change of population $= \frac{826 - 1{,}483}{1997 - 1995} = \frac{-657}{2} = -328.5$ persons/yr.

(c) The population was increasing from 1990 to 1994.

(d) The population was decreasing from 1994 to 1999.

35. (a) Average rate of change of sales $= \frac{584 - 512}{1999 - 1989} = \frac{72}{10} = 7.2$ units/yr.

(b) Average rate of change of sales $= \frac{520 - 512}{1990 - 1989} = \frac{8}{1} = 8$ units/yr.

(c) Average rate of change of sales $= \frac{410 - 520}{1992 - 1990} = \frac{-110}{2} = -55$ units/yr.

(d)

Year	CD players sold	Change in sales from previous year
1989	512	−
1990	520	8
1991	413	−107
1992	410	−3
1993	468	58
1994	510	42
1995	590	80
1996	607	17
1997	732	125
1998	612	−120
1999	584	−28

Sales *increased* most quickly between 1996 and 1997. Sales *decreased* most quickly between 1997 and 1998.

35. (a)

Time (s)	Average speed (ft/s)
0-2	17
2-4	18
4-6	63
6-8	147
8-10	237

The average speed is increasing, so the car is accelerating. From the shape of the graph, we see that the slope continues to get steeper.

(b)

Time (s)	Average speed (ft/s)
30-32	263
32-34	144
34-36	91
36-38	74
38-40	48

The average speed is decreasing, so the car is decelerating. From the shape of the graph, we see that the slope continues to get flatter.

37. (a) For all three runners, average rate of change $= \dfrac{d(10) - d(0)}{10 - 0} = \dfrac{100}{10} = 10.$

(b) Runner A gets a great jump out of the blocks but tires at the end of the race. Runner B runs a steady race. Runner C is slow at the beginning but accelerates down the track.

39. (a) $f(x)$ is always increasing, and
          $f(x) > 0$ for all x.

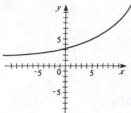

     (b) $f(x)$ is always decreasing, and
          $f(x) > 0$ for all x.

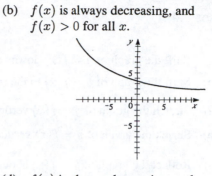

     (c) $f(x)$ is always increasing, and
          $f(x) < 0$ for all x.

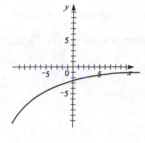

     (d) $f(x)$ is always decreasing, and
          $f(x) < 0$ for all x.

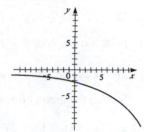

Exercises 3.4

1.   (a)   Shift the graph of $y = f(x)$ downward 3 units.

     (b)   Shift the graph of $y = f(x)$ to the right 3 units.

3.   (a)   Stretch the graph of $y = f(x)$ vertically by a factor of 3.

     (b)   Shrink the graph of $y = f(x)$ vertically by a factor of $\frac{1}{3}$.

5.   (a)   Reflect the graph of $y = f(x)$ about the x-axis, and stretch vertically by a factor of 2.

     (b)   Reflect the graph of $y = f(x)$ about the x-axis, and shrink vertically by a factor of $\frac{1}{2}$.

7.   (a)   Shift the graph of $y = f(x)$ to the right 4 units, and then upward $\frac{3}{4}$ of a unit.

     (b)   Shift the graph of $y = f(x)$ to the left 4 units, and then downward $\frac{3}{4}$ of a unit.

9.   (a)   Shrink the graph of $y = f(x)$ horizontally by a factor of $\frac{1}{4}$.

     (b)   Stretch the graph of $y = f(x)$ horizontally by a factor of 4.

11. $g(x) = f(x-2) = (x-2)^2 = x^2 - 4x + 4.$

13. $g(x) = f(x+1) + 2 = |x+1| + 2.$

15. $g(x) = -f(x+2) = -\sqrt{x+2}.$

17.   (a) $y = f(x-4)$ is graph #3.   (b) $y = f(x) + 3$ is graph #1.

     (c) $y = 2f(x+6)$ is graph #2.

19.   (a) $y = f(x-2)$ (b) $y = f(x) - 2$ (c) $y = 2f(x)$

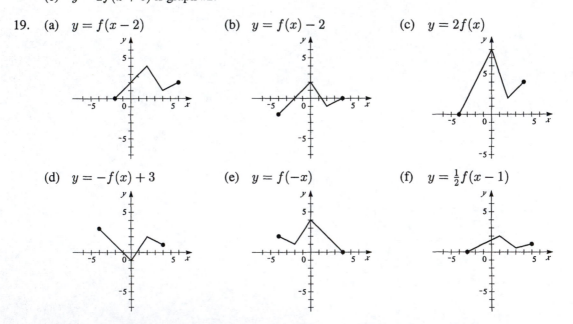

     (d) $y = -f(x) + 3$ (e) $y = f(-x)$ (f) $y = \frac{1}{2}f(x-1)$

21. (a) $f(x) = \dfrac{1}{x}$

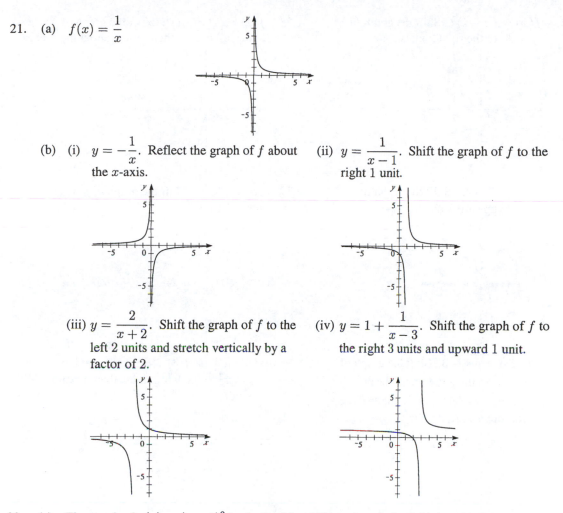

(b) (i) $y = -\dfrac{1}{x}$. Reflect the graph of f about the x-axis.

(ii) $y = \dfrac{1}{x-1}$. Shift the graph of f to the right 1 unit.

(iii) $y = \dfrac{2}{x+2}$. Shift the graph of f to the left 2 units and stretch vertically by a factor of 2.

(iv) $y = 1 + \dfrac{1}{x-3}$. Shift the graph of f to the right 3 units and upward 1 unit.

23. (a) The graph of $g(x) = (x+2)^2$ is obtained by shifting the graph of $f(x)$ to the left 2 units.

(b) The graph of $g(x) = x^2 + 2$ is obtained by shifting the graph of $f(x)$ upward 2 units.

25. (a) The graph of $g(x) = 2\sqrt{x}$ is obtained by stretching the graph of $f(x)$ vertically by a factor of 2.

(b) The graph of $g(x) = \frac{1}{2}\sqrt{x-2}$ is obtained by shifting the graph of $f(x)$ to the right 2 units, and then shrinking the graph vertically by a factor of $\frac{1}{2}$.

27. $y = f(x-2) + 3$. When $f(x) = x^2$, we get $y = (x-2)^2 + 3 = x^2 - 4x + 4 + 3 = x^2 - 4x + 7$.

29. $y = -5f(x+3)$. When $f(x) = \sqrt{x}$, we get $y = -5\sqrt{x+3}$

31. $y = 0.1f\left(x - \frac{1}{2}\right) - 2$. When $f(x) = |x|$, we get $y = 0.1\left|x - \frac{1}{2}\right| - 2$

33. $f(x) = (x - 2)^2$. Shift the graph of
     $y = x^2$ to the right 2 units.

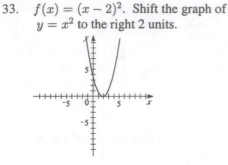

35. $f(x) = -(x + 1)^2$. Shift the graph of $y = x^2$
     to the left 1 unit, then reflect about the x-axis.

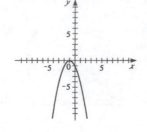

37. $f(x) = x^3 + 2$. Shift the graph of
     $y = x^3$ upward 2 units.

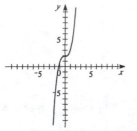

39. $y = 1 + \sqrt{x}$. Shift the graph of $y = \sqrt{x}$
     upward 1 unit.

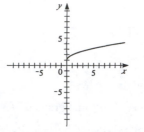

41. $y = \frac{1}{2}\sqrt{x + 4} - 3$. Shift the graph of
     $y = \sqrt{x}$ to the left 4 units, shrink
     vertically by a factor of $\frac{1}{2}$, and then
     shift it downward 3 units.

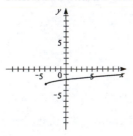

43. $y = 5 + (x + 3)^2$. Shift the graph of
     $y = x^2$ to the left 3 units, then upward
     5 units.

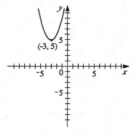

45. $y = |x| - 1$. Shift the graph of $y = |x|$
     downward 1 unit.

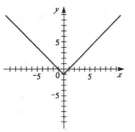

47. $y = |x + 2| + 2$. Shift the graph of $y = |x|$
     to the left 2 units and upward 2 units.

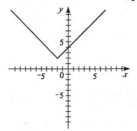

49.

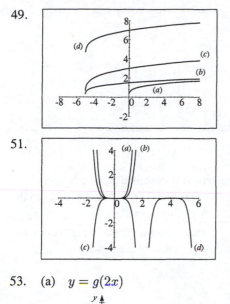

For part (b), shift the graph in (a) to the left 5 units; for part (c), shift the graph in (a) to the left 5 units, and stretch it vertically by a factor of 2; for part (d), shift the graph in (a) to the left 5 units, stretch it vertically by a factor of 2, and then shift it upward 4 units.

51.

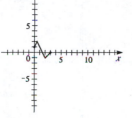

For part (b), shrink the graph in (a) vertically by a factor of $\frac{1}{3}$; for part (c), shrink the graph in (a) vertically by a factor of $\frac{1}{3}$, and reflect it about the x-axis; for part (d), shift the graph in (a) to the right 4 units, shrink vertically by a factor of $\frac{1}{3}$, and then reflect it about the x-axis.

53. (a) $y = g(2x)$ (b) $y = g(\frac{1}{2}x)$

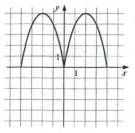   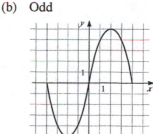

55. (a) Even   (b) Odd

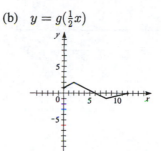

57. $y = [\![2x]\!]$

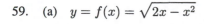

59. (a) $y = f(x) = \sqrt{2x - x^2}$ (b) $y = f(2x) = \sqrt{2(2x) - (2x)^2} = \sqrt{4x - 4x^2}$

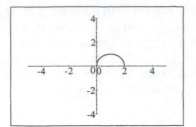

(c) $y = f(\frac{1}{2}x) = \sqrt{2\left(\frac{1}{2}x\right) - \left(\frac{1}{2}x\right)^2} = \sqrt{x - \frac{1}{4}x^2}$

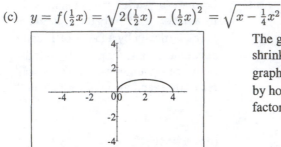

The graph in part (b) is obtained by horizontally shrinking the graph in part (a) by a factor of 2 (so the graph is $\frac{1}{2}$ as wide). The graph in part (c) is obtain by horizontally stretching the graph in part (a) by a factor of 2 (so the graph is twice as wide).

61. $f(-x) = (-x)^{-2} = x^{-2} = f(x)$. Thus $f(x)$ is even.

63. $f(-x) = (-x)^2 + (-x) = x^2 - x$. Thus $f(-x) \neq f(x)$. Also, $f(-x) \neq -f(x)$, so $f(x)$ is neither odd nor even.

65. $f(-x) = (-x)^3 - (-x) = -x^3 + x$
      $= -(x^3 - x) = -f(x)$. Thus $f(x)$ is odd.

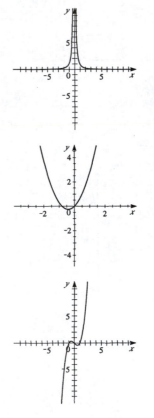

67. $f(-x) = 1 - \sqrt[3]{(-x)} = 1 + \sqrt[3]{x}$. Thus
     $f(-x) \neq f(x)$. Also $f(-x) \neq -f(x)$, so
     $f(x)$ is neither odd nor even.

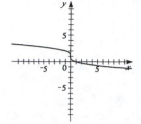

69. Since $f(x) = x^2 - 4 < 0$, for $-2 < x < 2$, the graph of $y = g(x)$ is found by sketching the graph of
     $y = f(x)$ for $x \leq -2$ and $x \geq 2$, then reflecting about the x-axis the part of the graph of $y = f(x)$
     for $-2 < x < 2$.

71. (a) $f(x) = 4x - x^2$ (b) $f(x) = |4x - x^2|$

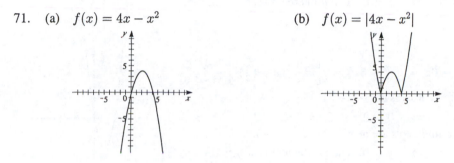

73. (a) The function $y = t^2$ must be shrunk vertically by a factor of 0.01 and shifted vertically 4 units
          up to obtain the function $y = f(t)$.

     (b) The function $y = f(t)$ must be shifted horizontally 10 units to the right to obtain the function
          $y = g(t)$. So $g(t) = f(t - 10) = 4 + 0.01(t - 10)^2 = 5 - 0.02t + 0.01t^2$.

75. f even implies $f(-x) = f(x)$; g even implies $g(-x) = g(x)$; f odd implies $f(-x) = -f(x)$; and
     g odd implies $g(-x) = -g(x)$

     If f and g are both even, then $(f + g)(-x) = f(-x) + g(-x) = f(x) + g(x) = (f + g)(x)$ and
     $f + g$ is even.

     If f and g are both odd, then $(f + g)(-x) = f(-x) + g(-x) = -f(x) - g(x) = -(f + g)(x)$ and
     $f + g$ is odd.

     If f odd and g even, then $(f + g)(-x) = f(-x) + g(-x) = -f(x) + g(x)$, which is neither odd
     nor even.

77. $f(x) = x^n$ is even when n is an even integer and $f(x) = x^n$ is odd when n is an odd integer.

     These names were chosen because polynomials with only terms with odd powers are odd functions,
     and polynomials with only terms with even powers are even functions.

Exercises 3.5

1.   (a)   Vertex: $(3, 4)$ (b)   Maximum value of f: 4.

3.   (a)   Vertex: $(1, -3)$ (b)   Minimum value of f: -3.

5. $y = x^2 - 6x$
     Vertex: $y = x^2 - 6x = x^2 - 6x + 9 - 9 = (x - 3)^2 - 9$.
     So the vertex is at $(3, -9)$.
     x-intercepts: $y = 0$ $\Rightarrow$ $0 = x^2 - 6x = x(x - 6)$. So $x = 0$ or
     $x = 6$. The x-intercepts are at $x = 0$ and $x = 6$.
     y-intercepts: $x = 0$ $\Rightarrow$ $y = 0$. The y-intercept is at $y = 0$.

7. $y = 2x^2 + 6x$
     Vertex: $y = 2x^2 - 6x = 2(x^2 + 3x) = 2\left[x^2 + 3x + \left(\frac{3}{2}\right)^2\right] - \frac{9}{2}$
     $= 2\left(x + \frac{3}{2}\right)^2 - \frac{9}{2}$. Vertex is at $\left(-\frac{3}{2}, -\frac{9}{2}\right)$.
     x-intercepts: $y = 0$ $\Rightarrow$ $0 = 2x^2 + 6x = 2x(x + 3)$ $\Rightarrow$ $x = 0$
     or $x = -3$. The x-intercepts are at $x = 0$ and $x = -3$.
     y-intercepts: $x = 0$ $\Rightarrow$ $y = 0$. The y-intercept is at $y = 0$.

9. $y = x^2 + 4x + 3$
     Vertex: $y = x^2 + 4x + 3 = x^2 + 4x + 4 - 4 + 3 = (x + 2)^2 - 1$.
     Vertex is at $(-2, -1)$.
     x-intercepts: $y = 0$ $\Rightarrow$ $0 = x^2 + 4x + 3 = (x + 1)(x + 3)$.
     So $x = -1$ or $x = -3$. The x-intercepts are at $x = -1$ and $x = -3$.
     y-intercepts: $x = 0$ $\Rightarrow$ $y = 3$. The y-intercept is at $y = 3$.

11. $y = -x^2 + 6x + 4$
     Vertex: $y = -x^2 + 6x + 4 = -(x^2 - 6x) + 4$
     $= -(x^2 - 6x + 9) + 4 + 9 = -(x - 3)^2 + 13$. Vertex is at $(3, 13)$.
     x-intercepts: $y = 0$ $\Rightarrow$ $0 = -(x - 3)^2 + 13$ $\Leftrightarrow$ $(x - 3)^2 = 13$
     $\Rightarrow$ $x - 3 = \pm\sqrt{13}$ $\Leftrightarrow$ $x = 3 \pm \sqrt{13}$.
     The x-intercepts are at $x = 3 - \sqrt{13}$ and $x = 3 + \sqrt{13}$
     y-intercepts: $x = 0$ $\Rightarrow$ $y = 4$. The y-intercept is at $y = 4$.

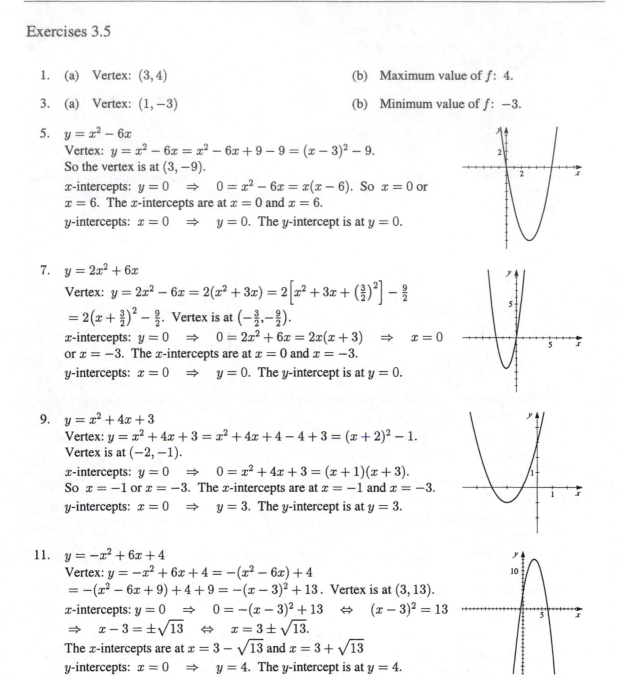

13. $y = 2x^2 + 4x + 3$

Vertex: $y = 2x^2 + 4x + 3 = 2(x^2 + 2x) + 3$
$= 2(x^2 + 2x + 1) + 3 - 2 = 2(x + 1)^2 + 1$. Vertex is at $(-1, 1)$.

x-intercepts: $y = 0 \quad \Rightarrow \quad 0 = 2x^2 + 4x + 3 = 2(x + 1)^2 + 1 \quad \Leftrightarrow$
$2(x + 1)^2 = -1$. Since this last equation has no real solution, there are
no x-intercepts.

y-intercepts: $x = 0 \quad \Rightarrow \quad y = 3$. The y-intercept is at $y = 3$.

15. $y = 2x^2 - 20x + 57$

Vertex: $y = 2x^2 - 20x + 57 = 2(x^2 - 10x) + 57$
$= 2(x^2 - 10x + 25) + 57 - 50 = 2(x - 5)^2 + 7$. Vertex is at $(5, 7)$.

x-intercepts: $y = 0 \quad \Rightarrow \quad 0 = 2x^2 - 20x + 57 = 2(x - 5)^2 + 7$
$\Leftrightarrow \quad 2(x - 5)^2 = -7$. Since this last equation has no real solution,
there are no x-intercepts.

y-intercepts: $x = 0 \quad \Rightarrow \quad y = 57$. The y-intercept is at $y = 57$.

17. $y = -4x^2 - 16x + 3$

Vertex: $y = -4x^2 - 16x + 3 = -4(x^2 + 4x) + 3 = -4(x^2 + 4x + 4) + 3 + 16$
$= -4(x + 2)^2 + 19$. Vertex is at $(-2, 19)$.

x-intercepts: $y = 0 \quad \Rightarrow \quad 0 = -4x^2 - 16x + 3 = -4(x + 2)^2 + 19$
$\Leftrightarrow \quad 4(x + 2)^2 = 19 \quad \Leftrightarrow \quad (x + 2)^2 = \frac{19}{4} \quad \Rightarrow$
$x + 2 = \pm\sqrt{\frac{19}{4}} = \pm\frac{\sqrt{19}}{2} \quad \Leftrightarrow \quad x = -2 \pm \frac{\sqrt{19}}{2}$. The x-intercepts are
at $x = -2 - \frac{\sqrt{19}}{2}$ and $x = -2 + \frac{\sqrt{19}}{2}$.

y-intercepts: $x = 0 \quad \Rightarrow \quad y = 3$. The y-intercept is at $y = 3$.

19. (a) $f(x) = 2x - x^2 = -(x^2 - 2x)$
$= -(x^2 - 2x + 1) + 1$
$= -(x - 1)^2 + 1$

(b)

(c) Therefore, the maximum value is $f(1) = 1$.

21. (a) $f(x) = x^2 + 2x - 1 = (x^2 + 2x) - 1$
$= (x^2 + 2x + 1) - 1 - 1$
$= (x + 1)^2 - 2$

(b)

(c) Therefore, the minimum value is $f(-1) = -2$.

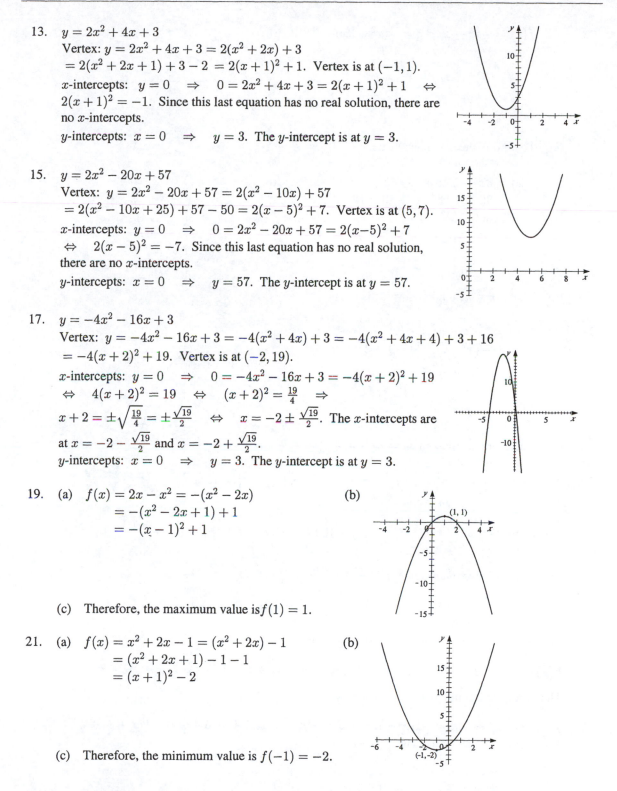

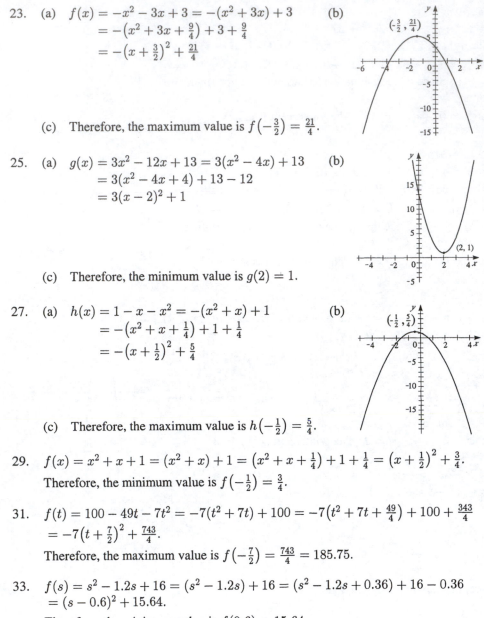

23. (a) $f(x) = -x^2 - 3x + 3 = -(x^2 + 3x) + 3$

      (b)

$$= -(x^2 + 3x + \tfrac{9}{4}) + 3 + \tfrac{9}{4}$$

$$= -(x + \tfrac{3}{2})^2 + \tfrac{21}{4}$$

     (c) Therefore, the maximum value is $f(-\tfrac{3}{2}) = \tfrac{21}{4}$.

25. (a) $g(x) = 3x^2 - 12x + 13 = 3(x^2 - 4x) + 13$

      (b)

$$= 3(x^2 - 4x + 4) + 13 - 12$$

$$= 3(x - 2)^2 + 1$$

     (c) Therefore, the minimum value is $g(2) = 1$.

27. (a) $h(x) = 1 - x - x^2 = -(x^2 + x) + 1$

      (b)

$$= -(x^2 + x + \tfrac{1}{4}) + 1 + \tfrac{1}{4}$$

$$= -(x + \tfrac{1}{2})^2 + \tfrac{5}{4}$$

     (c) Therefore, the maximum value is $h(-\tfrac{1}{2}) = \tfrac{5}{4}$.

29. $f(x) = x^2 + x + 1 = (x^2 + x) + 1 = (x^2 + x + \tfrac{1}{4}) + 1 + \tfrac{1}{4} = (x + \tfrac{1}{2})^2 + \tfrac{3}{4}$.

      Therefore, the minimum value is $f(-\tfrac{1}{2}) = \tfrac{3}{4}$.

31. $f(t) = 100 - 49t - 7t^2 = -7(t^2 + 7t) + 100 = -7(t^2 + 7t + \tfrac{49}{4}) + 100 + \tfrac{343}{4}$

      $= -7(t + \tfrac{7}{2})^2 + \tfrac{743}{4}$.

      Therefore, the maximum value is $f(-\tfrac{7}{2}) = \tfrac{743}{4} = 185.75$.

33. $f(s) = s^2 - 1.2s + 16 = (s^2 - 1.2s) + 16 = (s^2 - 1.2s + 0.36) + 16 - 0.36$

      $= (s - 0.6)^2 + 15.64$.

      Therefore, the minimum value is $f(0.6) = 15.64$.

35. $h(x) = \tfrac{1}{2}x^2 + 2x - 6 = \tfrac{1}{2}(x^2 + 4x) - 6 = \tfrac{1}{2}(x^2 + 4x + 4) - 6 - 2 = \tfrac{1}{2}(x + 2)^2 - 8$.

      Therefore, the minimum value is $h(-2) = -8$.

37. $f(x) = 3 - x - \tfrac{1}{2}x^2 = -\tfrac{1}{2}(x^2 + 2x) + 3 = -\tfrac{1}{2}(x^2 + 2x + 1) + 3 + \tfrac{1}{2} = -\tfrac{1}{2}(x + 1)^2 + \tfrac{7}{2}$.

      Therefore, the maximum value is $f(-1) = \tfrac{7}{2}$.

39. Since the vertex is at $(1, -2)$, the function is of the form $f(x) = a(x-1)^2 - 2$. Substituting the point $(4, 16)$, we get $16 = a(4-1)^2 - 2 \iff 16 = 9a - 2 \iff 9a = 18 \iff a = 2$. So the function is $f(x) = 2(x-1)^2 - 2 = 2x^2 - 4x$.

41. $f(x) = -x^2 + 4x - 3 = -(x^2 - 4x) - 3 = -(x^2 - 4x + 4) - 3 + 4 = -(x-2)^2 + 1$. So the domain of $f(x)$ is $(-\infty, \infty)$. Since $f(x)$ has a maximum value of 1, the range is $(-\infty, 1]$.

43. (a) $f(x) = x^2 + 1.79x - 3.21$ is shown in the viewing rectangle on the right. The minimum value is $f(x) \approx -4.01$.

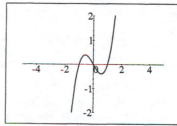

    (b) $f(x) = x^2 + 1.79x - 3.21$
$$= \left[x^2 + 1.79x + \left(\tfrac{1.79}{2} \right)^2 \right] - 3.21 - \left(\tfrac{1.79}{2} \right)^2$$
$$= (x + 0.895)^2 - 4.011025$$

    Therefore, the exact minimum of $f(x)$ is -4.011025.

45. Local maximum: 2 at $x = 0$. Local minimum: -1 at $x = -2$ and 0 at $x = 2$.

47. Local maximum: 0 at $x = 0$ and 1 at $x = 3$. Local minimum: -2 at $x = -2$ and -1 at $x = 1$.

49. In the viewing rectangle on the right, we see that $f(x) = x^3 - x$ has a local minimum and a local maximum. Smaller x and y ranges (shown in the viewing rectangles below) show that $f(x)$ has a local maximum of ≈ 0.38 when $x \approx -0.58$ and a local minimum of ≈ -0.38 when $x \approx 0.58$.

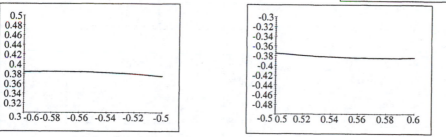

51. In the viewing rectangle on the right, we see that $g(x) = x^4 - 2x^3 - 11x^2$ has two local minimums and a local maximum. The local maximum is $g(x) = 0$ when $x = 0$. Smaller x and y ranges (shown in the viewing rectangles on the next page) show that local minima are $g(x) \approx -13.61$ when $x \approx -1.71$ and $g(x) \approx -73.32$. when $x \approx 3.21$

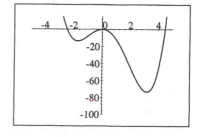

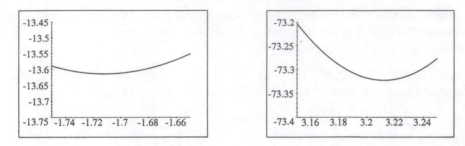

53. In the first viewing rectangle below, we see that $U(x) = x\sqrt{6-x}$ only has a local maximum. Smaller x and y ranges in the second viewing rectangle below show that $U(x)$ has a local maximum of ≈ 5.66 when $x \approx 4.00$.

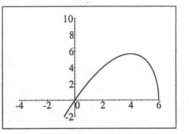

55. In the viewing rectangle on the right, we see that

$V(x) = \dfrac{1-x^2}{x^3}$ has a local minimum and a local maximum.

Smaller x and y ranges (shown in the viewing rectangles below) show that $V(x)$ has a local maximum of ≈ 0.38 when $x \approx -1.73$ and a local minimum of ≈ -0.38 when $x \approx 1.73$.

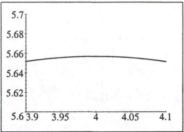

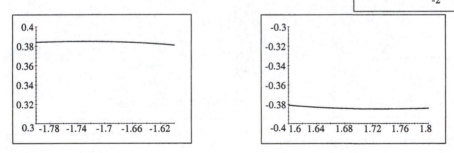

57. $y = f(t) = 40t - 16t^2 = -16\left(t^2 - \tfrac{5}{2}\right) = -16\left[t^2 - \tfrac{5}{2}t + \left(\tfrac{5}{4}\right)^2\right] + 16\left(\tfrac{5}{4}\right)^2 = -16\left(t - \tfrac{5}{4}\right)^2 + 25.$

Thus the maximum height attained by the ball is $f\left(\tfrac{5}{4}\right) = 25$ feet.

59. $R(x) = 80x - 0.4x^2 = -0.4(x^2 - 200x) = -0.4(x^2 - 200x + 10{,}000) + 4{,}000$
    $= -0.4(x - 100)^2 + 4{,}000.$

So revenue is maximized at \$4,000 when 100 units are sold.

61. $E(n) = \frac{2}{3}n - \frac{1}{90}n^2 = -\frac{1}{90}(n^2 - 60n) = -\frac{1}{90}(n^2 - 60n + 900) + 10 = -\frac{1}{90}(n - 30)^2 + 10$. Since
      the maximum of the function occurs when $n = 30$, the viewer should watch the commercial 30 times
      for maximum effectiveness.

63. Graphing $A(n) = n(900 - 9n)$ in the viewing rectangle
      $[0, 100]$ by $[0, 25{,}000]$ we see that maximum yield of apples
      occurs when there are 50 trees per acre

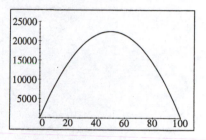

65. In the first viewing rectangle on the next page, we see the general location of the maximum of
      $N(s) = \dfrac{88s}{17 + 17\left(\frac{s}{20}\right)^2}$. In the second viewing rectangle we isolate the maximum, and from this

      graph we see that at the speed of 20 mi/h the largest number of cars that can use the highway safely
      is 52.

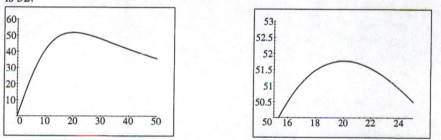

67. In the first viewing rectangle below, we see the general location of the maximum of
      $v(r) = 3.2(1 - r)r^2$ is around $r = 0.7$ cm. In the second viewing rectangle, we isolate the
      maximum, and from this graph we see that at the maximum velocity is 0.47 when $r \approx 0.67$ cm.

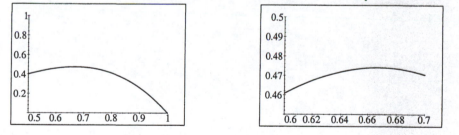

69. (a) If $x = a$ is a local maximum of $f(x)$ then $f(a) \geq f(x) \geq 0$ for all x around $x = a$. So
           $[g(a)]^2 \geq [g(x)]^2$ and thus $g(a) \geq g(x)$. Similarly, if $x = b$ is a local minimum of $f(x)$, then
           $f(x) \geq f(b) \geq 0$, for all x around $x = b$. So $[g(x)]^2 \geq [g(b)]^2$ and thus $g(x) \geq g(b)$.

      (b) Using the distance formula, $g(x) = \sqrt{(x - 3)^2 + (x^2 - 0)^2} = \sqrt{x^4 + x^2 - 6x + 9}$.

      (c) Let $f(x) = x^4 + x^2 - 6x + 9$. In the viewing rectangle we see that $f(x)$ has a minimum at
           $x = 1$. Thus $g(x)$ also has a minimum at $x = 1$ and this minimum value is
           $g(1) = \sqrt{1^4 + 1^2 - 6(1) + 9} = \sqrt{5}$.

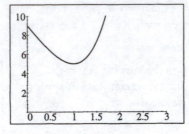

Exercises 3.6

1. $f(x) = x^2$ has domain $(-\infty, \infty)$. $g(x) = x + 2$ has domain $(-\infty, \infty)$. The intersection of the domains of f and g is $(-\infty, \infty)$.

    $(f + g)(x) = (x^2) + (x + 2) = x^2 + x + 2$, and the domain is $(-\infty, \infty)$.

    $(f - g)(x) = (x^2) - (x + 2) = x^2 - x - 2$, and the domain is $(-\infty, \infty)$.

    $(fg)(x) = (x^2)(x + 2) = x^3 + 2x^2$, and the domain is $(-\infty, \infty)$.

    $\left(\dfrac{f}{g}\right)(x) = \dfrac{x^2}{x + 2}$, and the domain is $\{\, x \mid x \neq -2 \,\}$.

3. $f(x) = \sqrt{1 + x^2}$, has domain $(-\infty, \infty)$. $g(x) = \sqrt{1 - x}$, has domain $(-\infty, 1]$. The intersection of the domains of f and g is $(-\infty, 1]$.

    $(f + g)(x) = \sqrt{1 + x^2} + \sqrt{1 - x}$, and the domain is $(-\infty, 1]$.

    $(f - g)(x) = \sqrt{1 + x^2} - \sqrt{1 - x}$, and the domain is $(-\infty, 1]$.

    $(fg)(x) = \sqrt{1 + x^2} \cdot \sqrt{1 - x} = \sqrt{1 - x + x^2 - x^3}$, and the domain is $(-\infty, 1]$.

    $\left(\dfrac{f}{g}\right)(x) = \dfrac{\sqrt{1 + x^2}}{\sqrt{1 - x}} = \sqrt{\dfrac{1 + x^2}{1 - x}}$, and the domain is $(-\infty, 1)$.

5. $f(x) = \dfrac{2}{x}$, has domain $x \neq 0$. $g(x) = -\dfrac{2}{x + 4}$, has domain $x \neq -4$. The intersection of the domains of f and g is $x \neq 0, -4$, in interval notation, this is $(-\infty, -4) \cup (-4, 0) \cup (0, \infty)$.

    $(f + g)(x) = \dfrac{2}{x} + \left(-\dfrac{2}{x + 4}\right) = \dfrac{2}{x} - \dfrac{2}{x + 4} = \dfrac{8}{x(x + 4)}$, and the domain is $(-\infty, -4) \cup (-4, 0) \cup (0, \infty)$.

    $(f - g)(x) = \dfrac{2}{x} - \left(-\dfrac{2}{x + 4}\right) = \dfrac{2}{x} + \dfrac{2}{x + 4} = \dfrac{4x + 8}{x(x + 4)}$, and the domain is $(-\infty, -4) \cup (-4, 0) \cup (0, \infty)$.

    $(fg)(x) = \dfrac{2}{x} \cdot \left(-\dfrac{2}{x + 4}\right) = -\dfrac{4}{x(x + 4)}$, and the domain is $(-\infty, -4) \cup (-4, 0) \cup (0, \infty)$.

    $\left(\dfrac{f}{g}\right)(x) = \dfrac{\dfrac{2}{x}}{-\dfrac{2}{x + 4}} = -\dfrac{x + 4}{x}$, and the domain is $(-\infty, -4) \cup (-4, 0) \cup (0, \infty)$.

7. $f(x) = \sqrt{x} + \sqrt{1 - x}$. The domain of $\sqrt{x}$ is $[0, \infty]]$, and the domain of $\sqrt{1 - x}$ is $(-\infty, 1]$. Thus the domain is $(-\infty, 1] \cap [0, \infty) = [0, 1]$.

9. $h(x) = (x - 3)^{-1/4} = \dfrac{1}{(x - 3)^{1/4}}$. Since $1/4$ is an even root and the denominator can not equal 0, $x - 3 > 0 \quad \Leftrightarrow \quad x > 3$. So the domain is $(3, \infty)$.

11.

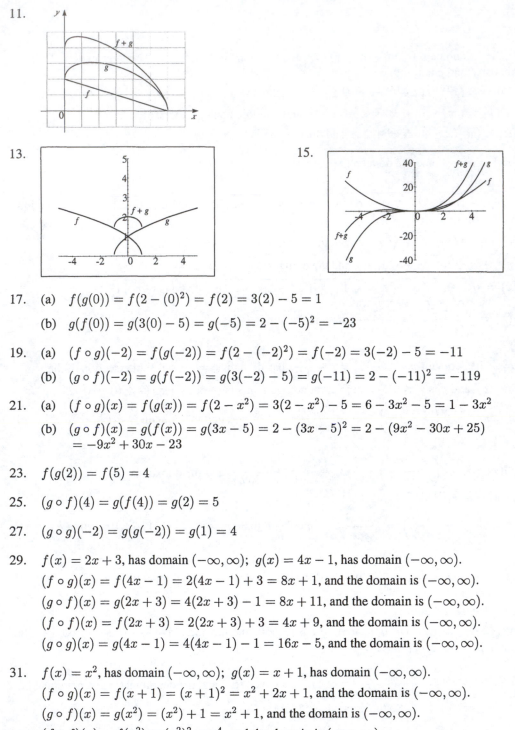

13.

15.

17. (a) $f(g(0)) = f(2 - (0)^2) = f(2) = 3(2) - 5 = 1$

(b) $g(f(0)) = g(3(0) - 5) = g(-5) = 2 - (-5)^2 = -23$

19. (a) $(f \circ g)(-2) = f(g(-2)) = f(2 - (-2)^2) = f(-2) = 3(-2) - 5 = -11$

(b) $(g \circ f)(-2) = g(f(-2)) = g(3(-2) - 5) = g(-11) = 2 - (-11)^2 = -119$

21. (a) $(f \circ g)(x) = f(g(x)) = f(2 - x^2) = 3(2 - x^2) - 5 = 6 - 3x^2 - 5 = 1 - 3x^2$

(b) $(g \circ f)(x) = g(f(x)) = g(3x - 5) = 2 - (3x - 5)^2 = 2 - (9x^2 - 30x + 25)$
    $= -9x^2 + 30x - 23$

23. $f(g(2)) = f(5) = 4$

25. $(g \circ f)(4) = g(f(4)) = g(2) = 5$

27. $(g \circ g)(-2) = g(g(-2)) = g(1) = 4$

29. $f(x) = 2x + 3$, has domain $(-\infty, \infty)$; $g(x) = 4x - 1$, has domain $(-\infty, \infty)$.
    $(f \circ g)(x) = f(4x - 1) = 2(4x - 1) + 3 = 8x + 1$, and the domain is $(-\infty, \infty)$.
    $(g \circ f)(x) = g(2x + 3) = 4(2x + 3) - 1 = 8x + 11$, and the domain is $(-\infty, \infty)$.
    $(f \circ f)(x) = f(2x + 3) = 2(2x + 3) + 3 = 4x + 9$, and the domain is $(-\infty, \infty)$.
    $(g \circ g)(x) = g(4x - 1) = 4(4x - 1) - 1 = 16x - 5$, and the domain is $(-\infty, \infty)$.

31. $f(x) = x^2$, has domain $(-\infty, \infty)$; $g(x) = x + 1$, has domain $(-\infty, \infty)$.
    $(f \circ g)(x) = f(x + 1) = (x + 1)^2 = x^2 + 2x + 1$, and the domain is $(-\infty, \infty)$.
    $(g \circ f)(x) = g(x^2) = (x^2) + 1 = x^2 + 1$, and the domain is $(-\infty, \infty)$.
    $(f \circ f)(x) = f(x^2) = (x^2)^2 = x^4$, and the domain is $(-\infty, \infty)$.
    $(g \circ g)(x) = g(x + 1) = (x + 1) + 1 = x + 2$, and the domain is $(-\infty, \infty)$.

33. $f(x) = \dfrac{1}{x}$, has domain $\{x \mid x \neq 0\}$; $g(x) = 2x + 4$, has domain $(-\infty, \infty)$.

$(f \circ g)(x) = f(2x + 4) = \dfrac{1}{2x + 4}$. $(f \circ g)(x)$ is defined for $2x + 4 \neq 0 \quad \Leftrightarrow \quad x \neq -2$. So the domain is $\{x \mid x \neq -2\} = (-\infty, -2) \cup (-2, \infty)$.

$(g \circ f)(x) = g\left(\dfrac{1}{x}\right) = 2\left(\dfrac{1}{x}\right) + 4 = \dfrac{2}{x} + 4$, the domain is $\{x \mid x \neq 0\} = (-\infty, 0) \cup (0, \infty)$.

$(f \circ f)(x) = f\left(\dfrac{1}{x}\right) = \dfrac{1}{\left(\dfrac{1}{x}\right)} = x$. $(f \circ f)(x)$ is defined whenever both $f(x)$ and $f(f(x))$ are defined; that is, whenever $\{x \mid x \neq 0\} = (-\infty, 0) \cup (0, \infty)$.

$(g \circ g)(x) = g(2x + 4) = 2(2x + 4) + 4 = 4x + 8 + 4 = 4x + 12$, and the domain is $(-\infty, \infty)$.

35. $f(x) = |x|$, has domain $(-\infty, \infty)$; $g(x) = 2x + 3$, has domain $(-\infty, \infty)$

$(f \circ g)(x) = f(2x + 4) = |2x + 3|$, and the domain is $(-\infty, \infty)$.

$(g \circ f)(x) = g(|x|) = 2|x| + 3$, and the domain is $(-\infty, \infty)$.

$(f \circ f)(x) = f(|x|) = ||x|| = |x|$, and the domain is $(-\infty, \infty)$.

$(g \circ g)(x) = g(2x + 3) = 2(2x + 3) + 3 = 4x + 6 + 3 = 4x + 9$. Domain is $(-\infty, \infty)$.

37. $f(x) = \dfrac{x}{x + 1}$, has domain $\{x \mid x \neq -1\}$; $g(x) = 2x - 1$, has domain $(-\infty, \infty)$

$(f \circ g)(x) = f(2x - 1) = \dfrac{2x - 1}{(2x - 1) + 1} = \dfrac{2x - 1}{2x}$, and the domain is
$\{x \mid x \neq 0\} = (-\infty, 0) \cup (0, \infty)$.

$(g \circ f)(x) = g\left(\dfrac{x}{x + 1}\right) = 2\left(\dfrac{x}{x + 1}\right) - 1 = \dfrac{2x}{x + 1} - 1$, and the domain is
$\{x \mid x \neq -1\} = (-\infty, -1) \cup (-1, \infty)$

$(f \circ f)(x) = f\left(\dfrac{x}{x + 1}\right) = \dfrac{\dfrac{x}{x + 1}}{\dfrac{x}{x + 1} + 1} \cdot \dfrac{x + 1}{x + 1} = \dfrac{x}{x + x + 1} = \dfrac{x}{2x + 1}$. $(f \circ f)(x)$ is defined

whenever both $f(x)$ and $f(f(x))$ are defined; that is, whenever $\{x \mid x \neq -1\}$ and $2x + 1 \neq 0 \quad \Rightarrow$
$\{x \mid x \neq -\frac{1}{2}\}$ which is $(-\infty, -1) \cup (-1, -\frac{1}{2}) \cup (-\frac{1}{2}, \infty)$.

$(g \circ g)(x) = g(2x - 1) = 2(2x - 1) - 1 = 4x - 2 - 1 = 4x - 3$, and the domain is $(-\infty, \infty)$.

39. $f(x) = \sqrt[3]{x}$, has domain $(-\infty, \infty)$; $g(x) = \sqrt[4]{x}$, has domain $[0, \infty)$.

$(f \circ g)(x) = f(\sqrt[4]{x}) = \sqrt[3]{\sqrt[4]{x}} = \sqrt[12]{x}$. $(f \circ g)(x)$ is defined whenever both $g(x)$ and $f(g(x))$ are defined. Since $f(x)$ has no restriction, the domain is $[0, \infty)$.

$(g \circ f)(x) = g(\sqrt[3]{x}) = \sqrt[4]{\sqrt[3]{x}} = \sqrt[12]{x}$. $(g \circ f)(x)$ is defined whenever both $f(x)$ and $g(f(x))$ are defined; that is, whenever $x \geq 0$. So the domain is $[0, \infty)$.

$(f \circ f)(x) = f(\sqrt[3]{x}) = \sqrt[3]{\sqrt[3]{x}} = \sqrt[9]{x}$. $(f \circ f)(x)$ is defined whenever both $f(x)$ and $f(f(x))$ are defined. Since $f(x)$ is defined everywhere, the domain is $(-\infty, \infty)$.

$(g \circ g)(x) = g(\sqrt[4]{x}) = \sqrt[4]{\sqrt[4]{x}} = \sqrt[16]{x}$. $(g \circ g)(x)$ is defined whenever both $g(x)$ and $g(g(x))$ are defined; that is, whenever $x \geq 0$. So the domain is $[0, \infty)$.

41. $(f \circ g \circ h)(x) = f(g(h(x))) = f(g(x - 1)) = f(\sqrt{x - 1}) = \sqrt{x - 1} - 1$

43. $(f \circ g \circ h)(x) = f(g(h(x))) = f(g(\sqrt{x})) = f(\sqrt{x} - 5) = (\sqrt{x} - 5)^4 + 1$

45. $F(x) = (x - 9)^5$. Let $f(x) = x^5$ and $g(x) = x - 9$, then $F(x) = (f \circ g)(x)$.

47. $G(x) = \dfrac{x^2}{x^2 + 4}$. Let $f(x) = \dfrac{x}{x + 4}$ and $g(x) = x^2$, then $G(x) = (f \circ g)(x)$.

49. $H(x) = |1 - x^3|$. Let $f(x) = |x|$ and $g(x) = 1 - x^3$, then $H(x) = (f \circ g)(x)$.

For Exercises 51 and 53 there are several possible solutions, only one of which is shown.

51. $F(x) = \dfrac{1}{x^2 + 1}$. Let $f(x) = \dfrac{1}{x}$, $g(x) = x + 1$, and $h(x) = x^2$, then $F(x) = (f \circ g \circ h)(x)$.

53. $G(x) = (4 + \sqrt[3]{x})^9$. Let $f(x) = x^9$, $g(x) = 4 + x$, and $h(x) = \sqrt[3]{x}$, then $G(x) = (f \circ g \circ h)(x)$.

55. Let r be the radius of the circular ripple in centimeters. Since the ripple travels at a speed of 60 cm/s, the distance traveled in t seconds is the radius, so $r = 60t$. Therefore, the area of the circle can be written as $A(t) = \pi r^2 = \pi(60t)^2 = 3600\pi t^2$ cm^2.

57. Let r be the radius of the spherical balloon in centimeters. Since the radius is increasing at a rate of 2 cm/s, the radius is $r = 2t$ after t seconds. Therefore, the surface areae of the balloon can be written as $S = 4\pi r^2 = 4\pi(2t)^2 = 4\pi(4t^2) = 16\pi t^2$.

59. (a) $f(x) = 0.90x$

(b) $g(x) = x - 100$

(c) $f \circ g = f(x - 100) = 0.90(x - 100) = 0.90x - 90$. $f \circ g$ represents applying the \$100 coupon, then the 10% discount.
$g \circ f = g(0.90x) = 0.90x - 100$. $g \circ f$ represents applying the 10% discount, then the \$100 coupon.
So applying the 10% discount, then the \$100 coupon gives the lower price.

61. $A(x) = 1.05x$. $(A \circ A)(x) = A(A(x)) = A(1.05x) = 1.05(1.05x) = (1.05)^2 x$.
$(A \circ A \circ A)(x) = A(A \circ A(x)) = A((1.05)^2 x) = 1.05[(1.05)^2 x] = (1.05)^3 x$.
$(A \circ A \circ A \circ A)(x) = A(A \circ A \circ A(x)) = A((1.05)^3 x) = 1.05[(1.05)^3 x] = (1.05)^4 x$. A represents the amount in the account after 1 year; $A \circ A$ represents the amount in the account after 2 years; $A \circ A \circ A$ represents the amount in the account after 3 years; and $A \circ A \circ A \circ A$ represents the amount in the account after 4 years. We can see that if we compose n copies of A, we get $(1.05)^n x$.

63. $g(x) = 2x + 1$ and $h(x) = 4x^2 + 4x + 7$

<u>Method 1:</u> Notice that $(2x + 1)^2 = 4x^2 + 4x + 1$. We see that adding 6 to this quantity gives $(2x + 1)^2 + 6 = 4x^2 + 4x + 1 + 6 = 4x^2 + 4x + 7$, which is $h(x)$. So let $f(x) = x^2 + 6$, and we have $(f \circ g)(x) = (2x + 1)^2 + 6 = h(x)$.

<u>Method 2:</u> Since $g(x)$ is linear and $h(x)$ is a second degree polynomial, $f(x)$ must be a second degree polynomial, that is, $f(x) = ax^2 + bx + c$ for some a, b, and c. Thus
$f(g(x)) = f(2x + 1) = a(2x + 1)^2 + b(2x + 1) + c \quad \Leftrightarrow$
$4ax^2 + 4ax + a + 2bx + b + c = 4ax^2 + (4a + 2b)x + (a + b + c) = 4x^2 + 4x + 7$. Comparing this with $f(g(x))$, we have $4a = 4$ (the x^2 coefficients), $4a + 2b = 4$ (the x coefficients), and $a + b + c = 7$ (the constant terms) $\quad \Leftrightarrow \quad a = 1$ and $2a + b = 2$ and $a + b + c = 7 \quad \Leftrightarrow \quad a = 1$, $b = 0, c = 6$. Thus $f(x) = x^2 + 6$.

$f(x) = 3x + 5$ and $h(x) = 3x^2 + 3x + 2$

Note since $f(x)$ is linear and $h(x)$ is quadratic, $g(x)$ must also be quadratic. We can then use trial and error to find $g(x)$. Another method is the following: We wish to find g so that $(f \circ g)(x) = h(x)$. Thus $f(g(x)) = 3x^2 + 3x + 2 \quad \Leftrightarrow \quad 3(g(x)) + 5 = 3x^2 + 3x + 2 \quad \Leftrightarrow$
$3(g(x)) = 3x^2 + 3x - 3 \quad \Leftrightarrow \quad g(x) = x^2 + x - 1$.

Exercises 3.7

1. By the Horizontal Line Test, f is not one-to-one.

3. By the Horizontal Line Test, f is one-to-one.

5. By the Horizontal Line Test, f is not one-to-one.

7. $f(x) = 3x + 1$. If $x_1 \neq x_2$, then $3x_1 \neq 3x_2$ and $3x_1 + 1 \neq 3x_2 + 1$. So f is a one-to-one function.

9. $g(x) = \sqrt{x}$. If $x_1 \neq x_2$, then $\sqrt{x_1} \neq \sqrt{x_2}$ because two different numbers cannot have the same square root. Therefore, g is a one-to-one function.

11. $h(x) = x^2 - 2x$. Since $h(0) = 0$ and $h(2) = (2) - 2(2) = 0$ we have $h(0) = h(2)$. So f is not a one-to-one function.

13. $f(x) = x^4 + 5$. Every nonzero number and its negative have the same fourth power. For example, $(-1)^4 = 1 = (1)^4$, so $f(-1) = f(1)$. Thus f is not a one-to-one function.

15. $f(x) = \dfrac{1}{x^2}$. Every nonzero number and its negative have the same square. For example, $\frac{1}{(-1)^2} = 1 = \frac{1}{(1)^2}$, so $f(-1) = f(1)$. Thus f is not a one-to-one function.

17. (a) $f(2) = 7$. Since f is one-to-one, $f^{-1}(7) = 2$.

    (b) $f^{-1}(3) = -1$. Since f is one-to-one, $f(-1) = 3$.

19. $f(x) = 5 - 2x$. Since f is one-to-one and $f(1) = 5 - 2(1) = 3$, then $f^{-1}(3) = 1$. (Find 1 by solving the equation $5 - 2x = 3$.)

21. $f(g(x)) = f(x - 3) = (x - 3) + 3 = x$, for all x.

    $g(f(x)) = g(x + 3) = (x + 3) - 3 = x$, for all x. Thus f and g are inverses of each other.

23. $f(g(x)) = f\left(\dfrac{x + 5}{2}\right) = 2\left(\dfrac{x + 5}{2}\right) - 5 = x + 5 - 5 = x$, for all x.

    $g(f(x)) = g(2x - 5) = \dfrac{(2x - 5) + 5}{2} = x$, for all x. Thus f and g are inverses of each other.

25. $f(g(x)) = f\left(\dfrac{1}{x}\right) = \dfrac{1}{1/x} = x$, for all x. Since $f(x) = g(x)$, we also have $g(f(x)) = x$. Thus f and g are inverses of each other.

27. $f(g(x)) = f(\sqrt{x + 4}) = (\sqrt{x + 4})^2 - 4 = x + 4 - 4 = x$, for all $x \geq -4$.

    $g(f(x)) = g(x^2 - 4) = \sqrt{(x^2 - 4) + 4} = \sqrt{x^2} = x$, for all $x \geq 0$. Thus f and g are inverses of each other.

29. $f(g(x)) = f\left(\dfrac{1}{x} + 1\right) = \dfrac{1}{\left(\dfrac{1}{x}+1\right)-1} = x$, for all $x \neq 0$.

$g(f(x)) = g\left(\dfrac{1}{x-1}\right) = \dfrac{1}{\left(\dfrac{1}{x-1}\right)} + 1 = (x-1) + 1 = x$, for all $x \neq 1$. Thus f and g are

inverses of each other.

31. $f(x) = 2x + 1$. $y = 2x + 1$ $\Leftrightarrow$ $2x = y - 1$ $\Leftrightarrow$ $x = \frac{1}{2}(y-1)$. So $f^{-1}(x) = \frac{1}{2}(x-1)$.

33. $f(x) = 4x + 7$. $y = 4x + 7$ $\Leftrightarrow$ $4x = y - 7$ $\Leftrightarrow$ $x = \frac{1}{4}(y-7)$. So $f^{-1}(x) = \frac{1}{4}(x-7)$.

35. $f(x) = \dfrac{x}{2}$. $y = \dfrac{x}{2}$ $\Leftrightarrow$ $x = 2y$. So $f^{-1}(x) = 2x$.

37. $f(x) = \dfrac{1}{x+2}$. $y = \dfrac{1}{x+2}$ $\Leftrightarrow$ $x + 2 = \dfrac{1}{y}$ $\Leftrightarrow$ $x = \dfrac{1}{y} - 2$. So $f^{-1}(x) = \dfrac{1}{x} - 2$.

39. $f(x) = \dfrac{1+3x}{5-2x}$. $y = \dfrac{1+3x}{5-2x}$ $\Leftrightarrow$ $y(5-2x) = 1 + 3x$ $\Leftrightarrow$ $5y - 2xy = 1 + 3x$ $\Leftrightarrow$

$3x + 2xy = 5y - 1$ $\Leftrightarrow$ $x(3+2y) = 5y - 1$ $\Leftrightarrow$ $x = \dfrac{5y-1}{2y+3}$. So $f^{-1}(x) = \dfrac{5x-1}{2x+3}$.

41. $f(x) = \sqrt{2+5x}, x \geq -\frac{2}{5}$. $y = \sqrt{2+5x}, y \geq 0$ $\Leftrightarrow$ $y^2 = 2 + 5x$ $\Leftrightarrow$ $5x = y^2 - 2$ $\Leftrightarrow$
$x = \frac{1}{5}(y^2 - 2)$ and $y \geq 0$. So $f^{-1}(x) = \frac{1}{5}(x^2 - 2), x \geq 0$.

43. $f(x) = 4 - x^2, x \geq 0$. $y = 4 - x^2$ $\Leftrightarrow$ $x^2 = 4 - y$ $\Leftrightarrow$ $x = \sqrt{4-y}$. So
$f^{-1}(x) = \sqrt{4-x}$. Note: $x \geq 0$ $\Rightarrow$ $f(x) \leq 4$.

45. $f(x) = 4 + \sqrt[3]{x}$. $y = 4 + \sqrt[3]{x}$ $\Leftrightarrow$ $\sqrt[3]{x} = y - 4$ $\Leftrightarrow$ $x = (y-4)^3$. So $f^{-1}(x) = (x-4)^3$.

47. $f(x) = 1 + \sqrt{1+x}$. $y = 1 + \sqrt{1+x}, y \geq 1$ $\Leftrightarrow$ $\sqrt{1+x} = y - 1$ $\Leftrightarrow$ $1 + x = (y-1)^2$
$\Leftrightarrow$ $x = (y-1)^2 - 1 = y^2 - 2y$. So $f^{-1}(x) = x^2 - 2x, x \geq 1$.

49. $f(x) = x^4, x \geq 0$. $y = x^4, y \geq 0$ $\Leftrightarrow$ $x = \sqrt[4]{y}$. So $f^{-1}(x) = \sqrt[4]{x}, x \geq 0$.

51. (a)   (b)

(c) $f(x) = 3x - 6$. $y = 3x - 6$ $\Leftrightarrow$ $3x = y + 6$ $\Leftrightarrow$ $x = \frac{1}{3}(y+6)$. So
$f^{-1}(x) = \frac{1}{3}(x+6)$.

53. (a)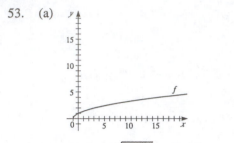

(b)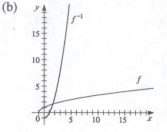

(c) $f(x) = \sqrt{x+1}$, $x \geq -1$. $y = \sqrt{x+1}$, $y \geq 0$ $\Leftrightarrow$ $y^2 = x+1$ $\Leftrightarrow$ $x = y^2 - 1$ and $y \geq 0$. So $f^{-1}(x) = x^2 - 1$, $x \geq 0$.

55. $f(x) = x^3 - x$. Using a graphing device and the Horizontal Line Test, we see that f is not a one-to-one function. For example, $f(0) = 0 = f(-1)$.

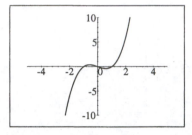

57. $f(x) = \dfrac{x+12}{x-6}$. Using a graphing device and the Horizontal Line Test, we see that f is a one-to-one function.

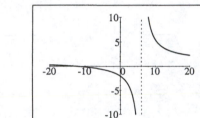

59. $f(x) = |x| - |x-6|$. Using a graphing device and the Horizontal Line Test, we see that f is not a one-to-one function. For example $f(0) = -6 = f(-2)$.

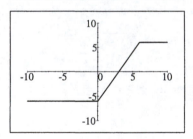

61. If we restrict the domain of $f(x)$ to $[0, \infty)$, then $y = 4 - x^2$ $\Leftrightarrow$ $x^2 = 4 - y$ $\Rightarrow$ $x = \sqrt{4-y}$ (since $x \geq 0$, we take the positive square root). So $f^{-1}(x) = \sqrt{4-x}$.
If we restrict the domain of $f(x)$ to $(-\infty, 0]$, then $y = 4 - x^2$ $\Leftrightarrow$ $x^2 = 4 - y$ $\Rightarrow$ $x = -\sqrt{4-y}$ (since $x \leq 0$, we take the negative square root). So $f^{-1}(x) = -\sqrt{4-x}$.

63. If we restrict the domain of $h(x)$ to $[-2, \infty)$, then $y = (x+2)^2$ $\Rightarrow$ $x+2 = \sqrt{y}$ (since $x \geq -2$, we take the positive square root) $\Leftrightarrow$ $x = -2 + \sqrt{y}$. So $h^{-1}(x) = -2 + \sqrt{x}$.
If we restrict the domain of $h(x)$ to $(-\infty, -2]$, then $y = (x+2)^2$ $\Rightarrow$ $x+2 = -\sqrt{y}$ (since $x \leq -2$, we take the negative square root) $\Leftrightarrow$ $x = -2 - \sqrt{y}$. So $h^{-1}(x) = -2 - \sqrt{x}$.

65.

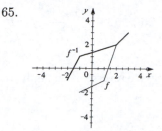

67. (a) $f(x) = 500 + 80x$.

   (b) $f(x) = 500 + 80x$. $y = 500 + 80x$ $\Leftrightarrow$ $80x = y - 500$ $\Leftrightarrow$ $x = \dfrac{y - 500}{80}$. So
   $f^{-1}(x) = \dfrac{x - 500}{80}$. f^{-1} represents the number of hours of investigation the investigate spends
   on a case for x dollars.

   (c) $f^{-1}(1220) = \dfrac{1220 - 500}{80} = \dfrac{720}{80} = 9$. The investigator spent 9 hours investigating this case.

69. (a) $v(r) = 18{,}500(0.25 - r^2)$. $t = 18{,}500(0.25 - r^2)$ $\Leftrightarrow$ $t = 4625 - 18{,}500r^2$ $\Leftrightarrow$
   $18500r^2 = 4625 - t$ $\Leftrightarrow$ $r^2 = \dfrac{4625 - t}{18{,}500}$ $\Rightarrow$ $r = \pm\sqrt{\dfrac{4625 - t}{18{,}500}}$. Since r represents a

   measure $r \geq 0$, so $v^{-1}(t) = \sqrt{\dfrac{4625 - t}{18{,}500}}$. v^{-1} represents the radius in the vein that has the
   velocity v.

   (b) $v^{-1}(30) = \sqrt{\dfrac{4625 - 30}{18{,}500}} \approx 0.498$ cm. The velocity is 30 at 0.498 cm from the center of the
   artery or vein.

71. (a) $F(x) = \frac{9}{5}x + 32$. $y = \frac{9}{5}x + 32$ $\Leftrightarrow$ $\frac{9}{5}x = y - 32$ $\Leftrightarrow$ $x = \frac{5}{9}(y - 32)$. So
   $F^{-1}(x) = \frac{5}{9}(x - 32)$. F^{-1} represents the Celsius temperature that corresponds to the
   Fahrenheit temperature of F.

   (b) $F^{-1}(86) = \frac{5}{9}(86 - 32) = \frac{5}{9}(54) = 30$. So 86° Fahrenheit is the same as 30° Celsius.

73. (a) $f(x) = \begin{cases} 0.1x, & \text{if } 0 \leq x \leq 20{,}000 \\ 2000 + 0.2(x - 20{,}000), & \text{if } x > 20{,}000 \end{cases}$

   (b) We will find the inverse of each piece of the function f.
   $f_1(x) = 0.1x$. $y = 0.1x$ $\Leftrightarrow$ $x = 10y$. So $f_1^{-1}(x) = 10x$.
   $f_2(x) = 2000 + 0.2(x - 20{,}000) = 0.2x - 2000$. $y = 0.2x - 2000$ $\Leftrightarrow$ $0.2x = y + 2000$
   $\Leftrightarrow$ $x = 5y + 10{,}000$. So $f_2^{-1}(x) = 5x + 10{,}000$.

   Since $f(0) = 0$ and $f(20{,}000) = 2000$ we have $f^{-1}(x) = \begin{cases} 10x, & \text{if } 0 \leq x \leq 2000 \\ 5x + 10{,}000 & \text{if } x > 2000 \end{cases}$

   It represents the taxpayer's income.

   (c) $f^{-1}(10{,}000) = 5(10{,}000) + 10{,}000 = 60{,}000$. The required income is $60,000.

75. $f(x) = 7 + 2x$. $y = 7 + 2x$ $\Leftrightarrow$ $2x = y - 7$ $\Leftrightarrow$ $x = \dfrac{y - 7}{2}$. So $f^{-1}(x) = \dfrac{x - 7}{2}$. f^{-1} is

the number of toppings on a pizza that costs x dollars.

77. (a) $f(x) = \dfrac{2x + 1}{5}$ is "multiply by 2, add 1, and then divide by 5". So the reverse is "multiply by

5, subtract 1 , and then divide by 2" or $f^{-1}(x) = \dfrac{5x - 1}{2}$. Check:

$$f \circ f^{-1}(x) = f\left(\frac{5x - 1}{2}\right) \qquad\qquad f^{-1} \circ f(x) = f\left(\frac{2x + 1}{5}\right)$$

$$= \frac{2\left(\dfrac{5x - 1}{2}\right) + 1}{5} \qquad\qquad = \frac{5\left(\dfrac{2x + 1}{5}\right) - 1}{2}$$

$$= \frac{5x - 1 + 1}{5} = \frac{5x}{5} = x \qquad\qquad = \frac{2x + 1 - 1}{2} = \frac{2x}{2} = x$$

(b) $f(x) = 3 - \dfrac{1}{x} = \dfrac{-1}{x} + 3$ is "take the negative reciprocal and add 3". Since the reverse of

"take the negative reciprocal" is "take the negative reciprocal ", $f^{-1}(x)$ is "subtract 3 and take

the negative reciprocal ", that is, $f^{-1}(x) = \dfrac{-1}{x - 3}$. Check:

$$f \circ f^{-1}(x) = f\left(\frac{-1}{x - 3}\right) \qquad\qquad f^{-1} \circ f(x) = f\left(3 - \frac{1}{x}\right)$$

$$= 3 - \frac{1}{\dfrac{-1}{x - 3}} \qquad\qquad = \frac{-1}{\left(3 - \dfrac{1}{x}\right) - 3}$$

$$= 3 - \left(1 \cdot \frac{x - 3}{-1}\right) \qquad\qquad = \frac{-1}{-\dfrac{1}{x}} = -1 \cdot \frac{x}{-1} = x$$

$$= 3 + x - 3 = x$$

(c) $f(x) = \sqrt{x^3 + 2}$ is "cube, add 2, and then take the square root 3". So the reverse is "square,

subtract 2, then take cube root " or $f^{-1}(x) = \sqrt[3]{x^2 - 2}$. Domain for $f(x)$ is $[-\sqrt[3]{2}, \infty)$;

domain for $f^{-1}(x)$ is $[0, \infty)$. Check:

$$f \circ f^{-1}(x) = f\left(\sqrt[3]{x^2 - 2}\right) \qquad\qquad f^{-1} \circ f(x) = f\left(\sqrt{x^3 + 2}\right)$$

$$= \sqrt{\left(\sqrt[3]{x^2 - 2}\right)^3 + 2} \qquad\qquad = \sqrt[3]{\left(\sqrt{x^3 + 2}\right)^2 - 2}$$

$$= \sqrt{x^2 - 2 + 2} \qquad\qquad = \sqrt[3]{x^3 + 2 - 2}$$

$$= \sqrt{x^2} = x \text{ (for the domains)} \qquad\qquad = \sqrt[3]{x^3} = x \text{ (for the domains)}$$

No; in a function like $f(x) = 3x - 2$, the variable occurs only once and it easy to see how to reverse

these operations step by step. But in $f(x) = \dfrac{3x - 2}{x + 7}$, you apply two different operations to the

variable x and then find the quotient of each of these values, so it is not possible to reverse the

operations step by step.

79. (a)   We find $g^{-1}(x)$: $y = 2x + 1$ $\Leftrightarrow$ $2x = y - 1$ $\Leftrightarrow$ $x = \frac{1}{2}(y - 1)$. So
$g^{-1}(x) = \frac{1}{2}(x - 1)$. Thus

$$f(x) = h \circ g^{-1}(x) = h\left(\tfrac{1}{2}(x - 1)\right) = 4\left[\tfrac{1}{2}(x - 1)\right]^2 + 4\left[\tfrac{1}{2}(x - 1)\right] + 7$$
$$= x^2 - 2x + 1 + 2x - 2 + 7 = x^2 + 6.$$

(b) $f \circ g = h$ $\Leftrightarrow$ $f^{-1} \circ f \circ g = f^{-1} \circ h$ $\Leftrightarrow$ $I \circ g = f^{-1} \circ h$ $\Leftrightarrow$ $g = f^{-1} \circ h$. Note
that we compose with f^{-1} on the left on each side of the equation. We find f^{-1}: $y = 3x + 5$
$\Leftrightarrow$ $3x = y - 5$ $\Leftrightarrow$ $x = \frac{1}{3}(y - 5)$. So $f^{-1}(x) = \frac{1}{3}(x - 5)$. Thus

$$g(x) = f^{-1} \circ h(x) = f^{-1}(3x^2 + 3x + 2) = \tfrac{1}{3}\left[(3x^2 + 3x + 2) - 5\right]$$
$$= \tfrac{1}{3}[3x^2 + 3x - 3] = x^2 + x - 1.$$

Review Exercises for Chapter 3

1. $f(x) = x^3 + 2x - 4$; $f(0) = (0)^3 + 2(0) - 4 = -4$; $f(2) = (2)^3 + 2(2) - 4 = 8$;
    $f(-2) = (-2)^3 + 2(-2) - 4 = -16$; $f(a) = (a)^3 + 2(a) - 4 = a^3 + 2a - 4$;
    $f(-a) = (-a)^3 + 2(-a) - 4 = -a^3 - 2a - 4$;
    $f(x+1) = (x+1)^3 + 2(x+1) - 4 = x^3 + 3x^2 + 3x + 1 + 2x + 2 - 4 = x^3 + 3x^2 + 5x - 1$;
    $f(2x) = (2x)^3 + 2(2x) - 4 = 8x^3 + 4x - 4$;
    $2f(x) - 2 = 2(x^3 + 2x - 4) - 2 = 2x^3 + 4x - 8 - 2 = 2x^3 + 4x - 10$.

3. (a) $f(-2) = -1$. $f(2) = 2$.

    (b) The domain of f is $[-4, 5]$.

    (c) The range of f is $[-4, 4]$.

    (d) f is increasing on $[-4, -2]$ and $[-1, 4]$; f is decreasing on $[-2, -1]$ and $[4, 5]$.

    (e) f is not a one-to-one, for example, $f(-2) = -1 = f(0)$. There are many more examples.

5. Domain: We must have $x + 3 \geq 0$ $\Leftrightarrow$ $x \geq -3$. In interval notation, the domain is $[-3, \infty)$.
    Range: For x in the domain of f, we have $x \geq -3$ $\Leftrightarrow$ $x + 3 \geq 0$ $\Leftrightarrow$ $\sqrt{x+3} \geq 0$ $\Leftrightarrow$
    $f(x) \geq 0$. So the range is $[0, \infty)$.

7. $f(x) = 7x + 15$. The domain is all real numbers, $(-\infty, \infty)$.

9. $f(x) = \sqrt{x + 4}$. We require $x + 4 \geq 0$ $\Leftrightarrow$ $x \geq -4$. Thus the domain is $[-4, \infty)$.

11. $f(x) = \dfrac{1}{x} + \dfrac{1}{x+1} + \dfrac{1}{x+2}$. The denominators cannot equal 0, therefore the domain is
    $\{x \mid x \neq 0, -1, -2\}$.

13. $h(x) = \sqrt{4 - x} + \sqrt{x^2 - 1}$. We require the expression inside the radicals be nonnegative. So
    $4 - x \geq 0$ $\Leftrightarrow$ $4 \geq x$; also $x^2 - 1 \geq 0$ $\Leftrightarrow$ $(x - 1)(x + 1) \geq 0$. Using the methods from
    Chapter 1, we have:

Interval	$(-\infty, -1)$	$(-1, 1)$	$(1, \infty)$
Sign of $x - 1$	$-$	$-$	$+$
Sign of $x + 1$	$-$	$+$	$+$
Sign of $(x - 1)(x + 1)$	$+$	$-$	$+$

    Thus the domain is $(-\infty, 4] \cap \{(-\infty, -1] \cup [1, \infty)\} = (-\infty, -1] \cup [1, 4]$.

15. $f(x) = 1 - 2x$ 17. $f(t) = 1 - \frac{1}{2}t^2$

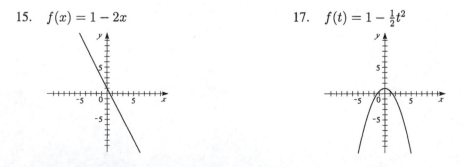

19. $f(x) = x^2 - 6x + 6$

21. $y = 1 - \sqrt{x}$

23. $y = \frac{1}{2}x^3$

25. $h(x) = \sqrt[3]{x}$

27. $g(x) = \frac{1}{x^2}$

29. $f(x) = \begin{cases} 1 - x & \text{if } x < 0 \\ 1 & \text{if } x \geq 0 \end{cases}$

31. $f(x) = \begin{cases} x + 6 & \text{if } x < -2 \\ x^2 & \text{if } x \geq -2 \end{cases}$

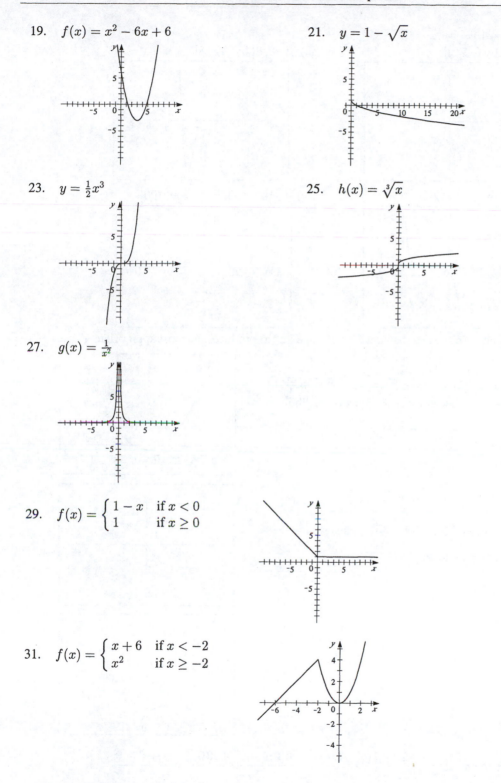

33. $f(x) = 6x^3 - 15x^2 + 4x - 1$

(i) $[-2, 2]$ by $[-2, 2]$

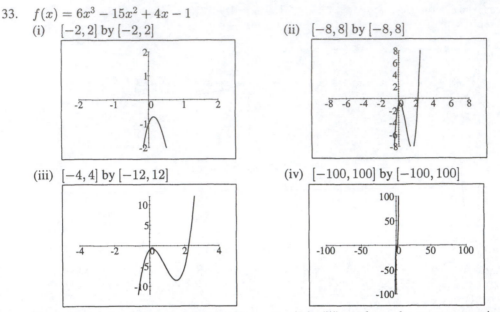

(ii) $[-8, 8]$ by $[-8, 8]$

(iii) $[-4, 4]$ by $[-12, 12]$

(iv) $[-100, 100]$ by $[-100, 100]$

From the graphs, we see that the viewing rectangle in (iii) produces the most appropriate graph.

35. $f(x) = x^2 + 25x + 173$

$= (x^2 + 25x + \frac{625}{4}) + 173 - \frac{625}{4} = (x + \frac{25}{2})^2 + \frac{67}{4}$.

We use the viewing rectangle $[-30, 5]$ by $[-20, 250]$.

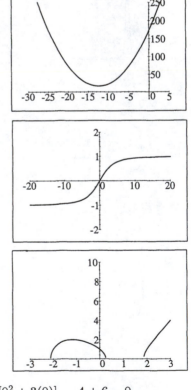

37. $y = \dfrac{x}{\sqrt{x^2 + 16}}$. Since $\sqrt{x^2 + 16} \geq \sqrt{x^2} = |x|$, it

follows that y should behave like $\dfrac{x}{|x|}$. Thus we use the

viewing rectangle $[-20, 20]$ by $[-2, 2]$.

39. $f(x) = \sqrt{x^3 - 4x + 1}$. The domain consists of all x
where $x^3 - 4x + 1 \geq 0$. Using a graphing device, the
domain is approximately $[-2.1, 0.2] \cup [1.9, \infty)$.

41. Average rate of change $= \dfrac{f(2) - f(0)}{2 - 0} = \dfrac{[(2)^2 + 3(2)] - [0^2 + 3(0)]}{2} = \dfrac{4 + 6 - 0}{2} = 5.$

43.   Average rate of change $= \dfrac{f(3+h) - f(3)}{(3+h) - 3} = \dfrac{\left[\frac{1}{3+h}\right] - \left[\frac{1}{3}\right]}{h} = \dfrac{\left[\frac{1}{3+h}\right] - \left[\frac{1}{3}\right]}{h} \cdot \dfrac{3(3+h)}{3(3+h)} = \dfrac{3 - (3+h)}{3h(3+h)}$

$= \dfrac{-h}{3h(3+h)} = \dfrac{-1}{3(3+h)}.$

45. $f(x) = x^3 - 4x^2$ is graphed in the viewing rectangle $[-5, 5]$ by $[-20, 10]$. $f(x)$ is increasing on $(-\infty, 0]$ and $[2.67, \infty)$. It is decreasing on $[0, 2.67]$.

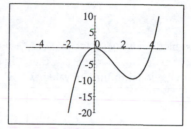

47.   (a) $y = f(x) + 8$. Shift the graph of $f(x)$ upward 8 units.

(b) $y = f(x + 8)$. Shift the graph of $f(x)$ to the left 8 units.

(c) $y = 1 + 2f(x)$. Stretch the graph of $f(x)$ vertically by a factor of 2, then shift it upward 1 unit.

(d) $y = f(x - 2) - 2$. Shift the graph of $f(x)$ to the right 2 units, then downward 2 units.

(e) $y = f(-x)$. Reflect the graph of $f(x)$ about the y-axis.

(f) $y = -f(-x)$. Reflect the graph of $f(x)$ first about the y-axis, then reflect about the x-axis.

(g) $y = -f(x)$. Reflect the graph of $f(x)$ about the x-axis.

(h) $y = f^{-1}(x)$. Reflect the graph of $f(x)$ about the line $y = x$.

49.   (a) $f(x) = 2x^5 - 3x^2 + 2$.

$f(-x) = 2(-x)^5 - 3(-x)^2 + 2 = -2x^5 - 3x^2 + 2$. Since $f(x) \neq f(-x)$, f is not even.

$-f(x) = -2x^5 + 3x^2 - 2$. Since $-f(x) \neq f(-x)$, f is not odd.

(b) $f(x) = x^3 - x^7$.

$f(-x) = (-x)^3 - (-x)^7 = -(x^3 - x^7) = -f(x)$, hence f is odd.

(c) $f(x) = \dfrac{1 - x^2}{1 + x^2}$. $f(-x) = \dfrac{1 - (-x)^2}{1 + (-x)^2} = \dfrac{1 - x^2}{1 + x^2} = f(x)$. Since $f(x) = f(-x)$, f is even.

(d) $f(x) = \dfrac{1}{x + 2}$. $f(-x) = \dfrac{1}{(-x) + 2} = \dfrac{1}{2 - x}$. $-f(x) = -\dfrac{1}{x + 2}$. Since $f(x) \neq f(-x)$, f is not even, and since $f(-x) \neq -f(x)$, f is not odd.

51. $f(x) = x^2 + 4x + 1 = (x^2 + 4x + 4) + 1 - 4 = (x + 2)^2 - 3$.

53. $g(x) = 2x^2 + 4x - 5 = 2(x^2 + 2x) - 5 = 2(x^2 + 2x + 1) - 5 - 2 = 2(x + 1)^2 - 7$. So the minimum value is $g(-1) = -7$.

55. $h(t) = -16t^2 + 48t + 32 = -16(t^2 - 3t) + 32 = -16(t^2 - 3t + \frac{9}{4}) + 32 + 36$

$= -16(t^2 - 3t + \frac{9}{4}) + 68 = -16(t - \frac{3}{2})^2 + 68$

The stone reaches a maximum height of 68 feet.

57. $f(x) = 3.3 + 1.6x - 2.5x^3$. In the first viewing rectangle, $[-2, 2]$ by $[-4, 8]$, we see that $f(x)$ has a local maximum and a local minimum.

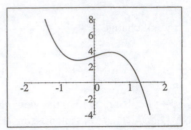

In the next viewing rectangle, $[0.4, 0.5]$ by $[3.78, 3.80]$, we isolate the local maximum value as approximately 3.79 when $x \approx 0.46$. In the last viewing rectangle, $[-0.5, -0.4]$ by $[2.80, 2.82]$, we isolate the local minimum value as 2.81 when $x \approx -0.46$.

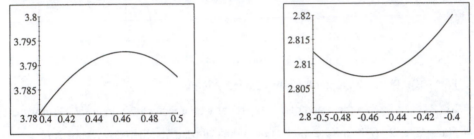

59. $f(x) = x^2 - 3x + 2$ and $g(x) = 4 - 3x$.

(a) $(f + g)(x) = (x^2 - 3x + 2) + (4 - 3x) = x^2 - 6x + 6$

(b) $(f - g)(x) = (x^2 - 3x + 2) - (4 - 3x) = x^2 - 2$

(c) $(fg)(x) = (x^2 - 3x + 2)(4 - 3x) = 4x^2 - 12x + 8 - 3x^3 + 9x^2 - 6x$
$= -3x^3 + 13x^2 - 18x + 8$

(d) $\left(\dfrac{f}{g}\right)(x) = \dfrac{x^2 - 3x + 2}{4 - 3x}, x \neq \frac{4}{3}$

(e) $(f \circ g)(x) = f(4 - 3x) = (4 - 3x)^2 - 3(4 - 3x) + 2 = 16 - 24x + 9x^2 - 12 + 9x + 2$
$= 9x^2 - 15x + 6$

(f) $(g \circ f)(x) = g(x^2 - 3x + 2) = 4 - 3(x^2 - 3x + 2) = -3x^2 + 9x - 2$

61. $f(x) = 3x - 1$ and $g(x) = 2x - x^2$.

$(f \circ g)(x) = f(2x - x^2) = 3(2x - x^2) - 1 = -3x^2 + 6x - 1$, and the domain is $(-\infty, \infty)$.

$(g \circ f)(x) = g(3x - 1) = 2(3x - 1) - (3x - 1)^2 = 6x - 2 - 9x^2 + 6x - 1 = -9x^2 + 12x - 3$, and the domain is $(-\infty, \infty)$

$(f \circ f)(x) = f(3x - 1) = 3(3x - 1) - 1 = 9x - 4$, and the domain is $(-\infty, \infty)$.

$(g \circ g)(x) = g(2x - x^2) = 2(2x - x^2) - (2x - x^2)^2 = 4x - 2x^2 - 4x^2 + 4x^3 - x^4$
$= -x^4 + 4x^3 - 6x^2 + 4x$, and domain is $(-\infty, \infty)$.

63. $f(x) = \sqrt{1 - x}$, $g(x) = 1 - x^2$ and $h(x) = 1 + \sqrt{x}$.

$(f \circ g \circ h)(x) = f(g(h(x))) = f\left(g(1 + \sqrt{x})\right) = f\left(1 - (1 + \sqrt{x})^2\right) = f(1 - (1 + 2\sqrt{x} + x) =$
$f(-x - 2\sqrt{x}) = \sqrt{1 - (-x - 2\sqrt{x})} = \sqrt{1 + 2\sqrt{x} + x} = \sqrt{(1 + \sqrt{x})^2} = 1 + \sqrt{x}$

65. $f(x) = 3 + x^3$. If $x_1 \neq x_2$, then $x_1^3 \neq x_2^3$ (unequal numbers have unequal cubes), and therefore $3 + x_1^3 \neq 3 + x_2^3$. Thus f is a one-to-one function.

67. $h(x) = \dfrac{1}{x^4}$. Since the fourth powers of a number and its negative are equal, h is not one-to-one. For example, $h(-1) = \frac{1}{(-1)^4} = 1$ and $h(1) = \frac{1}{(1)^4} = 1$, so $h(-1) = h(1)$.

69. $p(x) = 3.3 + 1.6x - 2.5x^3$. Using a graphing device and the Horizontal Line Test, we see that p is not a one-to-one function.

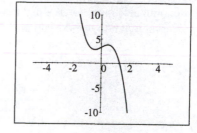

71. $f(x) = 3x - 2 \quad \Leftrightarrow \quad y = 3x - 2 \quad \Leftrightarrow \quad 3x = y + 2 \quad \Leftrightarrow \quad x = \frac{1}{3}(y + 2)$. So $f^{-1}(x) = \frac{1}{3}(x + 2)$.

73. $f(x) = (x + 1)^3 \quad \Leftrightarrow \quad y = (x + 1)^3 \quad \Leftrightarrow \quad x + 1 = \sqrt[3]{y} \quad \Leftrightarrow \quad x = \sqrt[3]{y} - 1$. So $f^{-1}(x) = \sqrt[3]{x} - 1$.

75. $f(x) = x^2 - 4,\ x \geq 0$.

(a)   (b)

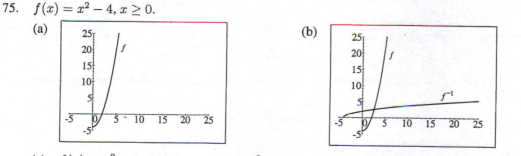

(c) $f(x) = x^2 - 4,\ x \geq 0 \quad \Leftrightarrow \quad y = x^2 - 4,\ y \geq -4 \quad \Leftrightarrow \quad x^2 = y + 4 \quad \Leftrightarrow \quad x = \sqrt{y + 4}$.
     So $f^{-1}(x) = \sqrt{x + 4},\ x \geq -4$.

Chapter 3 Test

1. By the Vertical Line Test, figures (a) and (b) are graphs of functions. By the Horizontal Line Test, only figure (a) is the graph of a one-to-one function.

2. (a) $f(3) = \frac{\sqrt{3+1}}{3} = \frac{\sqrt{4}}{3} = \frac{2}{3}$; $f(5) = \frac{\sqrt{5+1}}{5} = \frac{\sqrt{6}}{5}$; $f(a-+1) = \frac{\sqrt{(a-1)+1}}{a-1} = \frac{\sqrt{a}}{a-1}$.

   (b) $f(x) = \frac{\sqrt{x+1}}{x}$. Our restrictions are that the input to the radical is nonnegative, and the denominator must not be equal to zero. Thus $x + 1 \geq 0 \quad \Leftrightarrow \quad x \geq -1$ and $x \neq 0$. In interval notation, the domain is $[-1, 0) \cup (0, \infty)$.

3. Average rate of change $= \dfrac{f(2) - f(5)}{2 - 5} = \dfrac{[2^2 - 2(2)] - [5^2 - 2(5)]}{-3}$
$$= \frac{4 - 4 - (25 - 10)}{-3} = \frac{-15}{-3} = 5.$$

4. (a) $f(x) = x^3$

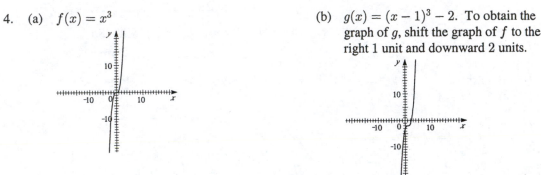

   (b) $g(x) = (x - 1)^3 - 2$. To obtain the graph of g, shift the graph of f to the right 1 unit and downward 2 units.

5. (a) $y = f(x - 3) + 2$. Shift the graph of $f(x)$ to the right 3 units, then shift the graph upward 2 units.

   (b) $y = f(-x)$. Reflect the graph of $f(x)$ about the y-axis.

6. (a) $f(x) = 2x^2 - 8x + 13$
$$= 2(x^2 - 4x) + 13$$
$$= 2(x^2 - 4x + 4) + 13 - 8$$
$$= 2(x - 2)^2 + 5.$$

   (b)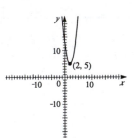

   (c) Since $f(x) = 2(x - 2)^2 + 5$ in standard form the minimum value of f is $f(2) = 5$.

7. (a) $f(-2) = 1 - (-2)^2 = 1 - 4 = -3$ (since $-2 \leq 0$).

    $f(1) = 2(1) + 1 = 2 + 1 = 3$ (since $1 > 0$).

    (b)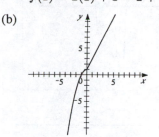

8. $f(x) = x^2 + 1$; $g(x) = x - 3$.

    (a) $(f \circ g)(x) = f(g(x)) = f(x - 3) = (x - 3)^2 + 1 = x^2 - 6x + 9 + 1 = x^2 - 6x + 10$

    (b) $(g \circ f)(x) = g(f(x)) = g(x^2 + 1) = (x^2 + 1) - 3 = x^2 - 2$

    (c) $f(g(2)) = f(-1) = (-1)^2 + 1 = 2$. (We have used the fact that $g(2) = (2) - 3 = -1$.)

    (d) $g(f(2)) = g(5) = 5 - 3 = 2$. (We have used the fact that $f(2) = 2^2 + 1 = 5$.)

    (e) $(g \circ g \circ g)(x) = g(g(g(x))) = g(g(x - 3)) = g(x - 6) = (x - 6) - 3 = x - 9$. (We have used the fact that $g(x - 3) = (x - 3) - 3 = x - 6$.)

9. (a) $f(x) = \sqrt{3 - x}$, $x \leq 3$ $\Leftrightarrow$

    $y = \sqrt{3 - x}$ $\Leftrightarrow$ $y^2 = 3 - x$ $\Leftrightarrow$

    $x = 3 - y^2$. Thus $f^{-1}(x) = 3 - x^2$,

    $x \geq 0$.

    (b) $f(x) = \sqrt{3 - x}$, $x \leq 3$ and

    $f^{-1}(x) = 3 - x^2$, $x \geq 0$

    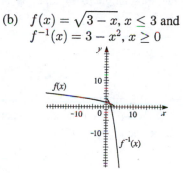

10. (a) The domain of f is $[0, 6]$, and the range of f is $[1, 7]$.

    (b)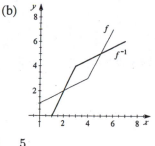

    (c) Average rate of change $= \dfrac{f(6) - f(2)}{6 - 2} = \dfrac{7 - 2}{4} = \dfrac{5}{4}$.

11. (a) $f(x) = 3x^4 - 14x^2 + 5x - 3$. The graph is shown in the viewing rectangle $[-10, 10]$ by $[-30, 10]$.

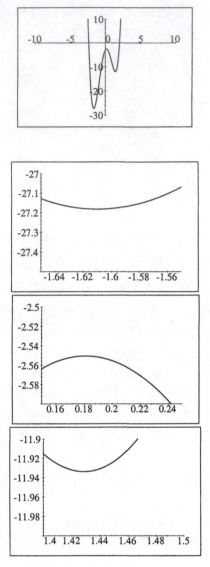

(b) No, by the Horizontal Line Test.

(c) The local minimum is approximately -27.18 when $x \approx -1.61$. Shown is the viewing rectangle $[-1.65, -1.55]$ by $[-27.5, -27]$.

The local maximum is approximately -2.55 when $x \approx 0.18$. Shown is the viewing rectangle $[0.15, 0.25]$ by $[-2.6, -2.5]$.

The local minimum is approximately -11.93 when $x \approx 1.43$. Shown is the viewing rectangle $[1.4, 1.5]$ by $[-12, -11.9]$.

(d) Using the graph in part (a) and the local minimum, -27.18, found in part (c), we see that the range is $[-27.18, \infty)$

(e) Using the information from part (c) and the graph in part (a), $f(x)$ is increasing on the intervals $[-1.61, 0.18]$ and $[1.43, \infty)$ and decreasing on the intervals $(-\infty, -1.61)$ and $[0.18, 1.43]$.

Focus on Modeling

1. Let w be the width of the building lot. Then the length of the lot is $3w$. So the area of the building lot is $A(w) = 3w^2$, $w > 0$.

3. Let w be the width of the base of the rectangle. Then the height of the rectangle is $\frac{1}{2}w$. Thus the volume of the box is given by the function $V(w) = \frac{1}{2}w^3$, $w > 0$.

5. Let P be the perimeter of the rectangle and y be the length of the other side. Since $P = 2x + 2y$ and the perimeter is 20, we have $2x + 2y = 20$ $\Leftrightarrow$ $x + y = 10$ $\Leftrightarrow$ $y = 10 - x$. Since area is $A = xy$, substituting gives $A(x) = x(10 - x) = 10x - x^2$, and since A must be positive, the domain is $0 < x < 10$.

7. Let h be the height of an altitude of the equilateral triangle whose side has length x, as shown in the diagram. Thus the area is given by $A = \frac{1}{2}xh$. By the Pythagorean Theorem, $h^2 + \left(\frac{1}{2}x\right)^2 = x^2$ $\Leftrightarrow$ $h^2 + \frac{1}{4}x^2 = x^2$ $\Leftrightarrow$ $h^2 = \frac{3}{4}x^2$ $\Leftrightarrow$ $h = \frac{\sqrt{3}}{2}x$. Substituting into the area of a triangle, we get $A(x) = \frac{1}{2}xh = \frac{1}{2}x\left(\frac{\sqrt{3}}{2}x\right) = \frac{\sqrt{3}}{4}x^2$, $x > 0$.

9. We solve for r in the formula for the area of a circle. This gives $A = \pi r^2$ $\Leftrightarrow$ $r^2 = \dfrac{A}{\pi}$ $\Rightarrow$ $r = \sqrt{\dfrac{A}{\pi}}$, so the model is $r(A) = \sqrt{\dfrac{A}{\pi}}$, $A > 0$.

11. Let h be the height of the box in feet. The volume of the box is $V = 60$. Then $x^2 h = 60$ $\Leftrightarrow$ $h = \dfrac{60}{x^2}$. The surface area, S, of the box is the sum of the area of the 4 sides and the area of the base and top. Thus $S = 4xh + 2x^2 = 4x\left(\dfrac{60}{x^2}\right) + 2x^2 = \dfrac{240}{x} + 2x^2$, so the model is $S(x) = \dfrac{240}{x} + 2x^2$, $x > 0$.

13. Let d_1 be the distance traveled south by the first ship and d_2 be the distance traveled east by the second ship. The first ship travels south for t hours at 5 mi/h, so $d_1 = 15t$ and, similarly, $d_2 = 20t$. Since the ships are traveling at right angles to each other, we can apply the Pythagorean Theorem to get
$D^2 = d_1^2 + d_2^2 = (15t)^2 + (20t)^2 = 225t^2 + 400t^2 = 625t^2$.
$D(t) = 25t$, $t \geq 0$.

15. Let b be the length of the base, l be the length of the equal sides, and h be the height in centimeters. Since the perimeter is 8, $2l + b = 8 \iff 2l = 8 - b \iff l = \frac{1}{2}(8 - b)$. By the Pythagorean Theorem, $h^2 + \left(\frac{1}{2}b\right)^2 = l^2 \iff h = \sqrt{l^2 - \frac{1}{4}b^2}$. Therefore the area of the triangle is

$$A = \tfrac{1}{2} \cdot b \cdot h = \tfrac{1}{2} \cdot b \sqrt{l^2 - \tfrac{1}{4}b^2} = \frac{b}{2}\sqrt{\tfrac{1}{4}(8 - b)^2 - \tfrac{1}{4}b^2}$$
$$= \frac{b}{4}\sqrt{64 - 16b + b^2 - b^2} = \frac{b}{4}\sqrt{64 - 16b} = \frac{b}{4} \cdot 4\sqrt{4 - b} = b\sqrt{4 - b},$$

so the model is $A(b) = b\sqrt{4 - b}, 0 < b < 4$.

17. Let w be the length of the rectangle. By the Pythagorean Theorem, $\left(\frac{1}{2}w\right)^2 + h^2 = 10^2 \iff$
$\dfrac{w^2}{4} + h^2 = 10^2 \iff w^2 = 4(100 - h^2) \iff w = 2\sqrt{100 - h^2}$ (since $w > 0$). Therefore, the area of the rectangle is $A = wh = 2h\sqrt{100 - h^2}$, so the model is $A(h) = 2h\sqrt{100 - h^2}$, $0 < h < 10$.

19. (a) We complete the table.

First number	Second number	Product
1	18	18
2	17	34
3	16	48
4	15	60
5	14	70
6	13	78
7	12	84
8	11	88
9	10	90
10	9	90
11	8	88

From the table we conclude that the numbers is still increasing, the numbers whose product is a maximum should both be 9.5.

(b) Let x be one number: then $19 - x$ is the other number, and so the product, p, is $p(x) = x(19 - x) = 19x - x^2$.

(c) $p(x) = 19x - x^2 = -(x^2 - 19x) = -\left[x^2 - 19x + \left(\frac{19}{2}\right)^2\right] + \left(\frac{19}{2}\right)^2 = -(x - 9.5)^2 + 90.25$. So the product is maximized when the numbers are both 9.5.

21. Let x and y be the two numbers. Since their sum is -24, we have $x + y = -24 \iff y = -x - 24$. The product of the two numbers is $P = xy = x(-x - 24) = -x^2 - 24x$, which we wish to maximize. So $P = -x^2 - 24x = -(x^2 + 24x) = -(x^2 + 24x + 144) + 144 = -(x + 12)^2 + 144$. Thus the maximum product is 144, and it occurs when $x = -12$ and $y = -(-12) - 24 = -12$. Thus the two numbers are -12 and -12.

23.   (a)   Let x be the width of the field (in feet) and l be the length of the field (in feet). Since the
            farmer has 2400 ft of fencing we must have $2x + l = 2400$.

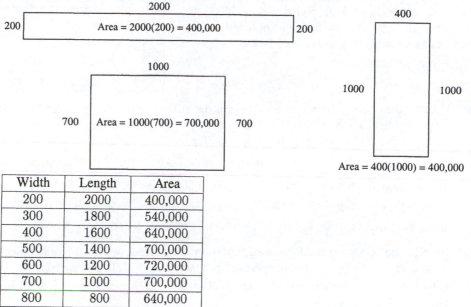

Width	Length	Area
200	2000	400,000
300	1800	540,000
400	1600	640,000
500	1400	700,000
600	1200	720,000
700	1000	700,000
800	800	640,000

It appears that the field of largest area is about 600 ft. × 1200 ft.

   (b)   Let x be the width of the field (in feet) and l be the length of the field (in feet). Since the
         farmer has 2400 ft of fencing we must have $2x + l = 2400 \quad \Leftrightarrow \quad l = 2400 - 2x$. The area
         of the fenced-in field is given by
         $A(x) = l \cdot x = (2400 - 2x)x = -2x^2 + 2400x = -2(x^2 - 1200x)$.

   (c)   The area is $A(x) = -2(x^2 - 1200x + 600^2) + 2(600^2) = -2(x - 600)^2 + 720000$. So the
         maximum area occurs when $x = 600$ feet and $l = 2400 - 2(600) = 1200$ feet.

25.   (a)   Let x be the length of the fence along the road. If the area is 1200, we have $1200 = x \cdot$ width,
            so the width of the garden is $\dfrac{1200}{x}$. Then the cost of the fence is given by the function

            $$C(x) = 5(x) + 3\left[x + 2 \cdot \frac{1200}{x}\right] = 8x + \frac{7200}{x}.$$

   (b)   We graph the function $y = C(x)$ in the viewing
         rectangle $[0, 75] \times [0, 800]$. From this we get
         the cost is minimized when $x = 30$ ft. Then the
         width is $\dfrac{1200}{30} = 40$ ft. So the length is 30 ft
         and the width is 40 ft.

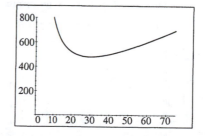

(c) We graph the function $y = C(x)$ and $y = 600$
in the viewing rectangle $[10, 65] \times [450, 650]$.
From this we get that the cost is at most 600
when $15 \le x \le 60$. So the range of lengths
he can fence along the road is 15 feet to 60 feet.

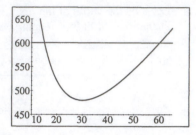

27. (a) Let p be the price of the ticket. So $10 - p$ is the difference in ticket price and therefore the
number of tickets sold is $27{,}000 + 3000(10 - p) = 57{,}000 - 3000p$. Thus the revenue is
$R(p) = p(57{,}000 - 3000p) = 57{,}000p - 3000p^2$.

(b) $R(p) = 0 = p(57{,}000 - 3000p)$. So $p = 0$ or $= \frac{57{,}000}{3000} = 19$. So at \$19 no one will come.

(c) We complete the square. $R(p) = 57{,}000p - 3000p^2 = -3000(p^2 - 19p)$
$= -3000\left(p^2 - 19p + \frac{19^2}{4}\right) + 270{,}750 = -3000\left(p - \frac{19}{2}\right)^2 + 270{,}750$. The revenue is
maximized when $p = \frac{19}{2}$, and so the price should be set at \$9.50.

29. (a) Let h be the height in feet of the straight portion of the window. The circumference of the
semicircle is $C = \frac{1}{2}\pi x$. Since the perimeter of the window is 30 feet, we have
$x + 2h + \frac{1}{2}\pi x = 30$. Solving for h, we get $2h = 30 - x - \frac{1}{2}\pi x$ $\Leftrightarrow$ $h = 15 - \frac{1}{2}x - \frac{1}{4}\pi x$.
The area of the window is
$A(x) = xh + \frac{1}{2}\pi\left(\frac{1}{2}x\right)^2 = x\left(15 - \frac{1}{2}x - \frac{1}{4}\pi x\right) + \frac{1}{8}\pi x^2 = 15x - \frac{1}{2}x^2 - \frac{1}{8}\pi x^2$.

(b) $A(x) = 15x - \frac{1}{2}x^2 - \frac{1}{8}\pi x^2 = 15x - \frac{1}{8}(\pi + 4)x^2 = -\frac{1}{8}(\pi + 4)\left[x^2 - \frac{120}{\pi+4}x\right]$
$= -\frac{1}{8}(\pi + 4)\left[x^2 - \frac{120}{\pi+4}x + \left(\frac{60}{\pi+4}\right)^2\right] + \frac{450}{\pi+4} = -\frac{1}{8}(\pi + 4)\left(x - \frac{60}{\pi+4}\right)^2 + \frac{450}{\pi+4}$. The area is
maximized when $x = \frac{60}{\pi+4} \approx 8.40$, and hence $h \approx 15 - \frac{1}{2}(8.40) - \frac{1}{4}\pi(8.40) \approx 4.20$.

31. (a) Let x be the length of one side of the base and let h be the height of the box in feet. Since the
volume of the box is $V = x^2h = 12$, we have $x^2h = 12$ $\Leftrightarrow$ $h = \dfrac{12}{x^2}$. The surface area, A,
of the box is sum of the area of the 4 sides and the area of the base. Thus the surface area of
the box is given by the formula $A(x) = 4xh + x^2 = 4x\left(\dfrac{12}{x^2}\right) + x^2 = \dfrac{48}{x} + x^2$, $x > 0$.

(b) The function $y = A(x)$ is shown in the first viewing rectangle below. In the second viewing
rectangle, we isolate the minimum, and we see that the amount of material is minimized when x
(the length and width) is 2.88 ft. Then the height is $h = \dfrac{12}{x^2} \approx 1.44$ ft.

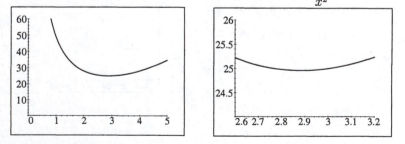

33. (a) Let w be the width of the pen and l be the length in meters. We use the area to establish a relationship between w and l. Since the area is 100 m², we have $l \cdot w = 100 \quad \Leftrightarrow \quad l = \dfrac{100}{w}$.

So the amount of fencing used is $F = 2l + 2w = 2\left(\dfrac{100}{w}\right) + 2w = \dfrac{200 + 2w^2}{w}$.

   (b) Using a graphing device, we first graph F in the viewing rectangle, $[0, 40]$ by $[0, 100]$, and locate the approximate location of the minimum value. In the second viewing rectangle, $[8, 12]$ by $[39, 41]$, we see that the minimum value of F occurs when $w = 10$. Therefore the pen should be a square with side 10 m.

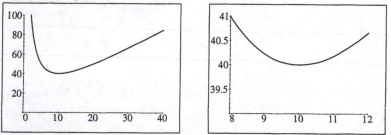

35. (a) Let x be the distance from point B to C, in miles. Then the distance from A to C is $\sqrt{x^2 + 25}$, and the energy used in flying from A to C then C to D is $f(x) = 14\sqrt{x^2 + 25} + 10(12 - x)$.

   (b) By using a graphing device, the energy expenditure is minimized when the distance from B to C is about 5.1 miles.

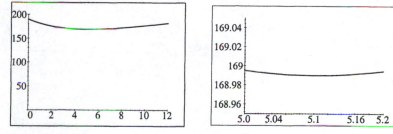

Chapter Four
Exercises 4.1

1. $P(x) = x^3 - 27$

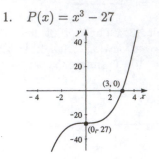

3. $P(x) = -(x + 2)^3$

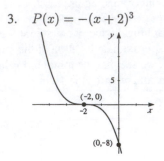

5. $P(x) = 2x^4 + 8$

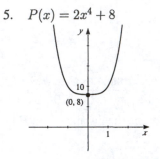

7. $P(x) = -(x - 1)^4 + 16$

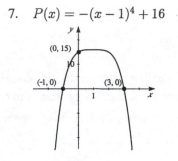

9. $P(x) = 2(x + 3)^5 - 64$

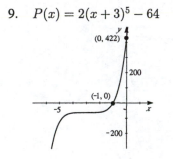

11. III

13. V

15. VI

17. $P(x) = (x - 1)(x + 2)$

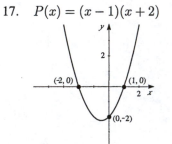

19. $P(x) = x(x - 3)(x + 2)$

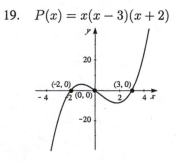

21. $P(x) = (x-3)(x+2)(3x-2)$

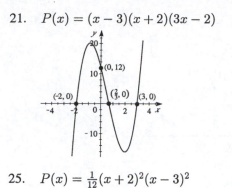

23. $P(x) = (x-1)^2(x-3)$

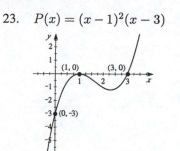

25. $P(x) = \frac{1}{12}(x+2)^2(x-3)^2$

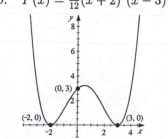

27. $P(x) = x^3(x+2)(x-3)^2$

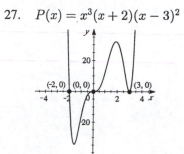

29. $P(x) = x^3 - x^2 - 6x$

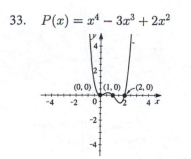

31. $P(x) = -x^3 + x^2 + 12x$

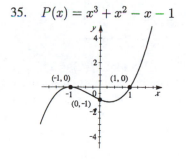

33. $P(x) = x^4 - 3x^3 + 2x^2$

35. $P(x) = x^3 + x^2 - x - 1$

37. $P(x) = 2x^3 - x^2 - 18x + 9$

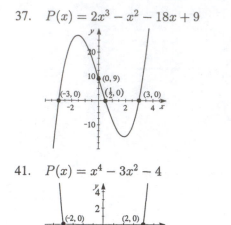

39. $P(x) = x^4 - 2x^3 - 8x + 16$

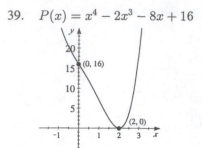

41. $P(x) = x^4 - 3x^2 - 4$

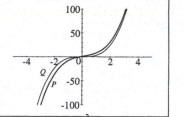

43. $P(x) = 3x^3 - x^2 + 5x + 1$; $Q(x) = 3x^3$.

Since P has odd degree and positive leading coefficient, it has the following end behavior:

$y \to \infty$ as $x \to \infty$ and $y \to -\infty$ as $x \to -\infty$.

On the large viewing rectangle, the graphs of P and Q look almost the same.

On the small viewing rectangle, the graphs of P and Q have different intercepts.

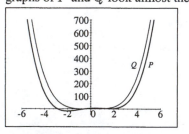

45. $P(x) = x^4 - 7x^2 + 5x + 5$; $Q(x) = x^4$.

Since P has even degree and positive leading coefficient, it has the following end behavior:

$y \to \infty$ as $x \to \infty$ and $y \to \infty$ as $x \to -\infty$.

On the large viewing rectangle, the graphs of P and Q look almost the same.

On the small viewing rectangle, the graphs of P and Q look very different and have different intercepts..

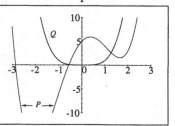

47. $P(x) = x^{11} - 9x^9$; $Q(x) = x^{11}$.

Since P has odd degree and positive leading coefficient, it has the following end behavior:

$y \to \infty$ as $x \to \infty$ and $y \to -\infty$ as $x \to -\infty$.

On the large viewing rectangle, the graphs of P and Q look like they have the same end behavior.

On the small viewing rectangle, the graphs of P and Q look very different and seem to have different end behavior.

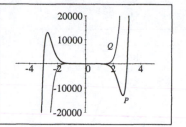   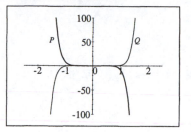

49. (a) x-intercepts at 0 and 4 y-intercept at 0.

     (b) Local maximum at $(2, 4)$. No local mimimum.

51. (a) x-intercepts at -2 and 1. y-intercept at -1.

     (b) Local maximum at $(1, 0)$. Local minimum at $(-1, -2)$.

53. $y = -x^2 + 8x$, $[-4, 12]$ by $[-50, 30]$.

No local minimum.

Local maximum at $(4, 16)$.

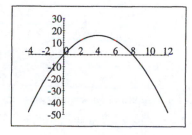

55. $y = x^3 - 12x + 9$, $[-5, 5]$ by $[-30, 30]$.

Local minimum at $(-2, 25)$,

Local maximum at $(2, -7)$.

57. $y = x^4 + 4x^3$, $[-5, 5]$ by $[-30, 30]$.

Local minimum at $(-3, -27)$.

No local maximum.

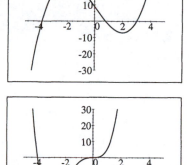

59. $y = 3x^5 - 5x^3 + 3$, $[-3, 3]$ by $[-5, 10]$.
Local maximum at $(-1, 5)$
Local minimum at $(1, 1)$.

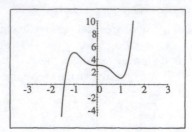

61. $y = -2x^2 + 3x + 5$.
One local maximum at $(0.75, 6.13)$.

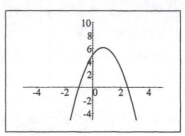

63. $y = x^3 - x^2 - x$.
One local maximum at $(-0.33, 0.19)$ and one
local minimum at $(1.00, -1.00)$.

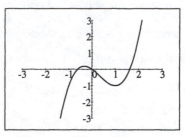

65. $y = x^4 - 5x^2 + 4$.
One local maximum at $(0, 4)$ and two local
minima at $(-1.58, -2.25)$ and $(1.58, -2.25)$.

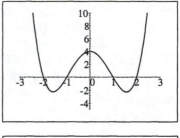

67. $y = (x - 2)^5 + 32$.
No extrema.

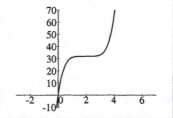

69. $y = x^8 - 3x^4 + x$.

One local maximum at $(0.44, 0.33)$ and two local minima at $(1.09, -1.15)$ and $(-1.12, -3.36)$.

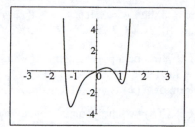

71. $y = cx^3$; $c = 1, 2, 5$, and $\frac{1}{2}$.

Increasing the value of c stretches the graph vertically.

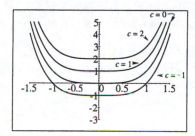

73. $P(x) = x^4 + c$; $c = -1, 0, 1$, and 2.

Increasing the value of c moves the graph up.

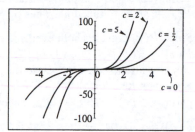

75. $P(x) = x^4 - cx$; $c = 0, 1, 8$, and 27.

Increasing the value of c causes a deeper dip in the graph, in the fourth quadrant, and moves the positive x-intercept to the right.

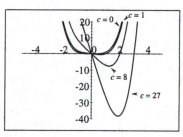

77. (a)

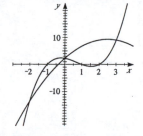

(b) The two graphs appear to intersect at 3 points.

(c) $x^3 - 2x^2 - x + 2 = -x^2 + 5x + 2 \iff x^3 - x^2 - 6x = 0 \iff x(x^2 - x - 6) = 0$
$\iff x(x - 3)(x + 2) = 0$. Then either $x = 0$, $x = 3$, or $x = -2$. If $x = 0$, then $y = 2$; if

$x = 3$ then $y = 8$; if $x = -2$, then $y = -12$. Hence the points where the two graphs intersect are $(0, 2)$, $(3, 8)$, and $(-2, -12)$.

79. (a) Let $P(x)$ be a polynomial containing only odd powers of x. Then each term of $P(x)$ can be written as Cx^{2n+1}, for some constant C and integer n. Since $C(-x)^{2n+1} = -Cx^{2n+1}$, each term of $P(x)$ is an odd function. Thus by part (a), $P(x)$ is an odd function.

     (b) Let $P(x)$ be a polynomial containing only even powers of x. Then each term of $P(x)$ can be written as Cx^{2n}, for some constant C and integer n. Since $C(-x)^{2n} = Cx^{2n}$, each term of $P(x)$ is an even function. Thus by part (b), $P(x)$ is an even function.

     (c) Since $P(x)$ contains both even and odd powers of x, we can write it in the form $P(x) = R(x) + Q(x)$, where $R(x)$ contains all the even-powered terms in $P(x)$ and $Q(x)$ contains all the odd-powered terms. By part (d), $Q(x)$ is an odd function, and by part (e), $R(x)$ is an even function. Thus, since neither $Q(x)$ nor $R(x)$ are constantly 0 (by assumption), by part (c), $P(x) = R(x) + Q(x)$ is neither even nor odd.

     (d) $P(x) = x^5 + 6x^3 - x^2 - 2x + 5 = (x^5 + 6x^3 - 2x) + (-x^2 + 5) = P_O(x) + P_E(x)$ where $P_O(x) = x^5 + 6x^3 - 2x$ and $P_E(x) = -x^2 + 5$. Since $P_O(x)$ contains only odd powers of x, it is an odd function, and since $P_E(x)$ contains only even powers of x, it is an even function.

81. (a) $P(x) = (x - 2)(x - 4)(x - 5)$.
          $P(x)$ has one local maximum and one local minimum.

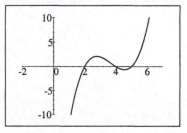

     (b) Since $P(a) = P(b) = 0$, and $P(x) > 0$ for $a < x < b$ (see the table below), the graph of P must first rise and then fall on the interval (a, b), and so P must have at least one local maximum between a and b. Using similar reasoning, the fact that $P(b) = P(c) = 0$ and $P(x) < 0$ for $b < x < c$ shows that P must have at least one local minimum between b and c. Thus P has at least two local extrema.

Interval	$(-\infty, a)$	(a, b)	(b, c)	(c, ∞)
Sign of $x - a$	$-$	$+$	$+$	$+$
Sign of $x - b$	$-$	$-$	$+$	$+$
Sign of $x - c$	$-$	$-$	$-$	$+$
Sign of $(x - a)(x - b)(x - c)$	$-$	$+$	$-$	$+$

83. $P(x) = 8x + 0.3x^2 - 0.0013x^3 - 372$

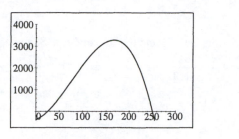

(a)   For the firm to break even, $P(x) = 0$. From the graph, we see that $P(x) = 0$ when $x \approx 25.2$. Of course, the firm cannot produce fractions of a blender, so the manufacturer must produce at least 26 blenders a year.

(b)   No, the profit does not increase indefinitely. The largest profit is approximately $3276.22, which occurs when the firm produces 166 blenders per year.

85.   (a)   The length of the bottom is $40 - 2x$, the width of the bottom is $20 - 2x$, and the height is x, so the volume of the box is $V = x(20 - 2x)(40 - 2x) = 4x^3 - 120x^2 + 800x$.

(b)   Since the height and width must be positive, we must have $x > 0$ and $20 - 2x > 0$, and so the domain of V is $0 < x < 10$.

(c)   Using the domain from part (b), we graph V in the viewing rectangle $[0, 10]$ by $[0, 1600]$. The maximum volume is $V \approx 1539.6$ when $x = 4.23$.

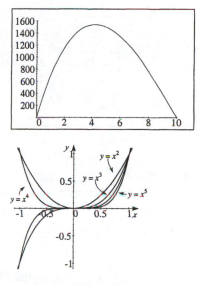

87.   The graph of $y = x^{100}$ would be close to the x-axis until just before $x = 1$ and would then pass through the points $(1, 1)$ and $(-1, 1)$. The graph of $y = x^{101}$ behaves similarly except that the y-values are negative for negative values of x.

84.   No, it is impossible. The end behavior of a third degree polynomial is the same as that of $y = kx^3$, and for this function, the values of y go off in opposite directions as $x \to \infty$ and $x \to -\infty$. But for a function with just one extremum, the values of y would head off in the same direction (either both up or both down) on either side of the extremum. An nth degree polynomial can have $n - 1$ extrema or $n - 3$ extrema or $n - 5$ extrema, and so on (decreasing by 2). An example of a polynomial that has six local extrema must be of degree 7 or higher. For example, $P(x) = (x - 1)(x - 2)(x - 3)(x - 4)(x - 5)(x - 6)(x - 7)$ has six local extrema.

Exercises 4.2

1. (a)

$$
\begin{array}{r}
3x - 4 \\
x + 3 \overline{\smash{\big)}\ 3x^2 + 5x - 4} \\
\underline{3x^2 + 9x} \\
-4x - 4 \\
\underline{-4x - 12} \\
8
\end{array}
$$

Thus the quotient is $3x - 4$ and the remainder is 8, and
$$
\frac{P(x)}{D(x)} = 3x - 4 + \frac{8}{x + 3}.
$$

(b) $P(x) = 3x^2 + 5x - 4 = (x + 3) \cdot (3x - 4) + 8$

3. (a)

$$
\begin{array}{r}
x^2 - x - 3 \\
x^2 + 3 \overline{\smash{\big)}\ x^4 - x^3 + 0x^2 + 4x + 2} \\
\underline{x^4 \qquad\quad + 3x^2} \\
-x^3 - 3x^2 + 4x \\
\underline{-x^2 \qquad - 3x} \\
-3x^2 + 7x + 2 \\
\underline{-3x^2 \qquad - 9} \\
7x + 11
\end{array}
$$

Thus the quotient is $x^2 - x - 3$ and the remainder is $7x + 11$, and
$$
\frac{P(x)}{D(x)} = x^2 - x - 3 + \frac{7x + 11}{x^2 + 3}.
$$

(b) $P(x) = x^3 + 4x^2 - 6x + 1 = (x^2 + 3) \cdot (x^2 - x - 3) + (7x + 11)$

5.

$$
\begin{array}{r}
x - 2 \\
x - 4 \overline{\smash{\big)}\ x^2 - 6x - 8} \\
\underline{x^2 - 4x} \\
-2x - 8 \\
\underline{-2x + 8} \\
-16
\end{array}
$$

Thus the quotient is $x - 2$ and the remainder is -16.

7.

$$
\begin{array}{r}
2x^2 \qquad\quad - 1 \\
2x + 1 \overline{\smash{\big)}\ 4x^3 + 2x^2 - 2x - 3} \\
\underline{4x^3 + 2x^2} \\
0x^2 - 2x - 3 \\
\underline{-2x - 1} \\
-2
\end{array}
$$

Thus the quotient is $2x^2 - 1$ and the remainder is -2.

9.

$$
\begin{array}{r}
x \quad + 2 \\
x^2 - 2x + 2 \overline{\smash{\big)}\ x^3 + 0x^2 + 6x + 3} \\
\underline{x^3 - 2x^2 + 2x} \\
2x^2 + 4x + 3 \\
\underline{2x^2 - 4x + 4} \\
8x - 1
\end{array}
$$

Thus the quotient is $x + 2$, and the remainder is $8x - 1$.

11.

$$\begin{array}{r} 3x \;\; +1 \\ 2x^2 + 0x + 5 \overline{\smash{\big)}\,6x^3 + 2x^2 + 22x + 0} \end{array}$$

$$\begin{array}{r} 6x^3 \qquad\quad +15x \\ \hline 2x^2 + 7x + 0 \\ 2x^2 \qquad +5 \\ \hline 7x - 5 \end{array}$$

Thus the quotient is $3x + 1$, and the remainder is $7x - 5$.

13.

$$\begin{array}{r} x^4 \qquad\qquad\qquad +1 \\ x^2 + 1 \overline{\smash{\big)}\,x^6 + 0x^5 + x^4 + 0x^3 + x^2 + 0x + 1} \end{array}$$

$$\begin{array}{r} x^6 \qquad + x^4 \\ \hline 0 \qquad + x^2 \qquad +1 \\ x^2 \qquad +1 \\ \hline 0 \end{array}$$

Thus the quotient is $x^4 + 1$, and the remainder is 0.

15. The synthetic division table for this problem takes the following form.

$$\begin{array}{r|rrr} 3 & 1 & -5 & 4 \\ & & 3 & -6 \\ \hline & 1 & -2 & -2 \end{array}$$

Thus the quotient is $x - 2$, and the remainder is -2.

17. The synthetic division table for this problem takes the following form.

$$\begin{array}{r|rrr} 6 & 3 & 5 & 0 \\ & & 18 & 138 \\ \hline & 3 & 23 & 138 \end{array}$$

Thus the quotient is $3x + 23$, and the remainder is 138.

19. Since $x + 2 = x - (-2)$, the synthetic division table for this problem takes the following form.

$$\begin{array}{r|rrrr} -2 & 1 & 2 & 2 & 1 \\ & & -2 & 0 & -4 \\ \hline & 1 & 0 & 2 & -3 \end{array}$$

Thus the quotient is $x^2 + 2$, and the remainder is -3.

21. Since $x + 3 = x - (-3)$ and $x^3 - 8x + 2 = x^3 + 0x^2 - 8x + 2$, the synthetic division table for this problem takes the following form.

$$\begin{array}{r|rrrr} -3 & 1 & 0 & -8 & 2 \\ & & -3 & 9 & -3 \\ \hline & 1 & -3 & 1 & -1 \end{array}$$

Thus the quotient is $x^2 - 3x + 1$, and the remainder is -1.

23. Since $x^5 + 3x^3 - 6 = x^5 + 0x^4 + 3x^3 + 0x^2 + 0x - 6$, the synthetic division table for this problem takes the following form.

$$\begin{array}{r|rrrrrr} 1 & 1 & 0 & 3 & 0 & 0 & -6 \\ & & 1 & 1 & 4 & 4 & 4 \\ \hline & 1 & 1 & 4 & 4 & 4 & -2 \end{array}$$

Thus the quotient is $x^4 + x^3 + 4x^2 + 4x + 4$, and the remainder is -2.

25. The synthetic division table for this problem takes the following form.

$$\begin{array}{r|rrrr} \tfrac{1}{2} & 2 & 3 & -2 & 1 \\ & & 1 & 2 & 0 \\ \hline & 2 & 4 & 0 & 1 \end{array}$$

Thus the quotient is $2x^2 + 4x$, and the remainder is 1.

27. Since $x^3 - 27 = x^3 + 0x^2 + 0x - 27$, the synthetic division table for this problem takes the following form.

$$
\begin{array}{r|rrrr}
3 & 1 & 0 & 0 & -27 \\
  &   & 3 & 9 & 27 \\
\hline
  & 1 & 3 & 9 & 0
\end{array}
$$
Thus the quotient is $x^2 + 3x + 9$, and the remainder is 0.

29. $P(x) = 4x^2 + 12x + 5$, $c = -1$

$$
\begin{array}{r|rrr}
-1 & 4 & 12 & 5 \\
   &   & -4 & -8 \\
\hline
   & 4 & 8 & -3
\end{array}
$$
Therefore by the Remainder Theorem, $P(-1) = -3$.

31. $P(x) = x^3 + 3x^2 - 7x + 6$, $c = 2$

$$
\begin{array}{r|rrrr}
2 & 1 & 3 & -7 & 6 \\
  &   & 2 & 10 & 6 \\
\hline
  & 1 & 5 & 3 & 12
\end{array}
$$
Therefore by the Remainder Theorem, $P(2) = 12$.

33. $P(x) = x^3 + 2x^2 - 7$, $c = -2$

$$
\begin{array}{r|rrrr}
-2 & 1 & 2 & 0 & -7 \\
   &   & -2 & 0 & 0 \\
\hline
   & 1 & 0 & 0 & -7
\end{array}
$$
Therefore by the Remainder Theorem, $P(-2) = -7$.

35. $P(x) = 5x^4 + 30x^3 - 40x^2 + 36x + 14$, $c = -7$

$$
\begin{array}{r|rrrrr}
-7 & 5 & 30 & -40 & 36 & 14 \\
   &   & -35 & 35 & 35 & -497 \\
\hline
   & 5 & -5 & -5 & 71 & -483
\end{array}
$$
Therefore by the Remainder Theorem, $P(-7) = -483$.

37. $P(x) = x^7 - 3x^2 - 1 = x^7 + 0x^6 + 0x^5 + 0x^4 + 0x^3 - 3x^2 + 0x - 1$, $c = 3$

$$
\begin{array}{r|rrrrrrrr}
3 & 1 & 0 & 0 & 0 & 0 & -3 & 0 & -1 \\
  &   & 3 & 9 & 27 & 81 & 243 & 720 & 2160 \\
\hline
  & 1 & 3 & 9 & 27 & 81 & 240 & 720 & 2159
\end{array}
$$
Therefore by the Remainder Theorem, $P(3) = 2159$.

39. $P(x) = 3x^3 + 4x^2 - 2x + 1$, $c = \frac{2}{3}$

$$
\begin{array}{r|rrrr}
\frac{2}{3} & 3 & 4 & -2 & 1 \\
            &   & 2 & 4 & \frac{4}{3} \\
\hline
            & 3 & 6 & 2 & \frac{7}{3}
\end{array}
$$
Therefore by the Remainder Theorem, $P\left(\frac{2}{3}\right) = \frac{7}{3}$.

41. $P(x) = x^3 + 2x^2 - 3x - 8$, $c = 0.1$

$$
\begin{array}{r|rrrr}
0.1 & 1 & 2 & -3 & -8 \\
    &   & 0.1 & 0.21 & -0.279 \\
\hline
    & 1 & 2.1 & -2.79 & -8.279
\end{array}
$$
Therefore by the Remainder Theorem, $P(0.1) = -8.279$.

43. $P(x) = x^3 - 3x^2 + 3x - 1$, $c = 1$

$$
\begin{array}{r|rrrr}
1 & 1 & -3 & 3 & -1 \\
  &   & 1 & -2 & 1 \\
\hline
  & 1 & -2 & 1 & 0
\end{array}
$$
Since the remainder is 0, $x - 1$ is a factor.

45. $P(x) = 2x^3 + 7x^2 + 6x - 5, c = \frac{1}{2}$

$$\frac{1}{2} \begin{array}{|rrrr} 2 & 7 & 6 & -5 \\ & 1 & 4 & 5 \\ \hline 2 & 8 & 10 & 0 \end{array}$$
Since the remainder is 0, $x - \frac{1}{2}$ is a factor.

47. $P(x) = x^3 - x^2 - 11x + 15, c = 3$

$$3 \begin{array}{|rrrr} 1 & -1 & -11 & 15 \\ & 3 & 6 & -15 \\ \hline 1 & 2 & -5 & 0 \end{array}$$

Since the remainder is 0, we know that 3 is a zero $x^3 - x^2 - 11x + 15 = (x - 3)(x^2 + 2x - 5)$.

Now $x^2 + 2x - 5 = 0$ when $x = \dfrac{-2 \pm \sqrt{2^2 + 4(1)(5)}}{2} = -1 \pm \sqrt{6}$. Hence, the zeros are $-1 - \sqrt{6}, -1 + \sqrt{6}$, and 3.

49. Since the zeros are $x = -1$, $x = 1$, and $x = 3$, the factors are $x + 1$, $x - 1$, and $x - 3$. Thus $P(x) = (x + 1)(x - 1)(x - 3) = x^3 - 3x^2 - x + 3$.

51. Since the zeros are $x = -1$, $x = 1$, $x = 3$, and $x = 5$, the factors are $x + 1$, $x - 1$, $x - 3$, and $x - 5$. Thus $P(x) = (x + 1)(x - 1)(x - 3)(x - 5) = x^4 - 8x^3 + 14x^2 + 8x - 15$.

53. Since the zeros of the polynomial are 1, -2, and 3, it follows that $P(x) = C(x - 1)(x + 2)(x - 3)$ $= C(x^3 - 2x^2 - 5x + 6) = Cx^3 - 2Cx^2 - 5Cx + 6C$. Since the coefficient of x^2 is to be 3, $-2C = 3$ so $C = -\frac{3}{2}$. Therefore, $P(x) = -\frac{3}{2}(x^3 - 2x^2 - 5x + 6) = -\frac{3}{2}x^3 + 3x^2 + \frac{15}{2}x - 9$ is the polynomial.

55. The y-intercept is 2 and the zeros of the polynomial are -1, 1, and 2. It follows that $P(x) = C(x + 1)(x - 1)(x - 2) = C(x^3 - 2x^2 - x + 2)$. Since $P(0) = 2$ we have $2 = C[(0)^3 - 2(0)^2 - (0) + 2] \quad \Leftrightarrow \quad 2 = 2C \quad \Leftrightarrow \quad C = 1$ and $P(x) = (x + 1)(x - 1)(x - 2) = x^3 - 2x^2 - x + 2$.

57. The y-intercept is 4 and the zeros of the polynomial are -2 and 1 both being degree two. It follows that $P(x) = C(x + 2)^2(x - 1)^2 = C(x^4 + 2x^3 - 3x^2 - 4x + 4)$. Since $P(0) = 4$ we have $4 = C[(0)^4 + 2(0)^3 - 3(0)^2 - 4(0) + 4] \quad \Leftrightarrow \quad 4 = 4C \quad \Leftrightarrow \quad C = 1$. Thus $P(x) = (x + 2)^2(x - 1)^2 = x^4 + 2x^3 - 3x^2 - 4x + 4$.

59. A   By the Remainder Theorem the remainder when $P(x) = 6x^{1000} - 17x^{562} + 12x + 26$ is divided by $x + 1$ is
$P(-1) = 6(-1)^{1000} - 17(-1)^{562} + 12(-1) + 26 = 6 - 17 - 12 + 26 = 3$.

   B   If $x - 1$ is a factor of $Q(x) = x^{567} - 3x^{400} + x^9 + 2$, then $Q(1)$ must equal 0.
$Q(1) = (1)^{567} - 3(1)^{400} + (1)^9 + 2 = 1 - 3 + 1 + 2 = 1 \neq 0$, so $x - 1$ is not a factor.

Exercises 4.3

1. $P(x) = x^3 - 4x^2 + 3$ has possible rational zeros ± 1 and ± 3.

3. $R(x) = 2x^5 + 3x^3 + 4x^2 - 8$ has possible rational zeros ± 1, ± 2, ± 4, ± 8, $\pm \frac{1}{2}$.

5. $T(x) = 4x^4 - 2x^2 - 7$ has possible rational zeros ± 1, ± 7, $\pm \frac{1}{2}$, $\pm \frac{7}{2}$, $\pm \frac{1}{4}$, $\pm \frac{7}{4}$.

7. (a) $P(x) = 5x^3 - x^2 - 5x + 1$ has possible rational zeros ± 1, $\pm \frac{1}{5}$.

    (b) From the graph, the actual zeroes are -1, $\frac{1}{5}$, and 1.

9. (a) $P(x) = 2x^4 - 9x^3 + 9x^2 + x - 3$ has possible rational zeros ± 1, ± 3, $\pm \frac{1}{2}$, $\pm \frac{3}{2}$.

    (b) From the graph, the actual zeroes are $-\frac{1}{2}$, 1, and 3.

11. $P(x) = x^3 + 3x^2 - 4$. The possible rational zeros are ± 1, ± 2, ± 4. $P(x)$ has 1 variation in sign and hence 1 positive real zero. $P(-x) = -x^3 + 3x^2 - 4$ has 2 variations in sign and hence 0 or 2 negative real zeros.

$$\begin{array}{r|rrr} 1 & 1 & 3 & 0 & -4 \\ & & 1 & 4 & 4 \\ \hline & 1 & 4 & 4 & 0 \end{array} \quad \Rightarrow \quad x = 1 \text{ is a zero.}$$

$P(x) = x^3 + 3x^2 - 4 = (x - 1)(x^2 + 4x + 4) = (x - 1)(x + 2)^2$. Therefore, the zeros are $x = -2, 1$.

13. $P(x) = x^3 - 3x - 2$. The possible rational zeros are ± 1, ± 2. $P(x)$ has 1 variation in sign and hence 1 positive real zero. $P(-x) = -x^3 + 3x - 2$ has 2 variations in sign and hence 0 or 2 negative real zeros.

$$\begin{array}{r|rrr} 1 & 1 & 0 & -3 & -2 \\ & & 1 & 1 & -2 \\ \hline & 1 & 1 & -2 & -4 \end{array} \quad \Rightarrow \quad x = 1 \text{ is not a zero.}$$

$$\begin{array}{r|rrr} 2 & 1 & 0 & -3 & -2 \\ & & 2 & 4 & 2 \\ \hline & 1 & 2 & 1 & 0 \end{array} \quad \Rightarrow \quad x = 2 \text{ is a zero.}$$

$P(x) = x^3 - 3x - 2 = (x - 2)(x^2 + 2x + 1) = (x - 2)(x + 1)^2$. Therefore, the zeros are $x = 2$, -1.

15. $P(x) = x^3 - 6x^2 + 12x - 8$. The possible rational zeros are ± 1, ± 2, ± 4, ± 8. $P(x)$ has 3 variations in sign and hence 1 or 3 positive real zeros. $P(-x) = -x^3 - 6x^2 - 12x - 8$ has no variations in sign and hence 0 negative real zeros.

$$\begin{array}{r|rrr} 1 & 1 & -6 & 12 & -8 \\ & & 1 & -5 & 7 \\ \hline & 1 & -5 & 7 & -1 \end{array} \quad \Rightarrow \quad x = 1 \text{ is not a zero.}$$

$$\begin{array}{r|rrr} 2 & 1 & -6 & 12 & -8 \\ & & 2 & -8 & 8 \\ \hline & 1 & -4 & 4 & 0 \end{array} \quad \Rightarrow \quad x = 2 \text{ is a zero.}$$

$P(x) = x^3 - 6x^2 + 12x - 8 = (x-2)(x^2 - 4x + 4) = (x-2)^3$. Therefore, the zero is $x = 2$.

17. $P(x) = x^3 - 4x^2 + x + 6$. The possible rational zeros are $\pm 1, \pm 2, \pm 3, \pm 6$. $P(x)$ has 2 variations in sign and hence 0 or 2 positive real zeros. $P(-x) = -x^3 - 4x^2 - x + 6$ has 1 variation in sign and hence 1 negative real zeros.

$$\begin{array}{r|rrrr} -1 & 1 & -4 & 1 & 6 \\ & & -1 & 5 & -6 \\ \hline & 1 & -5 & 6 & 0 \end{array} \Rightarrow \quad x + 1 \text{ is a factor.}$$

So $P(x) = x^3 - 4x^2 + x + 6 = (x+1)(x^2 - 5x + 6) = (x+1)(x-3)(x-2)$. Therefore, the zeros are $x = -1, 2, 3$.

19. $P(x) = x^3 + 3x^2 + 6x + 4$. The possible rational zeros are $\pm 1, \pm 2, \pm 4$. $P(x)$ has no variation in sign and hence no positive real zeros. $P(-x) = -x^3 + 3x^2 - 6x + 4$ has 3 variations in sign and hence 1 or 3 negative real zeros.

$$\begin{array}{r|rrrr} -1 & 1 & 3 & 6 & 4 \\ & & -1 & -2 & -4 \\ \hline & 1 & 2 & 4 & 0 \end{array} \Rightarrow \quad x + 1 \text{ is a factor.}$$

So $P(x) = x^3 + 3x^2 + 6x + 4 = (x+1)(x^2 + 2x + 4)$. Now, $Q(x) = x^2 + 2x + 4$ has no real zeros, since the discriminant of this quadratic is $b^2 - 4ac = (2)^2 - 4(1)(4) = -12 < 0$. Thus, the only real zero is $x = -1$.

21. Method #1: $P(x) = x^4 - 5x^2 + 4$. The possible rational zeros are $\pm 1, \pm 2, \pm 4$. $P(x)$ has 1 variation in sign and hence 1 positive real zero. $P(-x) = x^4 - 5x^2 + 4$ has 2 variations in sign and hence 0 or 2 negative real zeros.

$$\begin{array}{r|rrrrr} 1 & 1 & 0 & -5 & 0 & 4 \\ & & 1 & 1 & -4 & -4 \\ \hline & 1 & 1 & -4 & -4 & 0 \end{array} \Rightarrow \quad x = 1 \text{ is a zero.}$$

Thus $P(x) = x^4 - 5x^2 + 4 = (x-1)(x^3 + x^2 - 4x - 4)$. Continuing with the quotient we have:

$$\begin{array}{r|rrrr} -1 & 1 & 1 & -4 & -4 \\ & & -1 & 0 & 4 \\ \hline & 1 & 0 & -4 & 0 \end{array} \Rightarrow \quad x = -1 \text{ is a zero.}$$

$P(x) = x^4 - 5x^2 + 4 = (x-1)(x+1)(x^2 - 4) = (x-1)(x+1)(x-2)(x+2)$. Therefore, the zeros are $x = \pm 1, \pm 2$.

Method #2: Substituting $u = x^2$, the polynomial becomes $P(u) = u^2 - 5u + 4$, which factors: $u^2 - 5u + 4 = (u-1)(u-4) = (x^2 - 1)(x^2 - 4)$, so either $x^2 = 1$ or $x^2 = 4$. If $x^2 = 1$, then $x = \pm 1$; if $x^2 = 4$, then $x = \pm 2$. Therefore, the zeros are $x = \pm 1, \pm 2$.

23. $P(x) = x^4 + 6x^3 + 7x^2 - 6x - 8$. The possible rational zeros are $\pm 1, \pm 2, \pm 4, \pm 8$. $P(x)$ has 1 variation in sign and hence 1 positive real zero. $P(-x) = x^4 - 6x^3 + 7x^2 + 6x - 8$ has 3 variations in sign and hence 1 or 3 negative real zeros.

$$\begin{array}{r|rrrrr} 1 & 1 & 6 & 7 & -6 & -8 \\ & & 1 & 7 & 14 & 8 \\ \hline & 1 & 7 & 14 & 8 & 0 \end{array} \Rightarrow \quad x = 1 \text{ is a zero, and there are no other positive zeros.}$$

Thus $P(x) = x^4 + 6x^3 + 7x^2 - 6x - 8 = (x-1)(x^3 + 7x^2 + 14x + 8)$. Continuing by factoring the quotient, we have:

$$
\begin{array}{r|rrrr}
-1 & 1 & 7 & 14 & 8 \\
   &   & -1 & -6 & -8 \\
\hline
   & 1 & 6 & 8 & 0
\end{array}
\quad \Rightarrow \quad x = -1 \text{ is a zero.}
$$

So $P(x) = x^4 + 6x^3 + 7x^2 - 6x - 8 = (x - 1)(x + 1)(x^2 + 6x + 8)$
$= (x - 1)(x + 1)(x + 2)(x + 4)$. Therefore, the zeros are $x = -4, -2, \pm 1$.

25. $P(x) = 4x^4 - 25x^2 + 36$ has possible rational zeros $\pm 1, \pm 2, \pm 3, \pm 4, \pm 6, \pm 9, \pm 12, \pm 18, \pm 36,$
$\pm \frac{1}{2}, \pm \frac{1}{4}, \pm \frac{3}{2}, \pm \frac{3}{4}, \pm \frac{9}{2}, \pm \frac{9}{4}$. Since $P(x)$ has 2 variations in sign, there are 0 or 2 positive real zeros.
Since $P(-x) = 4x^4 - 25x^2 + 36$ has 2 variations in sign, there are 0 or 2 negative real zeros.

$$
\begin{array}{r|rrrrr}
1 & 4 & 0 & -25 & 0 & 36 \\
  &   & 4 & 4 & -21 & -21 \\
\hline
  & 4 & 4 & -21 & -21 & 15
\end{array}
$$

$$
\begin{array}{r|rrrrr}
2 & 4 & 0 & -25 & 0 & 36 \\
  &   & 8 & 16 & -18 & -36 \\
\hline
  & 4 & 8 & -9 & -18 & 0
\end{array}
\quad \Rightarrow \quad x = 2 \text{ is a zero.}
$$

$P(x) = (x - 2)(4x^3 + 8x^2 - 9x - 18)$

$$
\begin{array}{r|rrrr}
2 & 4 & 8 & -9 & -18 \\
  &   & 8 & 32 & 46 \\
\hline
  & 4 & 16 & 23 & 28
\end{array}
\quad \Rightarrow \quad \text{all positive, } x = 2 \text{ is an upper bound.}
$$

$$
\begin{array}{r|rrrr}
\frac{1}{2} & 4 & 8 & -9 & -18 \\
  &   & 2 & 5 & -2 \\
\hline
  & 4 & 10 & -4 & -20
\end{array}
\qquad
\begin{array}{r|rrrr}
\frac{1}{4} & 4 & 8 & -9 & -18 \\
  &   & 1 & \frac{9}{4} & -\frac{27}{16} \\
\hline
  & 4 & 9 & -\frac{27}{4} & -\frac{315}{16}
\end{array}
$$

$$
\begin{array}{r|rrrr}
\frac{3}{2} & 4 & 8 & -9 & -18 \\
  &   & 6 & 21 & 18 \\
\hline
  & 4 & 14 & 12 & 0
\end{array}
\quad \Rightarrow \quad x = \frac{3}{2} \text{ is a zero.}
$$

$P(x) = (x - 2)(2x - 3)(2x^2 + 7x + 6) = (x - 2)(2x - 3)(2x + 3)(x + 2)$. Therefore, the zeros
are $x = \pm 2, \pm \frac{3}{2}$.

Note: Since $P(x)$ has only even terms, factoring by substitution also works. Let $x^2 = u$; then
$P(u) = 4u^2 - 25u + 36 = (u - 4)(4u - 9) = (x^2 - 4)(4x^2 - 9)$, which yields the same results.

27. $P(x) = x^4 + 8x^3 + 24x^2 + 32x + 16$. The possible rational zeros are $\pm 1, \pm 2, \pm 4, \pm 8, \pm 16$. $P(x)$
has no variations in sign and hence no positive real zero. $P(-x) = x^4 - 8x^3 + 24x^2 - 32x + 16$
has 4 variations in sign and hence 0 or 2 or 4 negative real zeros.

$$
\begin{array}{r|rrrrr}
-1 & 1 & 8 & 24 & 32 & 16 \\
   &   & -1 & -7 & -17 & -15 \\
\hline
   & 1 & 7 & 17 & 15 & 1
\end{array}
\quad \Rightarrow \quad x = -1 \text{ is not a zero.}
$$

$$
\begin{array}{r|rrrrr}
-2 & 1 & 8 & 24 & 32 & 16 \\
   &   & -2 & -12 & -24 & -16 \\
\hline
   & 1 & 6 & 12 & 8 & 0
\end{array}
\quad \Rightarrow \quad x = -2 \text{ is a zero.}
$$

Thus $P(x) = x^4 + 8x^3 + 24x^2 + 32x + 16 = (x + 2)(x^3 + 6x^2 + 12x + 8)$. Continuing by
factoring the quotient, we have:

$$\begin{array}{r|rrrr}
-2 & 1 & 6 & 12 & 8 \\
& & -2 & -8 & -8 \\
\hline
& 1 & 4 & 4 & 0
\end{array}$$ $\Rightarrow$ $x = -2$ is a zero.

Thus $P(x) = (x+2)^2(x^2 + 4x + 4) = (x+2)^4$. Therefore, the zero is $x = -2$

29. Factoring by grouping can be applied to this exercise. $4x^3 + 4x^2 - x - 1 = 4x^2(x+1) - (x+1)$
$= (x+1)(4x^2 - 1) = (x+1)(2x+1)(2x-1)$. Therefore, the zeros are $x = -1, \pm\frac{1}{2}$.

31. $P(x) = 4x^3 - 7x + 3$. The possible rational zeros are $\pm 1, \pm 3, \pm\frac{1}{2}, \pm\frac{3}{2}, \pm\frac{1}{4}, \pm\frac{3}{4}$. Since $P(x)$ has 2 variations in sign, there are 0 or 2 positive zeros. Since $P(-x) = -4x^3 + 7x + 3$ has 1 variation in sign, there is 1 negative zero.

$$\begin{array}{r|rrrr}
\frac{1}{2} & 4 & 0 & -7 & 3 \\
& & 2 & 1 & -3 \\
\hline
& 4 & 2 & -6 & 0
\end{array}$$ $\Rightarrow$ $x = \frac{1}{2}$ is a zero.

$P(x) = (x - \frac{1}{2})(4x^2 + 2x - 6) = (2x-1)(2x^2 + x - 3) = (2x-1)(x-1)(2x+3) = 0$. Thus, the zeros are $x = -\frac{3}{2}, \frac{1}{2}, 1$.

33. $P(x) = 2x^4 - 7x^3 + 3x^2 + 8x - 4$. The possible rational zeros are $\pm 1, \pm 2, \pm 4, \pm\frac{1}{2}$. $P(x)$ has 3 variations in sign and hence 1 or 3 positive real zeros. $P(-x) = 2x^4 + 7x^3 + 3x^2 - 8x - 4$ has 1 variation in sign and hence 1 negative real zero.

$$\begin{array}{r|rrrrr}
1 & 2 & -7 & 3 & 8 & -4 \\
& & 2 & -5 & -2 & 6 \\
\hline
& 2 & -5 & -2 & 6 & 2
\end{array}$$ $\Rightarrow$ $x = 1$ is not a zero.

$$\begin{array}{r|rrrrr}
\frac{1}{2} & 2 & -7 & 3 & 8 & -4 \\
& & 1 & -3 & 0 & 4 \\
\hline
& 2 & -6 & 0 & 8 & 0
\end{array}$$ $\Rightarrow$ $x = \frac{1}{2}$ is a zero.

Thus $P(x) = 2x^4 - 7x^3 + 3x^2 + 8x - 4 = \left(x - \frac{1}{2}\right)(2x^3 - 6x^2 + 8)$. Continuing by factoring the quotient, we have:

$$\begin{array}{r|rrrr}
2 & 2 & -6 & 0 & 8 \\
& & 4 & -4 & -8 \\
\hline
& 2 & -2 & -4 & 0
\end{array}$$ $\Rightarrow$ $x = 2$ is a zero.

So $P(x) = \left(x - \frac{1}{2}\right)(x-2)(2x^2 - 2x - 4) = 2\left(x - \frac{1}{2}\right)(x-2)(x^2 - x - 2)$
$= 2\left(x - \frac{1}{2}\right)(x-2)^2(x+1)$. Thus, the zeros are $x = \frac{1}{2}, 2, -1$.

35. $P(x) = x^5 + 3x^4 - 9x^3 - 31x^2 + 36$. The possible rational zeros are $\pm 1, \pm 2, \pm 3, \pm 4, \pm 6, \pm 8,$ $\pm 9, \pm 12, \pm 18$. $P(x)$ has 2 variations in sign and hence 0 or 2 positive real zeros. $P(-x) = -x^5 + 3x^4 + 9x^3 - 31x^2 + 36$ has 3 variations in sign and hence 1 or 3 negative real zeros.

$$\begin{array}{r|rrrrrr}
1 & 1 & 3 & -9 & -31 & 0 & 36 \\
& & 1 & 4 & -5 & -36 & -36 \\
\hline
& 1 & 4 & -5 & -36 & -36 & 0
\end{array}$$ $\Rightarrow$ $x = 1$ is a zero.

So $P(x) = x^5 + 3x^4 - 9x^3 - 31x^2 + 36 = (x-1)(x^4 + 4x^3 - 5x^2 - 36x - 36)$. Continuing by factoring the quotient, we have:

$$
\begin{array}{r|rrrrr}
1 & 1 & 4 & -5 & -36 & -36 \\
  &   & 1 & 5 & 0 & -36 \\
\hline
  & 1 & 1 & 0 & -36 & -72 \\
\end{array}
\qquad
\begin{array}{r|rrrrr}
2 & 1 & 4 & -5 & -36 & -36 \\
  &   & 2 & 12 & 14 & -44 \\
\hline
  & 1 & 6 & 7 & -22 & -80 \\
\end{array}
$$

$$
\begin{array}{r|rrrrr}
3 & 1 & 4 & -5 & -36 & -36 \\
  &   & 3 & 21 & 48 & 36 \\
\hline
  & 1 & 7 & 16 & 12 & 0 \\
\end{array}
\Rightarrow \quad x = 3 \text{ is a zero.}
$$

So $P(x) = (x-1)(x-3)(x^3 + 7x^2 + 16x + 12)$. Since we have 2 positive zeros, there are no more positive zeros, so we continue by factoring the quotient with possible negative zeros.

$$
\begin{array}{r|rrrr}
-1 & 1 & 7 & 16 & 12 \\
   &   & -1 & -6 & -10 \\
\hline
   & 1 & 6 & 10 & 2 \\
\end{array}
\qquad
\begin{array}{r|rrrr}
-2 & 1 & 7 & 16 & 12 \\
   &   & -2 & -10 & -12 \\
\hline
   & 1 & 5 & 6 & 0 \\
\end{array}
\Rightarrow \quad x = -2 \text{ is a zero.}
$$

Then $P(x) = (x-1)(x-3)(x+2)(x^2 + 5x + 6) = (x-1)(x-3)(x+2)^2(x+3)$. Thus, the zeros are $x = 1, 3, -2, -3$.

37. $P(x) = 3x^5 - 14x^4 - 14x^3 + 36x^2 + 43x + 10$ has possible rational zeros $\pm 1, \pm 2, \pm 5, \pm 10, \pm\frac{1}{3},$ $\pm\frac{2}{3}, \pm\frac{5}{3}, \pm\frac{10}{3}$. Since $P(x)$ has 2 variations in sign, there are 0 or 2 positive real zeros. Since $P(-x) = -3x^5 - 14x^4 + 14x^3 + 36x^2 - 43x + 10$ has 3 variations in sign, there are 1 or 3 negative real zeros.

$$
\begin{array}{r|rrrrrr}
1 & 3 & -14 & -14 & 36 & 43 & 10 \\
  &   & 3 & -11 & -25 & 11 & 54 \\
\hline
  & 3 & -11 & -25 & 11 & 54 & 64 \\
\end{array}
$$

$$
\begin{array}{r|rrrrrr}
2 & 3 & -14 & -14 & 36 & 43 & 10 \\
  &   & 6 & -16 & -60 & -48 & -10 \\
\hline
  & 3 & -8 & -30 & -24 & -5 & 0 \\
\end{array}
\Rightarrow \quad x = 2 \text{ is a zero.}
$$

$P(x) = (x-2)(3x^4 - 8x^3 - 30x^2 - 24x - 5)$

$$
\begin{array}{r|rrrrr}
2 & 3 & -8 & -30 & -24 & -5 \\
  &   & 6 & -4 & -68 & -184 \\
\hline
  & 3 & -2 & -34 & -92 & -189 \\
\end{array}
$$

$$
\begin{array}{r|rrrrr}
5 & 3 & -8 & -30 & -24 & -5 \\
  &   & 15 & 35 & 25 & 5 \\
\hline
  & 3 & 7 & 5 & 1 & 0 \\
\end{array}
\Rightarrow \quad x = 5 \text{ is a zero.}
$$

$P(x) = (x-2)(x-5)(3x^3 + 7x^2 + 5x + 1)$

Since $3x^3 + 7x^2 + 5x + 1$ has no change in signs, there are no more positive zeros.

$$
\begin{array}{r|rrrr}
-1 & 3 & 7 & 5 & 1 \\
   &   & -3 & -4 & -1 \\
\hline
   & 3 & 4 & 1 & 0 \\
\end{array}
\Rightarrow \quad x = -1 \text{ is a zero.}
$$

$P(x) = (x-2)(x-5)(x+1)(3x^2 + 4x + 1) = (x-2)(x-5)(x+1)(x+1)(3x+1)$. Therefore, the zeros are $x = -1, -\frac{1}{3}, 2, 5$.

39. $P(x) = x^3 + 4x^2 + 3x - 2$. The possible rational zeros are $\pm 1, \pm 2$. $P(x)$ has 1 variation in sign and hence 1 positive real zero. $P(-x) = -x^3 + 4x^2 - 3x - 2$ has 2 variations in sign and hence 0 or 2 negative real zeros.

$$\begin{array}{r|rrr} 1 & 1 & 4 & 3 & -2 \\ & & 1 & 5 & 8 \\ \hline & 1 & 5 & 8 & 6 \end{array} \Rightarrow \quad x = 1 \text{ is an upper bound.}$$

$$\begin{array}{r|rrrr} -1 & 1 & 4 & 3 & -2 \\ & & -1 & -3 & 0 \\ \hline & 1 & 3 & 0 & -2 \end{array} \qquad \begin{array}{r|rrrr} -2 & 1 & 4 & 3 & -2 \\ & & -2 & -4 & 2 \\ \hline & 1 & 2 & -1 & 0 \end{array} \Rightarrow \quad x = -2 \text{ is a zero.}$$

So $P(x) = (x+2)(x^2 + 2x - 1)$. Using the quadratic formula on the second factor, we have:
$$x = \frac{-2 \pm \sqrt{2^2 - 4(1)(-1)}}{2(1)} = \frac{-2 \pm \sqrt{8}}{2} = \frac{-2 \pm 2\sqrt{2}}{2} = -1 \pm \sqrt{2}. \text{ Therefore, the zeros are}$$
$x = -2, -1 + \sqrt{2}, -1 - \sqrt{2}.$

41. $P(x) = x^4 - 6x^3 + 4x^2 + 15x + 4$. The possible rational zeros are $\pm 1, \pm 2, \pm 4$. $P(x)$ has 2 variations in sign and hence 0 or 2 positive real zeros. $P(-x) = x^4 + 6x^3 + 4x^2 - 15x + 4$ has 2 variations in sign and hence 0 or 2 negative real zeros.

$$\begin{array}{r|rrrrr} 1 & 1 & -6 & 4 & 15 & 4 \\ & & 1 & -5 & -1 & 14 \\ \hline & 1 & -5 & -1 & 14 & 18 \end{array} \qquad \begin{array}{r|rrrrr} 2 & 1 & -6 & 4 & 15 & 4 \\ & & 2 & -8 & -8 & 14 \\ \hline & 1 & -4 & -4 & 7 & 18 \end{array}$$

$$\begin{array}{r|rrrrr} 4 & 1 & -6 & 4 & 15 & 4 \\ & & 4 & -8 & -16 & -4 \\ \hline & 1 & -2 & -4 & -1 & 0 \end{array} \Rightarrow \quad x = 4 \text{ is a zero.}$$

So $P(x) = (x-4)(x^3 - 2x^2 - 4x - 1)$. Continuing by factoring the quotient, we have:

$$\begin{array}{r|rrrr} 4 & 1 & -2 & -4 & -1 \\ & & 4 & 8 & 16 \\ \hline & 1 & 2 & 4 & 15 \end{array} \Rightarrow \quad x = 4 \text{ is an upper bound.}$$

$$\begin{array}{r|rrrr} -1 & 1 & -2 & -4 & -1 \\ & & -1 & 3 & 1 \\ \hline & 1 & -3 & -1 & 0 \end{array} \Rightarrow \quad x = -1 \text{ is a zero.}$$

So $P(x) = (x-4)(x+1)(x^2 - 3x - 1)$. Using the quadratic formula on the third factor, we have:
$$x = \frac{-(-3) \pm \sqrt{(-3)^2 - 4(1)(-1)}}{2(1)} = \frac{3 \pm \sqrt{13}}{2}. \text{ Therefore, the zeros are } x = 4, -1, \frac{3 \pm \sqrt{13}}{2}.$$

43. $P(x) = x^4 - 7x^3 + 14x^2 - 3x - 9$. The possible rational zeros are $\pm 1, \pm 3, \pm 9$. $P(x)$ has 3 variations in sign and hence 1 or 3 positive real zeros. $P(-x) = x^4 + 7x^3 + 14x^2 + 3x - 4$ has 1 variation in sign and hence 1 negative real zero.

$$\begin{array}{r|rrrrr} 1 & 1 & -7 & 14 & -3 & -9 \\ & & 1 & -6 & 8 & 5 \\ \hline & 1 & -6 & 8 & 5 & 4 \end{array} \qquad \begin{array}{r|rrrrr} 3 & 1 & -7 & 14 & -3 & -9 \\ & & 3 & -12 & 6 & 9 \\ \hline & 1 & -4 & 2 & 3 & 0 \end{array} \Rightarrow \quad x = 3 \text{ is a zero.}$$

So $P(x) = (x-3)(x^3 - 4x^2 + 2x + 3)$. Since the constant term of the second term is 3, ± 9 are no longer possible zeros. Continuing by factoring the quotient, we have:

$$\begin{array}{r|rrrr} 3 & 1 & -4 & 2 & 3 \\ & & 3 & -3 & -3 \\ \hline & 1 & -1 & -1 & 0 \end{array} \Rightarrow \quad x = 3 \text{ is a zero again.}$$

So $P(x) = (x-3)^2(x^2 - x - 1)$. Using the quadratic formula on the second factor, we have:
$x = \dfrac{-(-1) \pm \sqrt{(-1)^2 - 4(1)(-1)}}{2(1)} = \dfrac{1 \pm \sqrt{5}}{2}$. Therefore, the zeros are $x = 3, \dfrac{1 \pm \sqrt{5}}{2}$.

45. $P(x) = 4x^3 - 6x^2 + 1$. The possible rational zeros are $\pm 1, \pm\frac{1}{2}, \pm\frac{1}{4}$. $P(x)$ has 2 variations in sign and hence 0 or 2 positive real zeros. $P(-x) = -4x^3 - 6x^2 + 1$ has 1 variation in sign and hence 1 negative real zero.

$$
\begin{array}{r|rrrr}
1 & 4 & -6 & 0 & 1 \\
  &   & 4 & -2 & -2 \\
\hline
  & 4 & -2 & -2 & -1
\end{array}
\qquad
\begin{array}{r|rrrr}
\frac{1}{2} & 4 & -6 & 0 & 1 \\
  &   & 2 & -2 & -1 \\
\hline
  & 4 & -4 & -2 & 0
\end{array}
\Rightarrow \quad x = \tfrac{1}{2} \text{ is a zero.}
$$

So $P(x) = \left(x - \frac{1}{2}\right)(4x^2 - 4x - 2)$. Using the quadratic formula on the second factor, we have:
$x = \dfrac{-(-4) \pm \sqrt{(-4)^2 - 4(4)(-2)}}{2(4)} = \dfrac{4 \pm \sqrt{48}}{8} = \dfrac{4 \pm 4\sqrt{3}}{8} = \dfrac{1 \pm \sqrt{3}}{2}$. Therefore, the zeros
are $x = \frac{1}{2}, \frac{1 \pm \sqrt{3}}{2}$.

47. $P(x) = 2x^4 + 15x^3 + 17x^2 + 3x - 1$. The possible rational zeros are $\pm 1, \pm\frac{1}{2}$. $P(x)$ has 1 variation in sign and hence 1 positive real zero. $P(-x) = 2x^4 - 15x^3 + 17x^2 - 3x - 1$ has 3 variations in sign and hence 1 or 3 negative real zeros.

$$
\begin{array}{r|rrrrr}
\frac{1}{2} & 2 & 15 & 17 & 3 & -1 \\
  &   & 1 & 8 & \frac{25}{2} & \frac{31}{4} \\
\hline
  & 2 & 16 & 25 & \frac{31}{2} & \frac{27}{4}
\end{array}
\Rightarrow \quad x = \tfrac{1}{2} \text{ is an upper bound.}
$$

$$
\begin{array}{r|rrrrr}
-\frac{1}{2} & 2 & 15 & 17 & 3 & -1 \\
  &   & -1 & -7 & -5 & 1 \\
\hline
  & 2 & 14 & 10 & -2 & 0
\end{array}
\Rightarrow \quad x = -\tfrac{1}{2} \text{ is a zero.}
$$

So $P(x) = \left(x + \frac{1}{2}\right)(2x^3 + 14x^2 + 10x - 2) = 2\left(x + \frac{1}{2}\right)(x^3 + 7x^2 + 5x - 1)$.

$$
\begin{array}{r|rrrr}
-1 & 1 & 7 & 5 & -1 \\
  &   & -1 & -6 & 1 \\
\hline
  & 1 & 6 & -1 & 0
\end{array}
\Rightarrow \quad x = -1 \text{ is a zero.}
$$

So $P(x) = \left(x + \frac{1}{2}\right)(2x^3 + 14x^2 + 10x - 2) = 2\left(x + \frac{1}{2}\right)(x + 1)(x^2 + 6x - 1)$

Using the quadratic formula on the third factor, we have: $x = \dfrac{-(6) \pm \sqrt{(6)^2 - 4(1)(-1)}}{2(1)}$
$= \frac{-6 \pm \sqrt{40}}{2} = \frac{-6 \pm 2\sqrt{10}}{2} = -3 \pm \sqrt{10}$. Therefore, the zeros are $x = -1, -\frac{1}{2}, -3 \pm \sqrt{10}$.

49. (a) $P(x) = x^3 - 3x^2 - 4x + 12$ has possible rational zeros $\pm 1, \pm 2, \pm 3, \pm 4, \pm 6, \pm 12$.

$$
\begin{array}{r|rrrr}
1 & 1 & -3 & -4 & 12 \\
  &   & 1 & -2 & -6 \\
\hline
  & 1 & -2 & -6 & 6
\end{array}
\qquad
\begin{array}{r|rrrr}
2 & 1 & -3 & -4 & 12 \\
  &   & 2 & -2 & -12 \\
\hline
  & 1 & -1 & -6 & 0
\end{array}
\Rightarrow \quad x = 2 \text{ is a zero.}
$$

So $P(x) = (x - 2)(x^2 - x - 6) = (x - 2)(x + 2)(x - 3)$. The real zeros of P are $-2, 2, 3$.

(b)

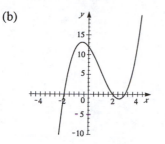

51. (a) $P(x) = -x^3 + 2x^2 + 15x - 36$ has possible rational zeros $\pm 1, \pm 2, \pm 3, \pm 4, \pm 6, \pm 9, \pm 12,$ ± 18.

$$
\begin{array}{r|rrrr}
1 & -1 & 2 & 15 & -36 \\
  &   & -1 & 1 & 16 \\
\hline
  & -1 & 1 & 16 & -20
\end{array}
\qquad
\begin{array}{r|rrrr}
2 & -1 & 2 & 15 & -36 \\
  &   & -2 & 0 & 30 \\
\hline
  & -1 & 0 & 15 & -6
\end{array}
$$

$$
\begin{array}{r|rrrr}
3 & -1 & 2 & 15 & -36 \\
  &   & -3 & -3 & -36 \\
\hline
  & -1 & -1 & 12 & 0
\end{array}
\quad \Rightarrow \quad x = 3 \text{ is a zero.}
$$

So $P(x) = (x-3)(-x^2 - x + 12) = (x-3)(-x+3)(x+4) = -(x-3)^2(x+4)$. The real zeros of P are $-4, 3$.

(b)

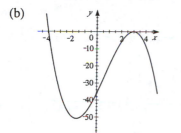

53. (a) $P(x) = x^4 - 5x^3 + 6x^2 + 4x - 8$ has possible rational zeros $\pm 1, \pm 2, \pm 4, \pm 8$.

$$
\begin{array}{r|rrrrr}
1 & 1 & -5 & 6 & 4 & -8 \\
  &   & 1 & -4 & 2 & 6 \\
\hline
  & 1 & -4 & 2 & 6 & -2
\end{array}
\qquad
\begin{array}{r|rrrrr}
2 & 1 & -5 & 6 & 4 & -8 \\
  &   & 2 & -6 & 0 & 8 \\
\hline
  & 1 & -3 & 0 & 4 & 0
\end{array}
\quad \Rightarrow \quad x = 2 \text{ is a zero.}
$$

So $P(x) = (x-2)(x^3 - 3x^2 + 4)$ and the possible rational zeros are restricted to $-1, \pm 2, \pm 4$.

$$
\begin{array}{r|rrrr}
2 & 1 & -3 & 0 & 4 \\
  &   & 2 & -2 & -4 \\
\hline
  & 1 & -1 & -2 & 0
\end{array}
\quad \Rightarrow \quad x = 2 \text{ is a zero again.}
$$

$P(x) = (x-2)^2(x^2 - x - 2) = (x-2)^2(x-2)(x+1) = (x-2)^3(x+1)$. So the real zeros of P are -1 and 2.

(b)

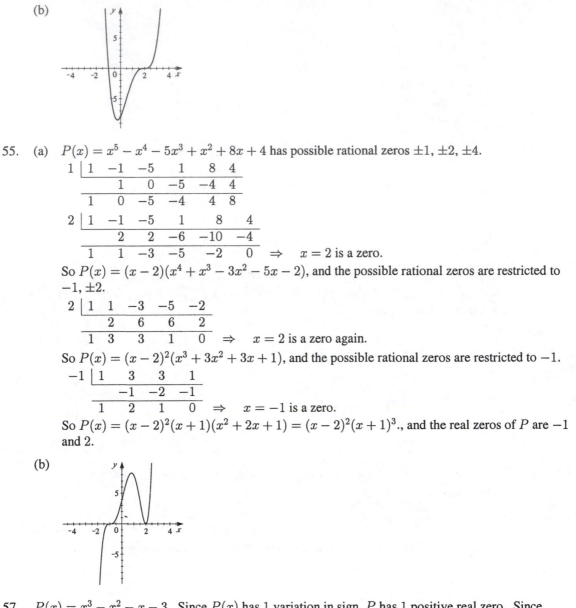

55. (a) $P(x) = x^5 - x^4 - 5x^3 + x^2 + 8x + 4$ has possible rational zeros $\pm 1, \pm 2, \pm 4$.

$$
\begin{array}{r|rrrrrr}
1 & 1 & -1 & -5 & 1 & 8 & 4 \\
  &   & 1 & 0 & -5 & -4 & 4 \\
\hline
  & 1 & 0 & -5 & -4 & 4 & 8
\end{array}
$$

$$
\begin{array}{r|rrrrrr}
2 & 1 & -1 & -5 & 1 & 8 & 4 \\
  &   & 2 & 2 & -6 & -10 & -4 \\
\hline
  & 1 & 1 & -3 & -5 & -2 & 0 \quad \Rightarrow \quad x = 2 \text{ is a zero.}
\end{array}
$$

So $P(x) = (x - 2)(x^4 + x^3 - 3x^2 - 5x - 2)$, and the possible rational zeros are restricted to $-1, \pm 2$.

$$
\begin{array}{r|rrrrr}
2 & 1 & 1 & -3 & -5 & -2 \\
  &   & 2 & 6 & 6 & 2 \\
\hline
  & 1 & 3 & 3 & 1 & 0 \quad \Rightarrow \quad x = 2 \text{ is a zero again.}
\end{array}
$$

So $P(x) = (x - 2)^2(x^3 + 3x^2 + 3x + 1)$, and the possible rational zeros are restricted to -1.

$$
\begin{array}{r|rrrr}
-1 & 1 & 3 & 3 & 1 \\
   &   & -1 & -2 & -1 \\
\hline
   & 1 & 2 & 1 & 0 \quad \Rightarrow \quad x = -1 \text{ is a zero.}
\end{array}
$$

So $P(x) = (x - 2)^2(x + 1)(x^2 + 2x + 1) = (x - 2)^2(x + 1)^3$., and the real zeros of P are -1 and 2.

(b)

57. $P(x) = x^3 - x^2 - x - 3$. Since $P(x)$ has 1 variation in sign, P has 1 positive real zero. Since $P(-x) = -x^3 - x^2 + x - 3$ has 2 variations in sign, P has 2 or 0 negative real zeros. Thus, P has 1 or 3 real zeros.

59. $P(x) = 2x^6 + 5x^4 - x^3 - 5x - 1$. Since $P(x)$ has 1 variation in sign, P has 1 positive real zero. Since $P(-x) = 2x^6 + 5x^4 + x^3 + 5x - 1$ has 1 variation in sign, P has 1 negative real zero. Therefore, P has 2 real zeros.

61. $P(x) = x^5 + 4x^3 - x^2 + 6x$. Since $P(x)$ has 2 variations in sign, P has 2 or 0 positive real zeros. Since $P(-x) = -x^5 - 4x^3 - x^2 - 6x$ has 0 variations in sign, P has 0 negative real zeros.

Therefore, P has a total of 1 or 3 real zeros (since $x = 0$ is a zero, but is neither positive nor negative).

63. $P(x) = 2x^3 + 5x^2 + x - 2;\ a = -3, b = 1$

$$
\begin{array}{r|rrrr}
-3 & 2 & 5 & 1 & -2 \\
   &   & -6 & 3 & -12 \\
\hline
   & 2 & -1 & 4 & -14
\end{array}
\qquad \text{alternating signs} \quad \Rightarrow \quad \text{lower bound.}
$$

$$
\begin{array}{r|rrrr}
1 & 2 & 5 & 1 & -2 \\
  &   & 2 & 7 & 8 \\
\hline
  & 2 & 7 & 8 & 6
\end{array}
\qquad \text{all nonnegative} \quad \Rightarrow \quad \text{upper bound.}
$$

Therefore $a = -3, b = 1$ are lower and upper bounds, respectively.

65. $P(x) = 8x^3 + 10x^2 - 39x + 9;\ a = -3, b = 2$

$$
\begin{array}{r|rrrr}
-3 & 8 & 10 & -39 & 9 \\
   &   & -24 & 42 & -9 \\
\hline
   & 8 & -14 & 3 & 0
\end{array}
\qquad \text{alternating signs} \quad \Rightarrow \quad \text{lower bound.}
$$

$$
\begin{array}{r|rrrr}
2 & 8 & 10 & -39 & 9 \\
  &   & 16 & 52 & 26 \\
\hline
  & 8 & 26 & 13 & 35
\end{array}
\qquad \text{all nonnegative} \quad \Rightarrow \quad \text{upper bound.}
$$

Therefore $a = -3, b = 2$ are lower and upper bounds, respectively. Note that $x = -3$ is also a zero.

There are many possible solutions to Exercises 67 and 69 since we are only asked to find 'an upper bound' and 'a lower bound'.

67. $P(x) = x^3 - 3x^2 + 4$ and use the Upper and Lower Bounds Theorem:

$$
\begin{array}{r|rrrr}
-1 & 1 & -3 & 0 & 4 \\
   &   & -1 & 4 & -4 \\
\hline
   & 1 & -4 & 4 & 0
\end{array}
\qquad \text{alternating signs} \quad \Rightarrow \quad \text{lower bound.}
$$

$$
\begin{array}{r|rrrr}
3 & 1 & -3 & 0 & 4 \\
  &   & 3 & 0 & 0 \\
\hline
  & 1 & 0 & 0 & 4
\end{array}
\qquad \text{all nonnegative} \quad \Rightarrow \quad \text{upper bound.}
$$

Therefore -1 is a lower bound (and a zero) and 3 is an upper bound.

69. $P(x) = x^4 - 2x^3 + x^2 - 9x + 2.$

$$
\begin{array}{r|rrrrr}
1 & 1 & -2 & 1 & -9 & 2 \\
  &   & 1 & -1 & 0 & -9 \\
\hline
  & 1 & -1 & 0 & -9 & -7
\end{array}
\qquad
\begin{array}{r|rrrrr}
2 & 1 & -2 & 1 & -9 & 2 \\
  &   & 2 & 0 & 2 & -14 \\
\hline
  & 1 & 0 & 1 & -7 & -12
\end{array}
$$

$$
\begin{array}{r|rrrrr}
3 & 1 & -2 & 1 & -9 & 2 \\
  &   & 3 & 3 & 12 & 9 \\
\hline
  & 1 & 1 & 4 & 3 & 11
\end{array}
\qquad \text{all positive} \quad \Rightarrow \quad \text{upper bound.}
$$

$$
\begin{array}{r|rrrrr}
-1 & 1 & -2 & 1 & -9 & 2 \\
   &   & -1 & 3 & -4 & 13 \\
\hline
   & 1 & -3 & 4 & -13 & 15
\end{array}
\qquad \text{alternating signs} \quad \Rightarrow \quad \text{lower bound.}
$$

Therefore -1 is a lower bound and 3 is an upper bound.

71. $P(x) = 2x^4 + 3x^3 - 4x^2 - 3x + 2$.

$$
\begin{array}{r|rrrrr}
1 & 2 & 3 & -4 & -3 & 2 \\
  &   & 2 & 5 & 1 & -2 \\
\hline
  & 2 & 5 & 1 & -2 & 0
\end{array}
\Rightarrow \quad x = 1 \text{ is a zero.}
$$

$P(x) = (x - 1)(2x^3 + 5x^2 + x - 2)$

$$
\begin{array}{r|rrrr}
-1 & 2 & 5 & 1 & -2 \\
   &   & -2 & -3 & 2 \\
\hline
   & 2 & 3 & -2 & 0
\end{array}
\Rightarrow \quad x = -1 \text{ is a zero.}
$$

$P(x) = (x-1)(x+1)(2x^2 + 3x - 2) = (x-1)(x+1)(2x-1)(x+2)$. Therefore, the zeros are $x = -2, \frac{1}{2}, \pm 1$.

73. Method #1: $P(x) = 4x^4 - 21x^2 + 5$ has 2 variations in sign, so by Descartes' rule of signs there are either 2 or 0 positive zeros. If we replace x with $(-x)$, the function does not change, so there are either 2 or 0 negative zeros. Possible rational zeros are $\pm 1, \pm\frac{1}{2}, \pm\frac{1}{4}, \pm 5, \pm\frac{5}{2}, \pm\frac{5}{4}$. By inspection, ± 1 and ± 5 are not zeros, so we must look for non-integer solutions:

$$
\begin{array}{r|rrrrr}
\frac{1}{2} & 4 & 0 & -21 & 0 & 5 \\
            &   & 2 & 1 & -10 & -5 \\
\hline
            & 4 & 2 & -20 & -10 & 0
\end{array}
\Rightarrow \quad x = \frac{1}{2} \text{ is a zero.}
$$

$P(x) = \left(x - \frac{1}{2}\right)(4x^3 + 2x^2 - 20x - 10)$, continuing with the quotient, we have:

$$
\begin{array}{r|rrrr}
-\frac{1}{2} & 4 & 2 & -20 & -10 \\
             &   & -2 & 0 & 10 \\
\hline
             & 4 & 0 & -20 & 0
\end{array}
\Rightarrow \quad x = -\frac{1}{2} \text{ is a zero.}
$$

$P(x) = \left(x - \frac{1}{2}\right)\left(x + \frac{1}{2}\right)(4x^2 - 20) = 0$. If $4x^2 - 20 = 0$, then $x = \pm\sqrt{5}$. Thus the zeros are $x = \pm\frac{1}{2}, \pm\sqrt{5}$.

Method #2: Substituting $u = x^2$, the equation becomes $4u^2 - 21u + 5 = 0$, which factors: $4u^2 - 21u + 5 = (4u - 1)(u - 5) = (4x^2 - 1)(x^2 - 5)$. Then either we have $x^2 = 5$, so that $x = \pm\sqrt{5}$, or we have $x^2 = \frac{1}{4}$, so that $x = \pm\sqrt{\frac{1}{4}} = \pm\frac{1}{2}$. Thus the zeros are $x = \pm\frac{1}{2}, \pm\sqrt{5}$.

75. $P(x) = x^5 - 7x^4 + 9x^3 + 23x^2 - 50x + 24$. The possible rational zeros are $\pm 1, \pm 2, \pm 3, \pm 4, \pm 6, \pm 8, \pm 12, \pm 24$. $P(x)$ has 4 variations in sign and hence 0, 2, or 4 positive real zeros. $P(-x) = -x^5 - 7x^4 - 9x^3 + 23x^2 + 50x + 24$ has 1 variation in sign, and hence 1 negative real zero.

$$
\begin{array}{r|rrrrrr}
1 & 1 & -7 & 9 & 23 & -50 & 24 \\
  &   & 1 & -6 & 3 & 26 & -24 \\
\hline
  & 1 & -6 & 3 & 26 & -24 & 0
\end{array}
\Rightarrow \quad x = 1 \text{ is a zero.}
$$

$P(x) = (x - 1)(x^4 - 6x^3 + 3x^2 + 26x - 24)$; continuing with the quotient, we try 1 again.

$$
\begin{array}{r|rrrrr}
1 & 1 & -6 & 3 & 26 & -24 \\
  &   & 1 & -5 & -2 & 24 \\
\hline
  & 1 & -5 & -2 & 24 & 0
\end{array}
\Rightarrow \quad x = 1 \text{ is a zero again.}
$$

$P(x) = (x - 1)^2(x^3 - 5x^2 - 2x + 24)$; continuing with the quotient, we start by trying 1 again.

$$
\begin{array}{r|rrrr}
1 & 1 & -5 & -2 & 24 \\
  &   & 1 & -4 & -6 \\
\hline
  & 1 & -4 & -6 & 18
\end{array}
\qquad\qquad
\begin{array}{r|rrrr}
2 & 1 & -5 & -2 & 24 \\
  &   & 2 & -6 & -16 \\
\hline
  & 1 & -3 & -8 & 8
\end{array}
$$

$$3 \enspace | \enspace \begin{array}{cccc} 1 & -5 & -2 & 24 \\ & 3 & -6 & -24 \\ \hline 1 & -2 & -8 & 0 \end{array} \enspace \Rightarrow \quad x = 3 \text{ is a zero.}$$

$P(x) = (x-1)^2(x-3)(x^2 - 2x - 8) = (x-1)^2(x-3)(x-4)(x+2)$. Therefore, the zeros are $x = -2, 1, 3, 4$.

77. $P(x) = x^3 - x - 2$. The only possible rational zeros of $P(x)$ are ± 1 and ± 2.

$$1 \enspace | \enspace \begin{array}{cccc} 1 & 0 & -1 & -2 \\ & 1 & 1 & 0 \\ \hline 1 & 1 & 0 & -2 \end{array} \qquad 2 \enspace | \enspace \begin{array}{cccc} 1 & 0 & -1 & -2 \\ & 2 & 4 & 6 \\ \hline 1 & 2 & 3 & 4 \end{array} \qquad -1 \enspace | \enspace \begin{array}{cccc} 1 & 0 & -1 & -2 \\ & -1 & 1 & 0 \\ \hline 1 & -1 & 0 & -2 \end{array}$$

Since the row that contains -1 alternates between nonnegative and nonpositive, -1 is a lower bound and there is no need to try -2. Therefore, $P(x)$ does not have any rational zeros.

79. $P(x) = 3x^3 - x^2 - 6x + 12$ has possible rational zeros $\pm 1, \pm 2, \pm 3, \pm 4, \pm 6, \pm 12, \pm \frac{1}{3}, \pm \frac{2}{3}, \pm \frac{4}{3}$.

$$\begin{array}{r|cccc} & 3 & -1 & -6 & 12 \\ \hline 1 & 3 & 2 & -4 & 8 \\ 2 & 3 & 5 & 4 & 20 \\ \frac{1}{3} & 3 & 0 & -6 & 10 \\ \frac{2}{3} & 3 & 1 & -\frac{16}{3} & \frac{76}{9} \\ \frac{4}{3} & 3 & 3 & -2 & \frac{28}{3} \\ -1 & 3 & -4 & -2 & 14 \\ -2 & 3 & -7 & 8 & -4 \\ -\frac{1}{3} & 3 & -2 & -\frac{16}{3} & \frac{124}{9} \\ -\frac{2}{3} & 3 & -3 & -4 & \frac{44}{3} \\ -\frac{4}{3} & 3 & -5 & \frac{2}{3} & \frac{100}{9} \end{array}$$

all positive $\quad \Rightarrow \quad x = 2$ is an upper bound

alternating signs $\quad \Rightarrow \quad x = -2$ is a lower bound

Therefore, there are no rational zeros.

81. $P(x) = x^3 - 3x^2 - 4x + 12$; $[-4, 4]$ by $[-15, 15]$.
The possible rational zeros are $\pm 1, \pm 2, \pm 3, \pm 4, \pm 6, \pm 12$.
By observing the graph of P, the rational zeros are $x = -2, 2, 3$.

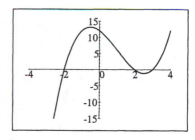

83. $P(x) = 2x^4 - 5x^3 - 14x^2 + 5x + 12$; $[-2, 5]$ by $[-40, 40]$.
The possible rational zeros are $\pm 1, \pm 2, \pm 3, \pm 4, \pm 6$, $\pm 12, \pm \frac{1}{2}, \pm \frac{3}{2}$.
By observing the graph of P, the zeros are $x = -\frac{3}{2}, -1, 1, 4$.

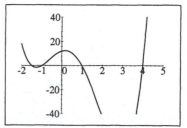

85. $x^4 - x - 4 = 0$. Possible rational solutions are $\pm 1, \pm 2, \pm 4$.

1	1	0	0	−1	−4
		1	1	1	0
	1	1	1	0	−4

2	1	0	0	−1	−4
		2	4	8	14
	1	2	4	7	10

$\Rightarrow \quad x = 2$ is an upper bound.

−1	1	0	0	−1	−4
		−1	1	−1	2
	1	−1	1	−2	−2

−2	1	0	0	−1	−4
		−2	4	−8	18
	1	−2	4	−9	14

$\Rightarrow \quad x = -2$ is a lower bound.

Therefore, we graph the function in the viewing rectangle $[-2, 2]$ by $[-5, 20]$ and see there are two solutions.

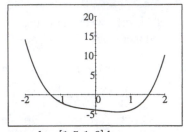

Viewing rectangle: $[-1.3, -1.25]$ by $[-0.1, 0.1]$. Solution $x \approx -1.28$.

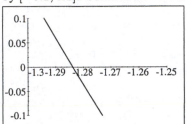

Viewing rectangle: $[1.5, 1.6]$ by $[-0.1, 0.1]$. Solution $x \approx 1.53$.

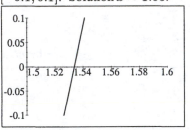

Thus the solutions are $x \approx -1.28, 1.53$.

87. $4.00x^4 + 4.00x^3 - 10.96x^2 - 5.88x + 9.09 = 0$.

1	4	4	−10.96	−5.88	9.09
		4	8	−2.96	−8.84
	4	8	−2.96	−8.84	0.25

2	4	4	−10.96	−5.88	9.09
		8	24	26.08	40.40
	4	12	13.04	20.2	49.49

$\Rightarrow \quad x = 2$ is an upper bound.

−2	4	4	−10.96	−5.88	9.09
		−8	8	5.92	−0.08
	4	−4	−2.96	0.04	9.01

−3	4	4	−10.96	−5.88	9.09
		−12	24	−39.12	135
	4	−8	13.04	−45	144.09

$\Rightarrow \quad x = -3$ is a lower bound.

Therefore, we graph the function in the viewing rectangle $[-3, 2]$ by $[-10, 40]$. There appear to be two solutions.

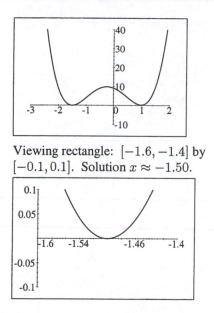

Viewing rectangle: $[-1.6, -1.4]$ by $[-0.1, 0.1]$. Solution $x \approx -1.50$.

Viewing rectangle: $[0.8, 1.2]$ by $[0, 1]$. The graph comes close but does not go through the x-axis. Thus there is no solution here.

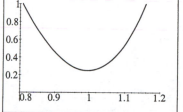

Therefore, the only solution is $x \approx -1.50$.

89. (a) Since $z > b$, we have $z - b > 0$. Since all the coefficients of $Q(x)$ are nonnegative, and since $z > 0$, we have $Q(z) > 0$ (being a sum of positive terms). Thus, $P(z) = (z - b) \cdot Q(z) + r > 0$, since the sum of a positive number and a nonnegative number.

(b) In part (a), we showed that if b satisfies the conditions of the first part of the Upper and Lower Bounds Theorem and $z > b$, then $P(z) > 0$. This means that no real zero of P can be larger than b, so b is an upper bound for the real zeros.

(c) Suppose $-b$ is a negative lower bound for the real zeros of $P(x)$. Then clearly b is an upper bound for $P_1(x) = P(-x)$. Thus, as in Part (a), we can write $P_1(x) = (x - b) \cdot Q(x) + r$, where $r > 0$ and the coefficients of Q are all nonnegative, and $P(x) = P_1(-x) = (-x - b) \cdot Q(-x) + r = (x + b) \cdot [-Q(-x)] + r$. Since the coefficients of $Q(x)$ are all nonnegative, the coefficients of $-Q(-x)$ will be alternately nonpositive and nonnegative, which proves the second part of the Upper and Lower Bounds Theorem.

91. Let $r =$ the radius of the silo. The volume of the hemispherical roof is $\frac{1}{2}\left(\frac{4}{3}\pi r^3\right) = \frac{2}{3}\pi r^3$. The volume of the cylindrical section is $\pi(r^2)(30) = 30\pi r^2$. Because the total volume of the silo is 15000 ft^3, we get the following equation: $\frac{2}{3}\pi r^3 + 30\pi r^2 = 15000 \iff \frac{2}{3}\pi r^3 + 30\pi r^2 - 15000 = 0 \iff \pi r^3 + 45\pi r^2 - 22500 = 0$. Using a graphing device, we first graph the polynomial in the viewing rectangle $[0, 15]$ by $[-10000, 10000]$. The solution, $r \approx 11.28$ ft., is shown in the viewing rectangle $[11.2, 11.4]$ by $[-1, 1]$.

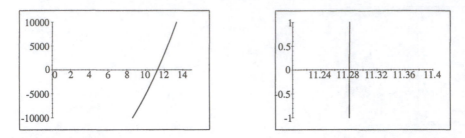

93. $h(t) = 11.60t - 12.41t^2 + 6.20t^3 - 1.58t^4 + 0.20t^5 - 0.01t^6$ is shown in the viewing rectangle
     $[0, 10]$ by $[0, 6]$.

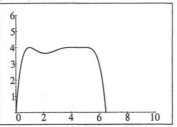

(a) It started to snow again.

(b) No, $h(t) \leq 4$.

(c) The function $h(t)$ is shown in the
     viewing rectangle $[6, 6.5]$ by $[0, 0.5]$.
     The x-intercept of the function is a little
     less than 6.5, which means that the snow
     melted just before midnight on Saturday
     night.

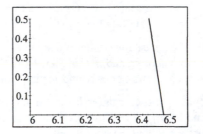

95. Let $r = $ the radius of the cone and cylinder. Let $h = $ the height of the cone. Since the height and
     diameter are equal, we get $h = 2r$. So the volume of the cylinder is $V_1 = \pi r^2 \cdot (cylinder\ height)$
     $= 20\pi r^2$, and the volume of the cone is $V_2 = \frac{1}{3}\pi r^2 h = \frac{1}{3}\pi r^2 (2r) = \frac{2}{3}\pi r^3$. Since the total volume is
     $\frac{500\pi}{3}$, it follows that $\frac{2}{3}\pi r^3 + 20\pi r^2 = \frac{500\pi}{3} \quad \Leftrightarrow \quad r^3 + 30r^2 - 250 = 0$. By Descartes' Rule of
     Signs, there is 1 positive zero.

r	$r^3 + 30r^2 - 250$	r	$r^3 + 30r^2 - 250$
1	-219	2.76	-2.33
2	-122	2.77	1.44
3	47	2.765	1.44
2.7	-11.62		
2.8	7.15		

Since r is between 2.76 and 2.765, the radius should be 2.76 m (correct to two decimals).

97. Let b be the width of the base, and let l be the length of the box. Then the length plus girth is
     $l + 4b = 108$, and the volume is $V = lb^2 = 2200$. Solving the first equation for l and substituting
     this value into the second equation yields $l = 108 - 4b \quad \Rightarrow V = (108 - 4b)b^2 = 2200 \quad \Leftrightarrow$

$4b^3 - 108b^2 + 2200 = 0 \quad \Leftrightarrow \quad 4(b^3 - 27b^2 + 550) = 0$. Now $P(b) = b^3 - 27b^2 + 550$ has two variations in sign, so there are 0 or 2 positive real zeros. We also observe that since $l > 0$, $b < 27$, so $b = 27$ is an upper bound. Thus the possible positive rational real zeros are 1, 2, 3, 10, 11, 22, 25.

$$
\begin{array}{r|rrrr}
1 & 1 & -27 & 0 & 550 \\
  &   & 1 & -26 & -26 \\
\hline
  & 1 & -26 & -26 & 524
\end{array}
\qquad
\begin{array}{r|rrrr}
2 & 1 & -27 & 0 & 550 \\
  &   & 2 & -50 & -100 \\
\hline
  & 1 & -25 & -50 & 450
\end{array}
$$

$$
\begin{array}{r|rrrr}
5 & 1 & -27 & 0 & 550 \\
  &   & 5 & -110 & -550 \\
\hline
  & 1 & -22 & -110 & 0
\end{array}
\qquad \Rightarrow \quad b = 5 \text{ is a zero.}
$$

$P(b) = (b - 5)(b^2 - 22b - 110)$. The other zeros are $b = \dfrac{22 \pm \sqrt{484 - 4(1)(-110)}}{2}$

$= \dfrac{22 \pm \sqrt{924}}{2} = \dfrac{22 \pm 30.397}{2}$. The positive answer from this factor is $b \approx 26.20$. Thus we have two possible solutions, $b = 5$ or $b \approx 26.20$. If $b = 5$, then $l = 108 - 4(5) = 88$; if $b \approx 26.20$, then $l = 108 - 4(26.20) = 3.20$. Thus the length of the box is either 88 in. or 3.20 in.

99. (a) Substituting $X - \frac{a}{3}$ for x we have

$$
\begin{aligned}
x^3 + ax^2 + bx + c &= \left(X - \tfrac{a}{3}\right)^3 + a\left(X - \tfrac{a}{3}\right)^2 + b\left(X - \tfrac{a}{3}\right) + c \\
&= X^3 - aX^2 + \tfrac{a^2}{3}X + \tfrac{a^3}{27} + a\left(X^2 - \tfrac{2a}{3}X + \tfrac{a^2}{9}\right) + bX - \tfrac{ab}{3} + c \\
&= X^3 - aX^2 + \tfrac{a^2}{3}X + \tfrac{a^3}{27} + aX^2 - \tfrac{2a^2}{3}X + \tfrac{a^3}{9} + bX - \tfrac{ab}{3} + c \\
&= X^3 + (-a + a)X^2 + \left(-\tfrac{a^2}{3} - \tfrac{2a^2}{3} + b\right)X + \left(\tfrac{a^3}{27} + \tfrac{a^3}{9} - \tfrac{ab}{3} + c\right) \\
&= X^3 + (b - a^2)X + \left(\tfrac{4a^3}{27} - \tfrac{ab}{3} + c\right).
\end{aligned}
$$

(b) $x^3 + 6x^2 + 9x + 4 = 0$. Setting $a = 6$, $b = 9$, and $c = 4$, we have:
$X^3 + (9 - 6^2)X + (32 - 18 + 4) = X^3 - 27X + 18$.

Exercises 4.4

1. (a) $x^4 + 4x^2 = 0$ $\Leftrightarrow$ $x^2(x^2 + 4) = 0$. So $x = 0$ or $x^2 + 4 = 0$. If
    $x^2 + 4 = 0$ then $x^2 = -4$ $\Leftrightarrow$ $x = \pm 2i$. Therefore, the solutions are $x = 0$ and $\pm 2i$.

    (b) To get the complete factorization, we factor the remaining quadratic factor
    $P(x) = x^2(x + 4) = x^2(x - 2i)(x + 2i)$.

3. (a) $x^3 - 2x^2 + 2x = 0$ $\Leftrightarrow$ $x(x^2 - 2x + 2) = 0$. So $x = 0$ or $x^2 - 2x + 2 = 0$. If
    $x^2 - 2x + 2 = 0$ then $x = \frac{-(-2) \pm \sqrt{(-2)^2 - 4(1)(2)}}{2} = \frac{2 \pm \sqrt{-4}}{2} = \frac{2 \pm 2i}{2} = 1 \pm i$. Therefore, the
    solutions are $x = 0, 1 \pm i$.

    (b) Since $1 - i$ and $1 + i$ are zeros, $x - (1 - i) = x - 1 + i$ and $x - (1 + i) = x - 1 - i$ are the
    factors of $x^2 - 2x + 2$. Thus the complete factorization is
    $P(x) = x(x^2 - 2x + 2) = x(x - 1 + i)(x - 1 - i)$.

5. (a) $x^4 + 2x^2 + 1 = 0$ $\Leftrightarrow$ $(x^2 + 1)^2 = 0$ $\Leftrightarrow$ $x^2 + 1 = 0$ $\Leftrightarrow$ $x^2 = -1$ $\Leftrightarrow$
    $x = \pm i$. Therefore the zeros of P are $x = \pm i$.

    (b) Since $-i$ and i are zeros, $x + i$ and $x - i$ are the factors of $x^2 + 1$. Thus the complete
    factorization is $P(x) = (x^2 + 1)^2 = [(x + i)(x - i)]^2 = (x + i)^2(x - i)^2$.

7. (a) $x^4 - 16 = 0$ $\Leftrightarrow$ $0 = (x^2 - 4)(x^2 + 4) = (x - 2)(x + 2)(x^2 + 4)$. So $x = \pm 2$ or
    $x^2 + 4 = 0$. If $x^2 + 4 = 0$ then $x^2 = -4$ $\Rightarrow$ $x = \pm 2i$. Therefore the zeros of P are
    $x = \pm 2, \pm 2i$.

    (b) Since $-i$ and i are zeros, $x + i$ and $x - i$ are the factors of $x^2 + 1$. Thus the complete
    factorization is $P(x) = (x - 2)(x + 2)(x^2 + 4) = (x - 2)(x + 2)(x - 2i)(x + 2i)$.

9. (a) $x^3 + 8 = 0$ $\Leftrightarrow$ $(x + 2)(x^2 - 2x + 4) = 0$. So $x = -2$ or $x^2 - 2x + 4 = 0$. If
    $x^2 - 2x + 4 = 0$ then $x = \frac{-(-2) \pm \sqrt{(-2)^2 - 4(1)(4)}}{2} = \frac{2 \pm \sqrt{-12}}{2} = \frac{2 \pm 2i\sqrt{3}}{2} = 1 \pm i\sqrt{3}$. Therefore,
    the zeros of P are $x = -2, 1 \pm i\sqrt{3}$.

    (b) Since $1 - i\sqrt{3}$ and $1 + i\sqrt{3}$ are the zeros from the $x^2 - 2x + 4 = 0$, $x - (1 - i\sqrt{3})$ and
    $x - (1 + i\sqrt{3})$ are the factors of $x^2 - 2x + 4$. Thus the complete factorization is
    $$P(x) = (x + 2)(x^2 - 2x + 4) = (x + 2)\left[x - (1 - i\sqrt{3})\right]\left[x - (1 + i\sqrt{3})\right]$$
    $$= (x + 2)(x - 1 + i\sqrt{3})(x - 1 - i\sqrt{3}).$$

11. (a) $x^6 - 1 = 0$ $\Leftrightarrow$ $0 = (x^3 - 1)(x^3 + 1) = (x - 1)(x^2 + x + 1)(x + 1)(x^2 - x + 1)$.
    Clearly, $x = \pm 1$ are solutions. If $x^2 + x + 1 = 0$, then $x = \frac{-1 \pm \sqrt{1 - 4(1)(1)}}{2} = \frac{-1 \pm \sqrt{-3}}{2}$
    $= -\frac{1}{2} \pm \frac{\sqrt{-3}}{2}$ so $x = -\frac{1}{2} \pm \frac{\sqrt{3}}{2}i$. And if $x^2 - x + 1 = 0$, then $x = \frac{1 \pm \sqrt{1 - 4(1)(1)}}{2}$
    $= \frac{1 \pm \sqrt{-3}}{2} = \frac{1}{2} \pm \frac{\sqrt{-3}}{2} = \frac{1}{2} \pm \frac{\sqrt{3}}{2}i$. Therefore, the zeros of P are $x = \pm 1, -\frac{1}{2} \pm \frac{\sqrt{3}}{2}i$,
    $\frac{1}{2} \pm \frac{\sqrt{3}}{2}i$.

(b) The zeros of $x^2 + x + 1 = 0$ are $-\frac{1}{2} - \frac{\sqrt{3}}{2}i$ and $-\frac{1}{2} + \frac{\sqrt{3}}{2}i$, so $x^2 + x + 1$ factors as

$\left[x - \left(-\frac{1}{2} - \frac{\sqrt{3}}{2}i\right)\right]\left[x - \left(-\frac{1}{2} + \frac{\sqrt{3}}{2}i\right)\right] = \left(x + \frac{1}{2} + \frac{\sqrt{3}}{2}i\right)\left(x + \frac{1}{2} - \frac{\sqrt{3}}{2}i\right)$. Similarly, since

the zeros of $x^2 - x + 1 = 0$ are $\frac{1}{2} - \frac{\sqrt{3}}{2}i$ and $\frac{1}{2} + \frac{\sqrt{3}}{2}i$, so $x^2 - x + 1$ factors as

$\left[x - \left(\frac{1}{2} - \frac{\sqrt{3}}{2}i\right)\right]\left[x - \left(\frac{1}{2} + \frac{\sqrt{3}}{2}i\right)\right] = \left(x - \frac{1}{2} + \frac{\sqrt{3}}{2}i\right)\left(x - \frac{1}{2} - \frac{\sqrt{3}}{2}i\right)$. Thus the complete

factorization is

$$P(x) = (x - 1)(x^2 + x + 1)(x + 1)(x^2 - x + 1)$$
$$= (x - 1)(x + 1)\left(x + \frac{1}{2} + \frac{\sqrt{3}}{2}i\right)\left(x + \frac{1}{2} - \frac{\sqrt{3}}{2}i\right)\left(x - \frac{1}{2} + \frac{\sqrt{3}}{2}i\right)\left(x - \frac{1}{2} - \frac{\sqrt{3}}{2}i\right).$$

13. $P(x) = x^2 + 25 = (x - 5i)(x + 5i)$. The zeros of P are $5i$ and $-5i$, both multiplicity 1.

15. $Q(x) = x^2 + 2x + 2$. Using the quadratic formula $x = \frac{-(2) \pm \sqrt{(2)^2 - 4(1)(2)}}{2(1)} = \frac{-2 \pm \sqrt{-4}}{2}$
$= \frac{-2 \pm 2i}{2} = -1 \pm i$. So $Q(x) = (x + 1 - i)(x + 1 + i)$. The zeros of Q are $-1 - i$ (multiplicity 1) and $-1 + i$ (multiplicity 1).

17. $P(x) = x^3 + 4x = x(x^2 + 4) = x(x - 2i)(x + 2i)$. The zeros of P are 0, $2i$, and $-2i$ (all multiplicity 1).

19. $Q(x) = x^4 - 1 = (x^2 - 1)(x^2 + 1) = (x - 1)(x + 1)(x^2 + 1) = (x - 1)(x + 1)(x - i)(x + i)$.
The zeros of Q are 1, -1, i, and $-i$ (all multiplicity 1).

21. $P(x) = 16x^4 - 81 = (4x^2 - 9)(4x^2 + 9) = (2x - 3)(2x + 3)(2x - 3i)(2x + 3i)$. The zeros of P are $\frac{3}{2}$, $-\frac{3}{2}$, $\frac{3}{2}i$, and $-\frac{3}{2}i$ (all multiplicity 1).

23. $P(x) = x^3 + x^2 + 9x + 9 = x^2(x + 1) + 9(x + 1) = (x + 1)(x^2 + 9) = (x + 1)(x - 3i)(x + 3i)$.
The zeros of P are -1, $3i$, and $-3i$ (all multiplicity 1).

25. $Q(x) = x^4 + 2x^2 + 1 = (x^2 + 1)^2 = (x - i)^2(x + i)^2$. The zeros of Q are i and $-i$ (both multiplicity 2).

27. $P(x) = x^4 + 3x^2 - 4 = (x^2 - 1)(x^2 + 4) = (x - 1)(x + 1)(x - 2i)(x + 2i)$. The zeros of P are 1, -1, $2i$, and $-2i$ (all multiplicity 1).

29. $P(x) = x^5 + 6x^3 + 9x = x(x^4 + 6x^2 + 9) = x(x^2 + 3)^2 = x(x - \sqrt{3}i)^2(x + \sqrt{3}i)^2$. The zeros of P are 0 (multiplicity 1), $\sqrt{3}i$ (multiplicity 2), and $-\sqrt{3}i$ (multiplicity 2).

31. Since $1 + i$ and $1 - i$ are conjugates, the factorization of the polynomial must be
$P(x) = a(x - [1 + i])(x - [1 - i]) = a(x^2 - 2x + 2)$. If we let $a = 1$, we get
$P(x) = x^2 - 2x + 2$.

33. Since $2i$ and $-2i$ are conjugates, the factorization of the polynomial must be
$Q(x) = b(x - 3)(x - 2i)(x + 2i]) = b(x - 3)(x^2 + 4) = b(x^3 - 3x^2 + 4x - 12)$. If we let $b = 1$,
we get $Q(x) = x^3 - 3x^2 + 4x - 12$.

35. Since i is a zero, by the Conjugate Roots Theorem, $-i$ is also a zero. So the factorization of the
polynomial must be $P(x) = a(x - 2)(x - i)(x + i) = a(x^3 - 2x^2 + x - 2)$. If we let $a = 1$, we
get $P(x) = x^3 - 2x^2 + x - 2$.

37. Since the zeros are $1 - 2i$ and 1 (with multiplicity two), by the Conjugate Roots Theorem, the other zero is $1 + 2i$. So a factorization is

$$R(x) = c(x - [1 - 2i])(x - [1 + 2i])(x - 1)^2 = c([x - 1] + 2i)([x - 1] - 2i)(x - 1)^2$$
$$= c([x - 1]^2 - [2i]^2)(x^2 - 2x + 1) = c(x^2 - 2x + 1 + 4)(x^2 - 2x + 1)$$
$$= c(x^2 - 2x + 5)(x^2 - 2x + 1) = c(x^4 - 2x^3 + x^2 - 2x^3 + 4x^2 - 2x + 5x^2 - 10x + 5)$$
$$= c(x^4 - 4x^3 + 10x^2 - 12x + 5). \text{ If we let } c = 1 \text{ we get } R(x) = x^4 - 4x^3 + 10x^2 - 12x + 5.$$

39. Since the zeros are i and $1 + i$, by the Conjugate Roots Theorem, the other zeros are $-i$ and $1 - i$. So a factorization is

$$T(x) = C(x - i)(x + i)(x - [1 + i])(x - [1 - i]) = C(x^2 - i^2)([x - 1] - i)([x - 1] + i)$$
$$= C(x^2 + 1)(x^2 - 2x + 1 - i^2) = C(x^2 + 1)(x^2 - 2x + 2) = C(x^4 - 2x^3 + 2x^2 + x^2 - 2x + 2)$$
$$= C(x^4 - 2x^3 + 3x^2 - 2x + 2) = Cx^4 - 2Cx^3 + 3Cx^2 - 2Cx + 2C.$$

Since the constant coefficient is 12, it follows that $2C = 12 \quad \Leftrightarrow \quad C = 6$, and so $T(x) = 6(x^4 - 2x^3 + 3x^2 - 2x + 2) = 6x^4 - 12x^3 + 18x^2 - 12x + 12.$

41. $P(x) = x^3 + 2x^2 + 4x + 8 = x^2(x + 2) + 4(x + 2) = (x + 2)(x^2 + 4)$
    $= (x + 2)(x - 2i)(x + 2i)$. Thus the zeros are -2 and $\pm 2i$.

43. $P(x) = x^3 - 2x^2 + 2x - 1$. By inspection, $P(1) = 1 - 2 + 2 - 1 = 0$, and hence $x = 1$ is a zero.

$$\begin{array}{r|rrrr} 1 & 1 & -2 & 2 & -1 \\ & & 1 & -1 & 1 \\ \hline & 1 & -1 & 1 & 0 \end{array}$$
Thus $P(x) = (x - 1)(x^2 - x + 1)$. So $x = 1$ or $x^2 - x + 1 = 0$.

Using the quadratic formula, we have $x = \dfrac{1 \pm \sqrt{1 - 4(1)(1)}}{2} = \dfrac{1 \pm i\sqrt{3}}{2}$. Hence, the zeros are 1 and $\dfrac{1 \pm i\sqrt{3}}{2}$.

45. $P(x) = x^3 - 3x^2 + 3x - 2$.

$$\begin{array}{r|rrrr} 2 & 1 & -3 & 3 & -2 \\ & & 2 & -2 & 2 \\ \hline & 1 & -1 & 1 & 0 \end{array}$$
Thus $P(x) = (x - 2)(x^2 - x + 1)$. So $x = 2$ or $x^2 - x + 1 = 0$

Using the quadratic formula we have $x = \dfrac{1 \pm \sqrt{1 - 4(1)(1)}}{2} = \dfrac{1 \pm i\sqrt{3}}{2}$. Hence, the zeros are 2, and $\frac{1 \pm i\sqrt{3}}{2}$

47. $P(x) = 2x^3 + 7x^2 + 12x + 9$ has possible rational zeros $\pm 1, \pm 3, \pm 9, \pm \frac{1}{2}, \pm \frac{3}{2}, \pm \frac{9}{2}$. Since all coefficients are positive, there are no positive real zeros.

$$\begin{array}{r|rrrr} -1 & 2 & 7 & 12 & 9 \\ & & -2 & -5 & -7 \\ \hline & 2 & 5 & 7 & 2 \end{array} \qquad \begin{array}{r|rrrr} -2 & 2 & 7 & 12 & 9 \\ & & -4 & -6 & -12 \\ \hline & 2 & 3 & 6 & -3 \end{array}$$

There is a zero between -1 and -2.

$$\begin{array}{r|rrrr} -\frac{3}{2} & 2 & 7 & 12 & 9 \\ & & -3 & -6 & -9 \\ \hline & 2 & 4 & 6 & 0 \end{array} \quad \Rightarrow \quad x = -\frac{3}{2} \text{ is a zero.}$$

$P(x) = \left(x + \frac{3}{2}\right)(2x^2 + 4x + 6) = 2\left(x + \frac{3}{2}\right)(x^2 + 2x + 3)$. Now $x^2 + 2x + 3$ has zeros

$$x = \frac{-2 \pm \sqrt{4 - 4(3)(1)}}{2} = \frac{-2 \pm 2\sqrt{-2}}{2} = -1 \pm i\sqrt{2}. \text{ Hence, the zeros are } -\frac{3}{2} \text{ and }$$
$-1 \pm i\sqrt{2}.$

49. $P(x) = x^4 + x^3 + 7x^2 + 9x - 18.$ Since $P(x)$ has one change in sign, we are guaranteed a positive zero, and since $P(-x) = x^4 - x^3 + 7x^2 - 9x - 18$, there are 1 or 3 negative zeros.

$$\begin{array}{r|rrrrr} 1 & 1 & 1 & 7 & 9 & -18 \\ & & 1 & 2 & 9 & 18 \\ \hline & 1 & 2 & 9 & 18 & 0 \end{array} \Rightarrow \quad \text{Therefore, } P(x) = (x-1)(x^3 + 2x^2 + 9x + 18).$$

Continuing with the quotient, we try negative zeros.

$$\begin{array}{r|rrrr} -1 & 1 & 2 & 9 & 18 \\ & & -1 & -1 & -8 \\ \hline & 1 & 1 & 8 & 10 \end{array}$$

$$\begin{array}{r|rrrr} -2 & 1 & 2 & 9 & 18 \\ & & -2 & 0 & -18 \\ \hline & 1 & 0 & 9 & 0 \end{array}$$

$P(x) = (x-1)(x+2)(x^2+9) = (x-1)(x+2)(x-3i)(x+3i).$ Therefore, the zeros are 1, -2, and $\pm 3i$.

51. We see a pattern and utilize it to factor by grouping. This gives
$P(x) = x^5 - x^4 + 7x^3 - 7x^2 + 12x - 12 = x^4(x-1) + 7x^2(x-1) + 12(x-1)$
$= (x-1)(x^4 + 7x^2 + 12) = (x-1)(x^2+3)(x^2+4)$
$= (x-1)(x - i\sqrt{3})(x + i\sqrt{3})(x - 2i)(x + 2i).$ Therefore, the zeros are 1, $\pm i\sqrt{3}$, and $\pm 2i$.

53. $P(x) = x^4 - 6x^3 + 13x^2 - 24x + 36$ has possible rational zeros $\pm 1, \pm 2, \pm 3, \pm 4, \pm 6, \pm 9, \pm 12, \pm 18.$ $P(x)$ has 4 variations in signs and $P(-x)$ has no variation in signs.

$$\begin{array}{r|rrrrr} 1 & 1 & -6 & 13 & -24 & 36 \\ & & 1 & -5 & 8 & -16 \\ \hline & 1 & -5 & 8 & -16 & 20 \end{array} \qquad \begin{array}{r|rrrrr} 2 & 1 & -6 & 13 & -24 & 36 \\ & & 2 & -8 & 10 & -28 \\ \hline & 1 & -4 & 5 & -14 & 8 \end{array}$$

$$\begin{array}{r|rrrrr} 3 & 1 & -6 & 13 & -24 & 36 \\ & & 3 & -9 & 12 & -36 \\ \hline & 1 & -3 & 4 & -12 & 0 \end{array} \Rightarrow \quad x = 3 \text{ is a zero.}$$

Continuing:

$$\begin{array}{r|rrrr} 3 & 1 & -3 & 4 & -12 \\ & & 3 & 0 & 12 \\ \hline & 1 & 0 & 4 & 0 \end{array} \Rightarrow \quad x = 3 \text{ is a zero.}$$

$P(x) = (x-3)^2(x^2+4) = (x-3)^2(x-2i)(x+2i).$ Therefore, the zeros are 3 (multiplicity 2) and $\pm 2i$.

55. $P(x) = x^5 - 3x^4 + 12x^3 - 28x^2 + 27x - 9$ has possible rational zeros $\pm 1, \pm 3, \pm 9.$ $P(x)$ has 4 variations in sign and $P(-x)$ has 1 variation in sign.

$$\begin{array}{r|rrrrrr} 1 & 1 & -3 & 12 & -28 & 27 & -9 \\ & & 1 & -2 & 10 & -18 & 9 \\ \hline & 1 & -2 & 10 & -18 & 9 & 0 \end{array}$$ $\Rightarrow$ $x = 1$ is a zero.

$$\begin{array}{r|rrrrr} 1 & 1 & -2 & 10 & -18 & 9 \\ & & 1 & -1 & 9 & -9 \\ \hline & 1 & -1 & 9 & -9 & 0 \end{array}$$ $\Rightarrow$ $x = 1$ is a zero.

$$\begin{array}{r|rrrr} 1 & 1 & -1 & 9 & -9 \\ & & 1 & 0 & 9 \\ \hline & 1 & 0 & 9 & 0 \end{array}$$ $\Rightarrow$ $x = 1$ is a zero.

$P(x) = (x-1)^3(x^2+9) = (x-1)^3(x-3i)(x+3i)$. Therefore, the zeros are 1 (multiplicity 3) and $\pm 3i$.

57. (a) $P(x) = x^3 - 5x^2 + 4x - 20 = x^2(x-5) + 4(x-5) = (x-5)(x^2+4)$.

     (b) $P(x) = (x-5)(x-2i)(x+2i)$.

59. (a) $P(x) = x^4 + 8x^2 - 9 = (x^2-1)(x^2+9) = (x-1)(x+1)(x^2+9)$.

     (b) $P(x) = (x-1)(x+1)(x-3i)(x+3i)$.

61. (a) $P(x) = x^6 - 64 = (x^3-8)(x^3+8) = (x-2)(x^2+2x+4)(x+2)(x^2-2x+4)$.

     (b) $P(x) = (x-2)(x+2)(x+1-i\sqrt{3})(x+1+i\sqrt{3})(x-1-i\sqrt{3})(x-1+i\sqrt{3})$.

63. (a) $x^4 - 2x^3 - 11x^2 + 12x = x(x^3 - 2x^2 - 11x + 12) = 0$. We first find the bounds for our viewing rectangle.

$$\begin{array}{r|rrrr} & 1 & -2 & -11 & 12 \\ 5 & 1 & 3 & 4 & 32 \\ -4 & 1 & -6 & 13 & -50 \end{array}$$ $\Rightarrow$ $x = 5$ is an upper bound.
 $\Rightarrow$ $x = -4$ is a lower bound.

The viewing rectangle $[-4, 5]$ by $[-50, 10]$ shows that $P(x) = x^4 - 2x^3 - 11x^2 + 12x$ has **4 real solutions**. Since this matches the degree of $P(x)$, $P(x)$ has **no imaginary solutions**.

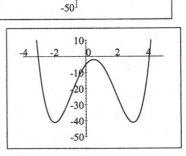

     (b) $x^4 - 2x^3 - 11x^2 + 12x - 5 = 0$. We use the same bounds for our viewing rectangle, $[-4, 5]$ by $[-50, 10]$, and see that $R(x) = x^4 - 2x^3 - 11x^2 + 12x - 5$ has **2 real solutions**. Since the degree of $R(x)$ is 4, $R(x)$ must have **2 imaginary solutions**.

(c) $x^4 - 2x^3 - 11x^2 + 12x + 40 = 0$. We use the same bounds for our viewing rectangle, $[-4, 5]$ by $[-10, 50]$, and see that $T(x) = x^4 - 2x^3 - 11x^2 + 12x + 40$ has no real solutions. Since the degree of $T(x)$ is 4, $T(x)$ must have 4 imaginary solutions.

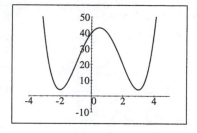

65. (a) $P(x) = x^2 - (1 + i)x + (2 + 2i)$. So
$P(2i) = (2i)^2 - (1+i)(2i) + 2 + 2i = -4 - 2i + 2 + 2 + 2i = 0$, and
$P(1-i) = (1-i)^2 - (1+i)(1-i) + (2+2i) = 1 - 2i - 1 - 1 - 1 + 2 + 2i = 0$.
Therefore, $2i$ and $1 - i$ are solutions of the equation $x^2 - (1+i)x + (2+2i) = 0$. However,
$P(-2i) = (-2i)^2 - (1+i)(-2i) + 2 + 2i = -4 + 2i - 2 + 2 + 2i = -4 + 4i$, and
$P(1+i) = (1+i)^2 - (1+i)(1+i) + 2 + 2i = 2 + 2i$. Since, $P(-2i) \neq 0$ and
$P(1+i) \neq 0$, $-2i$ and $1 + i$ are not solutions.

(b) This does not violate the Conjugate Roots Theorem because the coefficients of the polynomial $P(x)$ are not all real.

67. Since P has real coefficients, the imaginary zeros come in pairs: $a \pm bi$ (by the Conjugate Roots Theorem), where $b \neq 0$. Thus there must be an even number of imaginary zeros. Since P is of odd degree, it has an odd number of zeros (counting multiplicity). It follows that P has at least one real zero.

Exercises 4.5

1. $r(x) = \dfrac{x-1}{x+4}$. When $x = 0$, we have $r(0) = -\frac{1}{4}$, so the y-intercept is $-\frac{1}{4}$. The numerator is 0 when $x = 1$, so the x-intercept is 1.

3. $t(x) = \dfrac{x^2 - x - 2}{x - 6}$. When $x = 0$, we have $t(0) = \frac{-2}{-6} = \frac{1}{3}$, so the y-intercept is $\frac{1}{3}$. The numerator is 0 when $x^2 - x - 2 = (x-2)(x+1) = 0$ or when $x = 2$ or $x = -1$, so the x-intercepts are 2 and -1.

5. $r(x) = \dfrac{x^2 - 9}{x^2}$. Since 0 is not in the domain of $r(x)$, there is no y-intercept. The numerator is 0 when $x^2 - 9 = (x-3)(x+3) = 0$ or when $x = \pm 3$, so the x-intercepts are ± 3.

7. From the graph, the x-intercept is 3, the y-intercept is 3, the vertical asymptote is $x = 2$, and the horizontal asymptote is $y = 2$.

9. From the graph, the x-intercepts are -1 and 1, the y-intercept is about $\frac{1}{4}$, the vertical asymptotes are $x = -2$ and $x = 2$, and the horizontal asymptote is $y = 1$.

11. $r(x) = \dfrac{3}{x+2}$. There is a vertical asymptote where $x + 2 = 0 \quad \Leftrightarrow \quad x = -2$. We have

$$r(x) = \frac{3}{x+2} = \frac{\dfrac{3}{x}}{1 + \dfrac{2}{x}} \to 0 \text{ as } x \to \pm\infty \text{ , so the horizontal asymptote is } y = 0.$$

13. $t(x) = \dfrac{x^2}{x^2 - x - 6} = \dfrac{x^2}{(x-3)(x+2)} = \dfrac{1}{1 - \dfrac{1}{x} - \dfrac{6}{x^2}} \to 1 \text{ as } x \to \pm\infty.$ Hence, the

horizontal asymptote is $y = 1$. The vertical asymptotes occur when $(x-3)(x+2) = 0 \quad \Leftrightarrow \quad x = 3$ or $x = -2$, and so the vertical asymptotes are $x = 3$ and $x = -2$.

15. $s(x) = \dfrac{6}{x^2 + 2}$. There is no vertical asymptote since $x^2 + 2$ is never 0. Since

$$s(x) = \frac{6}{x^2 + 2} = \frac{\dfrac{6}{x^2}}{1 + \dfrac{2}{x^2}} \to 0 \text{ as } x \to \pm\infty, \text{ the horizontal asymptote is } y = 0.$$

17. $r(x) = \dfrac{6x - 2}{x^2 + 5x - 6}$. A vertical asymptote occurs when $x^2 + 5x - 6 = (x+6)(x-1) = 0 \quad \Leftrightarrow \quad x = 1$ or $x = -6$. Because the degree of the denominator is greater than the degree of the numerator, the horizontal asymptote is $y = 0$.

19. $y = \dfrac{x^2 + 2}{x - 1}$. A vertical asymptote occurs when $x - 1 = 0 \quad \Leftrightarrow \quad x = 1$. There are no horizontal asymptotes because the degree of the numerator is greater than the degree of the denominator.

In exercises 21-27, let $f(x) = \dfrac{1}{x}$.

21. $r(x) = \dfrac{1}{x-1} = f(x-1)$.

From this form we see that the graph of r is obtained from the graph of f by shifting 1 unit to the right. Thus r has vertical asymptote $x = 1$ and horizontal asymptote $y = 0$.

23. $s(x) = \dfrac{3}{x+1} = 3\left(\dfrac{1}{x+1}\right) = 3f(x+1)$.

From this form we see that the graph of s is obtained from the graph of f by shifting 1 unit to the left and stretching vertically by a factor of 3. Thus s has vertical asymptote $x = -1$ and horizontal asymptote $y = 0$.

25. $t(x) = \dfrac{2x-3}{x-2} = 2 + \dfrac{1}{x-2} = f(x-2) + 2$ (see long division below).

From this form we see that the graph of t is obtained from the graph of f by shifting 2 units to the right and 2 units vertically. Thus t has vertical asymptote $x = 2$ and horizontal asymptote $y = 2$.

$$\begin{array}{r} 2 \\ x-2 \overline{)\,2x-3} \\ \underline{2x-2} \\ 1 \end{array}$$

27. $r(x) = \dfrac{x+2}{x+3} = 1 - \dfrac{1}{x+3} = -f(x+3) + 1$ (see long division below).

From this form we see that the graph of r is obtained from the graph of f by shifting 3 units to the left, reflect about the x-axis, and then shifting vertically 1 unit. Thus r has vertical asymptote $x = -3$ and horizontal asymptote $y = 1$.

$$\begin{array}{r} 1 \\ x+3 \overline{)\,x+2} \\ \underline{x+3} \\ -1 \end{array}$$

29. $y = \dfrac{4x-4}{x+2}$. When $x = 0$, $y = -2$, so the y-intercept is -2. When $y = 0$, $4x - 4 = 0$ $\Leftrightarrow$ $x = 1$, so the x-intercept is 1. Since the degree of the numerator and denominator are the same the horizontal asymptote is $y = \frac{4}{1} = 4$. A vertical asymptote occurs when $x = -2$.

as $x \to$	-2^{+}	-2^{-}
sign of $y = \dfrac{4x-4}{x+2}$	$\dfrac{(-)}{(+)}$	$\dfrac{(-)}{(-)}$
$y \quad \to$	$-\infty$	∞

31. $s(x) = \dfrac{4 - 3x}{x + 7}$. When $x = 0$, $y = \frac{4}{7}$, so the y-intercept is $\frac{4}{7}$. The

x-intercepts occur when $y = 0$ $\Leftrightarrow$ $4 - 3x = 0$ $\Leftrightarrow$ $x = \frac{4}{3}$.
A vertical asymptote occurs when $x = -7$. Since the degree of the
numerator and denominator are the same the horizontal asymptote is

$y = \frac{-3}{1} = -3$.

33. $r(x) = \dfrac{18}{(x - 3)^2}$. When $x = 0$, $y = \frac{18}{9} = 2$, and so the y-intercept is

2. Since the numerator can never be zero, there is no x-intercept.
There is a vertical asymptote when $x - 3 = 0$ $\Leftrightarrow$ $x = 3$, and
because the degree of the asymptote is $y = 0$.

35. $s(x) = \dfrac{4x - 8}{(x - 4)(x + 1)}$. When $x = 0$, $y = \frac{-8}{(-4)(1)} = 2$, so the

y-intercept is 2. When $y = 0$, $4x - 8 = 0$ $\Leftrightarrow$ $x = 2$, so the
x-intercept is 2. The vertical asymptotes are $x = -1$ and $x = 4$, and
because the degree of the numerator is less than the degree of the
denominator, the horizontal asymptote is $y = 0$.

37. $s(x) = \dfrac{6}{x^2 - 5x - 6}$. When $x = 0$, $y = \frac{6}{-6} = -1$, so the y-intercept

is -1. Since the numerator is never zero, there is no x-intercept. The
vertical asymptotes occur when $x^2 - 5x - 6 = (x + 1)(x - 6)$ $\Leftrightarrow$
$x = -1$ and $x = 6$, and because the degree of the numerator is less
less than the degree of the denominator, the horizontal asymptote is
$y = 0$.

39. $t(x) = \dfrac{3x + 6}{x^2 + 2x - 8}$. When $x = 0$, $y = \frac{6}{-8} = -\frac{3}{4}$, so the y-intercept

y-intercept is $-\frac{3}{4}$. When $y = 0$, $3x + 6 = 0$ $\Leftrightarrow$ $x = -2$, so the
x-intercept is -2. The vertical asymptotes occur when
$x^2 + 2x - 8 = (x - 2)(x + 4) = 0$ $\Leftrightarrow$ $x = 2$ and $x = -4$.
Since the degree of the numerator is less than the degree of the
denominator, the horizontal asymptote is $y = 0$.

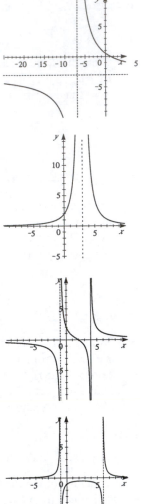

41. $r(x) = \dfrac{(x-1)(x+2)}{(x+1)(x-3)}$. When $x = 0$, $y = \frac{2}{3}$, so the y-intercept is $\frac{2}{3}$.
When $y = 0$, $(x-1)(x+2) = 0 \Rightarrow x = -2, 1$, so, the
x-intercepts are -2 and 1. The vertical asymptotes are $x = -1$ and
$x = 3$, and because the degree of the numerator and denominator are
the same the horizontal asymptote is $y = \frac{1}{1} = 1$.

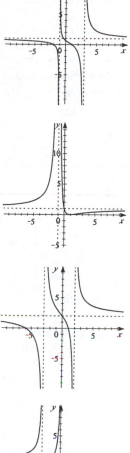

43. $r(x) = \dfrac{x^2 - 2x + 1}{x^2 + 2x + 1} = \dfrac{(x-1)^2}{(x+1)^2} = \left(\dfrac{x-1}{x+1}\right)^2$. When $x = 0$, $y = 1$,
so the y-intercept is 1. When $y = 0$, $x = 1$, so the x-intercept is 1. A
vertical asymptote occurs at $x + 1 = 0 \Leftrightarrow x = -1$. Because the
degree of the numerator and denominator are the same the horizontal
asymptote is $y = \frac{1}{1} = 1$.

45. $r(x) = \dfrac{2x^2 + 10x - 12}{x^2 + x - 6} = \dfrac{2(x-1)(x+6)}{(x-2)(x+3)}$. When $x = 0$,
$y = \dfrac{2(-1)(6)}{(-2)(3)} = 2$, so the y-intercept is 2. When $y = 0$,
$2(x-1)(x+6) = 0 \Rightarrow x = -6, 1$, so the x-intercepts are -6
and 1. Vertical asymptotes occur when $(x-2)(x+3) = 0 \Leftrightarrow$
$x = -3$ or $x = 2$. Because the degree of the numerator and
denominator are the same the horizontal asymptote is $y = \frac{2}{1} = 2$.

47. $y = \dfrac{x^2 - x - 6}{x^2 + 3x} = \dfrac{(x-3)(x+2)}{x(x+3)}$. The x-intercept occurs when $y = 0$
$\Leftrightarrow (x-3)(x+2) = 0 \Rightarrow x = -2, 3$, so the x-intercepts are
-2 and 3. There is no y-intercept because y is undefined when $x = 0$.
The vertical asymptotes are $x = 0$ and $x = -3$. Because the degree of
the numerator and denominator are the same, the horizontal asymptotes
is $y = \frac{1}{1} = 1$.

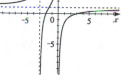

49. $r(x) = \dfrac{3x^2 + 6}{x^2 - 2x - 3} = \dfrac{3(x^2 + 2)}{(x-3)(x+1)}$. When $x = 0$, $y = -2$, so
the y-intercept is -2. Since the numerator can never equal zero, there
is no x-intercept. Vertical asymptotes occur when $x = -1, 3$. Because
the degree of the numerator and denominator are the same, the
horizontal asymptote is.$y = \frac{3}{1} = 3$.

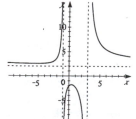

51. $s(x) = \dfrac{x^2 - 2x + 1}{x^3 - 3x^2} = \dfrac{(x-1)^2}{x^2(x-3)}$. Since $x = 0$ is not in the domain
of $s(x)$, there is no y-intercept. The x-intercept occurs when $y = 0$
$\Leftrightarrow \quad x^2 - 2x + 1 = (x-1)^2 = 0 \quad \Rightarrow \quad x = 1$, so the x-intercept
is 1. Vertical asymptotes occur when $x = 0, 3$. Since the degree of
the numerator is less than the degree of the denominator, the horizontal
asymptote is $y = 0$.

53. $r(x) = \dfrac{x^2}{x - 2}$. When $x = 0$, $y = 0$, so the graph passes through the
origin. There is a vertical asymptote when $x - 2 = 0 \quad \Leftrightarrow \quad x = 2$,
with $y \to \infty$ as $x \to 2^+$, and $y \to -\infty$ as $x \to 2^-$. Because the
degree of the numerator is greater than the degree of the denominator,
there is no horizontal asymptotes. By using long division, we see
that $y = x + 2 + \dfrac{4}{x - 2}$, so $y = x + 2$ is a slant asymptote.

55. $r(x) = \dfrac{x^2 - 2x - 8}{x} = \dfrac{(x-4)(x+2)}{x}$. The vertical asymptote is
$x = 0$, thus, there is no y-intercept. If $y = 0$, then $(x - 4)(x + 2) = 0$
$\Rightarrow \quad x = -2, 4$, so the x-intercepts are -2 and 4. Because the degree
f the numerator is greater than the degree of the denominator, there
are no horizontal asymptotes. By using long division, we see that
$y = x - 2 - \dfrac{8}{x}$, so $y = x - 2$ is a slant asymptote.

57. $r(x) = \dfrac{x^2 + 5x + 4}{x - 3} = \dfrac{(x+4)(x+1)}{x - 3}$. When $x = 0$, $y = -\frac{4}{3}$, so the
y-intercept is $-\frac{4}{3}$. When $y = 0$, $(x + 4)(x + 1) = 0 \quad \Leftrightarrow \quad x = -4$,
-1, so the two x-intercepts are -4 and -1. A vertical asymptote
occurs when $x = 3$, with $y \to \infty$ as $x \to 3^+$, and $y \to -\infty$
as $x \to 3^-$. Using long division, we see that $y = x + 8 + \dfrac{28}{x - 3}$,
so $y = x + 8$ is a slant asymptote.

59. $r(x) = \dfrac{x^3 + x^2}{x^2 - 4} = \dfrac{x^2(x+1)}{(x-2)(x+2)}$. When $x = 0$, $y = 0$, so the
graph passes through the origin. Moreover, when $y = 0$, we have
$x^2(x + 1) = 0 \quad \Rightarrow \quad x = 0, -1$, so the x-intercepts are 0 and -1.
Vertical asymptotes occur when $x = \pm 2$, and because the degree of
the numerator is greater than the degree of the dominator there are
no horizontal asymptotes. Using long division, we see that

$$y = x + 1 + \dfrac{4x + 4}{x^2 - 4}, \text{ so } y = x + 1 \text{ is a slant asymptote.}$$

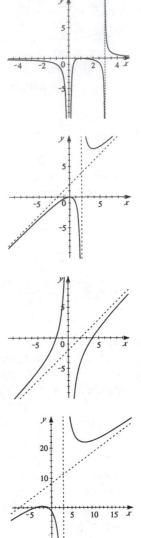

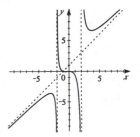

as $x \rightarrow$	-2^-	-2^+	2^-	2^+
sign of $y = \dfrac{x^2(x+1)}{(x-2)(x+2)}$	$\dfrac{(+)(-)}{(-)(-)}$	$\dfrac{(+)(-)}{(-)(+)}$	$\dfrac{(+)(+)}{(-)(+)}$	$\dfrac{(+)(+)}{(+)(+)}$
$y \rightarrow$	$-\infty$	∞	$-\infty$	∞

61. $f(x) = \dfrac{2x^2 + 6x + 6}{x + 3}$, $g(x) = 2x$. Vertical asymptote: $x = -3$.

$[-10, 5]$ by $[-20, 10]$ Graph of f. $[-20, 20]$ by $[-50, 50]$ Graph of f and g.

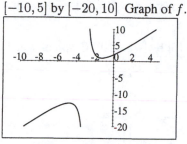

63. $f(x) = \dfrac{x^3 - 2x^2 + 16}{x - 2}$, $g(x) = x^2$. Vertical asymptote: $x = 2$.

$[-4, 8]$ by $[-30, 30]$ Graph of f. $[-20, 20]$ by $[-50, 50]$ Graph of f and g.

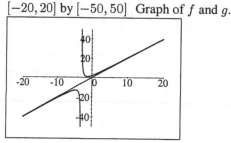

65. $f(x) = \dfrac{2x^2 - 5x}{2x + 3}$.

Vertical asymptote: $x = -1.5$
x-intercepts: $0, 2.5$
y-intercept: 0
Local maximum: $(-3.9, -10.4)$
Local minimum: $(0.9, -0.6)$

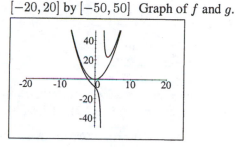

Using long division, we get $f(x) = x - 4 + \dfrac{12}{2x + 3}$. From the graph, we see that the end behavior of $f(x)$ is like the end behavior of $g(x) = x - 4$.

$[-10, 10]$ by $[-30, 30]$ Graph of f. $[-20, 20]$ by $[-50, 50]$ Graph of f and g.

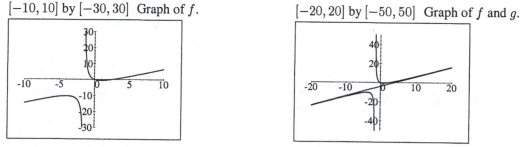

67. $f(x) = \dfrac{x^5}{x^3 - 1}$.

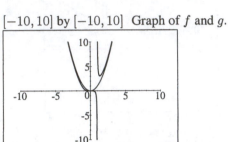

Vertical asymptote: $x = 1$
x-intercept: 0
y-intercept: 0
Local minimum: $(1.4, 3.1)$

Thus $y = x^2 + \dfrac{x^2}{x^3 - 1}$. From the graph we see that the end behavior of $f(x)$ is like the end behavior of $g(x) = x^2$.

$[-5, 5]$ by $[-10, 10]$ Graph of f.

$[-10, 10]$ by $[-10, 10]$ Graph of f and g.

69. $f(x) = \dfrac{x^4 - 3x^3 + 6}{x - 3}$.

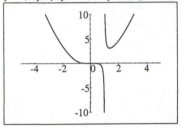

Vertical asymptote: $x = 3$
x-intercept: $1.6,\ 2.7$
y-intercept: -2
Local maximum: $(-0.4, -1.8),\ (2.4, 3.8)$
Local minimum: $(0.6, -2.3),\ (3.4, 54.3)$

Thus $y = x^3 + \dfrac{6}{x - 3}$. From the graphs on the next page we see that the end behavior of $f(x)$ is like the end behavior of $g(x) = x^3$.

$[-10, 10]$ by $[-20, 100]$ Graph of f.

$[-10, 10]$ by $[-20, 100]$ Graph of f and g.

71. (a)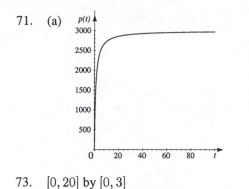

   (b) $p(t) = \dfrac{3000t}{t+1} = 3000 - \dfrac{3000}{t+1}.$

   So as $t \to \infty$, we have $p(t) \to 3000$.

73. [0, 20] by [0, 3]

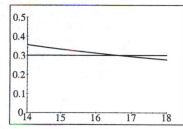

(a) The highest concentration of drug is 2.50 mg/L, and it is reached 1 hour after the drug is administered.

(b) The concentration of the drug in the bloodstream goes to 0.

(c) From the first viewing rectangle, we see that an approximate solution is near $t = 15$. Thus we graph $y = \dfrac{5t}{t^2 + 1}$ and $y = 0.3$ in the viewing rectangle $[14, 18]$ by $[0, 0.5]$. So it takes about 16.61 hours for the concentration to drop below 0.3 mg/L.

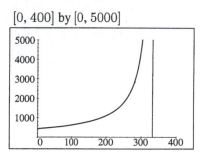

75. $P(v) = P_0\left(\dfrac{s_0}{s_0 - v}\right) \Rightarrow$

   $P(v) = 440\left(\dfrac{332}{332 - v}\right)$

   If the speed of the train approaches the speed of sound, the pitch of the whistle becomes very loud. This would be experienced as a "sonic boom"— an effect seldom heard with trains.

[0, 400] by [0, 5000]

77. Vertical asymptote $x = 3$: $p(x) = \dfrac{1}{x - 3}$

   Vertical asymptote $x = 3$ and horizontal asymptote $y = 2$: $r(x) = \dfrac{2x}{x - 3}$.

Vertical asymptotes $x = 1$ and $x = -1$, horizontal asymptote 0, and x-intercept 4:
$$q(x) = \frac{x - 4}{(x - 1)(x + 1)}.$$
Of course, other answers are possible.

79. (a) $r(x) = \dfrac{3x^2 - 3x - 6}{x - 2} = \dfrac{3(x - 2)(x + 1)}{x - 2} = 3(x + 1)$,
for $x \neq 2$. Therefore, $r(x) = 3x + 3$, $x \neq 2$. Since
$3(2) + 3 = 9$, the graph is the line $y = 3x + 3$ with
the point $(2, 9)$ removed.

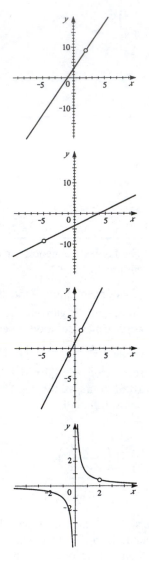

(b) $s(x) = \dfrac{x^2 + x - 20}{x + 5} = \dfrac{(x - 4)(x + 5)}{x + 5} = x - 4$, for
$x \neq -5$. Therefore, $s(x) = x - 4$, $x \neq -5$. Since
$(-5) - 4 = -9$, the graph is the line $y = x - 4$ with
the point $(-5, -9)$ removed.

$t(x) = \dfrac{2x^2 - x - 1}{x - 1} = \dfrac{(2x + 1)(x - 1)}{x - 1} = 2x + 1$, for
$x \neq 1$. Therefore, $t(x) = 2x + 1$, $x \neq 1$. Since
$2(1) + 1 = 3$, the graph is the line $y = 2x + 1$ with
the point $(1, 3)$ removed.

$u(x) = \dfrac{x - 2}{x^2 - 2x} = \dfrac{x - 2}{x(x - 2)} = \dfrac{1}{x}$, for $x \neq 2$.

Therefore, $u(x) = \dfrac{1}{x}$, $x \neq 2$. When $x = 2$, $\dfrac{1}{x} = \dfrac{1}{2}$,

the graph is the curve $y = \dfrac{1}{x}$ with the point $(2, \frac{1}{2})$

removed.

Review Exercises for Chapter 4

1. $P(x) = x^3 + 27$

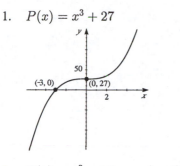

3. $P(x) = -(x-2)^4$

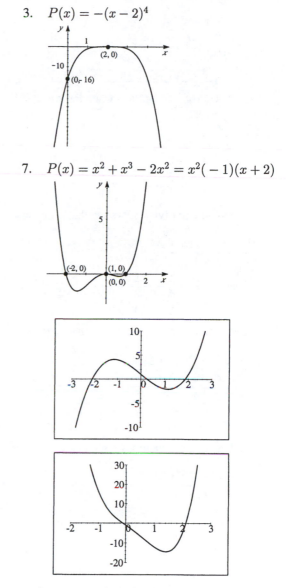

5. $P(x) = x^3 - 16x = x(x-4)(x+4)$

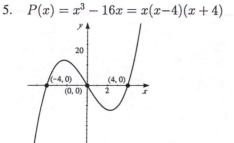

7. $P(x) = x^2 + x^3 - 2x^2 = x^2(-1)(x+2)$

9. $P(x) = x^3 - 4x + 1$.
   x-intercepts: -2.1, 0.3, and 1.9.
   y-intercept: 1
   Local maximum is $(-1.2, 4.1)$.
   Local minimum is $(1.2, -2.1)$.
   $y \to \infty$ as $x \to \infty$;
   $y \to -\infty$ as $x \to -\infty$.

11. $P(x) = 3x^4 - 4x^3 - 10x - 1$.
    x-intercepts: -0.1 and 2.1.
    y-intercept: -1
    Local maximum is $(1.4, -14.5)$.
    There is no local maximum.
    $y \to \infty$ as $x \to \pm\infty$.

13. (a) Use the Pythagorean Theorem and solving for y^2 we have, $x^2 + y^2 = 10^2 \iff$
    $y^2 = 100 - x^2$. Substituting we get $S = 13.8x(100 - x^2) = 1380x - 13.8x^3$.

    (b) Domain is $[0, 10]$.

(c)

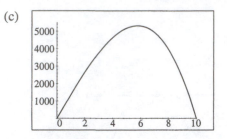

(d)   The strongest beam is when the width is 5.8 inches.

15. $\dfrac{x^2 - 3x + 5}{x - 2}$

$$2 \,\underline{|\ \begin{array}{rrr} 1 & -3 & 5 \\ & 2 & -2 \end{array}}$$
$$\begin{array}{rrr} 1 & -1 & 3 \end{array}$$

Using synthetic division, we see that $Q(x) = x - 1$ and $R(x) = 3$.

17. $\dfrac{x^3 - x^2 + 11x + 2}{x - 4}$

$$4 \,\underline{|\ \begin{array}{rrrr} 1 & -1 & 11 & 2 \\ & 4 & 12 & 92 \end{array}}$$
$$\begin{array}{rrrr} 1 & 3 & 23 & 94 \end{array}$$

Using synthetic division, we see that $Q(x) = x^2 + 3x + 23$ and $R(x) = 94$.

19. $\dfrac{x^4 - 8x^2 + 2x + 7}{x + 5}$

$$-5 \,\underline{|\ \begin{array}{rrrrr} 1 & 0 & -8 & 2 & 7 \\ & -5 & 25 & -85 & 415 \end{array}}$$
$$\begin{array}{rrrrr} 1 & -5 & 17 & -83 & 422 \end{array}$$

Using synthetic division, we see that $Q(x) = x^3 - 5x^2 + 17x - 83$ and $R(x) = 422$.

21. $\dfrac{2x^3 + x^2 - 8x + 15}{x^2 + 2x - 1}$

$$
\begin{array}{r}
2x - 3 \\
x^2 + 2x - 1 \,\overline{)\,2x^3 + x^2 - 8x + 15} \\
\underline{2x^3 + 4x^2 - 2x} \\
-3x^2 - 6x + 15 \\
\underline{-3x^2 - 6x + 3} \\
12
\end{array}
$$

Therefore, $Q(x) = 2x - 3$, and $R(x) = 12$.

23. $P(x) = 2x^3 - 9x^2 - 7x + 13$; find $P(5)$.

$$5 \,\underline{|\ \begin{array}{rrrr} 2 & -9 & -7 & 13 \\ & 10 & 5 & -10 \end{array}}$$
$$\begin{array}{rrrr} 2 & 1 & -2 & 3 \end{array} \qquad \text{Therefore, } P(5) = 3.$$

25.　$\frac{1}{2}$ is a zero of $P(x) = 2x^4 + x^3 - 5x^2 + 10x - 4$ if $P\left(\frac{1}{2}\right) = 0$.

$$
\begin{array}{r|rrrrr}
\frac{1}{2} & 2 & 1 & -5 & 10 & -4 \\
 & & 1 & 1 & -2 & 4 \\
\hline
 & 2 & 2 & -4 & 8 & 0
\end{array}
$$

Since $P\left(\frac{1}{2}\right) = 0$, $\frac{1}{2}$ is a zero of the polynomial.

27.　$P(x) = x^{500} + 6x^{201} - x^2 - 2x + 4$. The remainder from dividing $P(x)$ by $x - 1$ is
$P(1) = (1)^{500} + 6(1)^{201} - (1)^2 - 2(1) + 4 = 8$.

29.　(a)　$P(x) = x^5 - 6x^3 - x^2 + 2x + 18$ has possible rational zeros $\pm 1, \pm 2, \pm 3, \pm 6, \pm 9, \pm 18$.

　　(b)　Since $P(x)$ has 2 variations in sign, there are either 0 or 2 positive real zeros. Since
$P(-x) = -x^5 + 6x^3 - x^2 - 2x + 18$ has 3 variations in sign, there are 1 or 3 negative real
zeros.

31.　Since the zeros are $-\frac{1}{2}$, 2, and 3, a factorization is $P(x) = C\left(x + \frac{1}{2}\right)(x - 2)(x - 3)$
$= \frac{C}{2}(2x + 1)(x^2 - 5x + 6) = \frac{C}{2}(2x^3 - 10x^2 + 12x + x^2 - 5x + 6) = \frac{C}{2}(2x^3 - 9x^2 + 7x + 6)$.
Since the constant coefficient is 12, $\frac{C}{2}(6) = 12 \quad \Leftrightarrow \quad C = 4$, and so the polynomial is
$P(x) = 4x^3 - 18x^2 + 14x + 12$.

33.　No, there is no polynomial of degree 4 with integer coefficients that has zeros i, $2i$, $3i$ and $4i$. Since
the imaginary zeros of polynomial equations with real coefficients come in complex conjugate pairs,
there would have to be 8 zeros, which is impossible for a polynomial of degree 4.

35.　$P(x) = x^3 - 3x^2 - 13x + 15$ has possible rational zeros $\pm 1, \pm 3, \pm 5, \pm 15$.

$$
\begin{array}{r|rrrr}
1 & 1 & -3 & -13 & 15 \\
 & & 1 & -2 & -15 \\
\hline
 & 1 & -2 & -15 & 0
\end{array} \quad \Rightarrow \quad x = 1 \text{ is a zero.}
$$

So $P(x) = x^3 - 3x^2 - 13x + 15 = (x - 1)(x^2 - 2x - 15) = (x - 1)(x - 5)(x + 3)$. Therefore,
the zeros are -3, 1, and 5.

37.　$P(x) = x^4 + 6x^3 + 17x^2 + 28x + 20$ has possible rational zeros $\pm 1, \pm 2, \pm 4, \pm 5, \pm 10, \pm 20$.
Since all of the coefficients are positive, there are no positive real zeros.

$$
\begin{array}{r|rrrrr}
-1 & 1 & 6 & 17 & 28 & 20 \\
 & & -1 & -5 & -12 & -16 \\
\hline
 & 1 & 5 & 12 & 16 & 4
\end{array}
$$

$$
\begin{array}{r|rrrrr}
-2 & 1 & 6 & 17 & 28 & 20 \\
 & & -2 & -8 & -18 & -20 \\
\hline
 & 1 & 4 & 9 & 10 & 0
\end{array} \quad \Rightarrow \quad x = -2 \text{ is a zero.}
$$

$P(x) = x^4 + 6x^3 + 17x^2 + 28x + 20 = (x + 2)(x^3 + 4x^2 + 9x + 10)$. Continuing with the
quotient, we have

$$
\begin{array}{r|rrrr}
-2 & 1 & 4 & 9 & 10 \\
 & & -2 & -4 & -10 \\
\hline
 & 1 & 2 & 5 & 0
\end{array} \quad \Rightarrow \quad x = -2 \text{ is a zero.}
$$

Thus $P(x) = x^4 + 6x^3 + 17x^2 + 28x + 20 = (x + 2)^2(x^2 + 2x + 5)$. Now $x^2 + 2x + 5 = 0$ when
$x = \frac{-2 \pm \sqrt{4 - 4(5)(1)}}{2} = \frac{-2 \pm 4i}{2} = -1 \pm 2i$. Thus, the zeros are -2 (multiplicity 2) and $-1 \pm 2i$.

39. $P(x) = x^5 - 3x^4 - x^3 + 11x^2 - 12x + 4$ has possible rational zeros $\pm 1, \pm 2, \pm 4$.

$$
\begin{array}{r|rrrrrr}
1 & 1 & -3 & -1 & 11 & -12 & 4 \\
  &   & 1 & -2 & -3 & 8 & -4 \\
\hline
  & 1 & -2 & -3 & 8 & -4 & 0
\end{array}
\Rightarrow \quad x = 1 \text{ is a zero.}
$$

$P(x) = x^5 - 3x^4 - x^3 + 11x^2 - 12x + 4 = (x-1)(x^4 - 2x^3 - 3x^2 + 8x - 4)$. Continuing with the quotient, we have

$$
\begin{array}{r|rrrrr}
1 & 1 & -2 & -3 & 8 & -4 \\
  &   & 1 & -1 & -4 & 4 \\
\hline
  & 1 & -1 & -4 & 4 & 0
\end{array}
\Rightarrow \quad x = 1 \text{ is a zero.}
$$

So $x^5 - 3x^4 - x^3 + 11x^2 - 12x + 4 = (x-1)^2(x^3 - x^2 - 4x + 4) = (x-1)^3(x^2 - 4)$
$= (x-1)^3(x-2)(x+2)$. Therefore, the zeros are 1 (multiplicity 3), -2, and 2.

41. $P(x) = x^6 - 64 = (x^3 - 8)(x^3 + 8) = (x-2)(x^2 + 2x + 4)(x+2)(x^2 - 2x + 4)$. Now using the quadratic formula to find the zeros of $x^2 + 2x + 4$, we have

$x = \frac{-2 \pm \sqrt{4-4(4)(1)}}{2} = \frac{-2 \pm 2\sqrt{3}\,i}{2} = -1 \pm \sqrt{3}\,i$, and using the quadratic formula to find the zeros of

$x^2 - 2x + 4$, we have $x = \frac{2 \pm \sqrt{4-4(4)(1)}}{2} = \frac{2 \pm 2\sqrt{3}\,i}{2} = 1 \pm \sqrt{3}\,i$. Therefore, the zeros are 2, -2,

$1 \pm \sqrt{3}\,i$, and $-1 \pm \sqrt{3}\,i$.

43. $P(x) = 6x^4 - 18x^3 + 6x^2 - 30x + 36 = 6(x^4 - 3x^3 + x^2 - 5x + 6)$ has possible rational zeros $\pm 1, \pm 2, \pm 3, \pm 6$.

$$
\begin{array}{r|rrrrr}
1 & 6 & -18 & 6 & -30 & 36 \\
  &   & 6 & -12 & -6 & -36 \\
\hline
  & 6 & -12 & -6 & -36 & 0
\end{array}
\Rightarrow \quad x = 1 \text{ is a zero.}
$$

So $P(x) = 6x^4 - 18x^3 + 6x^2 - 30x + 36 = (x-1)(6x^3 - 12x^2 - 6x - 36)$
$= 6(x-1)(x^3 - 2x^2 - x - 6)$. Continuing with the quotient we have

$$
\begin{array}{r|rrrr}
1 & 1 & -2 & -1 & -6 \\
  &   & 1 & -1 & -2 \\
\hline
  & 1 & -1 & -2 & -8 \\
2 & 1 & -2 & -1 & -6 \\
  &   & 2 & 0 & -2 \\
\hline
  & 1 & 0 & -1 & -8 \\
3 & 1 & -2 & -1 & -6 \\
  &   & 3 & 3 & 6 \\
\hline
  & 1 & 1 & 2 & 0
\end{array}
\Rightarrow \quad x = 3 \text{ is a zero.}
$$

So, $P(x) = 6x^4 - 18x^3 + 6x^2 - 30x + 36 = 6(x-1)(x-3)(x^2 + x + 2)$. Now, $x^2 + x + 2 = 0$

when $x = \frac{-1 \pm \sqrt{1-4(1)(2)}}{2} = \frac{-1 \pm \sqrt{7}\,i}{2}$, and so the zeros are 1, 3, and $\frac{-1 \pm \sqrt{7}\,i}{2}$.

45. $2x^2 = 5x + 3 \quad \Leftrightarrow \quad 2x^2 - 5x - 3 = 0$.
The solutions are $x = -0.5, 3$.

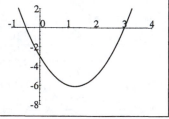

47. $x^4 - 3x^3 - 3x^2 - 9x - 2 = 0$ has solutions
     $x \approx -0.24, 4.24$.

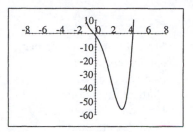

49. $r(x) = \dfrac{3x - 12}{x + 1}$. When $x = 0$, we have $r(0) = \frac{-12}{1} = -12$, so the
     y-intercept is -12. Since $y = 0$, when $3x - 12 = 0 \quad \Leftrightarrow \quad x = 4$,
     the x-intercept is 4. The vertical asymptote is $x = -1$. Because the
     degree of the denominator and numerator are the same, the
     horizontal asymptote is $y = \frac{3}{1} = 3$.

51. $r(x) = \dfrac{x - 2}{x^2 - 2x - 8} = \dfrac{x - 2}{(x + 2)(x - 4)}$. When $x = 0$, we have
     $r(0) = \frac{-2}{-8} = \frac{1}{4}$, so the y-intercept is $\frac{1}{4}$. When $y = 0$, we have
     $x - 2 = 0 \quad \Leftrightarrow \quad x = 2$, so the x-intercepts is 2. The vertical
     asymptotes occur when $x = -2$ and $x = 4$. The horizontal
     than the degree of the numerator.

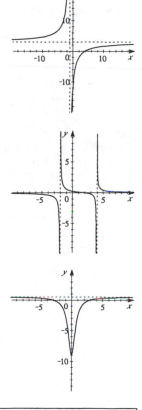

53. $r(x) = \dfrac{x^2 - 9}{2x^2 + 1} = \dfrac{(x + 3)(x - 3)}{2x^2 + 1}$. When $x = 0$, we have
     $r(0) = \frac{-9}{1}$, so the y-intercept is -9. When $y = 0$, we have
     $x^2 - 9 = 0 \quad \Leftrightarrow \quad x = \pm 3$ so the x-intercepts are -3 and 3. Since
     $2x^2 + 1 > 0$, the denominator is never zero so there are no vertical
     asymptotes. The horizontal asymptote is at $y = \frac{1}{2}$ because the
     degree of the denominator and numerator are the same.

55. $r(x) = \dfrac{x - 3}{2x + 6}$.
     From the graph we see that
     x-intercept: 3
     y-intercept: -0.5
     Vertical asymptote: $x = -3$
     Horizontal asymptote: $y = 0.5$
     No local extrema.

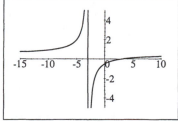

57. $r(x) = \dfrac{x^3 + 8}{x^2 - x - 2}.$

From the graph we see that

x-intercept: -2

y-intercept: -4

Vertical asymptote: $x = -1$, $x = 2$

Horizontal asymptote is $y = 0.5$.

Local maximum is $(0.425, -3.599)$.

Local minimum is $(4.216, 7.175)$.

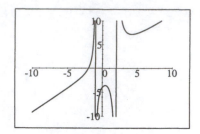

By using long division, we see that $f(x) = x + 1 + \dfrac{10 - x}{x^2 - x - 2}$, so f has a slant asymptote

$y = x + 1$.

59. The graphs $y = x^4 + x^2 + 24x$ and $y = 6x^3 + 20$ intersect when $x^4 + x^2 + 24x = 6x^3 + 20$ $\Leftrightarrow$

$x^4 - 6x^3 + x^2 + 24x - 20 = 0$. The possible rational zeros are $\pm 1, \pm 2, \pm 4, \pm 5, \pm 10, \pm 20$.

$$
\begin{array}{r|rrrrr}
1 & 1 & -6 & 1 & 24 & -20 \\
  &   & 1 & -5 & -4 & 20 \\
\hline
  & 1 & -5 & -4 & 20 & 0
\end{array}
\quad \Rightarrow \quad x = 1 \text{ is a zero.}
$$

So $x^4 - 6x^3 + x^2 + 24x - 20 = (x - 1)(x^3 - 5x^2 - 4x + 20) = 0$. Continuing with the quotient:

$$
\begin{array}{r|rrrr}
1 & 1 & -5 & -4 & 20 \\
  &   & 1 & -4 & -8 \\
\hline
  & 1 & -4 & -8 & 12
\end{array}
\qquad
\begin{array}{r|rrrr}
2 & 1 & -5 & -4 & 20 \\
  &   & 2 & -6 & -20 \\
\hline
  & 1 & -3 & -10 & 0
\end{array}
\quad \Rightarrow \quad x = 2 \text{ is a zero.}
$$

So $x^4 - 6x^3 + x^2 + 24x - 20 = (x - 1)(x - 2)(x^2 - 3x - 10)$

$= (x - 1)(x - 2)(x - 5)(x + 2) = 0$. Hence, the points of intersection are $(1, 26)$, $(2, 68)$,

$(5, 770)$, and $(-2, -28)$.

Chapter 4 Test

1. $f(x) = (x+1)^3 + 8$
    $= [(x+1) + 2][(x+1)^2 - 2(x+1) + 4]$
                                    Sum of cubes.
    $= (x+3)(x^2 + 2x + 1 - 2x - 2 + 4)$
    $= (x+3)(x^2 + 3)$

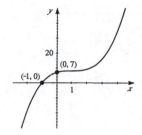

2. (a)

    $$\begin{array}{r|rrrrr} 2 & 1 & 0 & -4 & 2 & 5 \\ & & 2 & 4 & 0 & 4 \\ \hline & 1 & 2 & 0 & 2 & 9 \end{array}$$

    Therefore, quotient is $Q(x) = x^3 + 2x^2 + 2$, and the remainder is $R(x) = 9$.

    (b)
    $$\begin{array}{r} x^3 + 2x^2 \qquad + \frac{1}{2} \\ 2x^2 - 1 \overline{\smash{)}2x^5 + 4x^4 - x^3 - x^2 + 0x + 7} \\ \underline{2x^5 \qquad\quad -x^3} \\ 4x^4 \qquad\quad - x^2 \\ \underline{4x^4 \qquad\quad - 2x^2} \\ x^2 \qquad + 7 \\ \underline{x^2 \qquad - \frac{1}{2}} \\ 7\frac{1}{2} \text{ or } \frac{15}{2} \end{array}$$

    Therefore, quotient is $Q(x) = x^3 + 2x^2 + \frac{1}{2}$, and the remainder is $R(x) = \frac{15}{2}$.

3. (a) Possible rational zeros are: $\pm 1, \pm 3, \pm\frac{1}{2}, \pm\frac{3}{2}$.

    (b)
    $$\begin{array}{r|rrrr} -1 & 2 & -5 & -4 & 3 \\ & & -2 & 7 & -3 \\ \hline & 2 & -7 & 3 & 0 \end{array} \Rightarrow \quad x = -1 \text{ is a zero.}$$

    $P(x) = (x+1)(2x^2 - 7x + 3) = (x+1)(2x - 1)(x - 3) = 2(x+1)(x - \frac{1}{2})(x - 3)$

    (c) The zeros of P are $x = -1, 3, \frac{1}{2}$.

    (d)

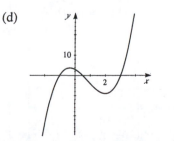

4. $P(x) = x^3 - x^2 - 4x - 6$. Possible rational zeros are: $\pm 1, \pm 2, \pm 3, \pm 6$.

$$\begin{array}{r|rrrr} 1 & 1 & -1 & -4 & -6 \\ & & 1 & 0 & -4 \\ \hline & 1 & 0 & -4 & -10 \end{array} \qquad \begin{array}{r|rrrr} 2 & 1 & -1 & -4 & -6 \\ & & 2 & 2 & -4 \\ \hline & 1 & 1 & -2 & -10 \end{array}$$

$$\begin{array}{r|rrrr} 3 & 1 & -1 & -4 & -6 \\ & & 3 & 6 & 6 \\ \hline & 1 & 2 & 2 & 0 \end{array} \quad \Rightarrow \quad x = 3 \text{ is a zero.}$$

So $P(x) = (x-3)(x^2 + 2x + 2)$. Using the quadratic formula on the second factor, we have

$$x = \frac{-2 \pm \sqrt{2^2 - 4(1)(2)}}{2(1)} = \frac{-2 \pm \sqrt{-4}}{2} = \frac{-2 \pm 2\sqrt{-1}}{2} = -1 \pm i. \text{ So zeros of } P(x) \text{ are } 3,$$

$-1 - i$, and $-1 + i$.

5. $P(x) = x^4 - 2x^3 + 5x^2 - 8x + 4$. The possible rational zeros of P are: $\pm 1, \pm 2,$ and ± 4. Since there are four changes in sign, P has 4, 2, or 0 positive real zeros.

$$\begin{array}{r|rrrrr} 1 & 1 & -2 & 5 & -8 & 4 \\ & & 1 & -1 & 4 & -4 \\ \hline & 1 & -1 & 4 & -4 & 0 \end{array}$$

So $P(x) = (x-1)(x^3 - x^2 + 4x - 4)$. Factoring the second factor by grouping we have:
$P(x) = (x-1)[x^2(x-1) + 4(x-1)] = (x-1)(x^2+4)(x-1) = (x-1)^2(x-2i)(x+2i)$.

6. Since $3i$ is a zero of $P(x)$, $-3i$ is also a zero of $P(x)$. And since -1 is a zero of multiplicity 2,
$P(x) = (x+1)^2(x-3i)(x+3i) = (x^2 + 2x + 1)(x^2 + 9) = x^4 + 2x^3 + 10x^2 + 18x + 9$.

7. $P(x) = 2x^4 - 7x^3 + x^2 - 18x + 3$.

    (a) Since $P(x)$ has 4 variations in sign, $P(x)$ can have 4, 2, or 0 positive real zeros. Since
    $P(-x) = 2x^4 + 7x^3 + x^2 + 18x + 3$ has no variations in sign, there are no negative real zeros.

    (b)
$$\begin{array}{r|rrrrr} 4 & 2 & -7 & 1 & -18 & 3 \\ & & 8 & 4 & 20 & 8 \\ \hline & 2 & 1 & 5 & 2 & 11 \end{array}$$

    Since the last row contains no negative entries, 4 is an upper bound for the real zeros of $P(x)$.

$$\begin{array}{r|rrrrr} -1 & 2 & -7 & 1 & -18 & 3 \\ & & -1 & 9 & -10 & 28 \\ \hline & 2 & -9 & 10 & -28 & 31 \end{array}$$

    Since the last row alternates in sign, -1 is a lower bound for the real zeros of $P(x)$.

    (c) Using the upper and lower limit from part b, we graph $P(x)$ in the viewing rectangle $[-1, 4]$ by $[-1, 1]$. The two real zeros are 0.17 and 3.93.

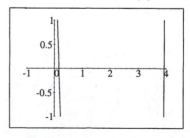

(d) Local minimum $(2.8, -70.3)$.

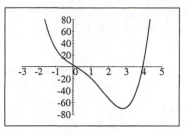

8. $r(x) = \dfrac{2x - 1}{x^2 - x - 2}$, $s(x) = \dfrac{x^3 + 27}{x^2 + 4}$, $t(x) = \dfrac{x^3 - 9x}{x + 2}$ and $u(x) = \dfrac{x^2 + x - 6}{x^2 - 25}$.

(a) $r(x)$ has the horizontal asymptote $y = 0$ because the degree of the denominator is greater than the degree of the numerator. $u(x)$ has the horizontal asymptote $y = \frac{1}{1} = 1$ because the degree of the numerator and the denominator are the same.

(b) The degree of the numerator of $s(x)$ is one more than the degree of the denominator, so $s(x)$ has a slant asymptote.

(c) The denominator of $s(x)$ is never 0, so $s(x)$ has no vertical asymptotes.

(d) $u(x) = \frac{x^2 + x - 6}{x^2 - 25} = \frac{(x+3)(x-2)}{(x-5)(x+5)}$. When $x = 0$, we have
$u(x) = \frac{-6}{-25} = \frac{6}{25}$, so the y-intercept is $y = \frac{6}{25}$. When $y = 0$,
we have $x = -3$ or $x = 2$, so the x-intercepts are -3 and 2. The
vertical asymptotes are $x = -5$ and $x = 5$. The horizontal
asymptote occurs at $y = \frac{1}{1} = 1$ because the degree of the
denominator and numerator are the same.

(e)
$$
\begin{array}{r}
x^2 \;\; - 2x \;\; - 5 \\
x + 2 \overline{\smash{\big)}\, x^3 + 0x^2 - 9x \;\; + 0} \\
\underline{x^3 + 2x^2 } \\
-2x^2 - 9x \\
\underline{-2x^2 - 4x } \\
-5x \;\; + 0 \\
\underline{-5x - 10} \\
-10
\end{array}
$$

Thus $P(x) = x^2 - 2x - 5$ and $t(x) = \dfrac{x^3 - 9x}{x + 2}$ have the same end behavior.

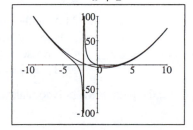

Principles of Modeling

1. (a) Using a graphing calculator, we obtain the quadratic polynomial
        $y = -0.275428x^2 + 19.7485x - 273.5523$.

    (b)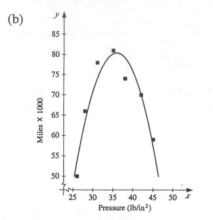

    (c) Moving the cursor along the path of the polynomial, we find that 35.85 lb/in^2 gives the longest tire life.

3. (a) Using a graphing calculator, we obtain the cubic polynomial
        $y = 0.00203708x^3 - 0.104521x^2 + 1.966206x + 1.45576$.

    (b)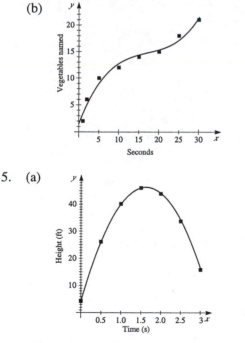

    (c) Moving the cursor along the path of the polynomial, we find that the subjects are estimated to name 43 vegetables in 40 seconds.

    (d) Moving the cursor along the path of the polynomial, we find that the subjects name 5 vegetables in about 2.0 seconds.

5. (a)

    From the data, a quadratic equation is an appropriate model.

    (b) Using a graphing calculator, we obtain the quadratic polynomial
        $y = -16.0x^2 + 51.8429x + 4.20714$.

    (c) Moving the cursor along the path of the polynomial, we find that the ball is 20 ft. above the ground 0.3 seconds and 2.9 seconds after it is thrown upward.

    (d) Again, moving the cursor along the path of the polynomial, we find that the maximum height is 46.2 ft.

Chapter Five

Exercises 5.1

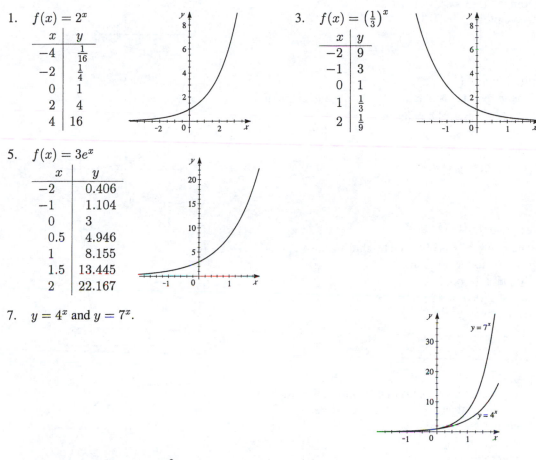

1. $f(x) = 2^x$

x	y
-4	$\frac{1}{16}$
-2	$\frac{1}{4}$
0	1
2	4
4	16

3. $f(x) = \left(\frac{1}{3}\right)^x$

x	y
-2	9
-1	3
0	1
1	$\frac{1}{3}$
2	$\frac{1}{9}$

5. $f(x) = 3e^x$

x	y
-2	0.406
-1	1.104
0	3
0.5	4.946
1	8.155
1.5	13.445
2	22.167

7. $y = 4^x$ and $y = 7^x$.

9. From the graph, $f(2) = a^2 = 9$, so $a = 3$. Thus $f(x) = 3^x$.

11. From the graph, $f(2) = a^2 = \frac{1}{16}$, so $a = \frac{1}{4}$. Thus $f(x) = \left(\frac{1}{4}\right)^x$.

13. III 15. I 17. II

19. $f(x) = -3^x$.
The graph of f is obtained by reflecting the graph of $y = 3^x$ about the x-axis.
Domain: $(-\infty, \infty)$.
Range: $(-\infty, 0)$.
Asymptote: $y = 0$.

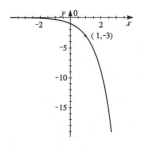

21. $g(x) = 2^x - 3$
     The graph of g is obtained by shifting the graph of $y = 2^x$
     downward 3 units.
     Domain: $(-\infty, \infty)$.
     Range: $(-3, \infty)$.
     Asymptote: $y = -3$.

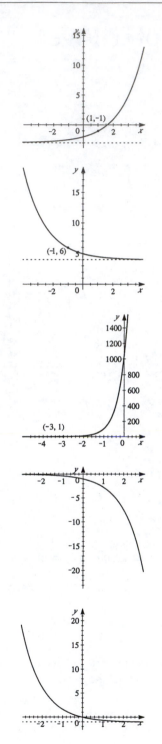

23. $h(x) = 4 + \left(\frac{1}{2}\right)^x$
     The graph of h is obtained by shifting the graph of
     $y = \left(\frac{1}{2}\right)^x$ upward 4 units.
     Domain: $(-\infty, \infty)$.
     Range: $(4, \infty)$.
     Asymptote: $y = 4$.

25. $f(x) = 10^{x+3}$
     The graph of f is obtained by shifting the graph of
     $y = 10^x$ to the left 3 units.
     Domain: $(-\infty, \infty)$.
     Range: $(0, \infty)$.
     Asymptote: $y = 0$.

27. $y = -e^x$
     The graph of $y = -e^x$ is obtained from the graph of
     $y = e^x$ by reflecting it about the x-axis.
     Domain: $(-\infty, \infty)$.
     Range: $(-\infty, 0)$.
     Asymptote: $y = 0$.

29. $y = e^{-x} - 1$
     The graph of $y = e^{-x} - 1$ is obtained from the graph of
     $y = e^x$ by reflecting it about the y-axis then shifting
     downward 1 unit.
     Domain: $(-\infty, \infty)$.
     Range: $(-1, \infty)$.
     Asymptote: $y = -1$.

31. $y = e^{x-2}$

The graph of $y = e^{x-2}$ is obtained from the graph of $y = e^x$ by shifting it to the right 2 units.

Domain: $(-\infty, \infty)$

Range: $(0, \infty)$

Asymptote: $y = 0$

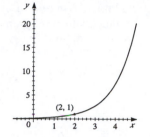

33. Using the points $(0, 3)$ and $(2, 12)$, we have $f(0) = Ca^0 = 3 \iff C = 3$. We also have $f(2) = 3a^2 = 12 \iff a^2 = 4 \iff a = 2$ (recall that for an exponential function $f(x) = a^x$ we require $a > 0$). Thus $f(x) = 3 \cdot 2^x$.

35. (a)

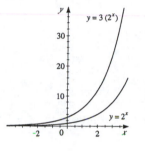

(b) Since $g(x) = 3(2^x) = 3f(x)$ and $f(x) > 0$, the height of the graph of $g(x)$ is always three times the height of the graph of $f(x) = 2^x$, so the graph of g is steeper than the graph of f.

37. $f(x) = 10^x$, so $\dfrac{f(x+h) - f(x)}{h} = \dfrac{10^{x+h} - 10^x}{h} = \dfrac{10^x \cdot 10^h - 10^x}{h} = 10^x \dfrac{10^h - 1}{h}$.

39.

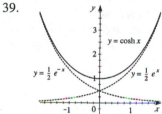

41. $\cosh(-x) = \dfrac{e^{-x} + e^{-(-x)}}{2} = \dfrac{e^{-x} + e^x}{2} = \dfrac{e^x + e^{-x}}{2} = \cosh(x)$.

43. $[\cosh(x)]^2 - [\sinh(x)]^2 = \left(\dfrac{e^x + e^{-x}}{2}\right)^2 - \left(\dfrac{e^x - e^{-x}}{2}\right)^2$

$= \frac{1}{4}(e^{2x} + 2 + e^{-2x}) - \frac{1}{4}(e^{2x} - 2 + e^{-2x}) = \frac{2}{4} + \frac{2}{4} = 1$.

45. (a) From the graphs in (i) - (iii) below, we see that the graph of f ultimately increases much more
          quickly than the graph of g.

     (i) $[0, 5]$ by $[0, 20]$ (ii) $[0, 25]$ by $[0, 10^7]$

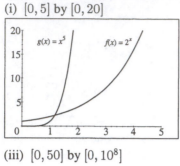

     (iii) $[0, 50]$ by $[0, 10^8]$

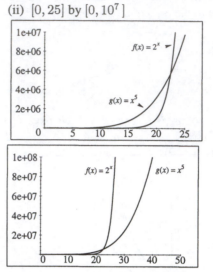

     (b) From the graphs in parts (i) and (ii), we see that the approximate solutions are $x \approx 1.2$ and
          $x \approx 22.4$.

47.

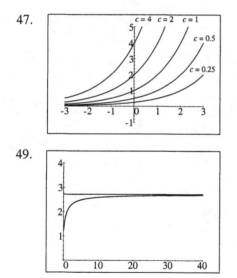

The larger the value of c, the more rapidly the
graph of $f(x) = c\,2^x$ increases. Also, some
students might notice that the graphs are just
shifted horizontally 1 unit. This is because of
our choice of c; each c in this exercise is of the
form 2^k. So $f(x) = 2^k \cdot 2^x = 2^{x+k}$.

49.

Note from the graph, that $y = \left(1 + \frac{1}{x}\right)^x$ approaches
e as x get large.

51. (a)

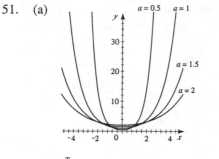

(b)   As a increases the curve $y = \dfrac{a}{2}\left(e^{x/a} + e^{-x/a}\right)$ flattens out and the y intercept increases.

53. $y = \dfrac{e^x}{x}$

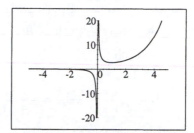

Vertical Asymptote: $x = 0$

Horizontal Asymptote: $y = 0$, left side only.

As $x \to -\infty$, $y \to 0$, and as $x \to \infty$,

$y \to \infty$.

55. $g(x) = e^x + e^{-3x}$

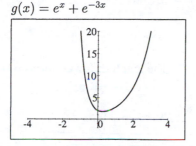

The graph of $g(x)$ is shown in the viewing rectangle $[-4, 4]$ by $[0, 20]$. From the graph, we see that there is a local minimum ≈ 1.75 when $x \approx 0.27$.

57. $y = xe^{-x}$

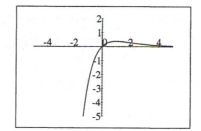

(a)   From the graph, we see that the function $f(x) = x\,e^x$ is *increasing* on $(-\infty, 1]$ and decreasing on $[1, \infty)$.

(b)   From the graph, we see that the range is approximately $(-\infty, 0.37)$.

59. $m(t) = 13\,e^{-0.015t}$

(a) $m(0) = 13$ kg.

(b) $m(45) = 13\,e^{-0.015(45)} = 13\,e^{-0.675} = 6.619$ kg. Thus the mass of the radioactive substance after 45 days is about 6.6 kg.

61. $v(t) = 80(1 - e^{-0.2t})$

(a) $v(0) = 80(1 - e^0) = 80(1 - 1) = 0$.

(b) $v(5) = 80(1 - e^{-0.2(5)}) \approx 80(0.632) = 50.57$ ft/s. So the velocity after 5 s is about 50.6 ft/s.

$v(10) = 80(1 - e^{-0.2(10)}) \approx 80(0.865) = 69.2$ ft/s. So the velocity after 10 s is about 69.2 ft/s.

(c)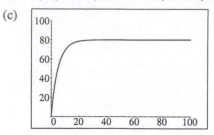

(d)   The terminal velocity is 80 ft/s.

63. $P(t) = \dfrac{1200}{1 + 11e^{-0.2t}}$.

(a) $P(0) = \dfrac{1200}{1 + 11e^{-0.2(0)}} = \dfrac{1200}{1 + 11} = 100$.

(b) $P(10) = \dfrac{1200}{1 + 11e^{-0.2(10)}} \approx 482$. $P(20) = \dfrac{1200}{1 + 11e^{-0.2(20)}} \approx 999$.

$P(30) = \dfrac{1200}{1 + 11e^{-0.2(30)}} \approx 1168$.

(c)   As $t \to \infty$ we have $e^{-0.2t} \to 0$, so $P(t) \to \dfrac{1200}{1 + 0} = 1200$. The graph shown confirms this.

65. $D(t) = \dfrac{5.4}{1 + 2.9e^{-0.01t}}$. So $D(20) = \dfrac{5.4}{1 + 2.9e^{-0.01(20)}} \approx 1.600$ ft.

67. $P = 10{,}000$, $r = 0.10$, and $n = 2$. So $A(t) = 10{,}000\left(1 + \dfrac{0.10}{2}\right)^{2t} = 10{,}000 \cdot 1.05^{2t}$.

(a) $A(5) = 10{,}000 \cdot 1.05^{10} \approx 16{,}288.95$, and so the value of the investment is \$16,288.95.

(b) $A(10) = 10{,}000 \cdot 1.05^{20} \approx 26{,}532.98$, and so the value of the investment is \$26,532.98.

(c) $A(15) = 10{,}000 \cdot 1.05^{30} \approx 43{,}219.42$, and so the value of the investment is \$43,219.42.

69. $P = 3000$ and $r = 0.09$. Then we have $A(t) = 3000\left(1 + \dfrac{0.09}{n}\right)^{nt}$, and so

$A(5) = 3000\left(1 + \dfrac{0.09}{n}\right)^{5n}$.

(a)   If $n = 1$, $A(5) = 3000\left(1 + \frac{0.09}{1}\right)^{5} = 3000 \cdot 1.09^{5} \approx \$4{,}615.87$.

(b)   If $n = 2$, $A(5) = 3000\left(1 + \frac{0.09}{2}\right)^{10} = 3000 \cdot 1.045^{10} \approx \$4{,}658.91$.

(c)   If $n = 12$, $A(5) = 3000\left(1 + \frac{0.09}{12}\right)^{60} = 3000 \cdot 1.0075^{60} \approx \$4{,}697.04$.

(d)   If $n = 52$, $A(5) = 3000\left(1 + \frac{0.09}{52}\right)^{260} \approx \$4{,}703.11$.

(e)   If $n = 365$, $A(5) = 3000\left(1 + \frac{0.09}{365}\right)^{1825} \approx \$4{,}704.68$.

(f)   If $n = 24 \cdot 365 = 8760$, $A(5) = 3000\left(1 + \frac{0.09}{8760}\right)^{43800} \approx \$4{,}704.93$.

(g)   If interest is compounded continuously, $A(5) = 3000 \cdot e^{0.45} \approx \$4{,}704.94$.

71.   We find the effective rate, with $P = 1$, and $t = 1$. So $A = \left(1 + \frac{r}{n}\right)^n$

(i) $n = 2, r = 0.085$; $A(2) = \left(1 + \frac{0.085}{2}\right)^2 = (1.0425)^2 \approx 1.0868$.

(ii) $n = 4, r = 0.0825$; $A(4) = \left(1 + \frac{0.0825}{4}\right)^4 = (1.020625)^4 \approx 1.0851$.

(iii)   continuous compounding, $r = 0.08$; $A(1) = e^{0.08} \approx 1.0833$.

Since (i) is larger than the others, the best investment is the one at 8.5% compounded semiannually.

73.   (a)   We must solve for P in the equation $10000 = P\left(1 + \frac{0.09}{2}\right)^{2(3)} = P(1.045)^6$ $\Leftrightarrow$
       $10000 = 1.3023P$ $\Leftrightarrow$ $P = 7678.96$. Thus the present value is \$7,678.96.

(b)   We must solve for P in the equation $100000 = P\left(1 + \frac{0.08}{12}\right)^{12(5)} = P(1.00667)^{60}$ $\Leftrightarrow$
       $100000 = 1.4898P$ $\Leftrightarrow$ $P = \$67{,}121.04$.

75.   Calculating the pay for method (b), we have $pay = 2 + 2^2 + 2^3 + \cdots + 2^{30} > 2^{30}$ cents
       $= \$10{,}737{,}418.24$. Since this is much more than method (a), method (b) is more profitable.

Exercises 5.2

1. (a) $5^2 = 25$ (b) $5^0 = 1$

3. (a) $8^{1/3} = 2$ (b) $2^{-3} = \frac{1}{8}$

5. (a) $e^x = 5$ (b) $e^5 = y$

7. (a) $\log_5 125 = 3$ (b) $\log_{10} 0.0001 = -4$

9. (a) $\log_8 \frac{1}{8} = -1$ (b) $\log_2 \left(\frac{1}{8}\right) = -3$

11. (a) $\ln 2 = x$ (b) $\ln y = 3$

13. (a) $\log_3 3 = 1$ (b) $\log_3 1 = \log_3 3^0 = 0$
    (c) $\log_3 3^2 = 2$

15. (a) $\log_6 36 = \log_6 6^2 = 2$ (b) $\log_9 81 = \log_9 9^2 = 2$
    (c) $\log_7 7^{10} = 10$

17. (a) $\log_3 \left(\frac{1}{27}\right) = \log_3 3^{-3} = -3$ (b) $\log_{10} \sqrt{10} = \log_{10} 10^{1/2} = \frac{1}{2}$
    (c) $\log_5 0.2 = \log_5 \left(\frac{1}{5}\right) = \log_5 5^{-1} = -1$

19. (a) $2^{\log_2 37} = 37$ (b) $3^{\log_3 8} = 8$
    (c) $e^{\ln \sqrt{5}} = \sqrt{5}$

21. (a) $\log_8 0.25 = \log_8 8^{-2/3} = -\frac{2}{3}$ (b) $\ln e^4 = 4$
    (c) $\ln \left(\frac{1}{e}\right) = \ln e^{-1} = -1$

23. (a) $\log_2 x = 5 \quad \Leftrightarrow \quad x = 2^5 = 32$ (b) $x = \log_2 16 = \log_2 2^4 = 4$

25. (a) $x = \log_3 243 = \log_3 3^5 = 5$ (b) $\log_3 x = 3 \quad \Leftrightarrow \quad x = 3^3 = 27$

27. (a) $\log_{10} x = 2 \quad \Leftrightarrow \quad x = 10^2 = 100$ (b) $\log_5 x = 2 \quad \Leftrightarrow \quad x = 5^2 = 25$

29. (a) $\log_x 16 = 4 \quad \Leftrightarrow \quad x^4 = 16 \quad \Leftrightarrow \quad x = 2$
    (b) $\log_x 8 = \frac{3}{2} \quad \Leftrightarrow \quad x^{3/2} = 8 \quad \Leftrightarrow \quad x = 8^{2/3} = 4$

31. (a) $\log 2 \approx 0.3010$ (b) $\log 35.2 \approx 1.5465$
    (c) $\log \left(\frac{2}{3}\right) \approx -0.1761$

33. (a) $\ln 5 \approx 1.6094$ (b) $\ln 25.3 \approx 3.2308$
    (c) $\ln \left(1 + \sqrt{3}\right) \approx 1.0051$

35. Since the point $(5, 1)$ is on the graph, we have $1 = \log_a 5 \quad \Leftrightarrow \quad a^1 = 5$. Thus the function is
$y = \log_5 x$.

37. Since the point $\left(3, \frac{1}{2}\right)$ is on the graph, we have $\frac{1}{2} = \log_a 3 \quad \Leftrightarrow \quad a^{1/2} = 3 \quad \Leftrightarrow \quad a = 9$. Thus the function is $y = \log_9 x$.

39. II   41. III   43. VI

45. The graph of $y = \log_4 x$ is obtained from the graph of $y = 4^x$ by reflecting it about the line $y = x$.

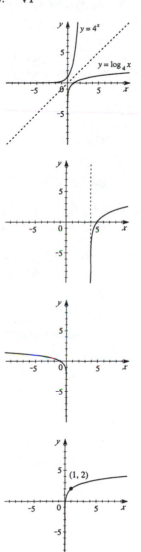

47. $f(x) = \log_2(x - 4)$
The graph of f is obtained from the graph of $y = \log_2 x$ by shifting it to the right 4 units.
Domain: $(4, \infty)$.
Range: $(-\infty, \infty)$.
Vertical asymptote: $x = 4$.

49. $g(x) = \log_5(-x)$
The graph of g is obtained from the graph of $y = \log_5 x$ by reflecting it about the y-axis.
Domain: $(-\infty, 0)$.
Range: $(-\infty, \infty)$.
Vertical asymptote: $x = 0$.

51. $y = 2 + \log_3 x$
The graph of $y = 2 + \log_3 x$ is obtained from the graph of $y = \log_3 x$ by shifting it upward 2 units.
Domain: $(0, \infty)$.
Range: $(-\infty, \infty)$.
Vertical asymptote: $x = 0$.

53. $y = 1 - \log_{10} x$
     The graph of $y = 1 - \log_{10} x$ is obtained from the graph of
     $y = \log_{10} x$ by reflecting it about the x-axis, and then shifting
     it upward 1 unit.
     Domain: $(0, \infty)$.
     Range: $(-\infty, \infty)$.
     Vertical asymptote: $x = 0$.

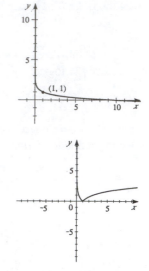

55. $y = |\ln x|$
     The graph of $y = |\ln x|$ is obtained from the graph of $y = \ln x$
     by reflecting the part of the graph for $0 < x < 1$ about the
     x-axis.
     Domain: $(0, \infty)$.
     Range: $[0, \infty)$.
     Vertical asymptote: $x = 0$.

57. $f(x) = \log_{10}(x + 3)$. We require that $x + 3 > 0$ $\Leftrightarrow$ $x > -3$, so the domain is $(-3, \infty)$.

59. $g(x) = \log_3(x^2 - 1)$. We require that $x^2 - 1 > 0$ $\Leftrightarrow$ $x^2 > 1$ $\Rightarrow$ $x < -1$ or $x > 1$, so the
     domain is $(-\infty, -1) \cup (1, \infty)$.

61. $h(x) = \ln x + \ln(2 - x)$. We require that $x > 0$ and $2 - x > 0$ $\Leftrightarrow$ $x > 0$ and $x < 2$ $\Leftrightarrow$
     $0 < x < 2$, so the domain is $(0, 2)$.

63. $y = \log_{10}(1 - x^2)$
     Domain: $(-1, 1)$
     Vertical asymptote: $x = -1$ and $x = 1$
     Local maximum $y = 0$ at $x = 0$

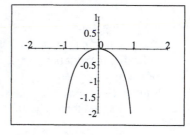

65. $y = x + \ln x$
     Domain: $(0, \infty)$
     Vertical asymptote: $x = 0$
     No local extrema.

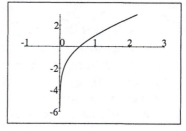

67. $y = \dfrac{\ln x}{x}$

Domain: $(0, \infty)$

Vertical asymptote: $x = 0$

Horizontal asymptote: $y = 0$

Local maximum $y \approx 0.37$ at $x \approx 2.72$

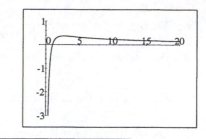

69. The graph of $g(x) = \sqrt{x}$ grows faster than the graph of $f(x) = \ln x$.

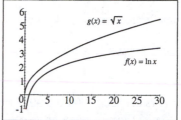

71. (a)

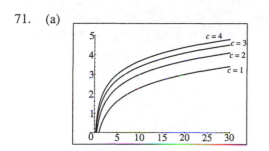

(b) Notice that $f(x) = \log(cx) = \log c + \log x$, so as c increases, the graph of $f(x) = \log(cx)$ is shifted upward $\log c$ units.

73. (a) $f(x) = \log_2(\log_{10} x)$. Since the domain of $\log_2 x$ is the positive real numbers, we have: $\log_{10} x > 0 \iff x > 10^0 = 1$. Thus the domain of $f(x)$ is $(1, \infty)$.

(b) $y = \log_2(\log_{10} x) \iff 2^y = \log_{10} x \iff 10^{2^y} = x$. Thus $f^{-1}(x) = 10^{2^x}$.

75. (a) $f(x) = \dfrac{2^x}{1 + 2^x}$. $y = \dfrac{2^x}{1 + 2^x} \iff y + y\,2^x = 2^x \iff y = 2^x - y\,2^x = 2^x(1 - y) \iff$

$2^x = \dfrac{y}{1 - y} \iff x = \log_2\left(\dfrac{y}{1 - y}\right)$. Thus $f^{-1}(x) = \log_2\left(\dfrac{x}{1 - x}\right)$.

(b) $\dfrac{x}{1 - x} > 0$. Solving this using the methods from Chapter 1, we start with the endpoints of the potential intervals, 0 and 1.

Interval	$(-\infty, 0)$	$(0, 1)$	$(1, \infty)$
Sign of x	$-$	$+$	$+$
Sign of $1 - x$	$+$	$+$	$-$
Sign of $\dfrac{x}{1 - x}$	$-$	$+$	$-$

Thus the domain of $f^{-1}(x)$ is $(0, 1)$.

77. Using $D = 0.73 D_0$ we have $A = -8267 \ln\left(\dfrac{D}{D_0}\right) = -8267 \ln 0.73 \approx 2601$ years.

79.   When $r = 6\%$ we have $t = \dfrac{\ln 2}{0.06} \approx 11.6$ years. When $r = 7\%$ we have $t = \dfrac{\ln 2}{0.07} \approx 9.9$ years.

And when $r = 8\%$ we have $t = \dfrac{\ln 2}{0.08} \approx 8.7$ years.

81.   Using $A = 100$ and $W = 5$ we find the ID $= \dfrac{\log(2A/W)}{\log 2} = \dfrac{\log(2 \cdot 100/5)}{\log 2} = \dfrac{\log 40}{\log 2} \approx 5.32$.

Using $A = 100$ and $W = 10$ we find the ID $= \dfrac{\log(2A/W)}{\log 2} = \dfrac{\log(2 \cdot 100/10)}{\log 2} = \dfrac{\log 20}{\log 2} \approx 4.32$.

So the smaller icon is $\frac{5.23}{4.32} \approx 1.23$ times harder.

83. $\log(\log 10^{100}) = \log 100 = 2$

$\log(\log(\log 10^{\text{googol}})) = \log(\log(\text{googol})) = \log(\log 10^{100}) = \log(100) = 2$

85.   The numbers between 1000 and 9999 (inclusive) each have 4 digits, while $\log 1000 = 3$ and $\log 10{,}000 = 4$. Since $[\![\, \log x \,]\!] = 3$ for all integers x where $1000 \leq x < 10000$, the number of digits is $[\![\, \log x \,]\!] + 1$. Likewise, if x is an integer where $10^{n-1} \leq x < 10^n$, then x would have n digits and $[\![\, \log x \,]\!] = n - 1$. Since $[\![\, \log x \,]\!] = n - 1 \quad \Leftrightarrow \quad n = [\![\, \log x \,]\!] + 1$, the number of digits in x is $[\![\, \log x \,]\!] + 1$.

Exercises 5.3

1. $\log_5 \sqrt{125} = \log_5 5^{3/2} = \frac{3}{2}$

3. $\log 2 + \log 5 = \log 10 = 1$

5. $\log_4 192 - \log_4 3 = \log_4 \frac{192}{3} = \log_4 64 = \log_4 4^3 = 3$

7. $\log_2 6 - \log_2 15 + \log_2 20 = \log_2 \frac{6}{15} + \log_2 20 = \log_2(\frac{2}{5} \cdot 20) = \log_2 8 = \log_2 2^3 = 3$

9. $\log_4 16^{100} = \log_4 (4^2)^{100} = \log_4 4^{200} = 200$

11. $\log(\log 10^{10,000}) = \log(10{,}000 \log 10) = \log(10{,}000 \cdot 1) = = \log(10{,}000) = \log 10^4$
     $= 4 \log 10 = 4$

13. $\log_2 2x = \log_2 2 + \log_2 x = 1 + \log_2 x$

15. $\log_2 [x(x-1)] = \log_2 x + \log_2 (x-1)$

17. $\log 6^{10} = 10 \log 6$

19. $\log_2 (AB^2) = \log_2 A + \log_2 B^2 = \log_2 A + 2 \log_2 B$

21. $\log_3 (x\sqrt{y}) = \log_3 x + \log_3 \sqrt{y} = \log_3 x + \frac{1}{2} \log_3 y$

23. $\log_5 \sqrt[3]{x^2 + 1} = \frac{1}{3} \log_5 (x^2 + 1)$

25. $\ln \sqrt{ab} = \frac{1}{2} \ln ab = \frac{1}{2} (\ln a + \ln b)$

27. $\log \left(\dfrac{x^3 y^4}{z^6} \right) = \log (x^3 y^4) - \log z^6 = 3 \log x + 4 \log y - 6 \log z$

29. $\log_2 \left(\dfrac{x(x^2 + 1)}{\sqrt{x^2 - 1}} \right) = \log_2 x + \log_2 (x^2 + 1) - \frac{1}{2} \log_2 (x^2 - 1)$

31. $\ln \left(x \sqrt{\dfrac{y}{z}} \right) = \ln x + \frac{1}{2} \ln \left(\dfrac{y}{z} \right) = \ln x + \frac{1}{2} (\ln y - \ln z)$

33. $\log \sqrt[4]{x^2 + y^2} = \frac{1}{4} \log(x^2 + y^2)$

35. $\log \sqrt{\dfrac{x^2 + 4}{(x^2 + 1)(x^3 - 7)^2}} = \frac{1}{2} \log \dfrac{x^2 + 4}{(x^2 + 1)(x^3 - 7)^2} = \frac{1}{2}[\log(x^2 + 4) - \log(x^2 + 1)(x^3 - 7)^2]$

     $= \frac{1}{2}[\log(x^2 + 4) - \log(x^2 + 1) - 2\log(x^3 - 7)]$

37. $\ln \dfrac{x^3 \sqrt{x-1}}{3x + 4} = \ln \left(x^3 \sqrt{x-1} \right) - \ln(3x + 4) = 3 \ln x + \frac{1}{2} \ln(x-1) - \ln(3x + 4)$

39. $\log_3 5 + 5\log_3 2 = \log_3 5 + \log_3 2^5 = \log_3(5 \cdot 2^5) = \log_3 160$

41. $\log_2 A + \log_2 B - 2\log_2 C = \log_2(AB) - \log_2(C^2) = \log_2\left(\dfrac{AB}{C^2}\right)$

43. $4\log x - \frac{1}{3}\log(x^2+1) + 2\log(x-1) = \log x^4 - \log\sqrt[3]{x^2+1} + \log(x-1)^2$

$\quad = \log\left(\dfrac{x^4}{\sqrt[3]{x^2+1}}\right) + \log(x-1)^2 = \log\left(\dfrac{x^4(x-1)^2}{\sqrt[3]{x^2+1}}\right)$

45. $\ln 5 + 2\ln x + 3\ln(x^2+5) = \ln(5x^2) + \ln(x^2+5)^3 = \ln[5x^2(x^2+5)^3]$

47. $\frac{1}{3}\log(2x+1) + \frac{1}{2}[\log(x-4) - \log(x^4-x^2-1)] =$

$\quad \log\sqrt[3]{2x+1} + \dfrac{1}{2}\log\dfrac{x-4}{x^4-x^2-1} = \log\left(\sqrt[3]{2x+1} \cdot \sqrt{\dfrac{x-4}{x^4-x^2-1}}\right)$

49. $\log_2 5 = \frac{\log 5}{\log 2} \approx 2.321928$

51. $\log_3 16 = \frac{\log 16}{\log 3} \approx 2.523719$

53. $\log_7 2.61 = \frac{\log 2.61}{\log 7} \approx 0.493008$

55. $\log_4 125 = \frac{\log 125}{\log 4} \approx 3.482892$

57. $\log_3 x = \dfrac{\log_e x}{\log_e 3} = \dfrac{\ln x}{\ln 3} = \dfrac{1}{\ln 3}\ln x$

The graph of $y = \dfrac{1}{\ln 3}\ln x$ is shown in the
viewing rectangle $[-1, 4]$ by $[-3, 2]$.

59. $\log e = \dfrac{\ln e}{\ln 10} = \dfrac{1}{\ln 10}$

61. $-\ln(x - \sqrt{x^2-1}) = \ln\left(\dfrac{1}{x-\sqrt{x^2-1}}\right) = \ln\left(\dfrac{1}{x-\sqrt{x^2-1}} \cdot \dfrac{x+\sqrt{x^2-1}}{x+\sqrt{x^2-1}}\right)$

$\quad = \ln\left(\dfrac{x+\sqrt{x^2-1}}{x^2-(x^2-1)}\right) = \ln\left(x+\sqrt{x^2-1}\right)$

63. (a) $\log P = \log c - k\log W \quad\Leftrightarrow\quad \log P = \log c - \log W^k \quad\Leftrightarrow\quad \log P = \log\left(\dfrac{c}{W^k}\right) \quad\Leftrightarrow$

$\quad P = \dfrac{c}{W^k}.$

(b)   Using $k = 2.1$ and $c = 8000$, when $W = 2$ we have $P = \dfrac{8000}{2^{2.1}} \approx 1866$ and when $W = 10$ we

have $P = \dfrac{8000}{10^{2.1}} \approx 64$.

65.   (a) $M = -2.5 \log\left(\dfrac{B}{B_0}\right) = -2.5 \log B + 2.5 \log B_0$.

(b)   Suppose B_1 and B_2 are the brightness of two stars such that $B_1 < B_2$ and let M_1 and M_2 be
their respective magnitudes. Since log is an increasing function, we have $\log B_1 < \log B_2$.
Then $\log B_1 < \log B_2 \quad \Leftrightarrow \quad \log B_1 - \log B_0 < \log B_2 - \log B_0 \quad \Leftrightarrow$

$\log\left(\dfrac{B_1}{B_0}\right) < \log\left(\dfrac{B_2}{B_0}\right) \quad \Leftrightarrow \quad -2.5 \log\left(\dfrac{B_1}{B_0}\right) > -2.5 \log\left(\dfrac{B_2}{B_0}\right) \quad \Leftrightarrow \quad M_1 > M_2$. Thus

the brighter star has less magnitudes.

(c)   Let B_1 be the brightness of the star Albiero. Then $100B_1$ is the brightness of Betelgeuse. The

magnitude of Betelgeuse is $M = -2.5 \log\left(\dfrac{100B_1}{B_0}\right) = -2.5\left[\log 100 + \log\left(\dfrac{B_1}{B_0}\right)\right]$

$= -2.5\left[2 + \log\left(\dfrac{B_1}{B_0}\right)\right] = -5 - 2.5 \log\left(\dfrac{B_1}{B_0}\right) = -5 + \text{magnitude of Albiero}$.

67.   The error is on the first line: $\log 0.1 < 0$, so $2 \log 0.1 < \log 0.1$.

Exercises 5.4

1. $e^x = 16 \quad \Leftrightarrow \quad \ln e^x = \ln 16 \quad \Leftrightarrow \quad x \ln e = \ln 16 \quad \Leftrightarrow \quad x = 2.7726$

3. $10^{2x} = 5 \quad \Leftrightarrow \quad \log 10^{2x} = \log 5 \quad \Leftrightarrow \quad 2x \log 10 = \log 5 \quad \Leftrightarrow \quad 2x = \log 5 \quad \Leftrightarrow$
   $x = \frac{1}{2} \log 5 = 0.3495$

5. $2^{1-x} = 3 \quad \Leftrightarrow \quad \log 2^{1-x} = \log 3 \quad \Leftrightarrow \quad (1-x) \log 2 = \log 3 \quad \Leftrightarrow \quad 1 - x = \frac{\log 3}{\log 2} \quad \Leftrightarrow$
   $x = 1 - \frac{\log 3}{\log 2} \approx -0.5850$

7. $3e^x = 10 \quad \Leftrightarrow \quad e^x = \frac{10}{3} \quad \Leftrightarrow \quad x = \ln\left(\frac{10}{3}\right) \approx 1.2040$

9. $e^{1-4x} = 2 \quad \Leftrightarrow \quad 1 - 4x = \ln 2 \quad \Leftrightarrow \quad -4x = -1 + \ln 2 \quad \Leftrightarrow \quad x = \frac{1-\ln 2}{4} = 0.0767$

11. $4 + 3^{5x} = 8 \quad \Leftrightarrow \quad 3^{5x} = 4 \quad \Leftrightarrow \quad \log 3^{5x} = \log 4 \quad \Leftrightarrow \quad 5x \log 3 = \log 4 \quad \Leftrightarrow \quad 5x = \frac{\log 4}{\log 3}$
    $\Leftrightarrow \quad x = \frac{\log 4}{5 \log 3} \approx 0.2524$

13. $8^{0.4x} = 5 \quad \Leftrightarrow \quad \log 8^{0.4x} = \log 5 \quad \Leftrightarrow \quad 0.4x \log 8 = \log 5 \quad \Leftrightarrow \quad 0.4x = \frac{\log 5}{\log 8} \quad \Leftrightarrow$
    $x = \frac{\log 5}{0.4 \log 8} \approx 1.9349$

15. $5^{-x/100} = 2 \quad \Leftrightarrow \quad \log 5^{-x/100} = \log 2 \quad \Leftrightarrow \quad -\frac{x}{100} \log 5 = \log 2 \quad \Leftrightarrow$
    $x = -\frac{100 \log 2}{\log 5} \approx -43.0677$

17. $e^{2x+1} = 200 \quad \Leftrightarrow \quad 2x + 1 = \ln 200 \quad \Leftrightarrow \quad 2x = -1 + \ln 200 \quad \Leftrightarrow \quad x = \frac{-1+\ln 200}{2} \approx 2.1492$

19. $5^x = 4^{x+1} \quad \Leftrightarrow \quad \log 5^x = \log 4^{x+1} \quad \Leftrightarrow \quad x \log 5 = (x+1) \log 4 = x \log 4 + \log 4 \quad \Leftrightarrow$
    $x \log 5 - x \log 4 = \log 4 \quad \Leftrightarrow \quad x(\log 5 - \log 4) = \log 4 \quad \Leftrightarrow \quad x = \frac{\log 4}{\log 5 - \log 4} \approx 6.2126$

21. $2^{3x+1} = 3^{x-2} \quad \Leftrightarrow \quad \log 2^{3x+1} = \log 3^{x-2} \quad \Leftrightarrow \quad (3x+1) \log 2 = (x-2) \log 3 \quad \Leftrightarrow$
    $3x \log 2 + \log 2 = x \log 3 - 2 \log 3 \quad \Leftrightarrow \quad 3x \log 2 - x \log 3 = -\log 2 - 2 \log 3 \quad \Leftrightarrow$
    $x(3 \log 2 - \log 3) = -(\log 2 + 2 \log 3) \quad \Leftrightarrow \quad s = -\frac{\log 2 + 2 \log 3}{3 \log 2 - \log 3} \approx -2.9469$

23. $\frac{50}{1 + e^{-x}} = 4 \quad \Leftrightarrow \quad 50 = 4 + 4e^{-x} \quad \Leftrightarrow \quad 46 = 4e^{-x} \quad \Leftrightarrow \quad 11.5 = e^{-x} \quad \Leftrightarrow \quad \ln 11.5 = -x$
    $\Leftrightarrow \quad x = -\ln 11.5 \approx -2.4423$

25. $100(1.04)^{2t} = 300 \quad \Leftrightarrow \quad 1.04^{2t} = 3 \quad \Leftrightarrow \quad \log 1.04^{2t} = \log 3 \quad \Leftrightarrow \quad 2t \log 1.04 = \log 3 \quad \Leftrightarrow$
    $t = \frac{\log 3}{2 \log 1.04} \approx 14.0055$

27. $x^2 2^x - 2^x = 0 \quad \Leftrightarrow \quad 2^x(x^2 - 1) = 0 \quad \Rightarrow \quad 2^x = 0$ (never) or $x^2 - 1 = 0$. If $x^2 - 1 = 0$, then
    $x^2 = 1 \quad \Rightarrow \quad x = \pm 1$. So the only solutions are $x = \pm 1$.

29. $4x^3 e^{-3x} - 3x^4 e^{-3x} = 0 \quad \Leftrightarrow \quad x^3 e^{-3x}(4 - 3x) = 0 \quad \Rightarrow \quad x = 0$ or $e^{-3x} = 0$ (never) or
    $4 - 3x = 0$. If $4 - 3x = 0$, then $3x = 4 \quad \Leftrightarrow \quad x = \frac{4}{3}$. So the solutions are $x = 0$ and $x = \frac{4}{3}$.

31. $e^{2x} - 3e^x + 2 = 0 \quad \Leftrightarrow \quad (e^x - 1)(e^x - 2) = 0 \quad \Rightarrow \quad e^x - 1 = 0$ or $e^x - 2 = 0$. If $e^x - 1 = 0$, then $e^x = 1 \quad \Leftrightarrow \quad x = \ln 1 = 0$. If $e^x - 2 = 0$, then $e^x = 2 \quad \Leftrightarrow \quad x = \ln 2 \approx 0.6931$. So the solutions are $x = 0$ and $x \approx 0.6931$.

33. $e^{4x} + 4e^{2x} - 21 = 0 \quad \Leftrightarrow \quad (e^{2x} + 7)(e^{2x} - 3) = 0 \quad \Rightarrow \quad e^{2x} = -7$ or $e^{2x} = 3$. Now $e^{2x} = -7$ has no solution, since $e^{2x} > 0$ for all x. But we <u>can</u> solve $e^{2x} = 3 \quad \Leftrightarrow \quad 2x = \ln 3 \quad \Leftrightarrow \quad x = \frac{1}{2}\ln 3 \approx 0.5493$. So the only solution is $x \approx 0.5493$.

35. $\ln x = 10 \quad \Leftrightarrow \quad x = e^{10} \approx 22026$

37. $\log x = -2 \quad \Leftrightarrow \quad x = 10^{-2} = 0.01$

39. $\log(3x + 5) = 2 \quad \Leftrightarrow \quad 3x + 5 = 10^2 = 100 \quad \Leftrightarrow \quad 3x = 95 \quad \Leftrightarrow \quad x = \frac{95}{3} \approx 31.6667$

41. $2 - \ln(3 - x) = 0 \quad \Leftrightarrow \quad 2 = \ln(3 - x) \quad \Leftrightarrow \quad e^2 = 3 - x \quad \Leftrightarrow \quad x = 3 - e^2 \approx -4.3891$

43. $\log_2 3 + \log_2 x = \log_2 5 + \log_2(x - 2) \quad \Leftrightarrow \quad \log_2(3x) = \log_2(5x - 10) \quad \Leftrightarrow \quad 3x = 5x - 10$
    $\Leftrightarrow \quad 2x = 10 \quad \Leftrightarrow \quad x = 5$

45. $\log x + \log(x - 1) = \log(4x) \quad \Leftrightarrow \quad \log[x(x - 1)] = \log(4x) \quad \Leftrightarrow \quad x^2 - x = 4x \quad \Leftrightarrow$
    $x^2 - 5x = 0 \quad \Leftrightarrow \quad x(x - 5) = 0 \quad \Rightarrow \quad x = 0$ or $x = 5$. So the *possible* solutions are $x = 0$ and $x = 5$. However, when $x = 0$, $\log x$ is undefined. Thus the only solution is $x = 5$.

47. $\log_5(x + 1) - \log_5(x - 1) = 2 \quad \Leftrightarrow \quad \log_5\left(\dfrac{x + 1}{x - 1}\right) = 2 \quad \Leftrightarrow \quad \dfrac{x + 1}{x - 1} = 5^2 \quad \Leftrightarrow$
    $x + 1 = 25x - 25 \quad \Leftrightarrow \quad 24x = 26 \quad \Leftrightarrow \quad x = \frac{13}{12}$

49. $\log_9(x - 5) + \log_9(x + 3) = 1 \quad \Leftrightarrow \quad \log_9[(x - 5)(x + 3)] = 1 \quad \Leftrightarrow \quad (x - 5)(x + 3) = 9^1$
    $\Leftrightarrow \quad x^2 - 2x - 24 = 0 \quad \Leftrightarrow \quad (x - 6)(x + 4) = 0 \quad \Rightarrow \quad x = 6$ or -4. However, $x = -4$ is inadmissible, so $x = 6$ is the only solution.

51. $\log(x + 3) = \log x + \log 3 \quad \Leftrightarrow \quad \log(x + 3) = \log(3x) \quad \Leftrightarrow \quad x + 3 = 3x \quad \Leftrightarrow \quad 2x = 3 \quad \Leftrightarrow$
    $x = \frac{3}{2}$

53. $2^{2/\log_5 x} = \frac{1}{16} \quad \Leftrightarrow \quad \log_2 2^{2/\log_5 x} = \log_2\left(\frac{1}{16}\right) \quad \Leftrightarrow \quad \dfrac{2}{\log_5 x} = -4 \quad \Leftrightarrow \quad \log_5 x = -\frac{1}{2} \quad \Leftrightarrow$
    $x = 5^{-1/2} = \frac{1}{\sqrt{5}} \approx 0.4472$

55. $\ln x = 3 - x \quad \Leftrightarrow \quad \ln x + x - 3 = 0$. Let $f(x) = \ln x + x - 3$. We need to solve the equation $f(x) = 0$. From the graph of f, we get $x \approx 2.21$.

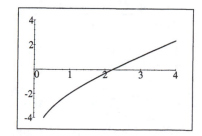

57. $x^3 - x = \log_{10}(x + 1)$ $\Leftrightarrow$ $x^3 - x - \log_{10}(x + 1) = 0$.
     Let $f(x) = x^3 - x - \log_{10}(x + 1)$. We need to solve the
     equation $f(x) = 0$. From the graph of f, we get $x = 0$
     or $x \approx 1.14$.

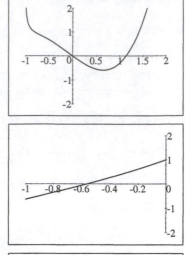

59. $e^x = -x$ $\Leftrightarrow$ $e^x + x = 0$. Let $f(x) = e^x + x$.
     We need to solve the equation $f(x) = 0$. From the graph
     of f, we get $x \approx -0.57$.

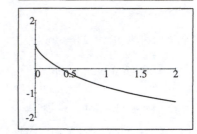

61. $4^{-x} = \sqrt{x}$ $\Leftrightarrow$ $4^{-x} - \sqrt{x} = 0$. Let $f(x) = 4^{-x} - \sqrt{x}$.
     We need to solve the equation $f(x) = 0$. From the graph
     of f, we get $x \approx 0.36$.

63. $\log(x - 2) + \log(9 - x) < 1$ $\Leftrightarrow$ $\log[(x - 2)(9 - x)] < 1$ $\Leftrightarrow$ $\log(-x^2 + 11x - 18) < 1$
     $\Rightarrow$ $-x^2 + 11x - 18 < 10^1$ $\Leftrightarrow$ $0 < x^2 - 11x + 28$ $\Leftrightarrow$ $0 < (x - 7)(x - 4)$. Also, since
     the domain of a logarithm is positive we must have $0 < -x^2 + 11x - 18$ $\Leftrightarrow$
     $0 < (x - 2)(9 - x)$. Using the methods from Chapter 1 with the endpoints 2, 4, 7, 9 for the
     intervals, we have:

Interval	$(-\infty, 2)$	$(2, 4)$	$(4, 7)$	$(7, 9)$	$(9, \infty)$
Sign of $x - 7$	$-$	$-$	$-$	$+$	$+$
Sign of $x - 4$	$-$	$-$	$+$	$+$	$+$
Sign of $x - 2$	$-$	$+$	$+$	$+$	$+$
Sign of $9 - x$	$+$	$+$	$+$	$+$	$-$
Sign of $(x - 7)(x - 4)$	$+$	$+$	$-$	$+$	$+$
Sign of $(x - 2)(9 - x)$	$-$	$+$	$+$	$+$	$-$

     Thus the solution is $(2, 4) \cup (7, 9)$.

65. $2 < 10^x < 5$ $\Leftrightarrow$ $\log 2 < x < \log 5$ $\Leftrightarrow$ $0.3010 < x < 0.6990$. Hence the solution to the
     inequality is approximately the interval $(0.3010, 0.6990)$.

67. (a) $A(3) = 5000\left(1 + \frac{0.085}{4}\right)^{4(3)} = 5000(1.02125^{12}) = 6435.09$. Thus the amount after 3 years is
          \$6,435.09.

(b) $10000 = 5000\left(1 + \frac{0.085}{4}\right)^{4t} = 5000(1.02125^{4t})$ $\Leftrightarrow$ $2 = 1.02125^{4t}$ $\Leftrightarrow$
$\log 2 = 4t \log 1.02125$ $\Leftrightarrow$ $t = \frac{\log 2}{4 \log 1.02125} \approx 8.24$ years. Thus the investment will double
in about 8.24 years.

69. $8000 = 5000\left(1 + \frac{0.075}{4}\right)^{4t} = 5000(1.01875^{4t})$ $\Leftrightarrow$ $1.6 = 1.01875^{4t}$ $\Leftrightarrow$
$\log 1.6 = 4t \log 1.01875$ $\Leftrightarrow$ $t = \frac{\log 1.6}{4 \log 1.01875} \approx 6.33$ years. The investment will increase to
$8000 in approximately 6 years and 4 months.

71. $2 = e^{0.085t}$ $\Leftrightarrow$ $\ln 2 = 0.085t$ $\Leftrightarrow$ $t = \frac{\ln 2}{0.085} \approx 8.15$ years. Thus the investment will double in
about 8.15 years.

73. $r_{\text{eff}} = \left(1 + \frac{r}{n}\right)^n - 1$. Here $r = 0.08$ and $n = 12$, so $r_{\text{eff}} = \left(1 + \frac{0.08}{12}\right)^{12} - 1 = (1.0066667)^{12} - 1$
$= 8.30\%$.

75. $15e^{-0.087t} = 5$ $\Leftrightarrow$ $e^{-0.087t} = \frac{1}{3}$ $\Leftrightarrow$ $-0.087t = \ln\left(\frac{1}{3}\right) = -\ln 3$ $\Leftrightarrow$ $t = \frac{\ln 3}{0.087} \approx 12.6277$.
So only 5 grams remain after approximately 13 days.

77.   (a) $P(3) = \dfrac{10}{1 + 4e^{-0.8(3)}} = 7.337$ So there are approximately 7337 fish after 3 years.

(b)   We solve for t. $\dfrac{10}{1 + 4e^{-0.8t}} = 5$ $\Leftrightarrow$ $1 + 4e^{-0.8t} = \dfrac{10}{5} = 2$ $\Leftrightarrow$ $4e^{-0.8t} = 1$ $\Leftrightarrow$
$e^{-0.8t} = 0.25$ $\Leftrightarrow$ $-0.8t = \ln 0.25$ $\Leftrightarrow$ $t = \frac{\ln 0.25}{-0.8} = 1.73$.   So the population will reach
5000 fish in about 1 year and 9 months.

79.   (a) $\ln\left(\dfrac{P}{P_0}\right) = -\dfrac{h}{k}$ $\Leftrightarrow$ $\dfrac{P}{P_0} = e^{-h/k}$ $\Leftrightarrow$ $P = P_0\,e^{-h/k}$. Substituting $k = 7$ and $P_0 = 100$
we get $P = 100e^{-h/7}$.

(b)   When $h = 4$ we have $P = 100e^{-4/7} \approx 56.47$ kPa.

81.   (a) $I = \frac{60}{13}\left(1 - e^{-13t/5}\right)$ $\Leftrightarrow$ $\frac{13}{60}I = 1 - e^{-13t/5}$ $\Leftrightarrow$ $e^{-13t/5} = 1 - \frac{13}{60}I$ $\Leftrightarrow$
$-\frac{13}{5}t = \ln\left(1 - \frac{13}{60}I\right)$ $\Leftrightarrow$ $t = -\frac{5}{13}\ln\left(1 - \frac{13}{60}I\right)$.

(b)   Substituting $I = 2$, we have $t = -\frac{5}{13}\ln\left[1 - \frac{13}{60}(2)\right] \approx 0.218$ seconds.

83.   Since $9^1 = 9$, $9^2 = 81$, and $9^3 = 729$, the solution of $9^x = 20$ must be between 1 and 2 (because 20
is between 9 and 81), whereas the solution to $9^x = 100$ must be between 2 and 3 (because 100 is
between 81 and 729).

Exercises 5.5

1. (a) $n(0) = 500$.

   (b) The relative growth rate is $0.45 = 45\%$.

   (c) $n(3) = 500e^{0.45(3)} \approx 1929$.

   (d) $10000 = 500\,e^{0.45t}$ $\Leftrightarrow$ $20 = e^{0.45t}$ $\Leftrightarrow$ $0.45t = \ln 20$ $\Leftrightarrow$ $t = \frac{\ln 20}{0.45} \approx 6.66$ hours, or 6 hours 40 minutes.

3. (a) $r = 0.08$ and $n(0) = 18000$. Thus the population is given by the formula $n(t) = 18{,}000\,e^{0.08t}$.

   (b) $t = 2008 - 2000 = 8$. Then we have $n(8) = 18000\,e^{0.08(8)} = 18000\,e^{0.64} \approx 34{,}137$. Thus there should be 34,137 foxes in the region by the year 2008.

   (c)
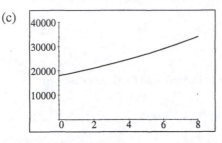

5. (a) $n(t) = 112{,}000\,e^{0.04t}$.

   (b) $t = 2000 - 1994 = 6$ and $n(6) = 112{,}000\,e^{0.04(6)} \approx 142380$. The projected population is about 142,000.

   (c) $200{,}000 = 112{,}000\,e^{0.04t}$ $\Leftrightarrow$ $\frac{25}{14} = e^{0.04t}$ $\Leftrightarrow$ $0.04t = \ln\left(\frac{25}{14}\right)$ $\Leftrightarrow$

   $t = 25\ln\left(\frac{25}{14}\right) \approx 14.5$. Since $1994 + 14.5 = 2008.5$, the population will reach 200,000 during the year 2008.

7. (a) The deer population in 1996 was 20,000.

   (b) Using the model $n(t) = 20000\,e^{rt}$ and the point $(4, 31000)$, we have $31000 = 20000\,e^{4r}$ $\Leftrightarrow$ $1.55 = e^{4r}$ $\Leftrightarrow$ $4r = \ln 1.55$ $\Leftrightarrow$ $r = \frac{1}{4}\ln 1.55 \approx 0.1096$. Thus $n(t) = 20000\,e^{0.1096t}$

   (c) $n(8) = 20000\,e^{0.1096(8)} \approx 48218$, so the projected deer population in 2004 is about 48,000.

   (d) $100000 = 20000\,e^{0.1096t}$ $\Leftrightarrow$ $5 = e^{0.1096t}$ $\Leftrightarrow$ $0.1096t = \ln 5$ $\Leftrightarrow$ $t = \frac{\ln 5}{0.1096}$ ≈ 14.63. Since $1996 + 14.63 = 2010.63$, the deer population will reach 100,000 during the year 2010.

9. (a) Using the formula $n(t) = n_0\,e^{rt}$ with $n_0 = 8600$ and $n(1) = 10000$, we solve for r, giving $10000 = n(1) = 8600\,e^r$ $\Leftrightarrow$ $\frac{50}{43} = e^r$ $\Leftrightarrow$ $r = \ln\left(\frac{50}{43}\right) \approx 0.1508$. Thus $n(t) = 8600\,e^{0.1508\,t}$.

   (b) $n(2) = 8600\,e^{0.1508(2)} \approx 11627$. Thus the number of bacteria after two hours is about 11,600.

(c) $17200 = 8600e^{0.1508t}$ ⟺ $2 = e^{0.1508t}$ ⟺ $0.1508t = \ln 2$ ⟺ $t = \frac{\ln 2}{0.1508} \approx 4.596$.
Thus the number of bacteria will double in about 4.6 hours.

11. (a) $2n_0 = n_0 e^{0.02t}$ ⟺ $2 = e^{0.02t}$ ⟺ $0.02t = \ln 2$ ⟺ $t = 50 \ln 2 \approx 34.65$. So we have
$t = 1995 + 34.65 = 2029.65$, and hence at the current growth rate the population will double
by the year 2029.

(b) $3n_0 = n_0 e^{0.02t}$ ⟺ $3 = e^{0.02t}$ ⟺ $0.02t = \ln 3$ ⟺ $t = 50 \ln 3 \approx 54.93$. So we
have $t = 1995 + 54.93 = 2049.93$, and hence at the current growth rate the population will
triple by the year 2049.

13. $n(t) = n_0 e^{2t}$. When $n_0 = 1$, the critical level is $n(24) = e^{2(24)} = e^{48}$. We solve the equation
$e^{48} = n_0 e^{2t}$, where $n_0 = 10$. This gives $e^{48} = 10\,e^{2t}$ ⟺ $48 = \ln 10 + 2t$ ⟺
$2t = 48 - \ln 10$ ⟺ $t = \frac{1}{2}(48 - \ln 10) \approx 22.85$ hours.

15. (a) Using $m(t) = m_0\,e^{-rt}$ with $m_0 = 10$ and $h = 30$, we have $r = \dfrac{\ln 2}{h} = \dfrac{\ln 2}{30} \approx 0.0231$. Thus
$m(t) = 10e^{-0.0231t}$.

(b) $m(80) = 10e^{-0.0231(80)} \approx 1.6$ grams.

(c) $2 = 10e^{-0.0231t}$ ⟺ $\frac{1}{5} = e^{-0.0231t}$ ⟺ $\ln\left(\frac{1}{5}\right) = -0.0231t$ ⟺ $t = \frac{-\ln 5}{-0.0231} \approx 70$
years.

17. By the formula in the text, $m(t) = m_0\,e^{-rt}$ where $r = \frac{\ln 2}{h}$, so $m(t) = 50e^{-\frac{\ln 2}{28}\cdot t}$. We need to solve
for t in the equation $32 = 50e^{-\frac{\ln 2}{28}t}$. This gives $e^{-\frac{\ln 2}{28}t} = \dfrac{32}{50}$ ⟺ $-\dfrac{\ln 2}{28}t = \ln\left(\dfrac{32}{50}\right)$ ⟺
$t = -\frac{28}{\ln 2}\cdot\ln\left(\frac{32}{50}\right) \approx 18.03$, so it takes about 18 years.

19. By the formula for radioactive decay, we have $m(t) = m_0\,e^{-rt}$, where $r = \frac{\ln 2}{h}$, in other words
$m(t) = m_0 e^{-\frac{\ln 2}{h}\cdot t}$. In this exercise we have to solve for h in the equation $200 = 250e^{-\frac{\ln 2}{h}\cdot 48}$
⟺ $0.8 = e^{-\frac{\ln 2}{h}\cdot 48}$ ⟺ $\ln(0.8) = -\frac{\ln 2}{h}\cdot 48$ ⟺ $h = -\frac{\ln 2}{\ln 0.8}\cdot 48 \approx 149.1$ hours. So the
half-life is approximately 149 hours.

21. By the formula in the text, $m(t) = m_0\,e^{-\frac{\ln 2}{h}\cdot t}$, so we have $0.65 = 1\cdot e^{-\frac{\ln 2}{5730}\cdot t}$ ⟺
$\ln(0.65) = -\dfrac{\ln 2}{5730}t$ ⟺ $t = -\dfrac{5730 \ln 0.65}{\ln 2} \approx 3561$. Thus the artifact is about 3560 years old.

23. (a) $T(0) = 65 + 145\,e^{-0.05(0)} = 65 + 145 = 210°F$.

(b) $T(10) = 65 + 145\,e^{-0.05(10)} \approx 152.9$. Thus the temperature after 10 minutes is about 153°F.

(c) $100 = 65 + 145\,e^{-0.05t}$ ⟺ $35 = 145\,e^{-0.05t}$ ⟺ $0.2414 = e^{-0.05t}$ ⟺
$\ln 0.2414 = -0.05t$ ⟺ $t = -\dfrac{\ln 0.2414}{0.05} \approx 28.4$. Thus the temperature will be 100°F in
about 28 minutes.

25. Using Newton's Law of Cooling, $T(t) = T_s + D_0\,e^{-kt}$ with $T_s = 75$ and $D_0 = 185 - 75 = 110$. So
$T(t) = 75 + 110\,e^{-kt}$.

(a)   Since $T(30) = 150$, we have $T(30) = 75 + 110\,e^{-30k} = 150$ $\Leftrightarrow$ $110\,e^{-30k} = 75$ $\Leftrightarrow$
$e^{-30k} = \frac{15}{22}$ $\Leftrightarrow$ $-30k = \ln\left(\frac{15}{22}\right)$ $\Leftrightarrow$ $k = -\frac{1}{30}\ln\left(\frac{15}{22}\right)$. Thus we have
$T(45) = 75 + 110\,e^{(45/30)\ln(15/22)} \approx 136.9$, and so the temperature of the turkey after 45
minutes is about $137°\,\text{F}$.

(b)   The temperature will be $100°\,\text{F}$ when $75 + 110\,e^{(t/30)\ln(15/22)} = 100$ $\Leftrightarrow$
$e^{(t/30)\ln(15/22)} = \frac{25}{110} = \frac{5}{22}$ $\Leftrightarrow$ $\left(\frac{t}{30}\right)\ln\left(\frac{15}{22}\right) = \ln\left(\frac{5}{22}\right)$ $\Leftrightarrow$ $t = 30\dfrac{\ln\left(\frac{5}{22}\right)}{\ln\left(\frac{15}{22}\right)} \approx 116.1$. So the

temperature will be $100°\,\text{F}$ after 116 minutes.

27.   (a) $\text{pH} = -\log[\text{H}^+] = -\log(5.0 \times 10^{-3}) \approx 2.3$

      (b) $\text{pH} = -\log[\text{H}^+] = -\log(3.2 \times 10^{-4}) \approx 3.5$

      (c) $\text{pH} = -\log[\text{H}^+] = -\log(5.0 \times 10^{-9}) \approx 8.3$

29.   (a) $\text{pH} = -\log[\text{H}^+] = 3.0$ $\Leftrightarrow$ $[\text{H}^+] = 10^{-3}\,\text{M}$

      (b) $\text{pH} = -\log[\text{H}^+] = 6.5$ $\Leftrightarrow$ $[\text{H}^+] = 10^{-6.5} \approx 3.2 \times 10^{-7}\,\text{M}$

31. $4.0 \times 10^{-7} \le [\text{H}^+] \le 1.6 \times 10^{-5}$ $\Leftrightarrow$ $\log(4.0 \times 10^{-7}) \le \log[\text{H}^+] \le \log(1.6 \times 10^{-5})$ $\Leftrightarrow$
      $-\log(4.0 \times 10^{-7}) \ge \text{pH} \ge -\log(1.6 \times 10^{-5})$ $\Leftrightarrow$ $6.4 \ge \text{pH} \ge 4.8$. Therefore the range of pH
      readings for cheese is approximately 4.8 to 6.4.

33.   Let I_0 be the intensity of the smaller earthquake and I_1 the intensity of the larger earthquake. Then
      $I_1 = 20\,I_0$. Notice that $M_0 = \log\left(\frac{I_0}{S}\right) = \log I_0 - \log S$ and $M_1 = \log\left(\frac{I_1}{S}\right) = \log\left(\frac{20\,I_0}{S}\right)$
      $= \log 20 + \log I_0 - \log S$. Then $M_1 - M_0 = \log 20 + \log I_0 - \log S - \log I_0 + \log S$
      $= \log 20 \approx 1.3$. Therefore the magnitude is 1.3 times larger.

35.   Let the subscript A represent the Alaska earthquake and SF represent the San Francisco earthquake.
      Then $M_A = \log\left(\frac{I_A}{S}\right) = 8.6$ $\Leftrightarrow$ $I_A = S \cdot 10^{8.6}$; also, $M_{SF} = \log\left(\frac{I_{SF}}{S}\right) = 8.3$ $\Leftrightarrow$
      $I_{SF} = S \cdot 10^{8.6}$. So $\dfrac{I_A}{I_{SF}} = \dfrac{S \cdot 10^{8.6}}{S \cdot 10^{8.3}} = 10^{0.3} \approx 1.995$, and hence the Alaskan earthquake was
      roughly twice as intense as the San Francisco earthquake.

37.   Let the subscript MC represent the Mexico City earthquake, and T represent the Tangshan
      earthquake. We have $\dfrac{I_T}{I_{MC}} = 1.26$ $\Leftrightarrow$ $\log 1.26 = \log\dfrac{I_T}{I_{MC}} = \log\dfrac{I_T/S}{I_{MC}/S} = \log\dfrac{I_T}{S} - \log\dfrac{I_{MC}}{S}$
      $= M_T - M_{MC}$. Therefore $M_T = M_{MC} + \log 1.26 \approx 8.1 + 0.1 = 8.2$. Thus the magnitude of the
      Tangshan earthquake was roughly 8.2.

39. $98 = 10\log\left(\dfrac{I}{10^{-12}}\right)$ $\Leftrightarrow$ $\log(I \cdot 10^{12}) = 9.8$ $\Leftrightarrow$ $\log I = 9.8 - \log 10^{12} = -2.2$ $\Leftrightarrow$
      $I = 10^{-2.2} \approx 6.3 \times 10^{-3}$. So the intensity was 6.3×10^{-3} watts/m^2.

41.   (a) $\beta_1 = 10\log\left(\dfrac{I_1}{I_0}\right)$ and $I_1 = \dfrac{k}{d_1^2}$ $\Leftrightarrow$ $\beta_1 = 10\log\left(\dfrac{k}{d_1^2\,I_0}\right) = 10\left[\log\left(\dfrac{k}{I_0}\right) - 2\log d_1\right]$

$$= 10 \log\left(\frac{k}{I_0}\right) - 20 \log d_1. \text{ Similarly, } \beta_2 = 10 \log\left(\frac{k}{I_0}\right) - 20 \log d_2. \text{ Substituting the}$$

expression for β_1, gives $\beta_2 = 10 \log\left(\frac{k}{I_0}\right) - 20 \log d_1 + 20 \log d_1 - 20 \log d_2$

$$= \beta_1 + 20 \log d_1 - 20 \log d_2 = \beta_1 + 20 \log\left(\frac{d_1}{d_2}\right).$$

(b) $\beta_1 = 120$, $d_1 = 2$, and $d_2 = 10$. Then $\beta_2 = \beta_1 + 20 \log\left(\frac{d_1}{d_2}\right) = 120 + 20 \log\left(\frac{2}{10}\right)$
$= 120 + 20 \log 0.2 \approx 106$, and so the intensity level at 10 m is approximately 106 dB.

Review Exercises for Chapter 5

1. $f(x) = 3^{-x}$.
   Domain: $(-\infty, \infty)$
   Range: $(0, \infty)$
   Asymptote: $y = 0$.

3. $y = 5 - 10^x$.
   Domain: $(-\infty, \infty)$
   Range: $(-\infty, 5)$
   Asymptote: $y = 5$.

5. $f(x) = \log_3(x - 1)$.
   Domain: $(1, \infty)$
   Range: $(-\infty, \infty)$
   Asymptote: $x = 1$.

7. $y = 2 - \log_2 x$.
   Domain: $(0, \infty)$
   Range: $(-\infty, \infty)$
   Asymptote: $x = 0$.

9. $F(x) = e^x - 1$.
   Domain: $(-\infty, \infty)$
   Range: $(-1, \infty)$
   Asymptote: $y = -1$.

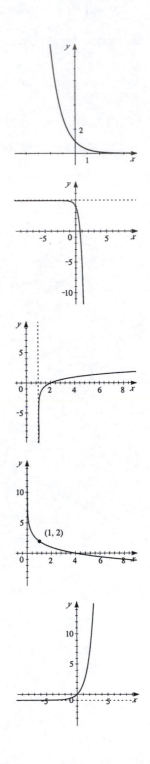

11. $y = 2 \ln x.$
     Domain: $(0, \infty)$
     Range: $(-\infty, \infty)$
     Asymptote: $x = 0.$

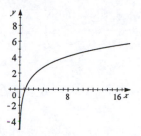

13. $f(x) = 10^{x^2} + \log(1 - 2x).$ Since $\log u$ is defined only for $u > 0$, we require $1 - 2x > 0$ $\Leftrightarrow$
     $-2x > -1$ $\Leftrightarrow$ $x < \frac{1}{2}$, and so the domain is $\left(-\infty, \frac{1}{2}\right).$

15. $h(x) = \ln(x^2 - 4).$ We must have $x^2 - 4 > 0$ (since $\ln y$ is defined only for $y > 0$) $\Leftrightarrow$
     $x^2 - 4 > 0$ $\Leftrightarrow$ $(x - 2)(x + 2) > 0.$ The endpoints of the intervals are -2 and $2.$

Interval	$(-\infty, -2)$	$(-2, 2)$	$(2, \infty)$
Sign of $x - 2$	$-$	$-$	$+$
Sign of $x + 2$	$-$	$+$	$+$
Sign of $(x - 2)(x + 2)$	$+$	$-$	$+$

     Thus the domain is $(-\infty, -2) \cup (2, \infty).$

17. $\log_2 1024 = 10$ $\Leftrightarrow$ $2^{10} = 1024$

19. $\log x = y$ $\Leftrightarrow$ $10^y = x$

21. $2^6 = 64$ $\Leftrightarrow$ $\log_2 64 = 6$

23. $10^x = 74$ $\Leftrightarrow$ $\log_{10} 74 = x$ $\Leftrightarrow$ $\log 74 = x$

25. $\log_2 128 = \log_2(2^7) = 7$

27. $10^{\log 45} = 45$

29. $\ln(e^6) = 6$

31. $\log_3 \frac{1}{27} = \log_3 3^{-3} = -3$

33. $\log_5 \sqrt{5} = \log_5 5^{1/2} = \frac{1}{2}$

35. $\log 25 + \log 4 = \log(25 \cdot 4) = \log 10^2 = 2$

37. $\log_2(16^{23}) = \log_2(2^4)^{23} = \log_2 2^{92} = 92$

39. $\log_8 6 - \log_8 3 + \log_8 2 = \log_8\left(\frac{6}{3} \cdot 2\right) = \log_8 4 = \log_8 8^{2/3} = \frac{2}{3}$

41. $\log(AB^2 C^{\,3}) = \log A + 2\log B + 3\log C$

43. $\ln\sqrt{\dfrac{x^2 - 1}{x^2 + 1}} = \frac{1}{2} \ln\left(\dfrac{x^2 - 1}{x^2 + 1}\right) = \frac{1}{2}[\ln(x^2 - 1) - \ln(x^2 + 1)]$

45. $\log_5\left(\dfrac{x^2(1-5x)^{3/2}}{\sqrt{x^3-x}}\right) = \log_5 x^2(1-5x)^{3/2} - \log_5\sqrt{x(x^2-1)}$

$= 2\log_5 x + \frac{3}{2}\log_5(1-5x) - \frac{1}{2}\log_5(x^3-x)$

47. $\log 6 + 4\log 2 = \log 6 + \log 2^4 = \log(6\cdot 2^4) = \log 96$

49. $\frac{3}{2}\log_2(x-y) - 2\log_2(x^2+y^2) = \log_2(x-y)^{3/2} - \log_2(x^2+y^2)^2 = \log_2\left(\dfrac{(x-y)^{3/2}}{(x^2+y^2)^2}\right)$

51. $\log(x-2) + \log(x+2) - \frac{1}{2}\log(x^2+4) = \log[(x-2)(x+2)] - \log\sqrt{x^2+4} =$

$\log\left(\dfrac{x^2-4}{\sqrt{x^2+4}}\right)$

53. $\log_2(1-x) = 4 \quad\Leftrightarrow\quad 1-x = 2^4 \quad\Leftrightarrow\quad x = 1 - 2^4 = -15$

55. $5^{5-3x} = 26 \quad\Leftrightarrow\quad \log_5 26 = 5 - 3x \quad\Leftrightarrow\quad 3x = 5 - \log_5 26 \quad\Leftrightarrow\quad x = \frac{1}{3}(5 - \log_5 26) \approx 0.99$

57. $e^{3x/4} = 10 \quad\Leftrightarrow\quad \ln e^{3x/4} = \ln 10 \quad\Leftrightarrow\quad \frac{3x}{4} = \ln 10 \quad\Leftrightarrow\quad x = \frac{4}{3}\ln 10 \approx 3.07$

59. $\log x + \log(x+1) = \log 12 \quad\Leftrightarrow\quad \log[x(x+1)] = \log 12 \quad\Leftrightarrow\quad x(x+1) = 12 \quad\Leftrightarrow$
$x^2 + x - 12 = 0 \quad\Leftrightarrow\quad (x+4)(x-3) = 0 \quad\Rightarrow\quad x = 3$ or -4. Because $\log x$ and $\log(x+1)$
are undefined at $x = -4$, it follows that $x = 3$ is the only solution.

61. $x^2 e^{2x} + 2x\,e^{2x} = 8\,e^{2x} \quad\Leftrightarrow\quad e^{2x}(x^2 + 2x - 8) = 0 \quad\Leftrightarrow\quad x^2 + 2x - 8 = 0$ (since $e^{2x} > 0$ for
all x) $\quad\Leftrightarrow\quad (x+4)(x-2) = 0 \quad\Leftrightarrow\quad x = 2, -4$

63. $5^{-2x/3} = 0.63 \quad\Leftrightarrow\quad \dfrac{-2x}{3}\log 5 = \log 0.63 \quad\Leftrightarrow\quad x = -\dfrac{3\log 0.63}{2\log 5} \approx 0.430618$

65. $5^{2x+1} = 3^{4x-1} \quad\Leftrightarrow\quad (2x+1)\log 5 = (4x-1)\log 3 \quad\Leftrightarrow\quad 2x\log 5 + \log 5 = 4x\log 3 - \log 3$

$\Leftrightarrow\quad x(2\log 5 - 4\log 3) = -\log 3 - \log 5 \quad\Leftrightarrow\quad x = \dfrac{\log 3 + \log 5}{4\log 3 - 2\log 5} \approx 2.303600$

67. $y = e^{x/(x+2)}$.
Vertical Asymptote: $x = -2$
Horizontal Asymptote: $y = 2.72$
No maximum or minimum.

69. $y = \log(x^3 - x)$.
    Vertical Asymptotes: $x = -1$, $x = 0$, $x = 1$
    Horizontal Asymptote: none
    Local maximum ≈ -0.41 when $x \approx -0.58$

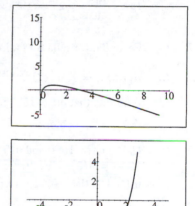

71. $3 \log x = 6 - 2x$.
    We graph $y = 3 \log x$ and $y = 6 - 2x$ in the same
    viewing rectangle. The solution occurs where the
    two graphs intersect. From the graphs, we see that the
    solution is $x \approx 2.42$.

73. $\ln x > x - 2$. We graph the function $f(x) = \ln x - x + 2$,
    and we see that the graph lies above the x-axis for
    $0.16 < x < 3.15$. So the approximate solution of the
    given inequality is $0.16 < x < 3.15$.

75. $f(x) = e^x - 3e^{-x} - 4x$.
    We graph the function $f(x)$, and we see that the
    function is increasing on $(-\infty, 0]$ and $[1.10, \infty)$
    and that it is decreasing on $[0, 1.10]$.

77. $\log_4 15 = \frac{\log 15}{\log 4} = 1.953445$

79. Notice that $\log_4 258 > \log_4 256 = \log_4 4^4 = 4$ and so $\log_4 258 > 4$. Also
    $\log_5 620 < \log_5 625 = \log_5 5^4 = 4$ and so $\log_5 620 < 4$. Then $\log_4 258 > 4 > \log_5 620$ and so
    $\log_4 258$ is larger.

81. $P = 12{,}000$, $r = 0.10$, and $t = 3$. Then $A = P(1 + \frac{r}{n})^{nt}$.

    (a) For $n = 2$, $A = 12{,}000(1 + \frac{0.10}{2})^{2(3)} = 12{,}000(1.05^6) \approx \$16{,}081.15$.

    (b) For $n = 12$, $A = 12{,}000(1 + \frac{0.10}{12})^{12(3)} \approx \$16{,}178.18$.

    (c) For $n = 365$, $A = 12{,}000(1 + \frac{0.10}{365})^{365(3)} \approx \$16{,}197.64$.

    (d) For $n = \infty$, $A = P e^{rt} = 12{,}000\, e^{0.10(3)} \approx \$16{,}198.31$.

83. (a) Using the model $n(t) = n_0 e^{rt}$, with $n_0 = 30$ and $r = 0.15$, we have the formula

$n(t) = 30 e^{0.15t}$.

(b) $n(4) = 30 e^{0.15(4)} \approx 55$.

(c) $500 = 30 e^{0.15t}$ $\Leftrightarrow$ $\frac{50}{3} = e^{0.15t}$ $\Leftrightarrow$ $0.15t = \ln\left(\frac{50}{3}\right)$ $\Leftrightarrow$ $t = \frac{1}{0.15} \ln\left(\frac{50}{3}\right) \approx 18.76$.
So the stray cat population will reach 500 in about 19 years.

85. (a) From the formula for radioactive decay, we have $m(t) = 10e^{-rt}$, where $r = -\dfrac{\ln 2}{2.7 \times 10^5}$. So
after 1000 years the amount remaining is $m(1000) = 10 \cdot e^{[-\ln 2/(2.7 \times 10^5)] \cdot 1000}$
$= 10e^{-(\ln 2)/(2.7 \times 10^2)} = 10e^{-(\ln 2)/270} \approx 9.97$. Therefore the amount remaining is about 9.97
mg.

(b) We solve for t in the equation $7 = 10e^{-[\ln 2/(2.7 \times 10^5)] \cdot t}$. We have $7 = 10 \, e^{-[\ln 2/(2.7 \times 10^5)] \cdot t}$ $\Leftrightarrow$

$0.7 = e^{-[\ln 2/(2.7 \times 10^5)] \cdot t}$ $\Leftrightarrow$ $\ln 0.7 = -\dfrac{\ln 2}{2.7 \times 10^5} \cdot t$ $\Leftrightarrow$

$t = -\dfrac{\ln 0.7}{\ln 2} \cdot 2.7 \times 10^5 \approx 138{,}934.75$. Thus it takes about 139,000 years.

87. (a) From the formula for radioactive decay, $r = \dfrac{\ln 2}{1590} \approx 0.0004359$ and $n(t) = 150 \cdot e^{-0.0004359 t}$.

(b) $n(1000) = 150 \cdot e^{-0.0004359 \cdot 1000} \approx 97.00$, and so the amount remaining is about 97.00 mg.

(c) Find t so that $50 = 150 \cdot e^{-0.0004359 t}$. We have $50 = 150 \cdot e^{-0.0004359 t}$ $\Leftrightarrow$ $\frac{1}{3} = e^{-0.0004359 t}$
$\Leftrightarrow$ $t = -\frac{1}{0.0004359} \ln\left(\frac{1}{3}\right) \approx 2520$. Thus only 50 mg remain after about 2520 years.

89. (a) Using $n_0 = 1500$ and $n(5) = 3200$ in the formula $n(t) = n_0 e^{rt}$, we have
$3200 = n(5) = 1500 e^{5r}$ $\Leftrightarrow$ $e^{5r} = \frac{32}{15}$ $\Leftrightarrow$ $5r = \ln\left(\frac{32}{15}\right)$ $\Leftrightarrow$ $r = \frac{1}{5}$
$\ln\left(\frac{32}{15}\right) \approx 0.1515$. Thus $n(t) = 1500 \cdot e^{0.1515t}$.

(b) We have $t = 1999 - 1988 = 11$ so $n(11) = 1500e^{0.1515 \cdot 11} \approx 7940$. Thus in 1999 the bird
population should be about 7940 birds.

91. $[H^+] = 1.3 \times 10^{-8}$ M. Then $pH = -\log[H^+] = -\log(1.3 \times 10^{-8}) \approx 7.9$, and so fresh egg whites
are basic.

93. Let I_0 be the intensity of the smaller earthquake and I_1 be the intensity of the larger earthquake.
Then $I_1 = 35 I_0$. Since $M = \log\left(\frac{I}{S}\right)$, we have $M_0 = \log\left(\frac{I_0}{S}\right) = 6.5$ and $M_1 = \log\left(\frac{I_1}{S}\right)$
$= \log\left(\frac{35 I_0}{S}\right) = \log 35 + \log\left(\frac{I_0}{S}\right) = \log 35 + M_0 = \log 35 + 6.5 \approx 8.04$. So the magnitude on the
Richter scale of the larger earthquake is approximately 8.0.

Chapter 5 Test

1. $y = 2^x$ and $y = \log_2 x$.

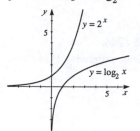

2. $f(x) = \log(x + 2)$.
   Domain: $(-2, \infty)$
   Range: $(-\infty, \infty)$
   Vertical asymptote: $x = -2$.

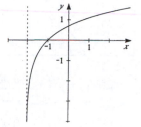

3. (a) $\log_5 \sqrt{125} = \log_5 (5^3)^{\frac{1}{2}} = \log_5 5^{3/2} = \frac{3}{2}$

   (b) $\log_2 56 - \log_2 7 = \log_2 \left(\frac{56}{7}\right) = \log_2 8 = \log_2 2^3 = 3$

   (c) $\log_8 4 = \log_8 8^{2/3} = \frac{2}{3}$

   (d) $\log_6 4 + \log_6 9 = \log_6 (4 \cdot 9) = \log_6 6^2 = 2$

4. $\log \sqrt{\dfrac{x + 2}{x^4(x^2 + 4)}} = \frac{1}{2} \log \left(\dfrac{x + 2}{x^4(x^2 + 4)}\right) = \frac{1}{2} \left[\log(x + 2) - (4 \log x + \log(x^2 + 4))\right]$

   $= \frac{1}{2}\log(x + 2) - 2 \log x - \frac{1}{2} \log(x^2 + 4)$

5. $\ln x - 2\ln(x^2 + 1) + \frac{1}{2} \ln(3 - x^4) = \ln\left(x\sqrt{3 - x^4}\right) - \ln(x^2 + 1)^2 = \ln\left(\dfrac{x\sqrt{3 - x^4}}{(x^2 + 1)^2}\right)$

6. (a) $2^{x-1} = 10 \quad \Leftrightarrow \quad \log 2^{x-1} = \log 10 = 1 \quad \Leftrightarrow \quad (x - 1)\log 2 = 1 \quad \Leftrightarrow \quad x - 1 = \frac{1}{\log 2}$
   $\Leftrightarrow \quad x = 1 + \frac{1}{\log 2} \approx 4.32$

   (b) $5 \ln(3 - x) = 4 \quad \Leftrightarrow \quad \ln(3 - x) = \frac{4}{5} \quad \Leftrightarrow \quad e^{\ln(3-x)} = e^{4/5} \quad \Leftrightarrow \quad 3 - x = e^{4/5} \quad \Leftrightarrow$
   $x = 3 - e^{4/5} \approx 0.77$

   (c) $10^{x+3} = 6^{2x} \quad \Leftrightarrow \quad \log 10^{x+3} = \log 6^{2x} \quad \Leftrightarrow \quad x + 3 = 2x \log 6 \quad \Leftrightarrow \quad 2x \log 6 - x = 3$
   $\Leftrightarrow \quad x(2 \log 6 - 1) = 3 \quad \Leftrightarrow \quad x = \dfrac{3}{2 \log 6 - 1} \approx 5.39$

   (d) $\log_2(x + 2) + \log_2(x - 1) = 2 \quad \Leftrightarrow \quad \log_2((x + 2)(x - 1)) = 2 \quad \Leftrightarrow \quad x^2 + x - 2 = 2^2$
   $\Leftrightarrow \quad x^2 + x - 6 = 0 \quad \Leftrightarrow \quad (x + 3)(x - 2) = 0 \quad \Rightarrow \quad x = -3 \text{ or } x = 2.$ However, both

logarithms are undefined at $x = -3$, so the only solution is $x = 2$.

7. (a) From the formula for population growth, we have $8000 = 1000e^{r \cdot 1}$ $\Leftrightarrow$ $8 = e^r$ $\Leftrightarrow$
        $r = \ln 8 \approx 2.07944$. Thus $n(t) = 1000e^{2.07944t}$.

    (b) $n(1.5) = 1000e^{2.07944(1.5)} \approx 22{,}627$

    (c) $15000 = 1000e^{2.07944t}$ $\Leftrightarrow$ $15 = e^{2.07944t}$ $\Leftrightarrow$ $\ln 15 = 2.07944t$ $\Leftrightarrow$
        $t = \frac{\ln 15}{2.07944} \approx 1.3$. Thus the population will reach 15,000 after approximately 1.3 hours.

    (d)

8. (a) $A(t) = 12{,}000\left(1 + \dfrac{0.056}{12}\right)^{12t}$, where t is in years.

    (b) $A(t) = 12{,}000\left(1 + \dfrac{0.056}{365}\right)^{365t}$. So $A(3) = 12{,}000\left(1 + \dfrac{0.056}{365}\right)^{365(3)} = \$14{,}195.06$.

    (c) $A(t) = 12{,}000\left(1 + \dfrac{0.056}{2}\right)^{2t} = 12{,}000(1.028)^{2t}$. So $20{,}000 = 12{,}000(1.028)^{2t}$ $\Leftrightarrow$
        $1.6667 = 1.028^{2t}$ $\Leftrightarrow$ $\ln 1.6667 = \ln 1.028^{2t}$ $\Leftrightarrow$ $\ln 1.6667 = 2t \ln 1.028$ $\Leftrightarrow$
        $t = \frac{\ln 1.6667}{2 \ln 1.028} \approx 9.25$ years.

9. $f(x) = \dfrac{e^x}{x^3}$

    (a)   (b) Vertical asymptote: $x = 0$; Horizontal
                                                      asymptote: $y = 0$.

                                                 (c) Local minimum ≈ 0.74 when $x \approx 3.00$

                                                 (d) Range $\approx (-\infty, 0) \cup [0.74, \infty)$

    (e) $\dfrac{e^x}{x^3} = 2x + 1$.

        We see that the graphs intersect at
        $x \approx -0.85$, 0.96, and 9.92.

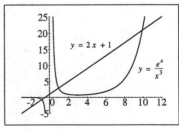

Focus on Modeling

1. (a)

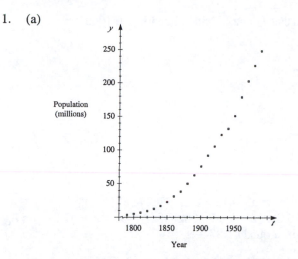

(b) Using a graphing calculator, we obtain the model $y = ab^t$, where $a = 4.041807 \times 10^{-16}$ and $b = 1.021003194$, and y is the population (in millions) in the year t.

(c) Substituting $t = 2000$ into the model of part (b), we get $y = ab^{2000} \approx 457.9$ million.

(d) According to the model, the population in 1965 should have been about $y = ab^{1965} \approx 221.2$ million.

(e) The values given by the model are clearly much too large. This means that an exponential model is NOT appropriate for these data.

3. (a) Yes.

(b)

Year	Health Expenditures (billions of dollars)	
t	E	$\ln E$
1970	74.3	4.30811
1980	251.1	5.52585
1985	434.5	6.07420
1987	506.2	6.22693
1989	623.9	6.43599
1990	696.6	6.54621
1992	820.3	6.70967
1994	937.2	6.84290
1996	1039.4	6.94640
1998	1150.0	7.04752
2000	1310.0	7.17778
2001	1424.5	7.26158

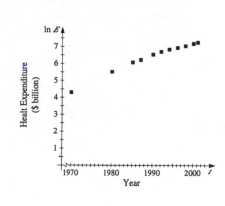

Yes, the scatter plot appears to be linear.

(c) Let t be the number of years elapsed since 1970. Then $\ln E = 4.494411 + 0.0970921464t$, where E is expenditure in billions of dollars.

(d) $E = e^{4.494411 + 0.0970921464t} = 89.51542887 \, e^{0.0970921464t}$.

(e) In 2004 we have $t = 2004 - 1970 = 34$, so the estimated 2004 health-care expenditures are $89.51542887\, e^{0.0970921464(34)} \approx 2429.8$ billion dollars.

5. (a) Using a graphing calculator, we obtain that $I_0 = 22.7586444$ and $k = 0.1062398$.

    (b)

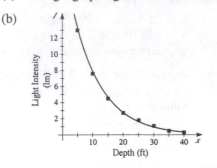

    (c) We solve $0.15 = 22.7586444\, e^{-0.1062398x}$ for x: $0.15 = 22.7586444\, e^{-0.1062398x}$ $\Leftrightarrow$ $0.006590902 = e^{-0.1062398x}$ $\Leftrightarrow$ $-5.022065 = -0.1062398x$ $\Leftrightarrow$ $x \approx 47.27$. So light intensity drops below 0.15 limuns below around 47.27 feet.

7. (a) Let t be the number of years elapsed since 1970, and let y be the number of millions of tons of lead emissions in year t. Using a graphing calculator, we obtain the exponential model $y = ab^t$, where $a = 301.813054$ and $b = 0.819745$.

    (b) Using a graphing calculator, we obtain the fourth-degree polynomial model $y = at^4 + bt^3 + ct^2 + dt + e$, where $a = -0.002430$, $b = 0.135159$, $c = -2.014322$, $d = -4.055294$, and $e = 199.092227$.

    (c) The exponential model is shown as a solid curve and the polynomial model as a dotted curve in the graph to the right. From the graph, the polynomial model appears to fit the data better.

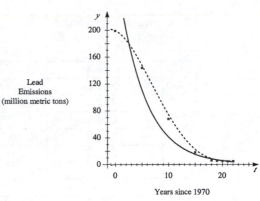

    (d) Exponential model: 1972 $(t = 2)$ $y = 202.8$
                              1982 $(t = 12)$ $y = 27.8$

         Polynomial model: 1972 $y = 184.0$
                              1982 $y = 43.5$

9. (a)

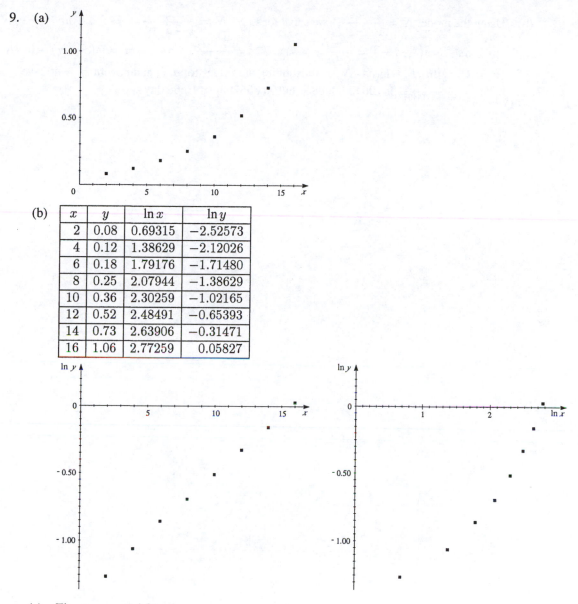

(b)

x	y	$\ln x$	$\ln y$
2	0.08	0.69315	−2.52573
4	0.12	1.38629	−2.12026
6	0.18	1.79176	−1.71480
8	0.25	2.07944	−1.38629
10	0.36	2.30259	−1.02165
12	0.52	2.48491	−0.65393
14	0.73	2.63906	−0.31471
16	1.06	2.77259	0.05827

(c) The exponential function.

(d) $y = a \cdot b^x$ where $a = 0.057697$ and $b = 1.200236$.

11. (a) Using the Logistic command on a TI-83 we get $y = \dfrac{c}{1 + ae^{-bx}}$ where $a = 49.10976596$, $b = 0.4981144989$, and $c = 500.855793$.

(b)   Using the model $N = \dfrac{c}{1 + ae^{-bt}}$ we solve for t. So $N = \dfrac{c}{1 + ae^{-bt}}$ $\Leftrightarrow$ $1 + ae^{-bt} = c/N$ $\Leftrightarrow$ $ae^{-bt} = \left(\dfrac{c}{N}\right) - 1 = \dfrac{c - N}{N}$ $\Leftrightarrow$ $e^{-bt} = \dfrac{c - N}{aN}$ $\Leftrightarrow$ $-bt = \ln(c - N) - \ln aN$ $\Leftrightarrow$ $t = \frac{1}{b}[\ln aN - \ln(c - N)]$. Substituting the values for a, b, and c, with $N = 400$ we have $t = \dfrac{1}{0.4981144989}[\ln 19643.90638 - \ln 100.855793] \approx 10.58$ days.

Chapter Six

Exercises 6.1

1. $\begin{cases} x - y = 2 \\ 2x + 3y = 9 \end{cases}$ Solving the first equation for x, we get $x = y + 2$ and substituting this into the

second equation, gives $2(y + 2) + 3y = 9$ $\Leftrightarrow$ $5y + 4 = 9$ $\Leftrightarrow$ $5y = 5$ $\Leftrightarrow$ $y = 1$. Substituting for y we get $x = y + 2 = (1) + 2 = 3$. Thus the solution is $(3, 1)$.

3. $\begin{cases} y = x^2 \\ y = x + 12 \end{cases}$ Substituting $y = x^2$ into the second equation gives $x^2 = x + 12$ $\Leftrightarrow$

$0 = x^2 - x - 12 = (x - 4)(x + 3)$ $\Rightarrow$ $x = 4$ or $x = -3$. So since $y = x^2$, the solutions are $(-3, 9)$ and $(4, 16)$.

5. $\begin{cases} x^2 + y^2 = 8 \\ x + y = 0 \end{cases}$ Solving the second equation for y gives $y = -x$, and substituting this into the first

equation gives $x^2 + (-x)^2 = 8$ $\Leftrightarrow$ $2x^2 = 8$ $\Leftrightarrow$ $x = \pm 2$. So since $y = -x$, the solutions are $(2, -2)$ and $(-2, 2)$.

7. $\begin{cases} x + y^2 = 0 \\ 2x + 5y^2 = 75 \end{cases}$ Solving the first equation for x gives $x = -y^2$, and substituting this into the

second equation gives $2(-y^2) + 5y^2 = 75$ $\Leftrightarrow$ $3y^2 = 75$ $\Leftrightarrow$ $y^2 = 25$ $\Leftrightarrow$ $y = \pm 5$. So since $x = -y^2$, the solutions are $(-25, -5)$ and $(-25, 5)$.

9. $\begin{cases} x + 2y = 5 \\ 2x + 3y = 8 \end{cases}$ Multiplying the first equation by 2 and the second by -1 gives the system

$\begin{cases} 2x + 4y = 10 \\ -2x - 3y = -8 \end{cases}$ Adding, we get $y = 2$, and substituting into the first equation in the original

system gives $x + 2(2) = 5$ $\Leftrightarrow$ $x + 4 = 5$ $\Leftrightarrow$ $x = 1$. The solution is $(1, 2)$.

11. $\begin{cases} x^2 - 2y = 1 \\ x^2 + 5y = 29 \end{cases}$ Subtracting the first equation from the second equation gives $7y = 28$ $\Rightarrow$

$y = 4$. Substituting $y = 4$ into the first equation of the original system gives $x^2 - 2(4) = 1$ $\Leftrightarrow$ $x^2 = 9$ $\Leftrightarrow$ $x = \pm 3$. The solutions are $(3, 4)$ and $(-3, 4)$.

13. $\begin{cases} 3x^2 - y^2 = 11 \\ x^2 + 4y^2 = 8 \end{cases}$ Multiplying the first equation by 4 gives the system $\begin{cases} 12x^2 - 4y^2 = 44 \\ x^2 + 4y^2 = 8 \end{cases}$.

Adding the equations gives $13x^2 = 52$ $\Leftrightarrow$ $x = \pm 2$. Substituting into the first equation we get $3(4) - y^2 = 11$ $\Leftrightarrow$ $y = \pm 1$. Thus, the solutions are $(2, 1)$, $(2, -1)$, $(-2, 1)$, and $(-2, -1)$.

15. $\begin{cases} x - y^2 + 3 = 0 \\ 2x^2 + y^2 - 4 = 0 \end{cases}$ Adding the two equations gives $2x^2 + x - 1 = 0$. Using the quadratic

formula we have $x = \frac{-1 \pm \sqrt{1 - 4(2)(-1)}}{2(2)} = \frac{-1 \pm \sqrt{9}}{4} = \frac{-1 \pm 3}{4}$. So $x = \frac{-1 - 3}{4} = -1$ or $x = \frac{-1 + 3}{4} = \frac{1}{2}$. Substituting $x = -1$ into the first equation gives $-1 - y^2 + 3 = 0$ $\Leftrightarrow$ $y^2 = 2$ $\Leftrightarrow$ $y = \pm \sqrt{2}$. Substituting $x = \frac{1}{2}$ into the first equation gives $\frac{1}{2} - y^2 + 3 = 0$ $\Leftrightarrow$ $y^2 = \frac{7}{2}$ $\Leftrightarrow$ $y = \pm \sqrt{\frac{7}{2}}$. Thus the solutions are $(-1, \pm \sqrt{2})$ and $(\frac{1}{2}, \pm \sqrt{\frac{7}{2}})$.

17. $\begin{cases} 2x + y = -1 \\ x - 2y = -8 \end{cases}$ By inspection of the graph, it appears that $(-2, 3)$ is the solution to the system. We check this in both equations to verify that it is a solution. $2(-2) + 3 = -4 + 3 = -1 \checkmark$ and $-2 - 2(3) = -2 - 6 = -8 \checkmark$. Since both equations are satisfied, the solution is $(-2, 3)$.

19. $\begin{cases} x^2 + y = 8 \\ x - 2y = -6 \end{cases}$ By inspection of the graph, it appears that $(2, 4)$ is a solution, but is difficult to get accurate values for the other point. Multiplying the first equation by 2 gives the system $\begin{cases} 2x^2 + 2y = 16 \\ x - 2y = -6 \end{cases}$. Adding the equations gives $2x^2 + x = 10 \quad \Leftrightarrow \quad 2x^2 + x - 10 = 0 \quad \Leftrightarrow$ $(2x + 5)(x - 2) = 0$. So $x = -\frac{5}{2}$ or $x = 2$. If $x = -\frac{5}{2}$, then $-\frac{5}{2} - 2y = -6 \quad \Leftrightarrow \quad -2y = -\frac{7}{2}$ $\Leftrightarrow \quad y = \frac{7}{4}$. And is $x = 2$ then $2 - 2y = -6 \quad \Leftrightarrow \quad -2y = -8 \quad \Leftrightarrow \quad y = 4$. Hence, the solutions are $\left(-\frac{5}{2}, \frac{7}{4}\right)$ and $(2, 4)$.

21. $\begin{cases} x^2 + y = 0 \\ x^3 - 2x - y = 0 \end{cases}$ By inspection of the graph, it appears that $(-2, -4)$, $(0, 0)$, and $(1, -1)$ are solutions to the system. We check each point in both equations to verify that it is a solution.
$(-2, -4)$: $\quad (-2)^2 + (-4) = 4 - 4 = 0 \checkmark \qquad (-2)^3 - 2(-2) - (-4) = -8 + 4 + 4 = 0 \checkmark$
$(0, 0)$: $\qquad (0)^2 + (0) = 0 \checkmark \qquad\qquad\qquad (0)^3 - 2(0) - (0) = 0 \checkmark$
$(1, -1)$: $\qquad (1)^2 + (-1) = 1 - 1 = 0 \checkmark \qquad (1)^3 - 2(1) - (-1) = 1 - 2 + 1 = 0 \checkmark$
Thus the solutions are $(-2, -4)$, $(0, 0)$, and $(1, -1)$.

23. $\begin{cases} y + x^2 = 4x \\ y + 4x = 16 \end{cases}$ Subtracting the second equation from the first equation gives $x^2 - 4x = 4x - 16$ $\Leftrightarrow \quad x^2 - 8x + 16 = 0 \quad \Leftrightarrow \quad (x - 4)^2 = 0 \quad \Leftrightarrow \quad x = 4$. Substituting this value for x into either of the original equations gives $y = 0$. Therefore, the solution is $(4, 0)$.

25. $\begin{cases} x - 2y = 2 \\ y^2 - x^2 = 2x + 4 \end{cases}$ Now $x - 2y = 2 \quad \Leftrightarrow \quad x = 2y + 2$. Substituting for x gives
$y^2 - x^2 = 2x + 4 \quad \Leftrightarrow \quad y^2 - (2y + 2)^2 = 2(2y + 2) + 4 \quad \Leftrightarrow$
$y^2 - 4y^2 - 8y - 4 = 4y + 4 + 4 \quad \Leftrightarrow \quad y^2 + 4y + 4 = 0 \quad \Leftrightarrow \quad (y + 2)^2 = 0 \quad \Leftrightarrow \quad y = -2$.
Since $x = 2y + 2$, we have $x = 2(-2) + 2 = -2$. Thus the solution is $(-2, -2)$.

27. $\begin{cases} x - y = 4 \\ xy = 12 \end{cases}$ Now $x - y = 4 \quad \Leftrightarrow \quad x = 4 + y$. Substituting for x gives $xy = 12 \quad \Leftrightarrow$
$(4 + y)y = 12 \quad \Leftrightarrow \quad y^2 + 4y - 12 = 0 \quad \Leftrightarrow \quad (y + 6)(y - 2) = 0 \quad \Leftrightarrow \quad y = -6, y = 2$.
Since $x = 4 + y$, the solutions are $(-2, -6)$ and $(6, 2)$.

29. $\begin{cases} x^2 y = 16 \\ x^2 + 4y + 16 = 0 \end{cases}$ Now $x^2 y = 16 \quad \Leftrightarrow \quad x^2 = \dfrac{16}{y}$. Substituting for x^2 gives
$\dfrac{16}{y} + 4y + 16 = 0 \quad \Rightarrow \quad 4y^2 + 16y + 16 = 0 \quad \Leftrightarrow \quad y^2 + 4y + 4 = 0 \quad \Leftrightarrow \quad (y + 2)^2 = 0$
$\Leftrightarrow \quad y = -2$. Therefore, $x^2 = \frac{16}{-2} = -8$, which has no real solution, and so the system has no solution.

31. $\begin{cases} x^2 + y^2 = 9 \\ x^2 - y^2 = 1 \end{cases}$ Adding the equations gives $2x^2 = 10 \quad \Leftrightarrow \quad x^2 = 5 \quad \Leftrightarrow \quad x = \pm\sqrt{5}$. Now

$x = \pm\sqrt{5} \quad \Rightarrow \quad y^2 = 9 - 5 = 4 \quad \Leftrightarrow \quad y = \pm 2$, and so the solutions are $(\sqrt{5}, 2)$, $(\sqrt{5}, -2)$, $(-\sqrt{5}, 2)$, and $(-\sqrt{5}, -2)$.

33. $\begin{cases} 2x^2 - 8y^3 = 19 \\ 4x^2 + 16y^3 = 34 \end{cases}$ Multiplying the first equation by 2 gives the system $\begin{cases} 4x^2 - 16y^3 = 38 \\ 4x^2 + 16y^3 = 34 \end{cases}$
Adding the two equations gives $8x^2 = 72 \quad \Leftrightarrow \quad x = \pm 3$, and then substituting into the first equation we have $2(9) - 8y^3 = 19 \quad \Leftrightarrow \quad y^3 = -\frac{1}{8} \quad \Leftrightarrow \quad y = -\frac{1}{2}$. Therefore, the solutions are $(3, -\frac{1}{2})$ and $(-3, -\frac{1}{2})$.

35. $\begin{cases} \dfrac{2}{x} - \dfrac{3}{y} = 1 \\ -\dfrac{4}{x} + \dfrac{7}{y} = 1 \end{cases}$ If we let $u = \dfrac{1}{x}$ and $v = \dfrac{1}{y}$, the system is equivalent to $\begin{cases} 2u - 3v = 1 \\ -4u + 7v = 1 \end{cases}$.

Multiplying the first equation by 4 gives the system $\begin{cases} 4u - 6v = 2 \\ -4u + 7v = 1 \end{cases}$. Adding the equations gives $v = 3$, and then substituting into the first equation gives $2u - 9 = 1 \quad \Leftrightarrow \quad u = 5$. Thus, the solution is $(\frac{1}{5}, \frac{1}{3})$.

37. $\begin{cases} y = 2x + 6 \\ y = -x + 5 \end{cases}$
The solution is approximately $(-0.33, 5.33)$.

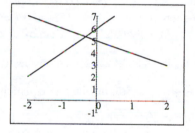

39. $\begin{cases} y = x^2 + 8x \\ y = 2x + 16 \end{cases}$
The solutions are $(-8, 0)$ and $(2, 20)$.

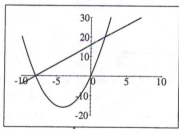

41. $\begin{cases} x^2 + y^2 = 25 \\ x + 3y = 2 \end{cases} \quad \Leftrightarrow \quad \begin{cases} y = \pm\sqrt{25 - x^2} \\ y = -\frac{1}{3}x + \frac{2}{3} \end{cases}$
The solutions are $(-4.51, 2.17)$ and $(4.91, -0.97)$.

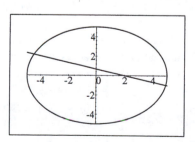

43. $\begin{cases} \frac{x^2}{9} + \frac{y^2}{18} = 1 \\ \qquad y = -x^2 + 6x - 2 \end{cases}$ ⇔

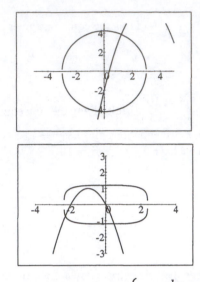

$\begin{cases} y = \pm\sqrt{18 - 2x^2} \\ y = -x^2 + 6x - 2 \end{cases}$

The solutions are $(1.23, 3.87)$ and $(-0.35, -4.21)$.

45. $\begin{cases} x^4 + 16y^4 = 32 \\ x^2 + 2x + y = 0 \end{cases}$ ⇔

$\begin{cases} y = \pm\dfrac{\sqrt[4]{32 - x^4}}{2} \\ y = -x^2 - 2x \end{cases}$

The solutions are $(-2.30, -0.70)$ and $(0.48, -1.19)$.

47. Let w and l be the lengths of the sides, in cm. Then we have the system $\begin{cases} lw = 180 \\ 2l + 2w = 54 \end{cases}$.

   We solve the second equation for w giving, $w = 27 - l$, and substitute into the first equation to get $l(27 - l) = 180$ ⇔ $l^2 - 27l + 180 = 0$ ⇔ $(l - 15)(l - 12) = 0$ ⇒ $l = 15$ or $l = 12$. If $l = 15$, then $w = 27 - 15 = 12$, and if $l = 12$, then $w = 27 - 12 = 15$. Therefore, the dimensions of the rectangle are 12 cm by 15 cm.

49. Let l and w be the length and width, respectively, of the rectangle. Then, the system of equations is $\begin{cases} 2l + 2w = 70 \\ \sqrt{l^2 + w^2} = 25 \end{cases}$. Solving the first equation for l, we have $l = 35 - w$, and substituting into the second yields $\sqrt{l^2 + w^2} = 25$ ⇔ $l^2 + w^2 = 625$ ⇔ $(35 - w)^2 + w^2 = 625$ ⇔ $1225 - 70w + w^2 + w^2 = 625$ ⇔ $2w^2 - 70w + 600 = 0$ ⇔ $(w - 15)(w - 20) = 0$ ⇒ $w = 15$ or $w = 20$. So the dimensions of the rectangle are 15 by 20.

51. At the points where the rocket path and the hillside meet, we have $\begin{cases} y = \frac{1}{2}x \\ y = -x^2 + 401x \end{cases}$. Substituting for y in the second equation gives $\frac{1}{2}x = -x^2 + 401x$ ⇔ $x^2 - \frac{801}{2}x = 0$ ⇔ $x\left(x - \frac{801}{2}\right) = 0$ ⇒ $x = 0, x = \frac{801}{2}$. When $x = 0$, the rocket has not left the pad. When $x = \frac{801}{2}$, then $y = \frac{1}{2}\left(\frac{801}{2}\right) = \frac{801}{4}$. So the rocket lands at the point $\left(\frac{801}{2}, \frac{801}{4}\right)$. The distance from the base of the hill is $\sqrt{\left(\frac{801}{2}\right)^2 + \left(\frac{801}{4}\right)^2} \approx 447.77$ meters.

53. The point P is at an intersection of the circle of radius 26 centered at $A\,(22, 32)$ and the circle of radius 20 centered at $B\,(28, 20)$. We have the system $\begin{cases} (x - 22)^2 + (y - 32)^2 = 26^2 \\ (x - 28)^2 + (y - 20)^2 = 20^2 \end{cases}$ ⇔

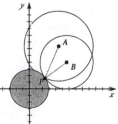

$\begin{cases} x^2 - 44x + 484 + y^2 - 64y + 1024 = 676 \\ x^2 - 56x + 784 + y^2 - 40y + 400 = 400 \end{cases}$ ⇔

$\begin{cases} x^2 - 44x + y^2 - 64y = -832 \\ x^2 - 56x + y^2 - 40y = -784 \end{cases}$. Subtracting the two equations, we get

$12x - 24y = -48 \quad \Leftrightarrow \quad x - 2y = -4$, which is the equation of a line. Solving for x, we have $x = 2y - 4$. Substituting into the first equation gives $(2y - 4)^2 - 44(2y - 4) + y^2 - 64y = -832$ $\Leftrightarrow \quad 4y^2 - 16y + 16 - 88y + 176 + y^2 - 64y = -832 \quad \Leftrightarrow \quad 5y^2 - 168y + 192 = -832 \quad \Leftrightarrow$ $5y^2 - 168y + 1024 = 0$. Using the quadratic formula, we have

$$y = \frac{168 \pm \sqrt{168^2 - 4(5)(1024)}}{2(5)} = \frac{168 \pm \sqrt{7744}}{10} = \frac{168 \pm 88}{10} \quad \Leftrightarrow \quad y = 8 \text{ or } y = 25.60.$$

Since the y coordinate of the point P must be less than that of point A, we have $y = 8$. Then $x = 2(8) - 4 = 12$. So the coordinates of P are $(12, 8)$.

To solve graphically, we must solve each equation for y. This gives $(x - 22)^2 + (y - 32)^2 = 26^2$ $\Leftrightarrow \quad (y - 32)^2 = 26^2 - (x - 22)^2 \quad \Rightarrow \quad y - 32 = \pm\sqrt{676 - (x - 22)^2} \quad \Leftrightarrow$ $y = 32 \pm \sqrt{676 - (x - 22)^2}$. We use the function $y = 32 - \sqrt{676 - (x - 22)^2}$ because the intersection we at interested in is below the point A. Likewise, solving the second equation for y, we would get the function $y = 20 - \sqrt{400 - (x - 28)^2}$.

In a 3 dimensional situation, you would need the minimum of 3 satellites, since a point on the earth can be uniquely specified as the intersection of 3 spheres centered at the satellites.

55. (a) $\begin{cases} \log x + \log y = \frac{3}{2} \\ 2\log x - \log y = 0 \end{cases}$ Adding the two equations gives $3\log x = \frac{3}{2} \quad \Leftrightarrow \quad \log x = \frac{1}{2}$

$\Leftrightarrow \quad x = \sqrt{10}$. Substituting into the second equation we get $2\log 10^{1/2} - \log y = 0 \quad \Leftrightarrow$ $\log 10 - \log y = 0 \quad \Leftrightarrow \quad \log y = 1 \quad \Leftrightarrow \quad y = 10$. Thus the solution is $\left(\sqrt{10}, 10\right)$.

(b) $\begin{cases} 2^x + 2^y = 10 \\ 4^x + 4^y = 68 \end{cases} \quad \Leftrightarrow \quad \begin{cases} 2^x + 2^y = 10 \\ 2^{2x} + 2^{2y} = 68 \end{cases}$

If we let $u = 2^x$ and $v = 2^y$, the system becomes $\begin{cases} u + v = 10 \\ u^2 + v^2 = 68 \end{cases}$.

Solving the first equation for u, and substituting this into the second equation gives $u + v = 10$ $\Leftrightarrow \quad u = 10 - v$, so $(10 - v)^2 + v^2 = 68 \quad \Leftrightarrow \quad 100 - 20v + v^2 + v^2 = 68 \quad \Leftrightarrow$ $v^2 - 10v + 16 = 0 \quad \Leftrightarrow \quad (v - 8)(v - 2) = 0 \quad \Rightarrow \quad v = 2 \text{ or } v = 8$. If $v = 2$, then $u = 8$, and so $y = 1$ and $x = 3$. If $v = 8$, then $u = 2$, and so $y = 3$ and $x = 1$. Thus, the solutions are $(1, 3)$ and $(3, 1)$.

(c) $\begin{cases} x - y = 3 \\ x^3 - y^3 = 387 \end{cases}$. Solving the first equation for x gives $x = 3 + y$ and using the hint, $x^3 - y^3 = 387 \quad \Leftrightarrow \quad (x - y)(x^2 + xy + y^2) = 387$. Next, substituting for x, we get $3[(3 + y)^2 + y(3 + y) + y^2] = 387 \quad \Leftrightarrow \quad 9 + 6y + y^2 + 3y + y^2 + y^2 = 129 \quad \Leftrightarrow$ $3y^2 + 9y + 9 = 129 \quad \Leftrightarrow \quad (y + 8)(y - 5) = 0 \quad \Rightarrow \quad y = -8 \text{ or } y = 5$. If $y = -8$, then $x = 3 + (-8) = -5$, and if $y = 5$, then $x = 3 + 5 = 8$. Thus the solutions are $(-5, -8)$ and $(8, 5)$.

(d) $\begin{cases} x^2 + xy = 1 \\ xy + y^2 = 3 \end{cases}$ Adding the equations gives

$x^2 + xy + xy + y^2 = 4 \quad \Leftrightarrow \quad x^2 + 2xy + y^2 = 4 \quad \Leftrightarrow \quad (x + y)^2 = 4 \quad \Rightarrow \quad x + y = \pm 2$. If $x + y = 2$, then from the first equation we get $x(x + y) = 1 \quad \Rightarrow \quad x \cdot 2 = 1 \quad \Rightarrow$ $x = \frac{1}{2}$, and so $y = 2 - \frac{1}{2} = \frac{3}{2}$. If $x + y = -2$, then from the first equation we get $x(x + y) = 1 \quad \Rightarrow \quad x \cdot (-2) = 1 \quad \Rightarrow \quad x = -\frac{1}{2}$, and so $y = -2 - \left(-\frac{1}{2}\right) = -\frac{3}{2}$. Thus the solutions are $\left(\frac{1}{2}, \frac{3}{2}\right)$ and $\left(-\frac{1}{2}, -\frac{3}{2}\right)$.

Exercises 6.2

1. $\begin{cases} x + y = 4 \\ 2x - y = 2 \end{cases}$

The solution is $x = 2$, $y = 2$.

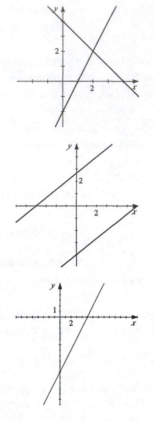

3. $\begin{cases} 2x - 3y = 12 \\ -x + \frac{3}{2}y = 4 \end{cases}$

No solution. The lines are parallel, so there is no intersection.

5. $\begin{cases} -x + \frac{1}{2}y = -5 \\ 2x - y = 10 \end{cases}$

Infinite number of solutions.

7. $\begin{cases} x + y = 4 \\ -x + y = 0 \end{cases}$ Adding the two equations gives $2y = 4$ $\Leftrightarrow$ $y = 2$. Substituting for y in the first equation gives $x + 2 = 4$ $\Leftrightarrow$ $x = 2$. Hence, the solution is $(2, 2)$.

9. $\begin{cases} 2x - 3y = 9 \\ 4x + 3y = 9 \end{cases}$ Adding the two equations gives $6x = 18$ $\Leftrightarrow$ $x = 3$. Substituting for x in the second equation gives $4(3) + 3y = 9$ $\Leftrightarrow$ $12 + 3y = 9$ $\Leftrightarrow$ $3y = -3$ $\Leftrightarrow$ $x = -1$. Hence, the solution is $(3, -1)$.

11. $\begin{cases} x + 3y = 5 \\ 2x - y = 3 \end{cases}$ Solving the first equation for x gives $x = -3y + 5$. Substituting for x in the second equation gives $2(-3y + 5) - y = 3$ $\Leftrightarrow$ $-6y + 10 - y = 3$ $\Leftrightarrow$ $-7y = -7$ $\Leftrightarrow$ $y = 1$. Then $x = -3(1) + 5 = 2$. Hence, the solution is $(2, 1)$.

13. $-x + y = 2$ $\Leftrightarrow$ $y = x + 2$. Substituting for y into $4x - 3y = -3$ gives $4x - 3(x + 2) = -3$ $\Leftrightarrow$ $4x - 3x - 6 = -3$ $\Leftrightarrow$ $x = 3$, and so $y = (3) + 2 = 5$. Hence, the solution is $(3, 5)$.

15. $x + 2y = 7$ $\Leftrightarrow$ $x = 7 - 2y$. Substituting for x into $5x - y = 2$ gives $5(7 - 2y) - y = 2$ $\Leftrightarrow$ $35 - 10y - y = 2$ $\Leftrightarrow$ $-11y = -33$ $\Leftrightarrow$ $y = 3$, and so $x = 7 - 2(3) = 1$. Hence, the solution is $(1, 3)$.

17. $\frac{1}{2}x + \frac{1}{3}y = 2$ $\Leftrightarrow$ $x + \frac{2}{3}y = 4$ $\Leftrightarrow$ $x = 4 - \frac{2}{3}y$. Substituting for x into $\frac{1}{5}x - \frac{2}{3}y = 8$ gives $\frac{1}{5}(4 - \frac{2}{3}y) - \frac{2}{3}y = 8$ $\Leftrightarrow$ $\frac{4}{5} - \frac{2}{15}y - \frac{10}{15}y = 8$ $\Leftrightarrow$ $12 - 2y - 10y = 120$ $\Leftrightarrow$ $y = -9$, and so $x = 4 - \frac{2}{3}(-9) = 10$. Hence, the solution is $(10, -9)$.

19. Adding gives $3x + 2y = 8$
$$\frac{x - 2y = 0}{4x \quad\quad = 8} \Leftrightarrow x = 2.$$
So $x - 2y = (2) - 2y = 0$ $\Leftrightarrow$ $2y = 2$ $\Leftrightarrow$ $y = 1$. Thus, the solution is $(2, 1)$.

21. $\begin{cases} x + 4y = 8 \\ 3x + 12y = 2 \end{cases}$ Adding -3 times the first equation to the second equation gives
$$\frac{\begin{aligned} -3x - 12y &= -24 \\ 3x + 12y &= 2 \end{aligned}}{0 = -22}$$, which is never true. Thus the system has no solution.

23. $\begin{cases} 2x - 6y = 10 \\ -3x + 9y = -15 \end{cases}$ Adding 3 times the first equation to 2 times the second equation gives
$$\frac{\begin{aligned} 6x - 18y &= 30 \\ -6x + 18y &= -30 \end{aligned}}{0 = 0}$$. Writing the equation in slope-intercept form, we have $2x - 6y = 10$ $\Leftrightarrow$ $-6y = -2x + 10$ $\Leftrightarrow$ $y = \frac{1}{3}x - \frac{5}{3}$, so a solution is any pair of the form $\left(x, \frac{1}{3}x - \frac{5}{3}\right)$, where x is any real number.

25. $\begin{cases} 6x + 4y = 12 \\ 9x + 6y = 18 \end{cases}$ Adding 3 times the first equation to -2 times the second equation gives
$$\frac{\begin{aligned} 18x + 12y &= 36 \\ -18x - 12y &= -36 \end{aligned}}{0 = 0}$$. Writing the equation in slope-intercept form, we have $6x + 4y = 12$ $\Leftrightarrow$ $4y = -6x + 12$ $\Leftrightarrow$ $y = -\frac{3}{2}x + 3$, so a solution is any pair of the form $\left(x, -\frac{3}{2}x + 3\right)$, where x is any real number.

27. $\begin{cases} 8s - 3t = -3 \\ 5s - 2t = -1 \end{cases}$. Adding 2 times the first equation to 3 times the second equation gives
$$\frac{\begin{aligned} 16s - 6t &= -6 \\ 15s - 6t &= -3 \end{aligned}}{s = -3}.$$ So $8(-3) - 3t = -3$ $\Leftrightarrow$ $-24 - 3t = -3$ $\Leftrightarrow$ $t = -7$. Thus, the solution is $(-3, -7)$.

29. $\begin{cases} \frac{1}{2}x + \frac{3}{5}y = 3 \\ \frac{5}{3}x + 2y = 10 \end{cases}$. Adding 10 times the first equation to -3 times the second equation gives

$5x + 6y = 30$
$-5x - 6y = -30$
_____
$\qquad 0 = 0\quad$. Writing the equation in slope-intercept form, we have $\frac{1}{2}x + \frac{3}{5}y = 3 \quad \Leftrightarrow$
$\frac{3}{5}y = -\frac{1}{2}x + 3 \quad \Leftrightarrow \quad y = -\frac{5}{6}x + 5$, so a solution is any pair of the form $\left(x, -\frac{5}{6}x + 5\right)$, where x is any real number.

31. $\begin{cases} 0.4x + 1.2y = 14 \\ 12x - 5y = 10 \end{cases}$. Adding 30 times the first equation to -1 times the second equation gives

$12x + 36y = 420$
$-12x + 5y = -10$
_____
$\qquad 41y = 410 \quad \Leftrightarrow \quad y = 10.$ So $12x - 5(10) = 10 \quad \Leftrightarrow \quad 12\text{x} = 60 \quad \Leftrightarrow \quad x = 5.$
Thus the solution is $(5, 10)$.

33. $\begin{cases} \frac{1}{3}x - \frac{1}{4}y = 2 \\ -8x + 6y = 10 \end{cases}$ Adding 24 times the first equation to the second equation gives

$8x - 6y = 48$
$-8x + 6y = 10$
_____
$\qquad 0 = 58$, which is never true. Thus the system has no solution.

35. $\begin{cases} 0.21x + 3.17y = 9.51 \qquad l_1 \\ 2.35x - 1.17y = 5.89 \qquad l_2 \end{cases}$
The solution is approximately $(3.87, 2.74)$.

37. $\begin{cases} 2371x - 6552y = 13{,}591 \qquad l_1 \\ 9815x + 992y = 618{,}555 \qquad l_2 \end{cases}$
The solution is approximately $(61.00, 20.00)$.

39. $x + y = 0 \qquad \times -1 \qquad -x - y = 0$
$x + ay = 1 \qquad \times 1 \quad \Rightarrow \quad x + ay = 1$
$\qquad\qquad\qquad\qquad ay - y = 1 \quad \Leftrightarrow \quad y(a - 1) = 1 \quad \Leftrightarrow \quad y = \dfrac{1}{a-1}, a \neq 1.$

So $x + \left(\dfrac{1}{a-1}\right) = 0 \quad \Leftrightarrow \quad x = \dfrac{1}{1-a} = -\dfrac{1}{a-1}.$ Thus the solution is $\left(-\dfrac{1}{a-1}, \dfrac{1}{a-1}\right).$

41. $ax + by = 1 \qquad \times -b \qquad -abx - b^2y = -b$
    $bx + ay = 1 \qquad \times a \qquad \Rightarrow \quad \underline{abx + a^2y = a}$

$$(a^2 - b^2)y = a - b \quad \Leftrightarrow \quad y = \frac{a - b}{a^2 - b^2} = \frac{1}{a + b}, a^2 - b^2 \neq 0.$$

So $ax + \dfrac{b}{a + b} = 1 \quad \Leftrightarrow \quad ax = \dfrac{a}{a + b} \quad \Leftrightarrow \quad x = \dfrac{1}{a + b}$. Thus the solution is $\left(\dfrac{1}{a + b}, \dfrac{1}{a + b} \right)$.

43. Let the two numbers be x and y. This gives $x + y = 34$
    $$\underline{x - y = 10}$$
    $$2x \qquad = 44 \quad \Leftrightarrow \quad x = 22.$$

So $22 + y = 34 \quad \Leftrightarrow \quad y = 12$. Therefore, the two numbers are 22 and 12.

45. Let d be the number of dimes and q be the number of quarters. This gives
    $d + q = 14 \qquad \times -1 \qquad -d - q = -14$
    $0.10d + 0.25q = 2.75 \qquad \times 10 \Rightarrow \underline{d + 2.5q = 27.5}$
    $$1.5q = 13.5 \quad \Leftrightarrow \quad q = 9$$

So $d + (9) = 14 \quad \Leftrightarrow \quad d = 5$. Thus the number of dimes is 5 and the number of quarters is 9.

47. Let x be the speed of the plane in still air and y be the speed of the wind. This gives
    $2x - 2y = 180 \qquad \times -6 \qquad 12x - 12y = 1080$
    $1.2x + 1.2y = 180 \qquad \times 10 \Rightarrow \underline{12x + 12y = 1800}$
    $$24x \qquad = 2880 \quad \Leftrightarrow \quad x = 120.$$

So $2(120) - 2y = 180 \quad \Leftrightarrow \quad -2y = -60 \quad \Leftrightarrow \quad y = 30$. Therefore, the speed of the plane is 120 mi/h and the wind speed is 30 mi/h.

49. Let x be the cycling speed and y be the running speed. (Remember to divide by 60 to convert minutes to decimal hours.) We have
    $0.5x + 0.5y = 12.5 \times -2 \qquad -x - y = -25$
    $0.75x + 0.2y = 16 \qquad \times 5 \Rightarrow \underline{3.75x + y = 80}$
    $$2.75x \qquad = 55 \quad \Leftrightarrow \quad x = 20.$$

So $20 + y = 25 \quad \Leftrightarrow \quad y = 5$. Thus, the cycling speed is 20 mi/h and the running speed is 5 mi/h.

51. Let a be the grams of food A and b be the grams of food B. This gives
    $0.12a + 0.20b = 32 \qquad \times -250 \qquad -30a - 50b = -8000$
    $100a + 50b = 22000 \qquad \times 1 \Rightarrow \underline{100a + 50b = 22000}$
    $$70a \qquad = 14000 \quad \Leftrightarrow \quad a = 200.$$

So $0.12(200) + 0.20b = 32 \quad \Leftrightarrow \quad 0.20b = 8 \quad \Leftrightarrow \quad b = 40$. Thus, she should use 200 grams of food A and 40 grams of food B.

53. Let x and y be the sulfuric acid concentrations in the first and second containers.
    $300x + 600y = 900(0.15) \qquad \times -1 \qquad -300x - 600y = -135$
    $100x + 500y = 600(0.125) \qquad \times 3 \Rightarrow \underline{300x + 1500y = 225}$
    $$900y = 90 \quad \Leftrightarrow \quad y = 0.10.$$

So $100x + 500(0.10) = 75 \quad \Leftrightarrow \quad x = 0.25$. Thus, the concentrations of sulfuric acid are 25% in the first container and 10% in the second.

55. Let x be the amount he invests at 6% and let y be the amount he invests at 10%.
The ratio of the amounts invested gives $x = 2y$. Then the interest earned is $0.06x + 0.10y = 3520$
$\Leftrightarrow$ $6x + 10y = 352{,}000$. Substituting gives $6(2y) + 10y = 352{,}000$ $\Leftrightarrow$ $22y = 352{,}000$
$\Leftrightarrow$ $y = 16{,}000$. Then $x = 2(16{,}000) = 32{,}000$. Thus, he invests \$32,000 at 6% and \$16,000 at
10%.

57. Let x be the tens digit and y be the ones digit of the number.

$$\begin{array}{llll} x + y = 7 & \times 9 & 9x + 9y = 63 & \\ 10y + x = 27 + 10x + y & \times 1 & \Rightarrow \quad \underline{9x - 9y = -27} & \\ & & 18x \qquad\; = 36 & \Leftrightarrow \quad x = 2. \end{array}$$

So $2 + y = 7$ $\Leftrightarrow$ $y = 5$. Thus, the number is 25.

59. $\displaystyle\sum_{k=1}^{n} x_k = 1 + 2 + 3 + 5 + 7 = 18,$

$\displaystyle\sum_{k=1}^{n} y_k = 3 + 5 + 6 + 6 + 9 = 29,$

$\displaystyle\sum_{k=1}^{n} x_k y_k = 1(3) + 2(5) + 3(6) + 5(6) + 7(9) = 124,$

$\displaystyle\sum_{k=1}^{n} x_k^2 = 1^2 + 2^2 + 3^2 + 5^2 + 7^2 = 88,\ n = 5.$ Thus we get the system

$$\begin{cases} 18a + 5b = 29 \\ 88a + 18b = 124 \end{cases} \quad \begin{array}{l} \times -18 \\ \times 5 \end{array} \quad \begin{array}{l} -324a - 90b = -522 \\ \underline{440a + 90b = 620} \\ 116a \qquad\quad = 98 \end{array} \quad \Leftrightarrow \quad a \approx 0.845.$$

Then $b = \frac{1}{5}[-18(0.845) + 29] \approx 2.758$. So, the regression line is $y = 0.845x + 2.758$.

Exercises 6.3

1. The equation $6x - \sqrt{3}y + \frac{1}{2}z = 0$ is linear.

3. The system $\begin{cases} xy - 3y + z = 5 \\ x - y^2 + 5z = 0 \\ 2x + yz = 3 \end{cases}$ is not a linear system, since the first equation contains a product of variables. In fact both the second and the third equation are not linear.

5. $\begin{cases} x - 2y + 4z = 3 \\ y + 2z = 7 \\ z = 2 \end{cases}$ Substituting $z = 2$ into the second equation gives $y + 2(2) = 7 \iff y = 3$.
    Substituting $z = 2$ and $y = 3$ into the first equation gives $x - 2(3) + 4(2) = 3 \iff x = 1$. Thus the solution is $(1, 3, 2)$.

7. $\begin{cases} x + 2y + z = 7 \\ -y + 3z = 9 \\ 2z = 6 \end{cases}$ Solving we get $2z = 6 \iff z = 3$. Substituting $z = 3$ into the second
    equation gives $-y + 3(3) = 9 \iff y = 0$. Substituting $z = 3$ and $y = 0$ into the first equation
    gives $x + 2(0) + 3 = 7 \iff x = 4$. Thus the solution is $(4, 0, 3)$.

9. $\begin{cases} 2x - y + 6z = 5 \\ y + 4z = 0 \\ -2z = 1 \end{cases}$ Solving we get $-2z = 1 \iff z = -\frac{1}{2}$. Substituting $z = -\frac{1}{2}$ into the
    second equation gives $y + 4(-\frac{1}{2}) = 0 \iff y = 2$. Substituting $z = -\frac{1}{2}$ and $y = 2$ into the first
    equation gives $2x - (2) + 6(-\frac{1}{2}) = 5 \iff x = 5$. Thus the solution is $(5, 2, -\frac{1}{2})$.

11. $\begin{cases} x - 2y - z = 4 \\ x - y + 3z = 0 \\ 2x + y + z = 0 \end{cases}$ Using the first equation we subtract the first equation from the second equation.
    $\begin{array}{r} -x + 2y + z = -4 \\ x - y + 3z = 0 \\ \hline y + 4z = -4 \end{array}$ This gives the system
    $\begin{cases} x - 2y - z = 4 \\ y + 4z = -4. \\ 2x + y + z = 0 \end{cases}$

    $\begin{cases} x - 2y - z = 4 \\ x - y + 3z = 0 \\ 2x + y + z = 0 \end{cases}$ Or using the third equation, we subtract $\frac{1}{2}$ times the third equation from the second equation.
    $\begin{array}{r} x - y + 3z = 0 \\ -x - \frac{1}{2}y - \frac{1}{2}z = 0 \\ \hline -\frac{3}{2}y + \frac{5}{2}z = 0 \end{array}$ This gives the
    system $\begin{cases} x - 2y - z = 4 \\ -\frac{3}{2}y + \frac{5}{2}z = 0. \\ 2x + y + z = 0 \end{cases}$

13. $\begin{cases} 2x - y + 3z = 2 \\ x + 2y - z = 4 \\ -4x + 5y + z = 10 \end{cases}$ Using the first equation we add 2 times the first equation to the third equation.
    $\begin{array}{r} 4x - 2y + 6z = 4 \\ -4x + 5y + z = 10 \\ \hline 3y + 7z = 14 \end{array}$ This gives the
    system $\begin{cases} 2x - y + 3z = 2 \\ x + 2y - z = 4 \\ 3y + 7z = 14 \end{cases}$.

$$\begin{cases} 2x - y + 3z = 2 \\ x + 2y - z = 4 \\ -4x + 5y + z = 10 \end{cases}$$
Or using the third equation, we add 4 times the second equation to the third equation.
$$\begin{aligned} 4x + 8y - 4z &= 16 \\ -4x + 5y + z &= 10 \\ \hline 13y - 3z &= 26 \end{aligned}$$
This gives

the system $$\begin{cases} 2x - y + 3z = 2 \\ x + 2y - z = 4 \\ 13y - 3z = 26 \end{cases}.$$

15. $$\begin{cases} x + y + z = 4 \\ x + 3y + 3z = 10 \\ 2x + y - z = 3 \end{cases}$$
Using the first equation we eliminate the x term from the second and the third equation.
$$\begin{aligned} -x - y - z &= -4 \\ x + 3y + 3z &= 10 \\ \hline 2y + 2z &= 6 \end{aligned}$$ and

$$\begin{aligned} -2x - 2y - 2z &= -8 \\ 2x + y - z &= 3 \\ \hline -y - 3z &= -5 \end{aligned} \quad \Leftrightarrow \quad y + 3z = 5$$

So we get the system $$\begin{cases} x + y + z = 4 \\ 2y + 2z = 6. \\ y + 3z = 5 \end{cases}$$
Which we can write as

$$\begin{cases} x + y + z = 4 \\ y + 3z = 5 \\ 2y + 2z = 6 \end{cases}$$ Eliminate the y term from the third equation,
$$\begin{aligned} -2y - 6z &= -10 \\ 2y + 2z &= 6 \\ \hline -4z &= -4 \end{aligned}$$ to get the system $$\begin{cases} x + y + z = 4 \\ y + 3z = 5 \\ -4z = -4 \end{cases}$$

So $z = 1$ and $y + 3(1) = 8 \quad \Leftrightarrow \quad y = 2$. Then $x + 2 + 1 = 4 \quad \Leftrightarrow \quad x = 1$. So the solution is $(1, 2, 1)$.

17. $$\begin{cases} x - 4z = 1 \\ 2x - y - 6z = 4 \\ 2x + 3y - 2z = 8 \end{cases}$$
Using the first equation we eliminate the x term from the second and the third equation.
$$\begin{aligned} -2x + 8z &= -2 \\ 2x - y - 6z &= 4 \\ \hline -y + 2z &= 2 \end{aligned}$$ and

$$\begin{aligned} -2x + 8z &= -2 \\ 2x + 3y - 2z &= 8 \\ \hline 3y + 6z &= 6 \end{aligned}$$ So we get the system $$\begin{cases} x - 4z = 1 \\ -y + 2z = 2 \\ 3y + 6z = 6 \end{cases}$$ Eliminate the y term from the third equation,
$$\begin{aligned} -3y + 6z &= 6 \\ 3y + 6z &= 6 \\ \hline 12z &= 12 \end{aligned}$$

To get the system $$\begin{cases} x - 4z = 1 \\ -y + 2z = 2 \\ 12z = 12 \end{cases}.$$ So $z = 1$ and $-y + 2(1) = 2 \quad \Leftrightarrow \quad y = 0$. Then $x - 4(1) = 1$

$\Leftrightarrow \quad x = 5$. So the solution is $(5, 0, 1)$.

19. $$\begin{cases} 2x + 4y - z = 2 \\ x + 2y - 3z = -4 \\ 3x - y + z = 1 \end{cases}$$
First rewrite as $$\begin{cases} x + 2y - 3z = -4 \\ 2x + 4y - z = 2 \\ 3x - y + z = 1 \end{cases}.$$
Using the first equation we eliminate the x term from the second and the third equation.

$$\begin{aligned} -2x - 4y + 6z &= 8 \\ 2x + 4y - z &= 2 \\ \hline 5z &= 10 \end{aligned}$$ and
$$\begin{aligned} -3x - 6y + 9z &= 12 \\ 3x - y + z &= 1 \\ \hline -7y + 10z &= 13 \end{aligned}$$ So we get the system $$\begin{cases} x + 2y - 3z = -4 \\ 5z = 10 \\ -7y + 10z = 13 \end{cases}.$$

which we can write as $$\begin{cases} x + 2y - 3z = -4 \\ -7y + 10z = 13 \\ 5z = 10 \end{cases}.$$ So $z = 2$ and $-7y + 10(2) = 13 \quad \Leftrightarrow \quad y = 1$. Then

$x + 2(1) - 3(2) = -4 \quad \Leftrightarrow \quad x = 0$. So the solution is $(0, 1, 2)$.

21. $$\begin{cases} y - 2z = 0 \\ 2x + 3y = 2 \\ -x - 2y + z = -1 \end{cases}$$
First rewrite as $$\begin{cases} -x - 2y + z = -1 \\ y - 2z = 0 \\ 2x + 3y = 2 \end{cases}.$$
Using the first equation we eliminate the x term from the third equation.

$$-2x - 4y + 2z = -2$$
$$\underline{2x + 3y \qquad = 2}$$
$$-y + 2z = 0$$

So we get the system

$$\begin{cases} -x - 2y + z = -1 \\ y - 2z = 0 \\ -y + 2z = 0 \end{cases}$$

Eliminate the y term from the third equation,

$$y - 2z = 0$$
$$\underline{-y + 2z = 0}$$
$$0 = 0$$

to get the system

$$\begin{cases} -x - 2y + z = -1 \\ y - 2z = 0 \\ 0 = 0 \end{cases}$$

. The system is dependent, so when $z = t$ we solve for y to get

$y - 2t = 0 \Leftrightarrow y = 2t$. Then $-x - 2(2t) + t = -1 \Leftrightarrow -x - 3t = -1 \Leftrightarrow x = -3t + 1$. So the solution is $(-3t + 1, 2t, t)$.

23. $\begin{cases} x + 2y - z = 1 \\ 2x + 3y - 4z = -3 \\ 3x + 6y - 3z = 4 \end{cases}$ Using the first equation we eliminate the x term from the second and the third equation.

$$-2x - 4y + 2z = -2$$
$$\underline{2x + 3y - 4z = -3}$$
$$-y - 2z = -5$$

and

$$-3x - 6y + 3z = -3$$
$$\underline{3x + 6y - 3z = 4}$$
$$0 = 1$$

Since $0 = 1$ is false, this system is inconsistent.

25. $\begin{cases} 2x + 3y - z = 1 \\ x + 2y = 3 \\ x + 3y + z = 4 \end{cases}$ First rewrite as

$\begin{cases} x + 2y = 3 \\ 2x + 3y - z = 1. \\ x + 3y + z = 4 \end{cases}$ Using the first equation we eliminate the x term from the second and the third equation.

$$-2x - 4y = -6$$
$$\underline{2x + 3y - z = 1}$$
$$-y - z = -5$$

and

$$-x - 2y = -3$$
$$\underline{x + 3y + z = 4}$$
$$y + z = 1$$

So we get the system

$\begin{cases} x + 2y = 3 \\ -y - z = -5. \\ y + z = 1 \end{cases}$

Eliminate the y term from the third equation.

$$-y - z = -5$$
$$\underline{y + z = 1}$$
$$0 = -4$$

Since $0 = -4$ is false, this system is inconsistent.

27. $\begin{cases} x + y - z = 0 \\ x + 2y - 3z = -3 \\ 2x + 3y - 4z = -3 \end{cases}$ Using the first equation we eliminate the x term from the second and the third equation.

$$-x - y + z = 0$$
$$\underline{x + 2y - 3z = -3}$$
$$y - 2z = -3$$

and

$$-2x - 2y - 2z = 0$$
$$\underline{2x + 3y - 4z = -3}$$
$$y - 2z = -3$$

So we get the system

$\begin{cases} x + y - z = 0 \\ y - 2z = -3 \\ y - 2z = -3 \end{cases}$ Eliminate the y term from the third equation.

$$-y + 2z = 3$$
$$\underline{y - 2z = -3}$$
$$0 = 0$$

So $z = t$ and $y - 2t = -3 \Leftrightarrow y = 2t - 3$. Then $x + (2t - 3) - t = 0 \Leftrightarrow x = -t + 3$. So the solution is $(-t + 3, 2t - 3, t)$.

29. $\begin{cases} x + 3y - 2z = 0 \\ 2x + 4z = 4 \\ 4x + 6y = 4 \end{cases}$ Using the first equation we eliminate the x term from the second and the third equation.

$$-2x - 6y + 4z = 0$$
$$\underline{2x + 4z = 4}$$
$$-6y + 8z = 4$$

and

$$-4x - 12y + 8z = 0$$
$$\underline{4x + 6y = 4}$$
$$-6y + 8z = 4$$

So we get the system

$\begin{cases} x + 3y - 2z = 0 \\ -6y + 8z = 4. \\ -6y + 8z = 4 \end{cases}$ Eliminate the y term from the third equation.

$$6y - 8z = -4$$
$$\underline{-6y + 8z = 4}$$
$$0 = 0$$

So $z = t$ and $-6y + 8t = 4 \Leftrightarrow -6y = -8t + 4 \Leftrightarrow y = \frac{4}{3}t - \frac{2}{3}$. Then $x + 3\left(\frac{4}{3}t - \frac{2}{3}\right) - 2t = 0 \Leftrightarrow x = -2t + 2$. So the solution is $\left(-2t + 2, \frac{4}{3}t - \frac{2}{3}, t\right)$.

31.
$$\begin{cases} x \quad\;\; + z + 2w = 6 \\ \quad y - 2z \quad\;\;\; = -3 \\ x + 2y - z \quad\;\;\; = -2 \\ 2x + y + 3z - 2w = 0 \end{cases}$$

Using the first equation we eliminate the x term from the third and the fourth equation.

$$\begin{aligned} -x \quad\quad\; - z - 2w &= -6 \\ x + 2y - z \quad\quad\; &= -2 \quad \text{and} \\ \hline 2y - 2z - 2w &= -8 \end{aligned}$$

$$\begin{aligned} -2x \quad\quad - 2z - 4w &= -12 \\ 2x + y + 3z - 2w &= 0 \\ \hline y + z - 6w &= -12 \end{aligned}$$

So we get the system

$$\begin{cases} x \quad\;\; + z + 2w = 6 \\ \quad y - 2z \quad\quad = -3 \\ \quad 2y - 2z - 2w = -8 \\ \quad\; y + z - 6w = -12 \end{cases}.$$

Using the first equation we eliminate the y term from the third and the fourth equation.

$$\begin{aligned} -2y + 4z \quad\quad &= 6 \\ 2y - 2z - 2w &= -8 \quad \text{and} \\ \hline 2z - 2w &= -2 \end{aligned}$$

$$\begin{aligned} -y + 2z \quad\quad &= 3 \\ y + z - 6w &= -12 \\ \hline 3z - 6w &= -9 \end{aligned}$$

This gives us the system

$$\begin{cases} x \;\; + z + 2w = 6 \\ y - 2z \quad\quad = -3 \\ \quad 2z - 2w = -2 \\ \quad 3z - 6w = -9 \end{cases}.$$

We divide the third equation by 2 to get this system .

$$\begin{cases} x \;\; + z + 2w = 6 \\ y - 2z \quad\quad = -3 \\ \quad z - 1w = -1 \\ \quad 3z - 6w = -9 \end{cases}.$$

Eliminate the z term from the fourth equation,

$$\begin{aligned} -3z + 3w &= 3 \\ 3z - 6w &= -9 \\ \hline -3w &= -6 \end{aligned}$$

to get the system

$$\begin{cases} x \;\; + z + 2w = 6 \\ y - 2z \quad\quad = -3 \\ \quad z \; - w = -1 \\ \quad\quad -3w = -6 \end{cases}.$$

So $w = 2$ and $z - 2 = -1 \;\Leftrightarrow\; z = 1$. Then $y - 2(1) = -3 \;\Leftrightarrow\; y = -1$ and $x + 1 + 2(2) = 6 \;\Leftrightarrow\; x = 1$. Thus the solution is $(1, -1, 1, 2)$.

33. Let x be the amount she invests at 4%, y be the amount she invests at 5%, and let z be the amount she invests at 6%. Modeling these we get the equations:

Total money $x + y + z = 100,000$

Annual income $0.04x + 0.05y + 0.06z = 0.051(100,000)$ $\Leftrightarrow$ $\begin{cases} x + y + z = 100,000 \\ 4x + 5y + 6z = 510,000. \\ x - y = 0 \end{cases}$

Equal amounts $x = y$

Using the first equation we eliminate the x term from the second and the third equation.

$$\begin{aligned} -4x - 4y - 4z &= -400,000 \\ 4x + 5y + 6z &= 510,000 \quad \text{and} \\ \hline y + 2z &= 110,000 \end{aligned}$$

$$\begin{aligned} -x - y - z &= -100,000 \\ x - y \quad\quad &= 0 \\ \hline -2y - z &= -100,000 \end{aligned}$$

So we get the system

$$\begin{cases} x + y + z = 100,000 \\ y + 2z = 110,000 \\ -2y - z = -100,000 \end{cases}$$

Eliminate the y term from the third equation.

$$\begin{aligned} 2y + 4z &= 220,000 \\ -2y - z &= -100,000 \\ \hline 3z &= 120,000 \end{aligned}$$

Finally, we get the system

$$\begin{cases} x + y + z = 100,000 \\ y + 2z = 110,000 \\ 3z = 120,000 \end{cases}$$

So $z = 40,000$ and $y + 2(40,000) = 110,000 \;\Leftrightarrow\; y = 30,000$. Since $x = y$, $x = 30,000$. She must invest $30,000 in short-term bonds, $30,000 in intermediate-term bonds, and $40,000 in long-term bonds.

35. Let a, b, and c be the ounce of Type A, Type B, and Type C pellets used. The requirements for the different vitamins gives the following system

$$\begin{cases} 2a + 3b + c = 9 \\ 3a + b + 3c = 14 \\ 8a + 5b + 7c = 32 \end{cases}$$
Using the first equation we eliminate the a term from the second and the third equation.

$$\begin{aligned} -6a - 9b - 3c &= -27 \\ 6a + 2b + 6c &= 28 \\ \hline -7b + 3c &= 1 \end{aligned}$$
and

$$\begin{aligned} -8a - 12b - 4c &= -36 \\ 8a + 5b + 7c &= 32 \\ \hline -7b + 3c &= -4 \end{aligned}$$
So we get the system
$$\begin{cases} 2a + 3b + c = 9 \\ -7b + 3c = 1 \\ -7b + 3c = -4 \end{cases}$$
Eliminate the b term from the third equation.

$$\begin{aligned} 7b - 3c &= -1 \\ -7b + 3c &= -4 \\ \hline 0 &= -5 \end{aligned}$$

But $0 = -5$ is always false, so there is no solution.

37. Let x, y, and z be the acres of land that is planted with corn, wheat, and soybean. Modeling these we get the equations:

Total acres $\quad\quad\quad x + y + z = 1200$
Market demand $\quad\quad\quad\quad 2x = y$
Total cost $\quad\quad 45x + 60y + 50z = 63{,}750$

Substituting $2x$ for y we get the system:

$$\begin{cases} x + 2x + z = 1200 \\ 2x = y \\ 45x + 60(2x) + 50z = 63{,}750 \end{cases}.$$

$$\Leftrightarrow \quad \begin{cases} 3x + z = 1200 \\ 2x = y \\ 165x + 50z = 63{,}750 \end{cases}$$
Using the first equation we eliminate the z term from the third equation.

$$\begin{aligned} -150x - 50z &= -60{,}000 \\ 165x + 50z &= 63{,}750 \\ \hline 15x &= 3{,}750 \end{aligned}$$
So we get the system

$$\begin{cases} 3x + z = 1200 \\ 2x = y \\ 15x = 3{,}750 \end{cases}$$

So $15x = 3{,}750 \quad \Leftrightarrow \quad x = 250$ and $y = 2(250) = 500$. Substituting into the original equation, we have $250 + 500 + z = 1200 \quad \Leftrightarrow \quad z = 450$. Thus the farmer should plant 250 acres of corn, 500 acres of wheat, and 450 acres of soybeans.

39. (a) We begin by substituting $\left(\dfrac{x_0 + x_1}{2}, \dfrac{y_0 + y_1}{2}, \dfrac{z_0 + z_1}{2} \right)$ into the left side of the first equation which gives: $a_1 \left(\dfrac{x_0 + x_1}{2} \right) + b_1 \left(\dfrac{y_0 + y_1}{2} \right) + c_1 \left(\dfrac{z_0 + z_1}{2} \right)$
$= \frac{1}{2}[(a_1 x_0 + b_1 y_0 + c_1 z_0) + (a_1 x_1 + b_1 y_1 + c_1 z_1)] = \frac{1}{2}[d_1 + d_1] = d_1$. Thus the given ordered triple satisfies the first equation. We can show that it satisfies the second and the third in exactly the same way. Thus it is a solution of the system.

(b) We have shown in part (a) that if the system has two different solutions, we can find a third one by averaging the two solutions. But then we can find a fourth and a fifth solution by averaging the new one with each of the previous two. Then we can find four more by repeating this process with these new solutions, and so on. Clearly this process can continue indefinitely, so there will be infinitely many solutions.

Exercises 6.4

1. $x < 3$

3. $y > x$

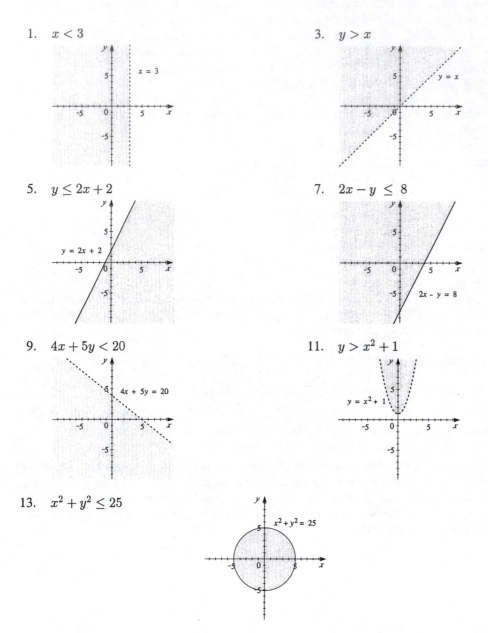

5. $y \leq 2x + 2$

7. $2x - y \leq 8$

9. $4x + 5y < 20$

11. $y > x^2 + 1$

13. $x^2 + y^2 \leq 25$

15. The boundary is a solid curve, so we have the inequality $y \leq \frac{1}{2}x - 1$. We take the test point $(0, -2)$ and verify that it satisfies the inequality: $-2 \leq \frac{1}{2}(0) - 1$ ✓.

17. The boundary is a broken curve, so we have the inequality $x^2 + y^2 > 4$. We take the test point $(0, 4)$ and verify that it satisfies the inequality: $0^2 + 4^2 \geq 4$ ✓.

19. $\begin{cases} x + y \le 4 \\ y \ge x \end{cases}$ The vertices occur where $\begin{cases} x + y = 4 \\ y = x \end{cases}$.
Substituting, we have $2x = 4 \iff x = 2$. Since $y = x$, the
vertex is $(2, 2)$, and the solution set is not bounded.

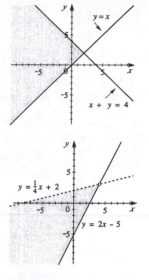

21. $\begin{cases} y < \frac{1}{4}x + 2 \\ y \ge 2x - 5 \end{cases}$ The vertex occurs where $\begin{cases} y = \frac{1}{4}x + 2 \\ y = 2x - 5 \end{cases}$.
Substituting for y gives $\frac{1}{4}x + 2 = 2x - 5 \iff \frac{7}{4}x = 7 \iff$
$x = 4$, so $y = 3$. Hence, the vertex is $(4, 3)$, and the solution is not
bounded.

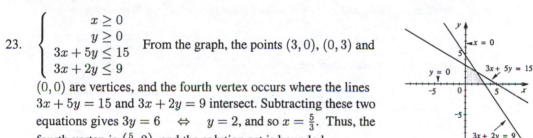

23. $\begin{cases} x \ge 0 \\ y \ge 0 \\ 3x + 5y \le 15 \\ 3x + 2y \le 9 \end{cases}$ From the graph, the points $(3, 0)$, $(0, 3)$ and
$(0, 0)$ are vertices, and the fourth vertex occurs where the lines
$3x + 5y = 15$ and $3x + 2y = 9$ intersect. Subtracting these two
equations gives $3y = 6 \iff y = 2$, and so $x = \frac{5}{3}$. Thus, the
fourth vertex is $\left(\frac{5}{3}, 2\right)$, and the solution set is bounded.

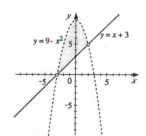

25. $\begin{cases} y < 9 - x^2 \\ y \ge x + 3 \end{cases}$ The vertices occur where $\begin{cases} y = 9 - x^2 \\ y = x + 3 \end{cases}$.
Substituting for y gives $9 - x^2 = x + 3 \iff x^2 + x - 6 = 0$
$\iff (x - 2)(x + 3) = 0 \implies x = -3, x = 2$. Therefore,
the vertices are $(-3, 0)$ and $(2, 5)$, and the solution set is bounded.

27. $\begin{cases} x^2 + y^2 \le 4 \\ x - y > 0 \end{cases}$ The vertices occur where $\begin{cases} x^2 + y^2 = 4 \\ x - y = 0 \end{cases}$.
Since $x - y = 0 \iff x = y$, substituting for x gives
$y^2 + y^2 = 4 \iff y^2 = 2 \implies y = \pm\sqrt{2}$, and $x = \pm\sqrt{2}$.
Therefore, the vertices are $(-\sqrt{2}, -\sqrt{2})$ and $(\sqrt{2}, \sqrt{2})$, and
the solution set is bounded.

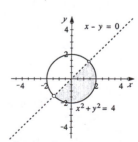

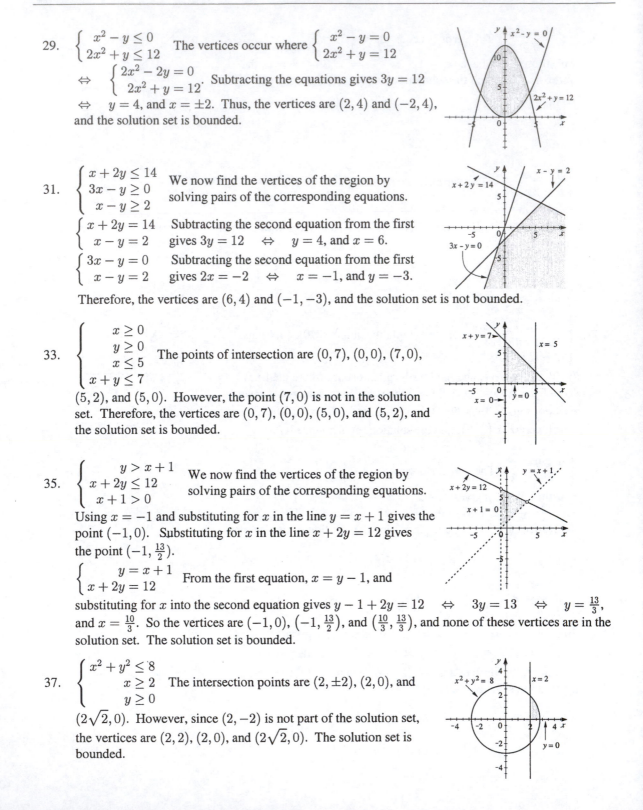

29. $\begin{cases} x^2 - y \leq 0 \\ 2x^2 + y \leq 12 \end{cases}$ The vertices occur where $\begin{cases} x^2 - y = 0 \\ 2x^2 + y = 12 \end{cases}$

$\Leftrightarrow \begin{cases} 2x^2 - 2y = 0 \\ 2x^2 + y = 12 \end{cases}$. Subtracting the equations gives $3y = 12$

$\Leftrightarrow \quad y = 4$, and $x = \pm 2$. Thus, the vertices are $(2, 4)$ and $(-2, 4)$, and the solution set is bounded.

31. $\begin{cases} x + 2y \leq 14 \\ 3x - y \geq 0 \\ x - y \geq 2 \end{cases}$ We now find the vertices of the region by solving pairs of the corresponding equations.

$\begin{cases} x + 2y = 14 \\ x - y = 2 \end{cases}$ Subtracting the second equation from the first gives $3y = 12 \quad \Leftrightarrow \quad y = 4$, and $x = 6$.

$\begin{cases} 3x - y = 0 \\ x - y = 2 \end{cases}$ Subtracting the second equation from the first gives $2x = -2 \quad \Leftrightarrow \quad x = -1$, and $y = -3$.

Therefore, the vertices are $(6, 4)$ and $(-1, -3)$, and the solution set is not bounded.

33. $\begin{cases} x \geq 0 \\ y \geq 0 \\ x \leq 5 \\ x + y \leq 7 \end{cases}$ The points of intersection are $(0, 7)$, $(0, 0)$, $(7, 0)$,

$(5, 2)$, and $(5, 0)$. However, the point $(7, 0)$ is not in the solution set. Therefore, the vertices are $(0, 7)$, $(0, 0)$, $(5, 0)$, and $(5, 2)$, and the solution set is bounded.

35. $\begin{cases} y > x + 1 \\ x + 2y \leq 12 \\ x + 1 > 0 \end{cases}$ We now find the vertices of the region by solving pairs of the corresponding equations.

Using $x = -1$ and substituting for x in the line $y = x + 1$ gives the point $(-1, 0)$. Substituting for x in the line $x + 2y = 12$ gives the point $(-1, \frac{13}{2})$.

$\begin{cases} y = x + 1 \\ x + 2y = 12 \end{cases}$ From the first equation, $x = y - 1$, and

substituting for x into the second equation gives $y - 1 + 2y = 12 \quad \Leftrightarrow \quad 3y = 13 \quad \Leftrightarrow \quad y = \frac{13}{3}$, and $x = \frac{10}{3}$. So the vertices are $(-1, 0)$, $(-1, \frac{13}{2})$, and $(\frac{10}{3}, \frac{13}{3})$, and none of these vertices are in the solution set. The solution set is bounded.

37. $\begin{cases} x^2 + y^2 \leq 8 \\ x \geq 2 \\ y \geq 0 \end{cases}$ The intersection points are $(2, \pm 2)$, $(2, 0)$, and

$(2\sqrt{2}, 0)$. However, since $(2, -2)$ is not part of the solution set, the vertices are $(2, 2)$, $(2, 0)$, and $(2\sqrt{2}, 0)$. The solution set is bounded.

39.
$$\begin{cases} x^2 + y^2 < 9 \\ x + y > 0 \\ x \le 0 \end{cases}$$

Substituting $x = 0$ into the equations $x^2 + y^2 = 9$

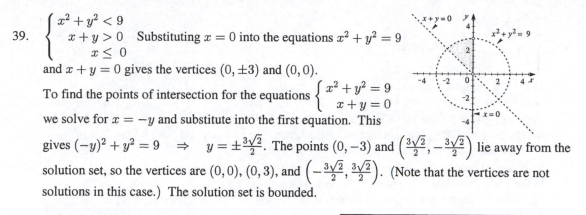

and $x + y = 0$ gives the vertices $(0, \pm 3)$ and $(0, 0)$.

To find the points of intersection for the equations $\begin{cases} x^2 + y^2 = 9 \\ x + y = 0 \end{cases}$

we solve for $x = -y$ and substitute into the first equation. This

gives $(-y)^2 + y^2 = 9 \Rightarrow y = \pm\frac{3\sqrt{2}}{2}$. The points $(0, -3)$ and $\left(\frac{3\sqrt{2}}{2}, -\frac{3\sqrt{2}}{2}\right)$ lie away from the

solution set, so the vertices are $(0, 0)$, $(0, 3)$, and $\left(-\frac{3\sqrt{2}}{2}, \frac{3\sqrt{2}}{2}\right)$. (Note that the vertices are not

solutions in this case.) The solution set is bounded.

41.
$$\begin{cases} y \ge x - 3 \\ y \ge -2x + 6 \\ y \le 8 \end{cases}$$

Using the graphing calculator we find the
region as shown. The vertices are $(3, 0)$, $(-1, 8)$,
and $(11, 8)$.

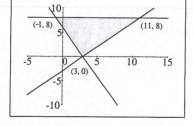

43.
$$\begin{cases} y \ge 6x - x^2 \\ x + y \ge 4 \end{cases}$$

Using the graphing calculator we find the
region as shown. The vertices are $(0.6, 3.4)$ and
$(6.4, -2.4)$.

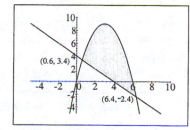

45. Let x be the number of fiction books published in a year, and y be
the number of nonfiction books. Then, the following system of
inequalities holds:

$$\begin{cases} x \ge 0 \\ y \ge 0 \\ x + y \le 100 \,. \\ y \ge 20 \\ x \ge y \end{cases}$$

From the graph, we see that the vertices are
$(50, 50)$, $(80, 20)$ and $(20, 20)$.

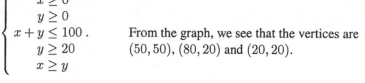

47. Let x be the number of Standard Blend packages and y be the
number of Deluxe Blend packages. Since there are 16 ounces per
pound, we get the following system of inequalities holds:

$$\begin{cases} x \ge 0 \\ y \ge 0 \\ \frac{1}{4}x + \frac{5}{8}y \le 80 \,. \\ \frac{3}{4}x + \frac{3}{8}y \le 90 \end{cases}$$

From the graph, we see that the vertices
are $(0, 0)$, $(120, 0)$, $(70, 100)$ and
$(0, 128)$.

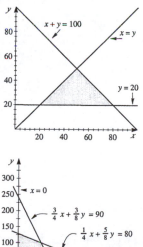

49.
$$\begin{cases} x + 2y > 4 \\ -x + y < 1 \\ x + 3y \le 9 \\ x < 3 \end{cases}.$$

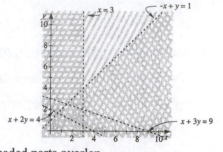

We solve the system using the first method, that is, shading the solution to each inequality. However, since this text is published in black and white, we shade the region with lines perpendicular to the boundary. As you can see, as the number of inequalities in the system increases, it gets harder to locate the region where all the shaded parts overlap. In the second method, if a region is shaded, then it does not satisfy at least one equality. As a result, the region that is left unshaded satisfies each inequality, and is the solution to the system of inequalities. This makes it easier to locate the solution set.

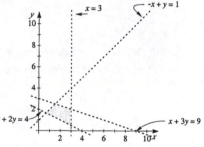

To finish, we find the vertices of the solution set. The line $x = 3$ intersects the line $x + 2y = 4$ at $\left(3, \frac{1}{2}\right)$ and the line $x + 3y = 9$ at $(3, 2)$. To find where the lines $-x + y = 1$ and $x + 2y = 4$ intersect, we add the two equations, which gives $3y = 5 \Leftrightarrow y = \frac{5}{3}$, and $x = \frac{2}{3}$. To find where the lines $-x + y = 1$ and $x + 3y = 9$ intersect, we add the two equations, which gives $4y = 10 \Leftrightarrow y = \frac{10}{4} = \frac{5}{2}$, and $x = \frac{3}{2}$. The vertices are $\left(3, \frac{1}{2}\right)$, $(3, 2)$, $\left(\frac{2}{3}, \frac{5}{3}\right)$, and $\left(\frac{3}{2}, \frac{5}{2}\right)$, and the solution set is bounded.

Exercises 6.5

1. $\dfrac{1}{(x-1)(x+2)} = \dfrac{A}{x-1} + \dfrac{B}{x+2}$

3. $\dfrac{x^2 - 3x + 5}{(x-2)^2(x+4)} = \dfrac{A}{x-2} + \dfrac{B}{(x-2)^2} + \dfrac{C}{x+4}$

5. $\dfrac{x^2}{(x-3)(x^2+4)} = \dfrac{A}{x-3} + \dfrac{Bx+C}{x^2+4}$

7. $\dfrac{x^3 - 4x^2 + 2}{(x^2+1)(x^2+2)} = \dfrac{Ax+B}{x^2+1} + \dfrac{Cx+D}{x^2+2}$

9. $\dfrac{x^3 + x + 1}{x(2x-5)^3(x^2+2x+5)^2} = \dfrac{A}{x} + \dfrac{B}{2x-5} + \dfrac{C}{(2x-5)^2} + \dfrac{D}{(2x-5)^3} +$

$$\dfrac{Ex+F}{x^2+2x+5} + \dfrac{Gx+H}{(x^2+2x+5)^2}$$

11. $\dfrac{2}{(x-1)(x+1)} = \dfrac{A}{x-1} + \dfrac{B}{x+1}$. Multiplying by $(x-1)(x+1)$, we get

$2 = A(x+1) + B(x-1) \quad \Leftrightarrow \quad 2 = Ax + A + Bx - B$. Thus

$\begin{cases} A + B = 0 \\ A - B = 2 \end{cases}$. Adding we get $2A = 2 \quad \Leftrightarrow \quad A = 1$. Now $A + B = 0 \quad \Leftrightarrow \quad B = -A$, so

$B = -1$. Thus, the required partial fraction decomposition is $\dfrac{2}{(x-1)(x+1)} = \dfrac{1}{x-1} - \dfrac{1}{x+1}$.

13. $\dfrac{5}{(x-1)(x+4)} = \dfrac{A}{x-1} + \dfrac{B}{x+4}$. Multiplying by $(x-1)(x+4)$, we get

$5 = A(x+4) + B(x-1) \quad \Leftrightarrow \quad 5 = Ax + 4A + Bx - B$. Thus

$\begin{cases} A + B = 0 \\ 4A - B = 5 \end{cases}$. Now $A + B = 0 \quad \Leftrightarrow \quad B = -A$, so substituting, we get $4A - (-A) = 5 \quad \Leftrightarrow$

$5A = 5 \quad \Leftrightarrow \quad A = 1$ and $B = -1$. The required partial fraction decomposition is

$\dfrac{5}{(x-1)(x+4)} = \dfrac{1}{x-1} - \dfrac{1}{x+4}$.

15. $\dfrac{12}{x^2-9} = \dfrac{12}{(x-3)(x+3)} = \dfrac{A}{x-3} + \dfrac{B}{x+3}$. Multiplying by $(x-3)(x+3)$, we get

$12 = A(x+3) + B(x-3) \quad \Leftrightarrow \quad 12 = Ax + 3A + Bx - 3B$. Thus

$\begin{cases} A + B = 0 \\ 3A - 3B = 12 \end{cases} \quad \Leftrightarrow \quad \begin{cases} A + B = 0 \\ A - B = 4 \end{cases}$. Adding we get $2A = 4 \quad \Leftrightarrow \quad A = 2$. So

$2 + B = 2 \quad \Leftrightarrow \quad$ and $B = -2$. The required partial fraction decomposition is

$\dfrac{12}{x^2-9} = \dfrac{2}{x-3} - \dfrac{2}{x+3}$.

17. $\dfrac{4}{x^2 - 4} = \dfrac{4}{(x-2)(x+2)} = \dfrac{A}{x-2} + \dfrac{B}{x+2}$. Multiplying by $x^2 - 4$, we get

$4 = A(x+2) + B(x-2) = (A+B)x + (2A - 2B)$, and so

$\begin{cases} A + B = 0 \\ 2A - 2B = 4 \end{cases} \Leftrightarrow \begin{cases} A + B = 0 \\ A - B = 2 \end{cases}$. Adding we get $2A = 2 \quad \Leftrightarrow \quad A = 1$, and $B = -1$.

Therefore, $\dfrac{4}{x^2 - 4} = \dfrac{1}{x-2} - \dfrac{1}{x+2}$.

19. $\dfrac{x+14}{x^2 - 2x - 8} = \dfrac{x+14}{(x-4)(x+2)} = \dfrac{A}{x-4} + \dfrac{B}{x+2}$. Hence,

$x + 14 = A(x+2) + B(x-4) = (A+B)x + (2A - 4B)$, and so

$\begin{cases} A + B = 1 \\ 2A - 4B = 14 \end{cases} \Leftrightarrow \begin{cases} 2A + 2B = 2 \\ A - 2B = 7 \end{cases}$. Adding, we get $3A = 9 \quad \Leftrightarrow \quad A = 3$. So

$(3) + B = 1 \quad \Leftrightarrow \quad B = -2$. Therefore, $\dfrac{x+14}{x^2 - 2x - 8} = \dfrac{3}{x-4} - \dfrac{2}{x+2}$.

21. $\dfrac{x}{8x^2 - 10x + 3} = \dfrac{x}{(4x-3)(2x-1)} = \dfrac{A}{4x-3} + \dfrac{B}{2x-1}$. Hence,

$x = A(2x-1) + B(4x-3) = (2A + 4B)x + (-A - 3B)$, and so

$\begin{cases} 2A + 4B = 1 \\ -A - 3B = 0 \end{cases} \Leftrightarrow \begin{cases} 2A + 4B = 1 \\ -2A - 6B = 0 \end{cases}$. Adding, we get $-2B = 1 \quad \Leftrightarrow \quad B = -\frac{1}{2}$, and

$A = \frac{3}{2}$. Therefore, $\dfrac{x}{8x^2 - 10x + 3} = \dfrac{\frac{3}{2}}{4x-3} - \dfrac{\frac{1}{2}}{2x-1}$.

23. $\dfrac{9x^2 - 9x + 6}{2x^3 - x^2 - 8x + 4} = \dfrac{9x^2 - 9x + 6}{(x-2)(x+2)(2x-1)} = \dfrac{A}{x-2} + \dfrac{B}{x+2} + \dfrac{C}{2x-1}$. Thus,

$\begin{aligned} 9x^2 - 9x + 6 &= A(x+2)(2x-1) + B(x-2)(2x-1) + C(x-2)(x+2) \\ &= A(2x^2 + 3x - 2) + B(2x^2 - 5x + 2) + C(x^2 - 4) \\ &= (2A + 2B + C)x^2 + (3A - 5B)x + (-2A + 2B - 4C). \end{aligned}$

This leads to the following system:

$\begin{cases} 2A + 2B + C = 9 & \text{Eq. 1: Coefficients of } x^2 \\ 3A - 5B = -9 & \text{Eq. 2: Coefficients of } x \\ -2A + 2B - 4C = 6 & \text{Eq. 3: Constant terms} \end{cases}$ Then

$\begin{array}{ll} \text{Eq. 1} & 2A + 2B + C = 9 \\ + \text{Eq. 3} & -2A + 2B - 4C = 6 \\ \hline \text{Eq. 4} & 4B - 3C = 15 \end{array}$

$\begin{array}{ll} & \text{Eq. 2} \quad 3A - 5B = -9 \\ \text{and} & + \text{Eq. 3} \quad -2A + 2B - 4C = 6 \\ \hline & \text{Eq. 5} \quad A - 3B - 4C = -3 \end{array} \Rightarrow \begin{cases} \text{Eq. 4} & 4B - 3C = 15 \\ \text{Eq. 5} & A - 3B - 4C = -3 \quad \text{Then} \\ \text{Eq. 3} & -2A + 2B - 4C = 6 \end{cases}$

$\begin{array}{ll} 2 \times \text{Eq. 5} & 2A - 6B - 8C = -6 \\ + \text{Eq. 3} & -2A + 2B - 4C = 6 \\ \hline \text{Eq. 6} & -4B - 12C = 0 \end{array}$ From which we get the system $\begin{cases} \text{Eq. 4} & 4B - 3C = 15 \\ \text{Eq. 5} \quad A - 3B - 4C = -3 \quad \text{Then} \\ \text{Eq. 6} & -4B - 12C = 0 \end{cases}$

$\begin{array}{ll} \text{Eq. 4} & 4B - 3C = 15 \\ + \text{Eq. 6} & -4B - 12C = 0 \\ \hline \text{Eq. 7} & -15C = 15 \end{array}$ Hence, $-15C = 15 \quad \Leftrightarrow \quad C = -1$; $4B + 3 = 15 \quad \Leftrightarrow \quad B = 3$;

and $A - 9 + 4 = -3 \quad \Leftrightarrow \quad A = 2$. Therefore, $\dfrac{9x^2 - 9x + 6}{2x^3 - x^2 - 8x + 4} = \dfrac{2}{x-2} + \dfrac{3}{x+2} - \dfrac{1}{2x-1}$.

25. $\dfrac{x^2+1}{x^3+x^2}=\dfrac{x^2+1}{x^2(x+1)}=\dfrac{A}{x}+\dfrac{B}{x^2}+\dfrac{C}{x+1}$. Hence,

$x^2+1=Ax(x+1)+B(x+1)+Cx^2=(A+C)x^2+(A+B)x+B$, and so $B=1$;

$A+1=0\Leftrightarrow A=-1$; and $-1+C=1\Leftrightarrow C=2$. Therefore, $\dfrac{x^2+1}{x^3+x^2}=\dfrac{-1}{x}+\dfrac{1}{x^2}+\dfrac{2}{x+1}$.

27. $\dfrac{2x}{4x^2+12x+9}=\dfrac{2x}{(2x+3)^2}=\dfrac{A}{2x+3}+\dfrac{B}{(2x+3)^2}$. Hence,

$2x=A(2x+3)+B=2Ax+(3A+B)$. So $2A=2\Leftrightarrow A=1$; and $3(1)+B=0\Leftrightarrow B=-3$.

Therefore, $\dfrac{2x}{4x^2+12x+9}=\dfrac{1}{2x+3}-\dfrac{3}{(2x+3)^2}$.

29. $\dfrac{4x^2-x-2}{x^4+2x^3}=\dfrac{4x^2-x-2}{x^3(x+2)}=\dfrac{A}{x}+\dfrac{B}{x^2}+\dfrac{C}{x^3}+\dfrac{D}{x+2}$. Hence,

$\begin{aligned}4x^2-x-2&=Ax^2(x+2)+Bx(x+2)+C(x+2)+Dx^3\\&=(A+D)x^3+(2A+B)x^2+(2B+C)x+2C.\end{aligned}$

So $2C=-2\Leftrightarrow C=-1$; $2B-1=-1\Leftrightarrow B=0$; $2A+0=4\Leftrightarrow A=2$; and $2+D=0\Leftrightarrow$

$D=-2$. Therefore, $\dfrac{4x^2-x-2}{x^4+2x^3}=\dfrac{2}{x}-\dfrac{1}{x^3}-\dfrac{2}{x+2}$.

31. $\dfrac{-10x^2+27x-14}{(x-1)^3(x+2)}=\dfrac{A}{x+2}+\dfrac{B}{x-1}+\dfrac{C}{(x-1)^2}+\dfrac{D}{(x-1)^3}$. Thus,

$\begin{aligned}-10x^2+27x-14&=A(x-1)^3+B(x+2)(x-1)^2+C(x+2)(x-1)+D(x+2)\\&=A(x^3-3x^2+3x-1)+B(x+2)(x^2-2x+1)+C(x^2+x-2)+D(x+2)\\&=A(x^3-3x^2+3x-1)+B(x^3-3x+2)+C(x^2+x-2)+D(x+2)\\&=(A+B)x^3+(-3A+C)x^2+(3A-3B+C+D)x+(-A+2B-2C+2D),\ \text{which leads}\end{aligned}$

to the following system of equations:

$$\begin{cases}A+\ B&=0\quad&\text{Eq. 1 Coefficients of }x^3\\-3A+\quad+\ C&=-10\quad&\text{Eq. 2: Coefficients of }x^2\\3A-3B+\ C+\ D&=27\quad&\text{Eq. 3: Coefficients of }x\\-A+2B-2C+2D&=-14\quad&\text{Eq. 4: Constant terms}\end{cases}$$

$\begin{array}{ll}3\times\text{Eq. 1}&3A+3B=0\\+\text{Eq. 2}&-3A++C=-10\\\hline\text{Eq. 5}&3B+C=-10\end{array}$

$\begin{array}{ll}-3\times\text{Eq. 1}&-3A-3B=0\\+\text{Eq. 3}&3A-3B+C+D=27\\\hline\text{Eq. 6}&-6B+C+D=27\end{array}$ and $\begin{array}{ll}\text{Eq. 1}&A+\ B=0\\+\text{Eq. 4}&-A+2B-2C+2D=-14\\\hline\text{Eq. 7}&3B-2C+2D=-14\end{array}$

Thus we get the system $\begin{cases}\text{Eq. 1}&A+B=0\\\text{Eq. 5:}&3B+\ C=-10\\\text{Eq. 6:}&-6B+\ C+\ D=27\\\text{Eq. 7:}&3B-2C+2D=-14\end{cases}$ Then $\begin{array}{ll}2\times\text{Eq. 5}&6B+2C=-20\\+\text{Eq. 6}&-6B+\ C+\ D=27\\\hline\text{Eq. 8}&3C+D=7\end{array}$

$$\begin{array}{ll} \text{Eq. 5} & 3B + C = -10 \\ -\text{Eq. 7} & \underline{-3B + 2C - 2D = 14} \\ \text{Eq. 9} & 3C -2D = 4 \end{array}$$
Thus we get the system
$$\begin{cases} \text{Eq. 1} & A + B = 0 \\ \text{Eq. 5:} & 3B + C = -10 \\ \text{Eq. 8:} & 3C + D = 7 \\ \text{Eq. 9:} & 3C - 2D = 4 \end{cases}$$

$$\begin{array}{ll} \text{Eq. 8} & 3C + D = 7 \\ -\text{Eq. 9} & \underline{-3C + 2D = -4} \\ \text{Eq. 10} & 3D = 3 \end{array}$$
Hence, $3D = 3 \iff D = 1;\ 3C + 1 = 7 \iff C = 2;$

$3B + 2 = -10 \iff B = -4;$ and $A - 4 = 0 \iff A = 4.$ Therefore,
$$\frac{-10x^2 + 27x - 14}{(x-1)^3(x+2)} = \frac{4}{x+2} - \frac{4}{x-1} + \frac{2}{(x-1)^2} + \frac{1}{(x-1)^3}.$$

33. $\dfrac{3x^3 + 22x^2 + 53x + 41}{(x+2)^2(x+3)^2} = \dfrac{A}{x+2} + \dfrac{B}{(x+2)^2} + \dfrac{C}{x+3} + \dfrac{D}{(x+3)^2}.$ Thus,

$3x^3 + 22x^2 + 53x + 41 = A(x+2)(x+3)^2 + B(x+3)^2 + C(x+2)^2(x+3) + D(x+2)^2$

$= A(x^3 + 8x^2 + 21x + 18) + B(x^2 + 6x + 9) + C(x^3 + 7x^2 + 16x + 12) + D(x^2 + 4x + 4)$

$= (A + C)x^3 + (8A + B + 7C + D)x^2 + (21A + 6B + 16C + 4D)x +$

$(18A + 9B + 12C + 4D),$ so we must solve the system:

$$\begin{cases} A + C = 3 & \text{Eq. 1 \ Coefficients of } x^3 \\ 8A + B + 7C + D = 22 & \text{Eq. 2: \ Coefficients of } x^2 \\ 21A + 6B + 16C + 4D = 53 & \text{Eq. 3: \ Coefficients of } x \\ 18A + 9B + 12C + 4D = 41 & \text{Eq. 4: \ Constant terms} \end{cases}$$
Then using Equation 1 we have

$$\begin{array}{ll} -8 \times \text{Eq. 1} & -8A - 8C = -24 \\ +\text{Eq. 2} & \underline{8A + B + 7C + D = 22} \\ \text{Eq. 5} & B - C + D = -2 \end{array}$$
and
$$\begin{array}{ll} -21 \times \text{Eq. 1} & -21A - 21C = -63 \\ +\text{Eq. 3} & \underline{21A + 6B + 16C + 4D = 53} \\ \text{Eq. 6} & 6B - 5C + 4D = -10 \end{array}$$

$$\begin{array}{ll} -18 \times \text{Eq. 1} & -18A - 18C = -54 \\ +\text{Eq. 4} & \underline{18A + 9B + 12C + 4D = 41} \\ \text{Eq. 7} & 9B - 6C + 4D = -13 \end{array}$$
Thus we get the following system.

$$\begin{cases} \text{Eq. 1} & A + C = 3 \\ \text{Eq. 5:} & B - C + D = -2 \\ \text{Eq. 6:} & 6B - 5C + 4D = -10 \\ \text{Eq. 7:} & 9B - 6C + 4D = -13 \end{cases}$$
Then
$$\begin{array}{ll} -6 \times \text{Eq. 5} & -6B + 6C - 6D = 12 \\ +\text{Eq. 6} & \underline{6B - 5C + 4D = -10} \\ \text{Eq. 8} & C - 2D = 2 \end{array}$$
and

$$\begin{array}{ll} -9 \times \text{Eq. 5} & -9B + 9C - 9D = 18 \\ +\text{Eq. 7} & \underline{9B - 6C + 4D = -13} \\ \text{Eq. 9} & 3C - 5D = 5 \end{array}$$
Thus we get the following system.
$$\begin{cases} \text{Eq. 1} & A + C = 3 \\ \text{Eq. 5:} & B - C + D = -2 \\ \text{Eq. 8:} & C - 2D = 2 \\ \text{Eq. 9:} & 3C - 5D = 5 \end{cases}$$

Finally,
$$\begin{array}{ll} -3 \times \text{Eq. 8} & -3C + 6D = -6 \\ +\text{Eq. 9} & \underline{3C - 5D = 5} \\ \text{Eq. 10} & D = -1 \end{array}$$
Hence, $D = -1;\ C + 2 = 2 \iff C = 0;$

$B - 0 - 1 = -2 \iff B = -1;$ and $A + 0 = 3 \iff A = 3.$ Therefore,
$$\frac{3x^3 + 22x^2 + 53x + 41}{(x+2)^2(x+3)^2} = \frac{3}{x+2} - \frac{1}{(x+2)^2} - \frac{1}{(x+3)^2}.$$

35. $\dfrac{x-3}{x^3+3x}=\dfrac{x-3}{x(x^2+3)}=\dfrac{A}{x}+\dfrac{Bx+C}{x^2+3}$. Hence, $x-3=A(x^2+3)+Bx^2+Cx$

$=(A+B)x^2+Cx+3A$. So $3A=-3\Leftrightarrow A=-1;\;\;C=1;\;$ and $-1+B=0\Leftrightarrow B=1$.

Therefore, $\dfrac{x-3}{x^3+3x}=-\dfrac{1}{x}+\dfrac{x+1}{x^2+3}$.

37. $\dfrac{2x^3+7x+5}{(x^2+x+2)(x^2+1)}=\dfrac{Ax+B}{x^2+x+2}+\dfrac{Cx+D}{x^2+1}$. Thus,

$2x^3+7x+5=(Ax+B)(x^2+1)+(Cx+D)(x^2+x+2)$

$=Ax^3+Ax+Bx^2+B+Cx^3+Cx^2+2Cx+Dx^2+Dx+2D$

$=(A+C)x^3+(B+C+D)x^2+(A+2C+D)x+(B+2D)$. We must solve the system:

$\begin{cases}A\ +\ C\ \ \ \ \ \ \ =2 & \text{Eq. 1\ \ Coefficients of }x^3\\ B+\ C+\ D=0 & \text{Eq. 2: Coefficients of }x^2\\ A\ +2C+\ D=7 & \text{Eq. 3: Coefficients of }x\\ B\ \ \ \ \ \ +2D=5 & \text{Eq. 4: Constant terms}\end{cases}$

Then $\begin{array}{l}-\text{Eq. 1}\ \ -A-\ \ C\ \ \ \ \ \ \ \ \ =-2\\ +\text{Eq. 3}\ \ \underline{\ \ A+2C+\ \ D=7}\\ \text{Eq. 5}\ \ \ \ \ \ \ \ \ \ \ \ \ \ C\ \ +D=5\end{array}$

and $\begin{array}{l}-\text{Eq. 2}\ \ -B-C-\ \ D=0\\ +\text{Eq. 4}\ \ \underline{\ \ B\ \ \ \ \ \ \ +2D=5}\\ \text{Eq. 6}\ \ \ \ \ \ \ \ \ -C+\ \ D=5\end{array}$ This leads to the following system $\begin{cases}\text{Eq. 1}\ \ A\ \ +C\ \ \ \ \ \ \ =2\\ \text{Eq. 2:}\ \ \ \ \ B+C+D=0\\ \text{Eq. 5:}\ \ \ \ \ \ \ \ \ \ C+D=5\\ \text{Eq. 6:}\ \ \ \ \ \ \ \ -C+D=5\end{cases}$

$\begin{array}{l}\text{Eq. 5}\ \ \ \ \ C+D=5\\ -\text{Eq. 6}\ \ \underline{-C+D=5}\\ \text{Eq. 7}\ \ \ \ \ \ \ \ \ 2D=10\end{array}$ Hence, $2D=10\quad\Leftrightarrow\quad D=5;\;\;C+5=5\quad\Leftrightarrow\quad C=0;$

$B+0+5=0\quad\Leftrightarrow\quad B=-5;$ and $A+0=2\quad\Leftrightarrow\quad A=2.$ Therefore,

$\dfrac{2x^3+7x+5}{(x^2+x+2)(x^2+1)}=\dfrac{2x-5}{x^2+x+2}+\dfrac{5}{x^2+1}$.

39. $\dfrac{x^4+x^3+x^2-x+1}{x(x^2+1)^2}=\dfrac{A}{x}+\dfrac{Bx+C}{x^2+1}+\dfrac{Dx+E}{(x^2+1)^2}$. Hence,

$x^4+x^3+x^2-x+1=A(x^2+1)^2+(Bx+C)x(x^2+1)+x(Dx+E)$

$=A(x^4+2x^2+1)+(Bx^2+Cx)(x^2+1)+Dx^2+Ex$

$=A(x^4+2x^2+1)+Bx^4+Bx^2+Cx^3+Cx+Dx^2+Ex$

$=(A+B)x^4+Cx^3+(2A+B+D)x^2+(C+E)x+A.$

Hence, $A=1;\;\;1+B=1\Leftrightarrow B=0;\;\;C=1;\;\;2+0+D=1\Leftrightarrow D=-1;\;$ and $1+E=-1$

$\Leftrightarrow E=-2.$ Therefore, $\dfrac{x^4+x^3+x^2-x+1}{x(x^2+1)^2}=\dfrac{1}{x}+\dfrac{1}{x^2+1}-\dfrac{x+2}{(x^2+1)^2}$.

41. We must first get a proper rational function. Using long division, we have:

$$\begin{array}{r}x^2\\ x^3-2x^2+x-2\overline{)\,x^5-2x^4+x^3+0x^2+x+5}\\ \underline{x^5-2x^4+x^3-2x^2}\\ 2x^2+x+5.\end{array}$$

Therefore, $\dfrac{x^5 - 2x^4 + x^3 + x + 5}{x^3 - 2x^2 + x - 2} = x^2 + \dfrac{2x^2 + x + 5}{x^3 - 2x^2 + x - 2} = x^2 + \dfrac{2x^2 + x + 5}{(x-2)(x^2+1)}$

$= x^2 + \dfrac{A}{x-2} + \dfrac{Bx+C}{x^2+1}$. Hence, $2x^2 + x + 5 = A(x^2 + 1) + (Bx + C)(x - 2)$

$= Ax^2 + A + Bx^2 + Cx - 2Bx - 2C = (A + B)x^2 + (C - 2B)x + (A - 2C)$. Equating coefficients, we get the system:

$$\begin{cases} A + B \quad\quad = 2 \\ \quad -2B + C = 1 \\ A \quad\quad - 2C = 5 \end{cases} \begin{array}{l} \text{Eq. 1: Coefficients of } x^2 \\ \text{Eq. 2: Coefficients of } x \\ \text{Eq. 3: Constant terms} \end{array} \quad \begin{array}{l} \text{Eq. 1} \\ -\text{Eq. 3} \\ \hline \text{Eq. 4} \end{array} \begin{array}{ll} A + B & = 2 \\ -A \quad + 2C = -5 \\ \hline B + 2C = -3 \end{array}$$

This leads to the following system $\begin{cases} \text{Eq. 1} \quad A + B \quad\quad = 2 \\ \text{Eq. 2:} \quad\quad -2B + C = 1 \\ \text{Eq. 4:} \quad\quad\quad B + 2C = -3 \end{cases}$ $\quad\begin{array}{l} \text{Eq. 2} \\ 2 \times \text{Eq. 4} \\ \hline \text{Eq. 5} \end{array} \begin{array}{l} -2B + C = 1 \\ 2B + 4C = -6. \\ \hline 5C = -5 \end{array}$

Therefore, $5C = -5 \quad\Leftrightarrow\quad C = -1, B - 2 = -3 \quad\Leftrightarrow\quad B = -1$, and $A - 1 = 2 \quad\Leftrightarrow$

$A = 3$. Therefore, $\dfrac{x^5 - 2x^4 + x^3 + x + 5}{x^3 - 2x^2 + x - 2} = x^2 + \dfrac{3}{x-2} - \dfrac{x+1}{x^2+1}$.

43. $\dfrac{ax + b}{x^2 - 1} = \dfrac{A}{x-1} + \dfrac{B}{x+1}$. Hence, $ax + b = A(x + 1) + B(x - 1) = (A + B)x + (A - B)$. So

$\begin{cases} A + B = a \\ A - B = b \end{cases}$. Adding, we get $2A = a + b \quad\Leftrightarrow\quad A = \dfrac{a+b}{2}$. Substituting, we get

$B = a - A = \dfrac{2a}{2} - \dfrac{a+b}{2} = \dfrac{a-b}{2}$. Therefore, $A = \dfrac{a+b}{2}$ and $B = \dfrac{a-b}{2}$.

45. (a) The expression $\dfrac{x}{x^2 + 1} + \dfrac{1}{x+1}$ is already a partial fraction decomposition. The denominator in the first term is a quadratic which can't be factored and the degree of the numerator is less than 2. The denominator of the second term is linear and the numerator is a constant.

(b) The term $\dfrac{x}{(x+1)^2}$ can be decomposed further, since the numerator and denominator both have linear factors. $\dfrac{x}{(x+1)^2} = \dfrac{A}{x+1} + \dfrac{B}{(x+1)^2}$. Hence, $x = A(x + 1) + B = Ax + (A + B)$.

So $A = 1, B = -1$, and $\dfrac{x}{(x+1)^2} = \dfrac{1}{x+1} + \dfrac{-1}{(x+1)^2}$.

(c) The expression $\dfrac{1}{x+1} + \dfrac{2}{(x+1)^2}$ is already a partial fraction decomposition, since each numerator is constant.

(d) The expression $\dfrac{x+2}{(x^2+1)^2}$ is already a partial fraction decomposition, since the denominator is the square of a quadratic which can't be factored and the degree of the numerator is less than 2.

Review Exercises for Chapter 6

1. $\begin{cases} 2x + 3y = 7 \\ x - 2y = 0 \end{cases}$ By inspection of the graph, it appears that $(2, 1)$ is the solution to the system. We check this in both equations to verify that it is the solution. $2(2) + 3(1) = 4 + 3 = 7\checkmark$ and $2 - 2(1) = 2 - 2 = 0\checkmark$. Since both equations are satisfied, the solution is $(2, 1)$.

3. $\begin{cases} x^2 + y = 2 \\ x^2 - 3x - y = 0 \end{cases}$ By inspection of the graph, it appears that $(2, -2)$ is a solution to the system, but is difficult to get accurate values for the other point. Adding the equations we get $2x^2 - 3x = 2$ $\Leftrightarrow$ $2x^2 - 3x - 2 = 0$ $\Leftrightarrow$ $(2x + 1)(x - 2) = 0$. So $2x + 1 = 0$ $\Leftrightarrow$ $x = -\frac{1}{2}$ or $x = 2$. If $x = -\frac{1}{2}$ then $(-\frac{1}{2})^2 + y = 2$ $\Leftrightarrow$ $y = \frac{7}{4}$. If $x = 2$ then $2^2 + y = 2$ $\Leftrightarrow$ $y = -2$. Thus the solutions are $(-\frac{1}{2}, \frac{7}{4})$ and $(2, -2)$.

5. $\begin{cases} 3x - y = 5 \\ 2x + y = 5 \end{cases}$ Adding, we get $5x = 10$ $\Leftrightarrow$ $x = 2$. So $2(2) + y = 5$ $\Leftrightarrow$ $y = 1$. Thus the solution is $(2, 1)$.

7. $\begin{cases} 2x - 7y = 28 \\ y = \frac{2}{7}x - 4 \end{cases}$ $\Leftrightarrow$ $\begin{cases} 2x - 7y = 28 \\ 2x - 7y = 28 \end{cases}$ Since these equations represent the same line, any point on this line will satisfy the system. Thus the solution are $\left(x, \frac{2}{7}x - 4\right)$, where x is any real number.

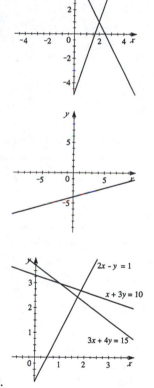

9. $\begin{cases} 2x - y = 1 \\ x + 3y = 10 \\ 3x + 4y = 15 \end{cases}$ Solving the first equation for y, we get $y = -2x + 1$. Substituting into the second equation gives $x + 3(-2x + 1) = 10$ $\Leftrightarrow$ $-5x = 7$ $\Leftrightarrow$ $x = -\frac{7}{5}$. So $y = -\left(-\frac{7}{5}\right) + 1 = \frac{12}{5}$. Checking the point $\left(-\frac{7}{5}, \frac{12}{5}\right)$ in the third equation we have $3\left(-\frac{7}{5}\right) + 4\left(\frac{12}{5}\right) \overset{?}{=} 15$ but $-\frac{21}{5} + \frac{48}{5} \neq 15$. Thus, there is no solution, and the lines do not intersect at one point.

11. $\begin{cases} y = x^2 + 2x \\ y = 6 + x \end{cases}$ Substituting for y gives $6 + x = x^2 + 2x$ $\Leftrightarrow$ $x^2 + x - 6 = 0$. Factoring we have $(x - 2)(x + 3) = 0$. Thus $x = 2$ or -3. If $x = 2$, then $y = 8$, and if $x = -3$, then $y = 3$. Thus the solutions are $(-3, 3)$ and $(2, 8)$.

13. $\begin{cases} 3x + \dfrac{4}{y} = 6 \\ x - \dfrac{8}{y} = 4 \end{cases}$ Adding twice the first equation to the second equation gives $7x = 16$ ⇔ $x = \frac{16}{7}$. So $\frac{16}{7} - \frac{8}{y} = 4$ ⇔ $16y - 56 = 28y$ ⇔ $-12y = 56$ ⇔ $y = -\frac{14}{3}$. Thus the solution is $\left(\frac{16}{7}, -\frac{14}{3}\right)$.

15. $\begin{cases} 0.32x + 0.43y = 0 \\ 7x - 12y = 341 \end{cases}$ ⇔ $\begin{cases} y = -\dfrac{32}{43}x \\ y = \dfrac{7x - 341}{12} \end{cases}$

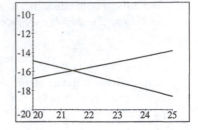

The solution is $(21.41, -15.93)$.

17. $\begin{cases} x - y^2 = 10 \\ x = \frac{1}{22}y + 12 \end{cases}$ ⇔ $\begin{cases} y = \pm\sqrt{x - 10} \\ y = 22(x - 12) \end{cases}$

The solutions are $(11.94, -1.39)$ and $(12.07, 1.44)$.

19. $\begin{cases} \text{Eq. 1:} & x + y + 2z = 6 \\ \text{Eq. 2:} & 2x \quad\;\; + 5z = 12 \\ \text{Eq. 3:} & x + 2y + 3z = 9 \end{cases}$ Then $\begin{array}{l} 2 \times \text{Eq. 1} \\ - \;\text{Eq. 2} \\ \hline \text{Eq. 4} \end{array} \begin{array}{l} 2x + 2y + 4z = 12 \\ -2x \qquad\;\; - 5z = -12 \\ \hline 2y - z = 0 \end{array}$ and

$\begin{array}{l} \text{Eq. 1} \\ - \;\text{Eq. 3} \\ \hline \text{Eq. 5} \end{array} \begin{array}{l} x + y + 2z = 6 \\ -x - 2y - 3z = -9 \\ \hline -y - z = -3 \end{array}$ This gives us the system $\begin{cases} \text{Eq. 1:} & x + y + 2z = 6 \\ \text{Eq. 4:} & 2y - z = 0 \\ \text{Eq. 5:} & -y - z = -3 \end{cases}$ Finally,

$\begin{array}{l} \text{Eq. 4} \\ + 2 \times \text{Eq. 5} \\ \hline \text{Eq. 6} \end{array} \begin{array}{l} 2y - z = 0 \\ -2y - 2z = -6 \\ \hline -3z = -6 \end{array}$ Therefore, $-3z = -6$ ⇔ $z = 2$; $-y - 2 = -3$ ⇔

$y = 1$; and $x + 1 + 2(2) = 6$ ⇔ $x = 1$. Hence, the solution is $(1, 1, 2)$.

21. $\begin{cases} \text{Eq. 1:} & x - 2y + 3z = 1 \\ \text{Eq. 2:} & 2x - y + z = 3 \\ \text{Eq. 3:} & 2x - 7y + 11z = 2 \end{cases}$ Then $\begin{array}{l} 2 \times \text{Eq. 1} \\ - \;\text{Eq. 2} \\ \hline \text{Eq. 4} \end{array} \begin{array}{l} 2x - 4y + 6z = 2 \\ -2x + y - z = -3 \\ \hline -3y + 5z = -1 \end{array}$ and

$\begin{array}{l} 2 \times \text{Eq. 1} \\ - \;\text{Eq. 3} \\ \hline \text{Eq. 5} \end{array} \begin{array}{l} 2x - 4y + 6z = 2 \\ -2x + 7y - 11z = -2 \\ \hline 3y - 5z = 0 \end{array}$ Thus we get the system $\begin{cases} \text{Eq. 1:} & x - 2y + 3z = 1 \\ \text{Eq. 4:} & -3y + 5z = -1 \\ \text{Eq. 5:} & 3y - 5z = 0 \end{cases}$ Then

$\begin{array}{l} \text{Eq. 4} \\ + \;\text{Eq. 5} \\ \hline \text{Eq. 6} \end{array} \begin{array}{l} -3y + 5z = -1 \\ 3y - 5z = 0 \\ \hline 0 = -1 \end{array}$ which is impossible. Therefore, the system has no solution.

23. $\begin{cases} \text{Eq. 1:} & x - 3y + z = 4 \\ \text{Eq. 2:} & 4x - y + 15z = 5 \end{cases}$ Then $\begin{array}{l} 4 \times \text{Eq. 1} \\ - \;\text{Eq. 2} \\ \hline \text{Eq. 3} \end{array} \begin{array}{l} 4x - 12y + 4z = 16 \\ -4x + y - 15z = -5 \\ \hline -11y - 11z = -11 \end{array}$

Dividing the last equation by -11 gives the system $\begin{cases} x - 3y + z = 4 \\ y + z = 1 \end{cases}$. Thus, the system has infinitely many solutions given by $z = t, y + t = -1 \quad \Leftrightarrow \quad y = -1 - t$, and $x + 3(1 + t) + t = 4$ $\Leftrightarrow \quad x = 1 - 4t$. Therefore, the solutions are $(1 - 4t, -1 - t, t)$, where t is any real number.

25. $\begin{cases} \text{Eq. 1: } -x + 4y + z = 8 \\ \text{Eq. 2: } 2x - 6y + z = -9 \\ \text{Eq. 3: } x - 6y - 4z = -15 \end{cases}$ Then $\begin{array}{l} 2 \times \text{Eq. 1} \quad -2x + 8y + 2z = 16 \\ \underline{+ \text{ Eq. 2} \quad\quad 2x - 6y + z = -9} \\ \text{Eq. 4} \quad\quad\quad\quad 2y + 3z = 7 \end{array}$ and

$\begin{array}{l} \text{Eq. 1} \quad -x + 4y + z = 8 \\ \underline{- \text{ Eq. 3} \quad\quad x - 6y - 4z = -15} \\ \text{Eq. 5} \quad\quad -2y - 3z = -7 \end{array}$ which gives us the system $\begin{cases} \text{Eq. 1: } -x + 4y + z = 8 \\ \text{Eq. 4: } \quad\quad 2y + 3z = 7 \\ \text{Eq. 5: } \quad -2y - 3z = -7 \end{cases}$ Finally,

$\begin{array}{l} \text{Eq. 4} \quad 2y + 3z = 7 \\ \underline{+ \text{ Eq. 5} \quad -2y - 3z = -7} \\ \text{Eq. 6} \quad\quad\quad 0 = 0 \end{array}$ Thus, the system has infinitely many solutions. Solve the first equation for x we get $x = -8 + 4y + z$ and solving the second for y gives $y = \frac{7}{2} - \frac{3}{2}z$. Let $z = t$, then $y = \frac{7}{2} - \frac{3}{2}t$, and $x = -8 + 4(\frac{7}{2} - \frac{3}{2}t) + t = -8 + 14 - 6t + t = 6 - 5t$. Therefore, the solutions are $(6 - 5t, \frac{7}{2} - \frac{3}{2}t, t)$, where t is any real number.

27. Let x be the amount in the 6% account, and y be the amount in the 7% account. Thus, the system is:
$\begin{cases} y = 2x \\ 0.06x + 0.07y = 600 \end{cases}$

Substituting gives $0.06x + 0.07(2x) = 600 \quad \Leftrightarrow \quad 0.2x = 600 \quad \Leftrightarrow \quad x = 3000$. So $y = 2(3000) = 6000$. Hence, the man has \$3,000 invested at 6% and \$6,000 invested at 7%.

29. The boundary is a solid curve, so we have the inequality $x + y^2 \le 4$. We take the test point $(0, 0)$ and verify that it satisfies the inequality: $0 + 0^2 \le 4 \checkmark$.

31.

33.

35.

37.

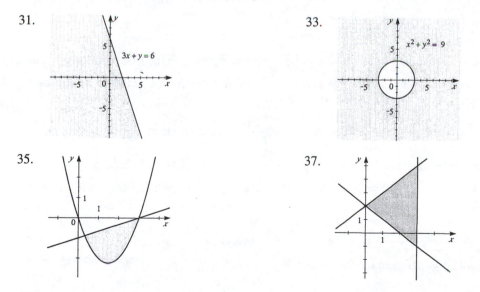

39. $\begin{cases} x^2 + y^2 < 9 \\ x + y < 0 \end{cases}$ The vertices occur where $y = -x$. By

substitution, $x^2 + x^2 = 9$ $\Leftrightarrow$ $x = \pm\frac{3}{\sqrt{2}}$, and so $y = \mp\frac{3}{\sqrt{2}}$.

Therefore, the vertices are $\left(\frac{3}{\sqrt{2}}, -\frac{3}{\sqrt{2}}\right)$ and $\left(-\frac{3}{\sqrt{2}}, \frac{3}{\sqrt{2}}\right)$ and

the solution set is bounded.

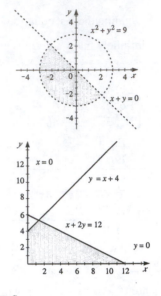

41. $\begin{cases} x \geq 0, \ y \geq 0 \\ x + 2y \leq 12 \\ y \leq x + 4 \end{cases}$ The intersection points are $(-4, 0)$, $(0, 4)$,

$\left(\frac{4}{3}, \frac{16}{3}\right)$, $(0, 6)$, $(0, 0)$, and $(12, 0)$. Since the points $(-4, 0)$ and

$(0, 6)$ are not in the solution set, the vertices are $(0, 4)$, $\left(\frac{4}{3}, \frac{16}{3}\right)$,

$(12, 0)$, and $(0, 0)$. The solution set is bounded.

43. $\begin{cases} -x + y + z = a \quad \text{Eq. 1:} \\ x - y + z = b \quad \text{Eq. 2:} \\ x + y - z = c \quad \text{Eq. 3:} \end{cases}$ Adding the first two equations $\begin{aligned} -x + y + z &= a \\ x - y + z &= b \\ \hline 2z &= a + b \end{aligned}$ so $z = \dfrac{a + b}{2}$.

Adding the Eq. 1 and Eq. 3, we get $\begin{aligned} -x + y + z &= a \\ x + y - z &= c \\ \hline 2y &= a + c \end{aligned}$ so $y = \dfrac{a + c}{2}$.

Finally adding the Eq. 2 and Eq. 3, we get $\begin{aligned} x - y + z &= b \\ x + y - z &= c \\ \hline 2x &= b + c \end{aligned}$ so $x = \dfrac{b + c}{2}$.

The solution is $\left(\dfrac{b + c}{2}, \dfrac{a + c}{2}, \dfrac{a + b}{2}\right)$.

45. Solving the second equation for y we have $y = kx$. Substituting for y in the first equation gives us

$x + kx = 12$ $\Leftrightarrow$ $(1 + k)x = 12$ $\Leftrightarrow$ $x = \dfrac{12}{k + 1}$. Substituting for y in the third equation

gives

us $kx - x = 2k$ $\Leftrightarrow$ $(k - 1)x = 2k$ $\Leftrightarrow$ $x = \dfrac{2k}{k - 1}$ These points of intersection are the

same when the x values are equal. Thus $\dfrac{12}{k + 1} = \dfrac{2k}{k - 1}$ $\Leftrightarrow$ $12(k - 1) = 2k(k + 1)$ $\Leftrightarrow$

$12k - 12 = 2k^2 + 2k$ $\Leftrightarrow$ $0 = 2k^2 - 10k + 12 = 2(k^2 - 5k + 6) = 2(k - 3)(k + 2)$. Hence

$k = 2$ or $k = 3$.

47. $\dfrac{3x + 1}{x^2 - 2x - 15} = \dfrac{3x + 1}{(x - 5)(x + 3)} = \dfrac{A}{x - 5} + \dfrac{B}{x + 3}$. Thus, $3x + 1 = A(x + 3) + B(x - 5)$

$= x(A + B) + (3A - 5B)$, and so

$$\begin{cases} A + B = 3 \\ 3A - 5B = 1 \end{cases} \Leftrightarrow \begin{cases} -3A - 3B = -9 \\ 3A - 5B = 1 \end{cases}.$$ Adding, we have $-8B = -8 \quad \Leftrightarrow \quad B = 1$, and

$A = 2$. Hence, $\dfrac{3x + 1}{x^2 - 2x - 15} = \dfrac{2}{x - 5} + \dfrac{1}{x + 3}$.

49. $\dfrac{2x - 4}{x(x - 1)^2} = \dfrac{A}{x} + \dfrac{B}{x - 1} + \dfrac{C}{(x - 1)^2}$. Then, $2x - 4 = A(x - 1)^2 + Bx(x - 1) + Cx$

$= Ax^2 - 2Ax + A + Bx^2 - Bx + Cx = x^2(A + B) + x(-2A - B + C) + A$. So $A = -4$;

$-4 + B = 0 \Leftrightarrow B = 4$; and $8 - 4 + C = 2 \Leftrightarrow C = -2$. Therefore,

$\dfrac{2x - 4}{x(x - 1)^2} = -\dfrac{4}{x} + \dfrac{4}{x - 1} - \dfrac{2}{(x - 1)^2}$.

51. $\dfrac{2x - 1}{x^3 + x} = \dfrac{2x - 1}{x(x^2 + 1)} = \dfrac{A}{x} + \dfrac{Bx + C}{x^2 + 1}$. Then $2x - 1 = A(x^2 + 1) + (Bx + C)x$

$= Ax^2 + A + Bx^2 + Cx = (A + B)x^2 + Cx + A$. So $A = -1$, $C = 2$ and $A + B = 0$ gives us

$B = 1$. Thus $\dfrac{2x - 1}{x^3 + x} = -\dfrac{1}{x} + \dfrac{x + 2}{x^2 + 1}$.

Chapter 6 Test

1. (a) Linear.

   (b) $\begin{cases} x + 3y = 7 \\ 5x + 2y = -4 \end{cases}$ Multiplying the first equation by -5 and then adding gives $-13y = -39$

   $\Leftrightarrow$ $y = 3$. So $x + 3(3) = 7$ $\Leftrightarrow$ $x = -2$. Thus, the solution is $(-2, 3)$.

2. (a) Nonlinear.

   (b) $\begin{cases} 6x + y^2 = 10 \\ 3x - y = 5 \end{cases}$ Multiplying the second equation by -2 gives $\begin{aligned} 6x + y^2 &= 10 \\ \underline{-6x + 2y} &= -10 \\ y^2 + 2y &= 0 \end{aligned}$. Thus

   $y^2 + 2y = y(y + 2) = 0$. Thus $y = 0$ or $y = -2$. When $y = 0$ then $3x = 5$ $\Leftrightarrow$ $x = \frac{5}{3}$ and
   when $y = -2$ then $3x - (-2) = 5$ $\Leftrightarrow$ $3x = 3$ $\Leftrightarrow$ $x = 1$ Thus the solutions are $\left(\frac{5}{3}, 0\right)$
   and $(1, -2)$.

3. $\begin{cases} x - 2y = 1 \\ y = x^3 - 2x^2 \end{cases}$

   The solutions are approximately $(-0.55, -0.78)$,
   $(0.43, -0.29)$, and $(2.12, 0.56)$.

   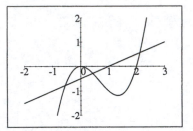

4. Let w be the speed of the wind and a be the speed of the airplane in still air, in km per hour. Then
   the speed of the of the plane flying against the wind is $a - w$, and the speed of the plane flying with
   the wind is $a + w$. Using *distance = time × rate*, we get the system

   $\begin{cases} 600 = 2.5(a - w) \\ 300 = \frac{50}{60}(a + w) \end{cases}$ $\Leftrightarrow$ $\begin{cases} 240 = a - w \\ 360 = a + w \end{cases}$.

   Adding the two equations, we get $600 = 2a$ $\Leftrightarrow$ $a = 300$. So $360 = 300 + w$ $\Leftrightarrow$ $w = 60$.
   Thus the speed of the airplane in still air is 300 km/h and the speed of the wind is 60 km/h.

5. (a) $\begin{cases} \text{Eq. 1:} & x + 2y + z = 3 \\ \text{Eq. 2:} & x + 3y + 2z = 3 \\ \text{Eq. 3:} & 2x + 3y - z = 8 \end{cases}$ $\begin{aligned} &\text{Subtracting Eq. 2} \\ &\text{from Eq. 1 we get} \end{aligned}$ $\begin{aligned} x + 2y + z &= 3 \\ \underline{-x - 3y - 2z} &= -3 \\ -y - z &= 0 \end{aligned}$

   $\begin{aligned} &\text{Then subtracting Eq. 3} \\ &\text{from } 2 \times \text{Eq. 1 we get} \end{aligned}$ $\begin{aligned} 2x + 4y + 2z &= 6 \\ \underline{-2x - 3y + z} &= -8 \\ y + 3z &= -2 \end{aligned}$ $\begin{aligned} &\text{Adding these two} \\ &\text{new equations gives} \end{aligned}$ $\begin{aligned} -y - z &= 0 \\ \underline{y + 3z} &= -2 \\ 2z &= -2 \end{aligned}$

   So $2z = -2$ $\Leftrightarrow$ $z = -1$. Since $-y - (-1) = 0$ $\Leftrightarrow$ $y = 1$. Thus
   $x + 2(1) + (-1) = 3$ $\Leftrightarrow$ $x = 2$. Thus the solution is $(2, 1, -1)$.

   (b) The system is neither inconsistent nor dependent.

6. (a) $\begin{cases} \text{Eq. 1:} & x - y + 9z = -8 \\ \text{Eq. 2:} & x \qquad - 4z = 7 \\ \text{Eq. 3:} & 3x - y + \ z = 5 \end{cases}$

Subtracting:
$\begin{array}{ll} & \text{Eq. 1} \\ - & \text{Eq. 2} \\ \hline & \text{Eq. 4} \end{array}$
$\begin{array}{l} x - y + \ 9z = -8 \\ -x \qquad + \ 4z = -7 \\ \hline -y + 13z = -15 \end{array}$

Then subtracting
$\begin{array}{ll} & 3 \times \text{Eq. 1} \\ - & \text{Eq. 3} \\ \hline & \text{Eq. 5} \end{array}$
$\begin{array}{l} 3x - 3y + 27z = -24 \\ -3x + \ y - \qquad z = -5 \\ \hline -2y + 26z = -29 \end{array}$

Thus we get the system

$\begin{cases} x - y + \ 9z = -8 \\ -y + 13z = -15 \\ -2y + 26z = -29 \end{cases}$
Finally subtracting these two new equations we get
$\begin{array}{l} 2y - 26z = 30 \\ -2y + 26z = -29 \\ \hline 0 = 1 \end{array}$

Which is always false. Thus there is no solution.

(b) The system is inconsistent.

7. (a) $\begin{cases} \text{Eq. 1:} & 2x - \ y + \ z = 0 \\ \text{Eq. 2:} & 3x + 2y - 3z = 1 \\ \text{Eq. 3:} & \ x - 4y + 5z = -1 \end{cases}$

Subtracting:
$\begin{array}{ll} & \text{Eq. 1} \\ - & 2 \times \text{Eq. 3} \\ \hline & \text{Eq. 4} \end{array}$
$\begin{array}{l} 2x - \ y + \quad z = 0 \\ -2x + 8y - 10z = 2 \\ \hline 7y - \quad 9z = 2 \end{array}$

Then subtracting
$\begin{array}{ll} & \text{Eq. 2} \\ - & 3 \times \text{Eq. 3} \\ \hline & \text{Eq. 5} \end{array}$
$\begin{array}{l} 3x + \ 2y - \ 3z = 1 \\ -3x + 12y - 15z = 3 \\ \hline 14y - 18z = 4 \end{array}$

Thus we get the system

$\begin{cases} \text{Eq. 4:} & 7y - \ 9z = 2 \\ \text{Eq. 5:} & 14y - 18z = 4 \\ \text{Eq. 3:} & x - 4y + \ 5z = -1 \end{cases}$
Finally subtracting
$\begin{array}{ll} & 2 \times \text{Eq. 4} \\ - & \text{Eq. 5} \\ \hline & \text{Eq. 5} \end{array}$
$\begin{array}{l} 14y - 18z = 4 \\ -14y + 18z = -4 \\ \hline 0 = 0 \end{array}$

Which is always true. So, $14y - 18z = 4 \iff y = \frac{2}{7} + \frac{9}{7}z$; $x - 4\left(\frac{2}{7} + \frac{9}{7}z\right) + 5z = -1$ $\iff x = \frac{1}{7} + \frac{1}{7}z$; and so the solutions are $\left(\frac{1}{7} + \frac{1}{7}z, \frac{2}{7} + \frac{9}{7}z, z\right)$ where z is any real number.

(b) The system is dependent.

8. Let x, y, and z represent the price in dollars for a coffee, juice, and donuts respectively. Then, the system of equations is

$\begin{cases} 2x + y + 2z = 6.25 \quad \text{Eq. 1: Anne's order} \\ x \qquad + 3z = 3.75 \quad \text{Eq. 2: Barry's order} \\ 3x + y + 4z = 9.25 \quad \text{Eq. 3: Cathy's order} \end{cases}$

Subtracting:
$\begin{array}{ll} & \text{Eq. 1} \\ - & 2 \times \text{Eq. 2} \\ \hline & \text{Eq. 4} \end{array}$
$\begin{array}{l} 2x + y + 2z = 6.25 \\ -2x \qquad - \ 6z = -7.50 \\ \hline y - 4z = -1.25 \end{array}$

and
$\begin{array}{ll} & \text{Eq. 3} \\ - & 3 \times \text{Eq. 2} \\ \hline & \text{Eq. 4} \end{array}$
$\begin{array}{l} 3x + y + 4z = 9.25 \\ -3x \qquad - \ 9z = -11.25 \\ \hline y - 5z = -2.00 \end{array}$

Thus we get the system
$\begin{cases} y - 4z = -1.25 \\ x \qquad + 3z = 3.75 \\ y - 5z = -2.00 \end{cases}$
Finally subtracting

$\begin{array}{l} y - 4z = -1.25 \\ -y + 5z = 2.00 \\ \hline z = 0.75 \end{array}$
Then $y - 4(0.75) = -1.25 \iff y = 1.75$ and $x + 3(0.75) = 3.75 \iff$

$x = 1.50$. Thus coffees are \$1.50, juices are \$1.75, and donuts are 75¢.

9. $\begin{cases} 2x + y \le 8 \\ x - y \ge -2 \\ x + 2y \ge 4 \end{cases}$

From the graph, the points $(4, 0)$ and $(0, 2)$ are vertices.
The third vertex occurs where the lines $2x + y = 8$ and
$x - y = -2$ intersect. Adding these two equations
gives $3x = 6 \quad \Leftrightarrow \quad x = 2$, and so $y = 8 - 2(2) = 4$.
Thus the third vertex is $(2, 4)$.

10. $\begin{cases} x^2 - 5 + y \le 0 \\ y \le 5 + 2x \end{cases}$

Substituting $y = 5 + 2x$ into the first equation gives
$x^2 - 5 + (5 + 2x) = 0 \quad \Leftrightarrow \quad x^2 + 2x = 0 \quad \Leftrightarrow$
$x(x + 2) = 0 \quad \Leftrightarrow \quad x = 0$ or $x = -2$. If $x = 0$, then
$y = 5 + 2(0) = 5$, and if $x = -2$, then $y = 5 + 2(-2) = 1$.
Thus the vertices are $(0, 5)$ and $(-2, 1)$.

11. $\dfrac{4x - 1}{(x - 1)^2 (x + 2)} = \dfrac{A}{x - 1} + \dfrac{B}{(x - 1)^2} + \dfrac{C}{x + 2}$. Thus,

$4x - 1 = A(x - 1)(x + 2) + B(x + 2) + C(x - 1)^2$
$= A(x^2 + x - 2) + B(x + 2) + C(x^2 - 2x + 1)$
$= (A + C)x^2 + (A + B - 2C)x + (-2A + 2B + C)$, which leads to the following system of

equations:

$\begin{cases} A \qquad + \ C = 0 \\ A + \ B - 2C = 4 \\ -2A + 2B + \ C = -1 \end{cases}$ Subtracting: $\begin{array}{l} \text{–Eq. 1} \\ \text{Eq. 2} \\ \hline \text{Eq. 4} \end{array}$ $\begin{array}{l} -A \quad - \ C = 0 \\ A + B - 2C = 4 \\ \hline B - 3C = 4 \end{array}$ and adding

$\begin{array}{l} 2 \times \text{Eq. 1} \\ + \text{Eq. 3} \\ \hline \text{Eq. 5} \end{array}$ $\begin{array}{l} 2A \qquad + 2C = 0 \\ -2A + 2B + \ C = -1 \\ \hline 2B + 3C = -1 \end{array}$ Then adding the new equations $\begin{array}{l} B - 3C = 4 \\ 2B + 3C = -1 \\ \hline 3B \qquad = 3 \end{array}$

Therefore, $3B = 3 \quad \Leftrightarrow \quad B = 1$; so $1 - 3C = 4 \quad \Leftrightarrow \quad C = -1$; and $A + (-1) = 0 \quad \Leftrightarrow$

$A = 1$. Therefore, $\dfrac{4x - 1}{(x - 1)^2 (x + 2)} = \dfrac{1}{x - 1} + \dfrac{1}{(x - 1)^2} - \dfrac{1}{x + 2}$.

12. $\dfrac{2x - 3}{x^3 + 3x} = \dfrac{2x - 3}{x(x^2 + 3)} = \dfrac{A}{x} + \dfrac{Bx + C}{x^2 + 3}$. Then $2x - 3 = A(x^2 + 3) + (Bx + C)x$

$= Ax^2 + 3A + Bx^2 + Cx = (A + B)x^2 + Cx + 3A$. So $3A = -3 \quad \Leftrightarrow \quad A = -1, C = 2$ and

$A + B = 0$ gives us $B = 1$. Thus $\dfrac{2x - 3}{x^3 + x} = -\dfrac{1}{x} + \dfrac{x + 2}{x^2 + 3}$.

Focus on Modeling

1.

Vertex	$M = 200 - x - y$
$(0, 2)$	$200 - (0) - (2) = 198$ ← maximum value
$(0, 5)$	$200 - (0) - (5) = 195$ ← minimum value
$(4, 0)$	$200 - (4) - (0) = 196$

Thus the maximum value is 198, and the minimum value is 195.

3. $\begin{cases} x \geq 0, y \geq 0 \\ 2x + y \leq 10 \\ 2x + 4y \leq 28 \end{cases}$

The objective function is $P = 140 - x + 3y$.

From the graph, the vertices are $(0, 0)$, $(5, 0)$, $(2, 6)$, and $(0, 7)$.

Vertex	$P = 140 - x + 3y$
$(0, 0)$	$140 - (0) + 3(0) = 140$
$(5, 0)$	$140 - (5) + 3(0) = 135$ ← minimum value
$(2, 6)$	$140 - (2) + 3(6) = 156$
$(0, 7)$	$140 - (0) + 3(7) = 161$ ← maximum value

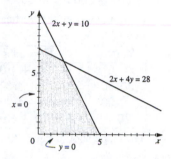

Thus the maximum value is 161, and the minimum value is 135.

5. Let t be the number of tables made daily and c be the number of chairs made daily. Then the data given can be summarized by the following table:

	Tables, t	Chairs, c	available time
carpentry	2 h	3 h	108 h
finishing	1 h	$\frac{1}{2}$ h	20 h
Profit	\$35	\$20	

Thus we wish to maximize the total profit, $P = 35t + 20c$, subject to the constraints:

$\begin{cases} 2t + 3c \leq 108 \\ t + \frac{1}{2}c \leq 20 \\ t \geq 0, c \geq 0. \end{cases}$

Thus we wish to maximize the total profit, $P = 35t + 20c$, subject to the constraints:

From the graph, the vertices occur at $(0, 0)$, $(20, 0)$, $(0, 36)$, and $(3, 34)$.

Vertex	$P = 35t + 20c$
$(0, 0)$	$35(0) + 20(0) = 0$
$(20, 0)$	$35(20) + 20(0) = 700$
$(0, 36)$	$35(0) + 20(36) = 720$
$(3, 34)$	$35(3) + 20(34) = 785$ ← maximum value

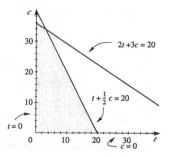

Hence, 3 tables and 34 chairs should be produced daily for a maximum profit of \$785.

7. Let x be the number of crates of oranges and y be the number of crates of grapefruit. Then the data given can be summarized by the following table:

	oranges	grapefruit	available
volume	4 ft^3	6 ft^3	300 ft^3
weight	80 lb	100 lb	5600 lb
Profit	$2.50	$4.00	

In addition, $x \geq y$. Thus we wish to maximize the total profit, $P = 2.5x + 4y$, subject to the constraints:

$$\begin{cases} x \geq 0, y \geq 0 \\ x \geq y \\ 4x + 6y \leq 300 \\ 80x + 100y \leq 5600. \end{cases}$$

From the graph, the vertices occur at $(0, 0)$, $(30, 30)$, $(45, 20)$, and $(70, 0)$.

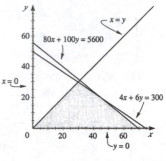

Vertex	$P = 2.5x + 4y$	
$(0,0)$	$2.5(0) + 4(0) = 0$	
$(30, 30)$	$2.5(30) + 4(30) = 195$	← maximum value
$(45, 20)$	$2.5(45) + 4(20) = 192.5$	
$(70, 0)$	$2.5(70) + 4(0) = 175$	

Thus, she should carry 30 crates of oranges and 30 crates of grapefruit for a maximum profit of $195.

9. Let x be the number of stereo sets shipped from Long Beach to Santa Monica and y be the number of stereo sets shipped from Long Beach to El Toro. Thus, $15 - x$ sets must be shipped to Santa Monica from Pasadena and $19 - y$ sets to El Toro from Pasadena. Thus $x \geq 0$, $y \geq 0$, $15 - x \geq 0$, $19 - y \geq 0$, $x + y \leq 24$, and $(15 - x) + (19 - y) \leq 18$. Simplifying, we get the inequalities (constraints):

$$\begin{cases} x \geq 0, y \geq 0 \\ x \leq 15, y \leq 19 \\ x + y \leq 24 \\ x + y \geq 16 \end{cases}$$

The objective function is the cost,
$C = 5x + 6y + 4(15 - x) + 5.5(19 - y) = x + 0.5y + 164.5$,
which we wish to minimize. From the graph, the vertices occur at $(0, 16)$, $(0, 19)$, $(5, 19)$, $(15, 9)$, and $(15, 1)$.

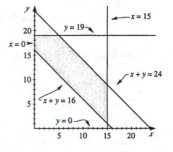

Vertex	$C = x + 0.5y + 164.5$	
$(0, 16)$	$(0) + 0.5(16) + 164.5 = 172.5$	← minimum value
$(0, 19)$	$(0) + 0.5(19) + 164.5 = 174$	
$(5, 19)$	$(5) + 0.5(19) + 164.5 = 179$	
$(15, 9)$	$(15) + 0.5(9) + 164.5 = 184$	
$(15, 1)$	$(15) + 0.5(1) + 164.5 = 180$	

The minimum cost is $172.50 and occurs when $x = 0$, $y = 16$. Hence, 0 sets should be shipped from Long Beach to Santa Monica, 16 from Long Beach to El Toro, 15 from Pasadena to Santa Monica, and 3 from Pasadena to El Toro.

11. Let x be the number of bags of standard mixtures and y be the number of bags of deluxe mixtures. Then the data can be summarized by the following table:

	standard	deluxe	available
cashews	100 g	150 g	15 kg
peanuts	200 g	50 g	20 kg
selling price	$1.95	$2.20	

Thus the total revenue, which we want to maximize, is given by $R = 1.95x + 2.25y$. We have the following constraints:

$$\begin{cases} x \geq 0, y \geq 0 \\ x \geq y \\ 0.1x + 0.15y \leq 15 \\ 0.2x + 0.05y \leq 20 \end{cases} \Leftrightarrow \begin{cases} x \geq 0, y \geq 0 \\ x \geq y \\ 10x + 15y \leq 1500 \\ 20x + 5y \leq 2000. \end{cases}$$

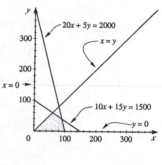

From the graph, the vertices occur at $(0,0)$, $(60,60)$, $(90,40)$, and $(100,0)$.

Vertex	$R = 1.95x + 2.25y$
$(0,0)$	$1.96(0) + 2.25(0) = 0$
$(60,60)$	$1.95(60) + 2.25(60) = 252$
$(90,40)$	$1.95(90) + 2.25(40) = 265.5$ ← maximum revenue
$(100,0)$	$1.95(100) + 2.25(0) = 195$

Hence, he should pack 90 bags of standard and 40 bags of deluxe mixture for a maximum revenue of $265.50.

13. Let x be the amount in municipal bonds and y be the amount in bank certificates, both in dollars. Then, $12000 - x - y$ is the amount in high-risk bonds. So our constraints can be stated as:

$$\begin{cases} x \geq 0, y \geq 0 \\ 12,000 - x - y \geq 0 \\ x \geq 3y \\ 12000 - x - y \leq 2000 \end{cases} \Leftrightarrow \begin{cases} x \geq 0, y \geq 0 \\ x + y \leq 12,000 \\ x \geq 3y \\ x + y \geq 10,000. \end{cases}$$

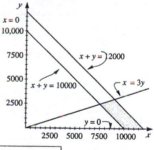

From the graph, the vertices occur at $(7500, 2500)$, $(10000, 0)$, $(12000, 0)$, and $(9000, 3000)$. The objective function is
$P = 0.07x + 0.08y + 0.12(12000 - x - y)$
$= 1440 - 0.05x - 0.04y$, which we wish to maximize.

Vertex	$P = 1440 - 0.05x - 0.04y$
$(7500, 2500)$	$1440 - 0.05(7500) - 0.04(2500) = 965$ ← maximum value
$(10000, 0)$	$1440 - 0.05(10000) - 0.04(0) = 940$
$(12000, 0)$	$1440 - 0.05(12000) - 0.04(0) = 840$
$(9000, 3000)$	$1440 - 0.05(9000) - 0.04(3000) = 870$

Hence, she should invest $7500 in municipal bonds, $2500 in bank certificates, and the remaining $2000 in high-risk bonds for a maximum yield of $965.

15. Let g be the number of games published and e be the number of educational programs published. Then the number of utility programs published is $36 - g - e$. Hence we wish to maximize profit,
$P = 5000g + 8000e + 6000(36 - g - e) = 216,000 - 1000g + 2000e$, subject to the constraints:

$$\begin{cases} g \geq 4,\, e \geq 0 \\ 36 - g - e \geq 0 \\ 36 - g - e \leq 2e \end{cases} \qquad \Leftrightarrow \qquad \begin{cases} g \geq 4,\, e \geq 0 \\ g + e \leq 36 \\ g + 3e \geq 36. \end{cases}$$

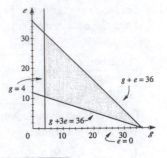

From the graph, the vertices are at $\left(4, \frac{32}{3}\right)$, $(4, 32)$, and $(36, 0)$. The objective function is $P = 216{,}000 - 1000g + 2000e$.

Vertex	$P = 216{,}000 - 1000g + 2000e$	
$\left(4, \frac{32}{3}\right)$	$216{,}000 - 1000(4) + 2000\left(\frac{32}{3}\right) = 233{,}333.33$	
$(4, 32)$	$216{,}000 - 1000(4) + 2000(32) = 276{,}000$	$\leftarrow$ maximum value
$(36, 0)$	$216{,}000 - 1000(36) + 2000(0) = 180{,}000$	

So, they should publish 4 games, 32 educational programs, and 0 utility programs for a maximum profit of \$276,000 annually.

Chapter Seven

Exercises 7.1

1. 3×2

3. 2×1

5. 1×3

7. (a) Yes, this matrix is in echelon form.

    (b) Yes, this matrix is in reduced echelon form.

    (c) $x = -3$
    $y = 5$

9. (a) Yes, this matrix is in echelon form.

    (b) No, this matrix is not in reduced echelon form, since the leading "one" in the second row does not have a zero above it.

    (c) $x + 2y + 8z = 0$
    $y + 3z = 2$
    $0 = 0$

11. (a) No, this matrix is not in echelon form, since the row of zeros is not at the bottom.

    (b) No, this matrix is not in reduced echelon form.

    (c) $x \qquad = 0$
    $0 = 0$
    $y + 5z = 1$

13. (a) Yes, this matrix is in echelon form.

    (b) Yes, this matrix is in reduced echelon form.

    (c) $x + 3y \quad - w = 0$ Notice that this system does not have a solution.
    $z + 2w = 0$
    $0 = 1$
    $0 = 0$

15.
$$\begin{bmatrix} 1 & -2 & 1 & 1 \\ 0 & 1 & 2 & 5 \\ 1 & 1 & 3 & 8 \end{bmatrix} \xrightarrow[R_3 - R_1 \to R_3]{} \begin{bmatrix} 1 & -2 & 1 & 1 \\ 0 & 1 & 2 & 5 \\ 0 & 3 & 2 & 7 \end{bmatrix} \xrightarrow[]{R_3 - 3R_2 \to R_3} \begin{bmatrix} 1 & -2 & 1 & 1 \\ 0 & 1 & 2 & 5 \\ 0 & 0 & -4 & -8 \end{bmatrix}$$
Thus, $-4z = -8 \quad \Leftrightarrow \quad z = 2; y + 2(2) = 5 \quad \Leftrightarrow \quad y = 1;$ and $x - 2(1) + (2) = 1 \quad \Leftrightarrow$
$x = 1.$ Therefore, the solution is $(1, 1, 2).$

17.
$$\begin{bmatrix} 1 & 1 & 1 & 2 \\ 2 & -3 & 2 & 4 \\ 4 & 1 & -3 & 1 \end{bmatrix} \xrightarrow[R_3 - 4R_1 \to R_3]{R_2 - 2R_1 \to R_2} \begin{bmatrix} 1 & 1 & 1 & 2 \\ 0 & -5 & 0 & 0 \\ 0 & -3 & -7 & -7 \end{bmatrix} \xrightarrow[]{R_3 - \frac{3}{5}R_2 \to R_3} \begin{bmatrix} 1 & 1 & 1 & 2 \\ 0 & -5 & 0 & 0 \\ 0 & 0 & -7 & -7 \end{bmatrix}$$

Thus, $-7z = -7$ $\Leftrightarrow$ $z = 1$; $-5y = 0$ $\Leftrightarrow$ $y = 0$; and $x + 0 + 1 = 2$ $\Leftrightarrow$ $x = 1$.
Therefore, the solution is $(1, 0, 1)$.

19. $\begin{bmatrix} 1 & 2 & -1 & -2 \\ 1 & 0 & 1 & 0 \\ 2 & -1 & -1 & -3 \end{bmatrix}$ $\xrightarrow[R_3-2R_1 \rightarrow R_3]{R_2-R_1 \rightarrow R_2}$ $\begin{bmatrix} 1 & 2 & -1 & -2 \\ 0 & -2 & 2 & 2 \\ 0 & -5 & 1 & 1 \end{bmatrix}$ $\xrightarrow{-\frac{1}{2}R_2}$

$\begin{bmatrix} 1 & 2 & -1 & -2 \\ 0 & 1 & 1 & 1 \\ 0 & -5 & 1 & 1 \end{bmatrix}$ $\xrightarrow{R_3+5R_2 \rightarrow R_3}$ $\begin{bmatrix} 1 & 2 & -1 & -2 \\ 0 & 1 & 1 & 1 \\ 0 & 0 & 6 & 6 \end{bmatrix}$

Thus, $6z = 6$ $\Leftrightarrow$ $z = 1$; $y + (1) = 1$ $\Leftrightarrow$ $y = 0$; and $x + 2(0) - (1) = -2$ $\Leftrightarrow$
$x = -1$. Therefore, the solution is $(-1, 0, 1)$.

21. $\begin{bmatrix} 1 & 2 & -1 & 9 \\ 2 & 0 & -1 & -2 \\ 3 & 5 & 2 & 22 \end{bmatrix}$ $\xrightarrow[R_3-3R_1 \rightarrow R_3]{R_2-2R_1 \rightarrow R_2}$ $\begin{bmatrix} 1 & 2 & -1 & 9 \\ 0 & -4 & 1 & -20 \\ 0 & -1 & 5 & -5 \end{bmatrix}$ $\xrightarrow{4R_3-R_2 \rightarrow R_3}$

$\begin{bmatrix} 1 & 2 & -1 & 9 \\ 0 & -4 & 1 & -20 \\ 0 & 0 & 19 & 0 \end{bmatrix}$

Thus, $19x_3 = 0$ $\Leftrightarrow$ $x_3 = 0$; $-4x_2 = -20$ $\Leftrightarrow$ $x_2 = 5$; and $x_1 + 2(5) = 9$ $\Leftrightarrow$
$x_1 = -1$. Therefore, the solution is $(-1, 5, 0)$.

23. $\begin{bmatrix} 2 & -3 & -1 & 13 \\ -1 & 2 & -5 & 6 \\ 5 & -1 & -1 & 49 \end{bmatrix}$ $\xrightarrow[2R_3-5R_1 \rightarrow R_3]{2R_2+R_1 \rightarrow R_2}$ $\begin{bmatrix} 2 & -3 & -1 & 13 \\ 0 & 1 & -11 & 25 \\ 0 & 13 & 3 & 33 \end{bmatrix}$ $\xrightarrow{R_3-13R_2 \rightarrow R_3}$

$\begin{bmatrix} 2 & -3 & -1 & 13 \\ 0 & 1 & -11 & 25 \\ 0 & 0 & 146 & -292 \end{bmatrix}$

Thus, $146z = -292$ $\Leftrightarrow$ $z = -2$; $y - 11(-2) = 25$ $\Leftrightarrow$ $y = 3$; and $2x - 3 \cdot 3 + 2 = 13$
$\Leftrightarrow$ $x = 10$. Therefore, the solution is $(10, 3, -2)$.

25. $\begin{bmatrix} 1 & 1 & 1 & 2 \\ 0 & 1 & -3 & 1 \\ 2 & 1 & 5 & 0 \end{bmatrix}$ $\xrightarrow{R_3-2R_1 \rightarrow R_3}$ $\begin{bmatrix} 1 & 1 & 1 & 2 \\ 0 & 1 & -3 & 1 \\ 0 & -1 & 3 & -4 \end{bmatrix}$ $\xrightarrow{R_3+R_2 \rightarrow R_3}$ $\begin{bmatrix} 1 & 1 & 1 & 3 \\ 0 & 1 & -3 & 1 \\ 0 & 0 & 0 & -3 \end{bmatrix}$
The third row of the matrix states $0 = -3$, which is impossible. Hence, the system is inconsistent,
and there is no solution.

27. $\begin{bmatrix} 2 & -3 & -9 & -5 \\ 1 & 0 & 3 & 2 \\ -3 & 1 & -4 & -3 \end{bmatrix}$ $\xrightarrow{R_1 \leftrightarrow R_2}$ $\begin{bmatrix} 1 & 0 & 3 & 2 \\ 2 & -3 & -9 & -5 \\ -3 & 1 & -4 & -3 \end{bmatrix}$ $\xrightarrow[R_3+3R_1 \rightarrow R_3]{R_2-2R_1 \rightarrow R_2}$

$\begin{bmatrix} 1 & 0 & 3 & 2 \\ 0 & -3 & -15 & -9 \\ 0 & 1 & 5 & 3 \end{bmatrix}$ $\xrightarrow{-\frac{1}{3}R_2}$ $\begin{bmatrix} 1 & 0 & 3 & 2 \\ 0 & 1 & 5 & 3 \\ 0 & 1 & 5 & 3 \end{bmatrix}$ $\xrightarrow{R_3-R_2 \rightarrow R_3}$

$\begin{bmatrix} 1 & 0 & 3 & 2 \\ 0 & 1 & 5 & 3 \\ 0 & 0 & 0 & 0 \end{bmatrix}$

Therefore, this system has infinitely many solutions, given by $x + 3t = 2 \quad \Leftrightarrow \quad x = 2 - 3t$, and $y + 5t = 3 \quad \Leftrightarrow \quad y = 3 - 5t$. Hence, the solutions are $(2 - 3t, 3 - 5t, t)$, where t is any real number.

29. $\begin{bmatrix} 1 & -1 & 3 & 3 \\ 4 & -8 & 32 & 24 \\ 2 & -3 & 11 & 4 \end{bmatrix} \xrightarrow[R_3 - 2R_1 \to R_3]{R_2 - 4R_1 \to R_2} \begin{bmatrix} 1 & -1 & 3 & 3 \\ 0 & -4 & 20 & 12 \\ 0 & -1 & 5 & -2 \end{bmatrix} \xrightarrow[R_3 + R_2 \to R_3]{-\frac{1}{4}R_2}$

$\begin{bmatrix} 1 & -1 & 3 & 3 \\ 0 & 1 & -5 & -3 \\ 0 & 0 & 0 & -5 \end{bmatrix}$

The third row of the matrix states $0 = -5$, which is impossible. Hence, the system is inconsistent, and there is no solution.

31. $\begin{bmatrix} 1 & 4 & -2 & -3 \\ 2 & -1 & 5 & 12 \\ 8 & 5 & 11 & 30 \end{bmatrix} \xrightarrow[R_3 - 8R_1 \to R_3]{R_2 - 2R_1 \to R_2} \begin{bmatrix} 1 & 4 & -2 & -3 \\ 0 & -9 & 9 & 18 \\ 0 & -27 & 27 & 54 \end{bmatrix} \xrightarrow{R_3 - 3R_2 \to R_3}$

$\begin{bmatrix} 1 & 4 & -2 & -3 \\ 0 & -9 & 9 & 18 \\ 0 & 0 & 0 & 0 \end{bmatrix}$

Therefore, this system has infinitely many solutions, given by $-9y + 9t = 18 \quad \Leftrightarrow \quad y = -2 + t$, and $x + 4(-2 + t) - 2t = -3 \quad \Leftrightarrow \quad x = 5 - 2t$. Hence, the solutions are $(5 - 2t, -2 + t, t)$, where t is any real number.

33. $\begin{bmatrix} 2 & 1 & -2 & 12 \\ -1 & -\frac{1}{2} & 1 & -6 \\ 3 & \frac{3}{2} & -3 & 18 \end{bmatrix} \xrightarrow[-R_1]{R_1 \leftrightarrow R_2} \begin{bmatrix} 1 & \frac{1}{2} & -1 & 6 \\ 2 & 1 & -2 & 12 \\ 3 & \frac{3}{2} & -3 & 18 \end{bmatrix} \xrightarrow[R_3 - 3R_1 \to R_3]{R_2 - 2R_1 \to R_2} \begin{bmatrix} 1 & \frac{1}{2} & -1 & 6 \\ 0 & 0 & 0 & 0 \\ 0 & 0 & 0 & 0 \end{bmatrix}$

Therefore, this system has infinitely many solutions, given by $x + \frac{1}{2}s - t = 6 \quad \Leftrightarrow$ $x = 6 - \frac{1}{2}s + t$. Hence, the solutions are $(6 - \frac{1}{2}s + t, s, t)$, where s and t are any real numbers.

35. $\begin{bmatrix} 4 & -3 & 1 & -8 \\ -2 & 1 & -3 & -4 \\ 1 & -1 & 2 & 3 \end{bmatrix} \xrightarrow{R_1 \leftrightarrow R_3} \begin{bmatrix} 1 & -1 & 2 & 3 \\ -2 & 1 & -3 & -4 \\ 4 & -3 & 1 & -8 \end{bmatrix} \xrightarrow[R_3 - 4R_1 \to R_3]{R_2 + 2R_1 \to R_2}$

$\begin{bmatrix} 1 & -1 & 2 & 3 \\ 0 & -1 & 1 & 2 \\ 0 & 1 & -7 & -20 \end{bmatrix} \xrightarrow{-R_2} \begin{bmatrix} 1 & -1 & 2 & 3 \\ 0 & 1 & -1 & -2 \\ 0 & 1 & -7 & -20 \end{bmatrix} \xrightarrow{R_3 - R_2 \to R_3} \begin{bmatrix} 1 & -1 & 2 & 3 \\ 0 & 1 & -1 & -2 \\ 0 & 0 & -6 & -18 \end{bmatrix}$

Therefore, $-6z = -18 \quad \Leftrightarrow \quad z = 3; y - (3) = -2 \quad \Leftrightarrow \quad y = 1;$ and $x - (1) + 2(3) = 3 \quad \Leftrightarrow$ $x = -2$. Hence, the solution is $(-2, 1, 3)$.

37. $\begin{bmatrix} 1 & 2 & -3 & -5 \\ -2 & -4 & -6 & 10 \\ 3 & 7 & -2 & -13 \end{bmatrix} \xrightarrow[R_3 - 3R_1 \to R_3]{R_2 + 2R_1 \to R_2} \begin{bmatrix} 1 & 2 & -3 & -5 \\ 0 & 0 & -12 & 0 \\ 0 & 1 & 7 & 2 \end{bmatrix} \xrightarrow{R_2 \leftrightarrow R_3}$

$\begin{bmatrix} 1 & 2 & -3 & -5 \\ 0 & 1 & 7 & 2 \\ 0 & 0 & -12 & 0 \end{bmatrix}$

Therefore, $-12z = 0 \quad \Leftrightarrow \quad z = 0; y + 7(0) = 2 \quad \Leftrightarrow \quad y = 2;$ and $x + 2(2) - 3(0) = -5$ $\Leftrightarrow \quad x = -9$. Hence, the solution is $(-9, 2, 0)$.

39. $\begin{bmatrix} -1 & 2 & 1 & -3 & 3 \\ 3 & -4 & 1 & 1 & 9 \\ -1 & -1 & 1 & 1 & 0 \\ 2 & 1 & 4 & -2 & 3 \end{bmatrix} \xrightarrow{-R_1} \begin{bmatrix} 1 & -2 & -1 & 3 & -3 \\ 3 & -4 & 1 & 1 & 9 \\ -1 & -1 & 1 & 1 & 0 \\ 2 & 1 & 4 & -2 & 3 \end{bmatrix}$

$\xrightarrow[\substack{R_2-3R_1 \to R_2,\ R_3+R_1 \to R_3 \\ R_4-2R_1 \to R_4}]{} \begin{bmatrix} 1 & -2 & -1 & 3 & -3 \\ 0 & 2 & 4 & -8 & 18 \\ 0 & -3 & 0 & 4 & -3 \\ 0 & 5 & 6 & -8 & 9 \end{bmatrix} \xrightarrow{\frac{1}{2}R_2}$

$\begin{bmatrix} 1 & -2 & -1 & 3 & -3 \\ 0 & 1 & 2 & -4 & 9 \\ 0 & -3 & 0 & 4 & -3 \\ 0 & 5 & 6 & -8 & 9 \end{bmatrix} \xrightarrow[\substack{R_3+3R_2 \to R_3 \\ R_4-5R_2 \to R_4}]{} \begin{bmatrix} 1 & -2 & -1 & 3 & -3 \\ 0 & 1 & 2 & -4 & 9 \\ 0 & 0 & 6 & -8 & 24 \\ 0 & 0 & -4 & 12 & -36 \end{bmatrix}$

$\xrightarrow{3R_4+2R_3 \to R_4} \begin{bmatrix} 1 & -2 & -1 & 3 & -3 \\ 0 & 1 & 2 & -4 & 9 \\ 0 & 0 & 6 & -8 & 24 \\ 0 & 0 & 0 & 20 & -60 \end{bmatrix}$

Therefore, $20w = -60 \Leftrightarrow w = -3; 6z + 24 = 24 \Leftrightarrow z = 0.$ Then $y + 12 = 9 \Leftrightarrow y = -3$ and $x + 6 - 9 = -3 \Leftrightarrow x = 0.$ Hence, the solution is $(0, -3, 0, -3).$

41. $\begin{bmatrix} 1 & 1 & 2 & -1 & -2 \\ 0 & 3 & 1 & 2 & 2 \\ 1 & 1 & 0 & 3 & 2 \\ -3 & 0 & 1 & 2 & 5 \end{bmatrix} \xrightarrow[\substack{R_3-R_1 \to R_3 \\ R_4+3R_1 \to R_4}]{} \begin{bmatrix} 1 & 1 & 2 & -1 & -2 \\ 0 & 3 & 1 & 2 & 2 \\ 0 & 0 & -2 & 4 & 4 \\ 0 & 3 & 7 & -1 & -1 \end{bmatrix} \xrightarrow{R_4-R_2 \to R_4}$

$\begin{bmatrix} 1 & 1 & 2 & -1 & -2 \\ 0 & 3 & 1 & 2 & 2 \\ 0 & 0 & -2 & 4 & 4 \\ 0 & 0 & 6 & -3 & -3 \end{bmatrix} \xrightarrow{R_4+3R_3 \to R_4} \begin{bmatrix} 1 & 1 & 2 & -1 & -2 \\ 0 & 3 & 1 & 2 & 2 \\ 0 & 0 & -2 & 4 & 4 \\ 0 & 0 & 0 & 9 & 9 \end{bmatrix}$

Therefore, $9w = 9 \Leftrightarrow w = 1; -2z + 4(1) = 4 \Leftrightarrow z = 0.$ Then $3y + (0) + 2(1) = 2 \Leftrightarrow y = 0$ and $x + (0) + 2(0) - (1) = -2 \Leftrightarrow x = -1.$ Hence, the solution is $(-1, 0, 0, 1).$

43. $\begin{bmatrix} 1 & 0 & 1 & 1 & 4 \\ 0 & 1 & -1 & 0 & -4 \\ 1 & -2 & 3 & 1 & 12 \\ 2 & 0 & -2 & 5 & -1 \end{bmatrix} \xrightarrow[\substack{R_3-R_1 \to R_3 \\ R_4-2R_1 \to R_4}]{} \begin{bmatrix} 1 & 0 & 1 & 1 & 4 \\ 0 & 1 & -1 & 0 & -4 \\ 0 & -2 & 2 & 0 & 8 \\ 0 & 0 & -4 & 3 & -9 \end{bmatrix} \xrightarrow{R_3+2R_2 \to R_3}$

$\begin{bmatrix} 1 & 0 & 1 & 1 & 4 \\ 0 & 1 & -1 & 0 & -4 \\ 0 & 0 & 0 & 0 & 0 \\ 0 & 0 & -4 & 3 & -9 \end{bmatrix} \xrightarrow{R_3 \leftrightarrow -R_4} \begin{bmatrix} 1 & 0 & 1 & 1 & 4 \\ 0 & 1 & -1 & 0 & -4 \\ 0 & 0 & 4 & -3 & 9 \\ 0 & 0 & 0 & 0 & 0 \end{bmatrix}$

Therefore, $4z - 3t = 9 \Leftrightarrow 4z = 9 + 3t \Leftrightarrow z = \frac{9}{4} + \frac{3}{4}t.$ Then we have
$y - \left(\frac{9}{4} + \frac{3}{4}t\right) = -4 \Leftrightarrow y = \frac{-7}{4} + \frac{3}{4}t$ and $x + \left(\frac{9}{4} + \frac{3}{4}t\right) + t = 4 \Leftrightarrow x = \frac{7}{4} - \frac{7}{4}t.$ Hence, the solutions are $\left(\frac{7}{4} - \frac{7}{4}t, \frac{-7}{4} + \frac{3}{4}t, \frac{9}{4} + \frac{3}{4}t, t\right),$ where t is any number.

45. $\begin{bmatrix} 1 & -1 & 0 & 1 & 0 \\ 3 & 0 & -1 & 2 & 0 \\ 1 & -4 & 1 & 2 & 0 \end{bmatrix} \xrightarrow[\substack{R_2-3R_1 \to R_2 \\ R_3-R_1 \to R_3}]{} \begin{bmatrix} 1 & -1 & 0 & 1 & 0 \\ 0 & 3 & -1 & -1 & 0 \\ 0 & -3 & 1 & 1 & 0 \end{bmatrix} \xrightarrow{R_3+R_2 \to R_3}$

$$\begin{bmatrix} 1 & -1 & 0 & 1 & 0 \\ 0 & 3 & -1 & -1 & 0 \\ 0 & 0 & 0 & 0 & 0 \end{bmatrix}$$

Therefore, the system has infinitely many solutions, given by $3y - s - t = 0 \quad \Leftrightarrow \quad y = \frac{1}{3}(s+t)$ and $x - \frac{1}{3}(s+t) + t = 0 \quad \Leftrightarrow \quad x = \frac{1}{3}(s - 2t)$. So the solutions are $\left(\frac{1}{3}[s - 2t], \frac{1}{3}[s+t], s, t\right)$, where s and t are any real numbers.

47. Let x, y, z represent the number of VitaMax, Vitron, and VitaPlus pills taken daily. The matrix representation for the system of equations is:

$$\begin{bmatrix} 5 & 10 & 15 & 50 \\ 15 & 20 & 0 & 50 \\ 10 & 10 & 10 & 50 \end{bmatrix} \xrightarrow[\frac{1}{5}R_3]{\frac{1}{5}R_1, \frac{1}{5}R_2} \begin{bmatrix} 1 & 2 & 3 & 10 \\ 3 & 4 & 0 & 10 \\ 2 & 2 & 2 & 10 \end{bmatrix} \xrightarrow[R_3-2R_1 \to R_3]{R_2-3R_1 \to R_2} \begin{bmatrix} 1 & 2 & 3 & 10 \\ 0 & -2 & -9 & -20 \\ 0 & -2 & -4 & -10 \end{bmatrix}$$

$$\xrightarrow{R_3-R_2 \to R_3} \begin{bmatrix} 1 & 2 & 3 & 10 \\ 0 & -2 & -9 & -20 \\ 0 & 0 & 5 & 10 \end{bmatrix}$$

Thus, $5z = 10 \quad \Leftrightarrow \quad z = 2; \; -2y - 18 = -20 \quad \Leftrightarrow \quad y = 1$; and $x + 2 + 6 = 10 \quad \Leftrightarrow \quad x = 2$. Hence, he should take 2 VitaMax, 1 Vitron, and 2 VitaPlus pills daily.

49. Let x, y, and z represent the distance, in miles, of the run, swim, and cycle parts of the race respectively. Then, since $time = \frac{distance}{speed}$, we get the following equations from the three contestants' race times:

$$\begin{cases} \left(\frac{x}{10}\right) + \left(\frac{y}{4}\right) + \left(\frac{z}{20}\right) = 2.5 \\ \left(\frac{x}{7.5}\right) + \left(\frac{y}{6}\right) + \left(\frac{z}{15}\right) = 3 \\ \left(\frac{x}{15}\right) + \left(\frac{y}{3}\right) + \left(\frac{z}{40}\right) = 1.75 \end{cases} \Leftrightarrow \begin{cases} 2x + 5y + z = 50 \\ 4x + 5y + 2z = 90 \\ 8x + 40y + 3z = 210 \end{cases},$$

which has the following matrix representation:

$$\begin{bmatrix} 2 & 5 & 1 & 50 \\ 4 & 5 & 2 & 90 \\ 8 & 40 & 3 & 210 \end{bmatrix} \xrightarrow[R_3-4R_1 \to R_3]{R_2-2R_1 \to R_2} \begin{bmatrix} 2 & 5 & 1 & 50 \\ 0 & -5 & 0 & -10 \\ 0 & 20 & -1 & 10 \end{bmatrix} \xrightarrow{R_3+4R_2 \to R_3}$$

$$\begin{bmatrix} 2 & 5 & 1 & 50 \\ 0 & -5 & 0 & -10 \\ 0 & 0 & -1 & -30 \end{bmatrix}.$$

Thus, $-z = -30 \quad \Leftrightarrow \quad z = 30; \; -5y = -10 \quad \Leftrightarrow \quad y = 2$; and $2x + 10 + 30 = 50 \quad \Leftrightarrow \quad x = 5$. So the race has a 5 mile run, 2 mile swim, and 30 mile cycle.

51. Let t be the number of tables produced, c the number of chairs, and a the number of armoires. Then, the system of equations is

$$\begin{cases} \frac{1}{2}t + c + a = 300 \\ \frac{1}{2}t + \frac{3}{2}c + a = 400 \\ t + \frac{3}{2}c + 2a = 590 \end{cases} \Leftrightarrow \begin{cases} t + 2c + 2a = 600 \\ t + 3c + 2a = 800 \\ 2t + 3c + 4a = 1180 \end{cases}$$ and the matrix representation is:

$$\begin{bmatrix} 1 & 2 & 2 & 600 \\ 1 & 3 & 2 & 800 \\ 2 & 3 & 4 & 1180 \end{bmatrix} \xrightarrow[R_3-2R_1 \to R_3]{R_2-R_1 \to R_2} \begin{bmatrix} 1 & 2 & 2 & 600 \\ 0 & 1 & 0 & 200 \\ 0 & -1 & 0 & -20 \end{bmatrix} \xrightarrow{R_3+R_2 \to R_3} \begin{bmatrix} 1 & 2 & 2 & 600 \\ 0 & 1 & 0 & 200 \\ 0 & 0 & 0 & 180 \end{bmatrix}.$$

The third row states $0 = 180$, which is impossible, and so the system is inconsistent. Therefore, it is impossible to use all of the available labor-hours.

53. Line containing the points $(0, 0)$, $(1, 12)$:

Using the general form of a line, $y = ax + b$, we substitute for x and y and solve for a and b. This gives $(0, 0)$ $0 = a(0) + b$ $\Rightarrow$ $b = 0$;
      $(1, 12)$ $12 = a(1) + b$ $\Rightarrow$ $a = 12$.

Since $a = 12$ and $b = 0$, the equation of the line is $y = 12x$.

Quadratic containing the points $(0, 0)$, $(1, 12)$, $(3, 6)$:

Using the general form of a quadratic, $y = ax^2 + bx + c$, we substitute for x and y and solve for a, b, and c. This gives

$(0,0)$ $0 = a(0)^2 + b(0) + c$ $\Rightarrow$ $c = 0$;
$(1,12)$ $12 = a(1)^2 + b(1) + c$ $\Rightarrow$ $a + b = 12$ $\times -3$ $-3a - 3b = -36$
$(3,6)$ $6 = a(3)^2 + b(3) + c$ $\Rightarrow$ $9a + 3b = 6$ $\underline{9a + 3b = 6}$
                                                                                              $6a = -30$ $\Leftrightarrow$

$a = -5$. So $a + b = 12$ $\Leftrightarrow$ $b = 12 - a$ $\Rightarrow$ $b = 17$. Since $a = -5$, $b = 17$, and $c = 0$, the equation of the quadratic is $y = -5x^2 + 17x$.

Cubic containing the points $(0, 0)$, $(1, 12)$, $(2, 40)$, $(3, 6)$:

Using the general form of a cubic, $y = ax^3 + bx^2 + cx + d$, we substitute for x and y and solve for a, b, c, and d. This gives

$(0,0)$ $0 = a(0)^3 + b(0)^2 + c(0) + d$ $\Rightarrow$ $d = 0$;
$(1,12)$ $12 = a(1)^3 + b(1)^2 + c(1) + d$ $\Rightarrow$ $a + b + c + d = 12$;
$(2,40)$ $40 = a(2)^3 + b(2)^2 + c(2) + d$ $\Rightarrow$ $8a + 4b + 2c + d = 40$;
$(3,6)$ $6 = a(3)^3 + b(3)^2 + c(3) + d$ $\Rightarrow$ $27a + 9b + 3c + d = 6$.

Since $d = 0$ the system reduces to $\begin{cases} a + b + c = 12 \\ 8a + 4b + 2c = 40 \\ 27a + 9b + 3c = 6 \end{cases}$

$\begin{bmatrix} 1 & 1 & 1 & 12 \\ 8 & 4 & 2 & 40 \\ 27 & 9 & 3 & 6 \end{bmatrix}$ $\xrightarrow[R_3 - 27R_1 \to R_3]{R_2 - 8R_1 \to R_2}$ $\begin{bmatrix} 1 & 1 & 1 & 12 \\ 0 & -4 & -6 & -56 \\ 0 & -18 & -24 & -318 \end{bmatrix}$ $\begin{matrix} -\frac{1}{2}R_2 \\ -\frac{1}{6}R_3 \end{matrix}$

$\begin{bmatrix} 1 & 1 & 1 & 12 \\ 0 & 2 & 3 & 28 \\ 0 & 3 & 4 & 53 \end{bmatrix}$ $\xrightarrow{2R_3 - 3R_2 \to R_3}$ $\begin{bmatrix} 1 & 1 & 1 & 12 \\ 0 & 2 & 3 & 28 \\ 0 & 0 & -1 & 22 \end{bmatrix}$

So $c = -22$ and back-substituting we have $2b + 3(-22) = 28$ $\Leftrightarrow$ $b = 47$ and $a + 47 + (-22) = 0$ $\Leftrightarrow$ $a = -13$. So the cubic is $y = -13x^3 + 47x^2 - 22x$

Fourth-degree polynomial containing the points $(0, 0)$, $(1, 12)$, $(2, 40)$, $(3, 6)$, $(-1, -14)$:

Using the general form of a fourth-degree polynomial, $y = ax^4 + bx^3 + cx^2 + dx + e$, we substitute for x and y and solve for a, b, c, d and d. This gives

$(0,0)$ $\qquad 0 = a(0)^4 + b(0)^3 + c(0)^2 + d(0) + e \quad \Rightarrow \quad e = 0$

$(1,12)$ $\qquad 12 = a(1)^4 + b(1)^3 + c(1)^2 + d(1) + e$

$(2,40)$ $\qquad 40 = a(2)^4 + b(2)^3 + c(2)^2 + d(2) + e$

$(3,6)$ $\qquad 6 = a(3)^4 + b(3)^3 + c(3)^2 + d(3) + e$

$(-1,-14)$ $\quad -14 = a(-1)^4 + b(-1)^3 + c(-1)^2 + d(-1) + e$

Since the first equation is $e = 0$, we eliminate e from the other equations to get the following system:

$$\begin{cases} a + b + c + d = 12 \\ 16a + 8b + 4c + 2d = 40 \\ 81a + 27b + 9c + 3d = 6 \\ a - b + c - d = -14 \end{cases}$$

$$\begin{bmatrix} 1 & 1 & 1 & 1 & 12 \\ 16 & 8 & 4 & 2 & 40 \\ 81 & 27 & 9 & 3 & 6 \\ 1 & -1 & 1 & -1 & -14 \end{bmatrix} \xrightarrow[\;R_4 - R_1 \to R_4\;]{R_2 - 16R_1 \to R_2,\; R_3 - 81R_1 \to R_3} \begin{bmatrix} 1 & 1 & 1 & 1 & 12 \\ 0 & -8 & -12 & -14 & -152 \\ 0 & -54 & -72 & -78 & -966 \\ 0 & -2 & 0 & -2 & -26 \end{bmatrix}$$

$$\xrightarrow[\;R_2 \to R_3, R_3 \to R_4\;]{-\frac{1}{2}R_4 \to R_2} \begin{bmatrix} 1 & 1 & 1 & 1 & 12 \\ 0 & 1 & 0 & 1 & 13 \\ 0 & -8 & -12 & -14 & -152 \\ 0 & -54 & -72 & -78 & -966 \end{bmatrix} \xrightarrow[\;R_4 + 54R_2 \to R_4\;]{R_3 + 8R_2 \to R_3} \begin{bmatrix} 1 & 1 & 1 & 1 & 12 \\ 0 & 1 & 0 & 1 & 13 \\ 0 & 0 & -12 & -6 & -48 \\ 0 & 0 & -72 & -24 & -264 \end{bmatrix}$$

$$\xrightarrow[\;R_4 - 6R_3 \to R_4\;]{} \begin{bmatrix} 1 & 1 & 1 & 1 & 12 \\ 0 & 1 & 0 & 1 & 13 \\ 0 & 0 & -12 & -6 & -48 \\ 0 & 0 & 0 & 12 & 24 \end{bmatrix}$$

So $d = 2$. Then $-12c - 6(2) = -48 \quad \Leftrightarrow \quad c = 3$ and $b + 2 = 13 \quad \Leftrightarrow \quad b = 11$. Finally, $a + 11 + 3 + 2 = 12 \quad \Leftrightarrow \quad a = -4$. So the fourth-degree polynomial containing these points is $y = -4x^4 + 11x^3 + 3x^2 + 2x$.

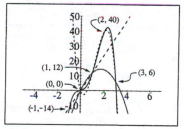

Exercises 7.2

1. The matrices have different dimensions, they cannot be equal.

3. $\begin{bmatrix} 2 & 6 \\ -5 & 3 \end{bmatrix} + \begin{bmatrix} -1 & -3 \\ 6 & 2 \end{bmatrix} = \begin{bmatrix} 1 & 3 \\ 1 & 5 \end{bmatrix}$

5. $3\begin{bmatrix} 1 & 2 \\ 4 & -1 \\ 1 & 0 \end{bmatrix} = \begin{bmatrix} 3 & 6 \\ 12 & -3 \\ 3 & 0 \end{bmatrix}$

7. $\begin{bmatrix} 2 & 6 \\ 1 & 3 \\ 2 & 4 \end{bmatrix}\begin{bmatrix} 1 & -2 \\ 3 & 6 \\ -2 & 0 \end{bmatrix}$ is undefined because these matrices have incompatible dimensions.

9. $\begin{bmatrix} 1 & 2 \\ -1 & 4 \end{bmatrix}\begin{bmatrix} 1 & -2 & 3 \\ 2 & 2 & -1 \end{bmatrix} = \begin{bmatrix} 5 & 2 & 1 \\ 7 & 10 & -7 \end{bmatrix}$

11. $2X + A = B \quad \Leftrightarrow \quad X = \frac{1}{2}(B - A) = \frac{1}{2}\left(\begin{bmatrix} 2 & 5 \\ 3 & 7 \end{bmatrix} - \begin{bmatrix} 4 & 6 \\ 1 & 3 \end{bmatrix} \right) = \frac{1}{2}\begin{bmatrix} -2 & -1 \\ 2 & 4 \end{bmatrix} = \begin{bmatrix} -1 & -\frac{1}{2} \\ 1 & 2 \end{bmatrix}.$

13. $2(B - X) = D$. Since B is a 2×2 matrix, $B - X$ is only defined when X is a 2×2 matrix, so $2(B - X)$ is a 2×2 matrix. But D is a 3×2 matrix. Thus no solution exists.

15. $\frac{1}{5}(X + D) = C \quad \Leftrightarrow \quad X + D = 5C \quad \Leftrightarrow \quad X = 5C - D = 5\begin{bmatrix} 2 & 3 \\ 1 & 0 \\ 0 & 2 \end{bmatrix} - \begin{bmatrix} 10 & 20 \\ 30 & 20 \\ 10 & 0 \end{bmatrix}$

$= \begin{bmatrix} 10 & 15 \\ 5 & 0 \\ 0 & 10 \end{bmatrix} - \begin{bmatrix} 10 & 20 \\ 30 & 20 \\ 10 & 0 \end{bmatrix} = \begin{bmatrix} 0 & -5 \\ -25 & -20 \\ -10 & 10 \end{bmatrix}$

In Exercises 17-- 37, the matrices $A, B, C, D, E, F,$ and G are defined as follows:

$$A = \begin{bmatrix} 2 & -5 \\ 0 & 7 \end{bmatrix} \quad B = \begin{bmatrix} 3 & \frac{1}{2} & 5 \\ 1 & -1 & 3 \end{bmatrix} \quad C = \begin{bmatrix} 2 & -\frac{5}{2} & 0 \\ 0 & 2 & -3 \end{bmatrix} \quad D = [7 \quad 3]$$

$$E = \begin{bmatrix} 1 \\ 2 \\ 0 \end{bmatrix} \quad F = \begin{bmatrix} 1 & 0 & 0 \\ 0 & 1 & 0 \\ 0 & 0 & 1 \end{bmatrix} \quad G = \begin{bmatrix} 5 & -3 & 10 \\ 6 & 1 & 0 \\ -5 & 2 & 2 \end{bmatrix}$$

17. $B + C = \begin{bmatrix} 3 & \frac{1}{2} & 5 \\ 1 & -1 & 3 \end{bmatrix} + \begin{bmatrix} 2 & -\frac{5}{2} & 0 \\ 0 & 2 & -3 \end{bmatrix} = \begin{bmatrix} 5 & -2 & 5 \\ 1 & 1 & 0 \end{bmatrix}$

19. $C - B = \begin{bmatrix} 2 & -\frac{5}{2} & 0 \\ 0 & 2 & -3 \end{bmatrix} - \begin{bmatrix} 3 & \frac{1}{2} & 5 \\ 1 & -1 & 3 \end{bmatrix} = \begin{bmatrix} -1 & -3 & -5 \\ -1 & 3 & -6 \end{bmatrix}$

21. $3B + 2C = 3\begin{bmatrix} 3 & \frac{1}{2} & 5 \\ 1 & -1 & 3 \end{bmatrix} + 2\begin{bmatrix} 2 & -\frac{5}{2} & 0 \\ 0 & 2 & -3 \end{bmatrix} = \begin{bmatrix} 13 & -\frac{7}{2} & 15 \\ 3 & 1 & 3 \end{bmatrix}$

23. $2C - 6B = 2\begin{bmatrix} 2 & -\frac{5}{2} & 0 \\ 0 & 2 & -3 \end{bmatrix} - 6\begin{bmatrix} 3 & \frac{1}{2} & 5 \\ 1 & -1 & 3 \end{bmatrix} = \begin{bmatrix} -14 & -8 & -30 \\ -6 & 10 & -24 \end{bmatrix}$

25. AD is undefined because A (2×2) and D (1×2) have incompatible dimensions.

27. $BF = \begin{bmatrix} 3 & \frac{1}{2} & 5 \\ 1 & -1 & 3 \end{bmatrix} \begin{bmatrix} 1 & 0 & 0 \\ 0 & 1 & 0 \\ 0 & 0 & 1 \end{bmatrix} = \begin{bmatrix} 3 & \frac{1}{2} & 5 \\ 1 & -1 & 3 \end{bmatrix}$

29. $(DA)B = [7 \;\; 3]\begin{bmatrix} 2 & -5 \\ 0 & 7 \end{bmatrix}\begin{bmatrix} 3 & \frac{1}{2} & 5 \\ 1 & -1 & 3 \end{bmatrix} = [14 \;\; -14]\begin{bmatrix} 3 & \frac{1}{2} & 5 \\ 1 & -1 & 3 \end{bmatrix} = [28 \;\; 21 \;\; 28]$

31. $GE = \begin{bmatrix} 5 & -3 & 10 \\ 6 & 1 & 0 \\ -5 & 2 & 2 \end{bmatrix}\begin{bmatrix} 1 \\ 2 \\ 0 \end{bmatrix} = \begin{bmatrix} -1 \\ 8 \\ -1 \end{bmatrix}$

33. $A^3 = \begin{bmatrix} 2 & -5 \\ 0 & 7 \end{bmatrix}\begin{bmatrix} 2 & -5 \\ 0 & 7 \end{bmatrix}\begin{bmatrix} 2 & -5 \\ 0 & 7 \end{bmatrix} = \begin{bmatrix} 4 & -45 \\ 0 & 49 \end{bmatrix}\begin{bmatrix} 2 & -5 \\ 0 & 7 \end{bmatrix} = \begin{bmatrix} 8 & -335 \\ 0 & 343 \end{bmatrix}$

35. B^2 is undefined because the dimensions (2×3 and 2×3) are incompatible.

37. $BF + FE$ is undefined because the dimensions, $(2 \times 3) \cdot (3 \times 3) = (2 \times 3)$ and $(3 \times 3) \cdot (3 \times 1)$ $= (3 \times 1)$, are incompatible.

39. $\begin{bmatrix} x & 2y \\ 4 & 6 \end{bmatrix} = \begin{bmatrix} 2 & -2 \\ 2x & -6y \end{bmatrix}$. Thus we must solve the system $\begin{cases} x = 2 \\ 2y = -2 \\ 4 = 2x \\ 6 = -6y \end{cases}$ So $x = 2$ and $2y = -2$

$\Leftrightarrow$ $y = -1$. Since these values for x and y also satisfy the last two equations, the solution is $x = 2, y = -1$.

41. $2\begin{bmatrix} x & y \\ x + y & x - y \end{bmatrix} = \begin{bmatrix} 2 & -4 \\ -2 & 6 \end{bmatrix}$. Since $2\begin{bmatrix} x & y \\ x + y & x - y \end{bmatrix} = \begin{bmatrix} 2x & 2y \\ 2(x + y) & 2(x - y) \end{bmatrix}$. Thus we

must solve the system $\begin{cases} 2x = 2 \\ 2y = -4 \\ 2(x + y) = -2 \\ 2(x - y) = 6 \end{cases}$ So $x = 1$ and $y = -2$. Since these values for x and y

also satisfy the last two equations, the solution is $x = 1, y = -2$.

43. $\begin{cases} 2x - 5y = 7 \\ 3x + 2y = 4 \end{cases}$ written as a matrix equation is $\begin{bmatrix} 2 & -5 \\ 3 & 2 \end{bmatrix}\begin{bmatrix} x \\ y \end{bmatrix} = \begin{bmatrix} 7 \\ 4 \end{bmatrix}$.

45. $\begin{cases} 3x_1 + 2x_2 - x_3 + x_4 = 0 \\ x_1 - x_3 = 5 \\ 3x_2 + x_3 - x_4 = 4 \end{cases}$ written as a matrix equation is $\begin{bmatrix} 3 & 2 & -1 & 1 \\ 1 & 0 & -1 & 0 \\ 0 & 3 & 1 & -1 \end{bmatrix}\begin{bmatrix} x_1 \\ x_2 \\ x_3 \\ x_4 \end{bmatrix} = \begin{bmatrix} 0 \\ 5 \\ 4 \end{bmatrix}$.

47. $A = \begin{bmatrix} 1 & 0 & 6 & -1 \\ 2 & \frac{1}{2} & 4 & 0 \end{bmatrix}$, $B = [1 \quad 7 \quad -9 \quad 2]$, and $C = \begin{bmatrix} 1 \\ 0 \\ -1 \\ -2 \end{bmatrix}$.

ABC is undefined because the dimensions of A (2×4) and B (1×4) are not compatible.

$ACB = \begin{bmatrix} -3 \\ -2 \end{bmatrix} [1 \quad 7 \quad -9 \quad 2] = \begin{bmatrix} -3 & -21 & 27 & -6 \\ -2 & -14 & 18 & -4 \end{bmatrix}$

BAC is undefined because the dimensions of B (1×4) and A (2×4) are not compatible.

BCA is undefined because the dimensions of C (4×1) and A (2×4) are not compatible.

CAB is undefined because the dimensions of C (4×1) and A (2×4) are not compatible.

CBA is undefined because the dimensions of B (1×4) and A (2×4) are not compatible.

49. (a) $BA = [\$0.90 \quad \$0.80 \quad \$1.10] \begin{bmatrix} 4000 & 1000 & 3500 \\ 400 & 300 & 200 \\ 700 & 500 & 9000 \end{bmatrix} = [\$4690 \quad \$1690 \quad \$13,210]$

(b) The entries in the product matrix represent the total food sales in Santa Monica, Long Beach, and Anaheim, respectively.

51. (a) $AB = [6 \quad 10 \quad 14 \quad 28] \begin{bmatrix} 2000 & 2500 \\ 3000 & 1500 \\ 2500 & 1000 \\ 1000 & 500 \end{bmatrix} = [105,000 \quad 58,000]$

(b) That day they canned 105,000 ounces of tomato sauce and 58,000 ounces of tomato paste.

53. (a) $\begin{bmatrix} 1 & 0 & 1 & 0 & 1 & 1 \\ 0 & 3 & 0 & 1 & 2 & 1 \\ 1 & 2 & 0 & 0 & 3 & 0 \\ 1 & 3 & 2 & 3 & 2 & 0 \\ 0 & 3 & 0 & 0 & 2 & 1 \\ 1 & 2 & 0 & 1 & 3 & 1 \end{bmatrix}$

(b) $\begin{bmatrix} 2 & 1 & 2 & 1 & 2 & 2 \\ 1 & 3 & 1 & 2 & 3 & 2 \\ 2 & 3 & 1 & 1 & 3 & 1 \\ 2 & 3 & 3 & 3 & 3 & 1 \\ 1 & 3 & 1 & 1 & 3 & 2 \\ 2 & 3 & 1 & 2 & 3 & 2 \end{bmatrix}$

(c) $\begin{bmatrix} 2 & 3 & 2 & 3 & 2 & 2 \\ 3 & 0 & 3 & 2 & 1 & 2 \\ 2 & 1 & 3 & 3 & 0 & 3 \\ 2 & 0 & 1 & 0 & 1 & 3 \\ 3 & 0 & 3 & 3 & 1 & 2 \\ 2 & 1 & 3 & 2 & 0 & 2 \end{bmatrix}$

(d) $\begin{bmatrix} 3 & 3 & 3 & 3 & 3 & 3 \\ 3 & 0 & 3 & 3 & 0 & 3 \\ 3 & 0 & 3 & 3 & 0 & 3 \\ 3 & 0 & 0 & 0 & 0 & 3 \\ 3 & 0 & 3 & 3 & 0 & 3 \\ 3 & 0 & 3 & 3 & 0 & 3 \end{bmatrix}$ which corresponds to this image

(e)

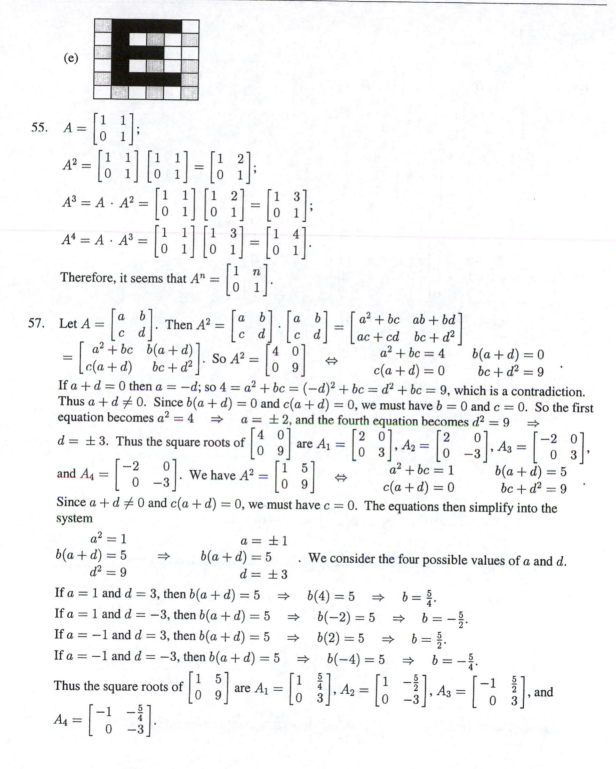

55. $A = \begin{bmatrix} 1 & 1 \\ 0 & 1 \end{bmatrix};$

$A^2 = \begin{bmatrix} 1 & 1 \\ 0 & 1 \end{bmatrix} \begin{bmatrix} 1 & 1 \\ 0 & 1 \end{bmatrix} = \begin{bmatrix} 1 & 2 \\ 0 & 1 \end{bmatrix};$

$A^3 = A \cdot A^2 = \begin{bmatrix} 1 & 1 \\ 0 & 1 \end{bmatrix} \begin{bmatrix} 1 & 2 \\ 0 & 1 \end{bmatrix} = \begin{bmatrix} 1 & 3 \\ 0 & 1 \end{bmatrix};$

$A^4 = A \cdot A^3 = \begin{bmatrix} 1 & 1 \\ 0 & 1 \end{bmatrix} \begin{bmatrix} 1 & 3 \\ 0 & 1 \end{bmatrix} = \begin{bmatrix} 1 & 4 \\ 0 & 1 \end{bmatrix}.$

Therefore, it seems that $A^n = \begin{bmatrix} 1 & n \\ 0 & 1 \end{bmatrix}.$

57. Let $A = \begin{bmatrix} a & b \\ c & d \end{bmatrix}.$ Then $A^2 = \begin{bmatrix} a & b \\ c & d \end{bmatrix} \cdot \begin{bmatrix} a & b \\ c & d \end{bmatrix} = \begin{bmatrix} a^2 + bc & ab + bd \\ ac + cd & bc + d^2 \end{bmatrix}$

$= \begin{bmatrix} a^2 + bc & b(a + d) \\ c(a + d) & bc + d^2 \end{bmatrix}.$ So $A^2 = \begin{bmatrix} 4 & 0 \\ 0 & 9 \end{bmatrix} \Leftrightarrow \begin{array}{ll} a^2 + bc = 4 & b(a + d) = 0 \\ c(a + d) = 0 & bc + d^2 = 9 \end{array}.$

If $a + d = 0$ then $a = -d$; so $4 = a^2 + bc = (-d)^2 + bc = d^2 + bc = 9$, which is a contradiction.
Thus $a + d \neq 0$. Since $b(a + d) = 0$ and $c(a + d) = 0$, we must have $b = 0$ and $c = 0$. So the first
equation becomes $a^2 = 4 \Rightarrow a = \pm 2$, and the fourth equation becomes $d^2 = 9 \Rightarrow$
$d = \pm 3$. Thus the square roots of $\begin{bmatrix} 4 & 0 \\ 0 & 9 \end{bmatrix}$ are $A_1 = \begin{bmatrix} 2 & 0 \\ 0 & 3 \end{bmatrix}$, $A_2 = \begin{bmatrix} 2 & 0 \\ 0 & -3 \end{bmatrix}$, $A_3 = \begin{bmatrix} -2 & 0 \\ 0 & 3 \end{bmatrix}$,

and $A_4 = \begin{bmatrix} -2 & 0 \\ 0 & -3 \end{bmatrix}.$ We have $A^2 = \begin{bmatrix} 1 & 5 \\ 0 & 9 \end{bmatrix} \Leftrightarrow \begin{array}{ll} a^2 + bc = 1 & b(a + d) = 5 \\ c(a + d) = 0 & bc + d^2 = 9 \end{array}.$

Since $a + d \neq 0$ and $c(a + d) = 0$, we must have $c = 0$. The equations then simplify into the
system

$\begin{array}{l} a^2 = 1 \\ b(a + d) = 5 \\ d^2 = 9 \end{array} \Rightarrow \begin{array}{l} a = \pm 1 \\ b(a + d) = 5 \\ d = \pm 3 \end{array}$. We consider the four possible values of a and d.

If $a = 1$ and $d = 3$, then $b(a + d) = 5 \Rightarrow b(4) = 5 \Rightarrow b = \frac{5}{4}.$

If $a = 1$ and $d = -3$, then $b(a + d) = 5 \Rightarrow b(-2) = 5 \Rightarrow b = -\frac{5}{2}.$

If $a = -1$ and $d = 3$, then $b(a + d) = 5 \Rightarrow b(2) = 5 \Rightarrow b = \frac{5}{2}.$

If $a = -1$ and $d = -3$, then $b(a + d) = 5 \Rightarrow b(-4) = 5 \Rightarrow b = -\frac{5}{4}.$

Thus the square roots of $\begin{bmatrix} 1 & 5 \\ 0 & 9 \end{bmatrix}$ are $A_1 = \begin{bmatrix} 1 & \frac{5}{4} \\ 0 & 3 \end{bmatrix}$, $A_2 = \begin{bmatrix} 1 & -\frac{5}{2} \\ 0 & -3 \end{bmatrix}$, $A_3 = \begin{bmatrix} -1 & \frac{5}{2} \\ 0 & 3 \end{bmatrix}$, and

$A_4 = \begin{bmatrix} -1 & -\frac{5}{4} \\ 0 & -3 \end{bmatrix}.$

Exercises 7.3

1. $A = \begin{bmatrix} 4 & 1 \\ 7 & 2 \end{bmatrix}; B = \begin{bmatrix} 2 & -1 \\ -7 & 4 \end{bmatrix}.$ $AB = \begin{bmatrix} 4 & 1 \\ 7 & 2 \end{bmatrix}\begin{bmatrix} 2 & -1 \\ -7 & 4 \end{bmatrix} = \begin{bmatrix} 1 & 0 \\ 0 & 1 \end{bmatrix}$ and

$BA = \begin{bmatrix} 2 & -1 \\ -7 & 4 \end{bmatrix}\begin{bmatrix} 4 & 1 \\ 7 & 2 \end{bmatrix} = \begin{bmatrix} 1 & 0 \\ 0 & 1 \end{bmatrix}.$

3. $A = \begin{bmatrix} 1 & 3 & -1 \\ 1 & 4 & 0 \\ -1 & -3 & 2 \end{bmatrix}; B = \begin{bmatrix} 8 & -3 & 4 \\ -2 & 1 & -1 \\ 1 & 0 & 1 \end{bmatrix}.$

$AB = \begin{bmatrix} 1 & 3 & -1 \\ 1 & 4 & 0 \\ -1 & -3 & 2 \end{bmatrix}\begin{bmatrix} 8 & -3 & 4 \\ -2 & 1 & -1 \\ 1 & 0 & 1 \end{bmatrix} = \begin{bmatrix} 1 & 0 & 0 \\ 0 & 1 & 0 \\ 0 & 0 & 1 \end{bmatrix}$ and

$BA = \begin{bmatrix} 8 & -3 & 4 \\ -2 & 1 & -1 \\ 1 & 0 & 1 \end{bmatrix}\begin{bmatrix} 1 & 3 & -1 \\ 1 & 4 & 0 \\ -1 & -3 & 2 \end{bmatrix} = \begin{bmatrix} 1 & 0 & 0 \\ 0 & 1 & 0 \\ 0 & 0 & 1 \end{bmatrix}$

5. $A = \begin{bmatrix} 7 & 4 \\ 3 & 2 \end{bmatrix} \Leftrightarrow A^{-1} = \frac{1}{14-12}\begin{bmatrix} 2 & -4 \\ -3 & 7 \end{bmatrix} = \begin{bmatrix} 1 & -2 \\ -\frac{3}{2} & \frac{7}{2} \end{bmatrix}.$

Then, $AA^{-1} = \begin{bmatrix} 7 & 4 \\ 3 & 2 \end{bmatrix}\begin{bmatrix} 1 & -2 \\ -\frac{3}{2} & \frac{7}{2} \end{bmatrix} = \begin{bmatrix} 1 & 0 \\ 0 & 1 \end{bmatrix},$

and $A^{-1}A = \begin{bmatrix} 1 & -2 \\ -\frac{3}{2} & \frac{7}{2} \end{bmatrix}\begin{bmatrix} 7 & 4 \\ 3 & 2 \end{bmatrix} = \begin{bmatrix} 1 & 0 \\ 0 & 1 \end{bmatrix}.$

7. $\begin{bmatrix} 5 & 3 \\ 3 & 2 \end{bmatrix}^{-1} = \frac{1}{10-9}\begin{bmatrix} 2 & -3 \\ -3 & 5 \end{bmatrix} = \begin{bmatrix} 2 & -3 \\ -3 & 5 \end{bmatrix}$

9. $\begin{bmatrix} 2 & 5 \\ -5 & -13 \end{bmatrix}^{-1} = \frac{1}{-26+25}\begin{bmatrix} -13 & -5 \\ 5 & 2 \end{bmatrix} = \begin{bmatrix} 13 & 5 \\ -5 & -2 \end{bmatrix}$

11. $\begin{bmatrix} 6 & -3 \\ -8 & 4 \end{bmatrix}^{-1} = \frac{1}{24-24}\begin{bmatrix} 4 & 3 \\ 8 & 6 \end{bmatrix},$ which is not defined, and so there is no inverse.

13. $\begin{bmatrix} 0.4 & -1.2 \\ 0.3 & 0.6 \end{bmatrix}^{-1} = \frac{1}{0.24+0.36}\begin{bmatrix} 0.6 & 1.2 \\ -0.3 & 0.4 \end{bmatrix} = \begin{bmatrix} 1 & 2 \\ -\frac{1}{2} & \frac{2}{3} \end{bmatrix}$

15. $\begin{bmatrix} 2 & 4 & 1 & 1 & 0 & 0 \\ -1 & 1 & -1 & 0 & 1 & 0 \\ 1 & 4 & 0 & 0 & 0 & 1 \end{bmatrix} \xrightarrow[2R_3-R_1 \to R_3]{2R_2+R_1 \to R_2} \begin{bmatrix} 2 & 4 & 1 & 1 & 0 & 0 \\ 0 & 6 & -1 & 1 & 2 & 0 \\ 0 & 4 & -1 & -1 & 0 & 2 \end{bmatrix}$

$\xrightarrow[3R_1-2R_2 \to R_1]{3R_3-2R_2 \to R_3} \begin{bmatrix} 6 & 0 & 5 & 1 & -4 & 0 \\ 0 & 6 & -1 & 1 & 2 & 0 \\ 0 & 0 & -1 & -5 & -4 & 6 \end{bmatrix} \xrightarrow[R_2-R_3 \to R_2]{R_1+5R_3 \to R_1}$

$$\begin{bmatrix} 6 & 0 & 0 & -24 & -24 & 30 \\ 0 & 6 & 0 & 6 & 6 & -6 \\ 0 & 0 & -1 & -5 & -4 & 6 \end{bmatrix} \xrightarrow[\frac{1}{6}R_2,\, -R_3]{\frac{1}{6}R_1} \begin{bmatrix} 1 & 0 & 0 & -4 & -4 & 5 \\ 0 & 1 & 0 & 1 & 1 & -1 \\ 0 & 0 & 1 & 5 & 4 & -6 \end{bmatrix}$$

Therefore, the inverse matrix is $\begin{bmatrix} -4 & -4 & 5 \\ 1 & 1 & -1 \\ 5 & 4 & -6 \end{bmatrix}$.

17. $\begin{bmatrix} 1 & 2 & 3 & 1 & 0 & 0 \\ 4 & 5 & -1 & 0 & 1 & 0 \\ 1 & -1 & -10 & 0 & 0 & 1 \end{bmatrix} \xrightarrow[R_3-R_1 \to R_3]{R_2-4R_1 \to R_2} \begin{bmatrix} 1 & 2 & 3 & 1 & 0 & 0 \\ 0 & -3 & -13 & -4 & 1 & 0 \\ 0 & -3 & -13 & -1 & 0 & 1 \end{bmatrix}$

$\xrightarrow{R_3-R_2 \to R_3} \begin{bmatrix} 1 & 2 & 3 & 1 & 0 & 0 \\ 0 & -3 & -13 & -4 & 1 & 0 \\ 0 & 0 & 0 & 3 & -1 & 1 \end{bmatrix}$

Since the left half of the last row consists entirely of zeros, there is no inverse matrix.

19. $\begin{bmatrix} 0 & -2 & 2 & 1 & 0 & 0 \\ 3 & 1 & 3 & 0 & 1 & 0 \\ 1 & -2 & 3 & 0 & 0 & 1 \end{bmatrix} \xrightarrow{R_1 \leftrightarrow R_3} \begin{bmatrix} 1 & -2 & 3 & 0 & 0 & 1 \\ 3 & 1 & 3 & 0 & 1 & 0 \\ 0 & -2 & 2 & 1 & 0 & 0 \end{bmatrix} \xrightarrow{R_2-3R_1 \to R_2}$

$\begin{bmatrix} 1 & -2 & 3 & 0 & 0 & 1 \\ 0 & 7 & -6 & 0 & 1 & -3 \\ 0 & -2 & 2 & 1 & 0 & 0 \end{bmatrix} \xrightarrow[R_2+3R_3 \to R_2]{R_1-R_3 \to R_1} \begin{bmatrix} 1 & 0 & 1 & -1 & 0 & 1 \\ 0 & 1 & 0 & 3 & 1 & -3 \\ 0 & -2 & 2 & 1 & 0 & 0 \end{bmatrix}$

$\xrightarrow{R_3+2R_2 \to R_3} \begin{bmatrix} 1 & 0 & 1 & -1 & 0 & 1 \\ 0 & 1 & 0 & 3 & 1 & -3 \\ 0 & 0 & 2 & 7 & 2 & -6 \end{bmatrix} \xrightarrow{\frac{1}{2}R_3} \begin{bmatrix} 1 & 0 & 1 & -1 & 0 & 1 \\ 0 & 1 & 0 & 3 & 1 & -3 \\ 0 & 0 & 1 & \frac{7}{2} & 1 & -3 \end{bmatrix}$

$\xrightarrow{R_1-R_3 \to R_1} \begin{bmatrix} 1 & 0 & 0 & -\frac{9}{2} & -1 & 4 \\ 0 & 1 & 0 & 3 & 1 & -3 \\ 0 & 0 & 1 & \frac{7}{2} & 1 & -3 \end{bmatrix}$

Therefore, the inverse matrix is $\begin{bmatrix} -\frac{9}{2} & -1 & 4 \\ 3 & 1 & -3 \\ \frac{7}{2} & 1 & -3 \end{bmatrix}$.

21. $\begin{bmatrix} 1 & 2 & 0 & 3 & 1 & 0 & 0 & 0 \\ 0 & 1 & 1 & 1 & 0 & 1 & 0 & 0 \\ 0 & 1 & 0 & 1 & 0 & 0 & 1 & 0 \\ 1 & 2 & 0 & 2 & 0 & 0 & 0 & 1 \end{bmatrix} \xrightarrow[R_4-R_1 \to R_4]{R_3-R_2 \to R_3} \begin{bmatrix} 1 & 2 & 0 & 3 & 1 & 0 & 0 & 0 \\ 0 & 1 & 1 & 1 & 0 & 1 & 0 & 0 \\ 0 & 0 & -1 & 0 & 0 & -1 & 1 & 0 \\ 0 & 0 & 0 & -1 & -1 & 0 & 0 & 1 \end{bmatrix}$

$\xrightarrow[-R_4]{-R_3} \begin{bmatrix} 1 & 2 & 0 & 3 & 1 & 0 & 0 & 0 \\ 0 & 1 & 1 & 1 & 0 & 1 & 0 & 0 \\ 0 & 0 & 1 & 0 & 0 & 1 & -1 & 0 \\ 0 & 0 & 0 & 1 & 1 & 0 & 0 & -1 \end{bmatrix} \xrightarrow[R_2-R_3 \to R_2]{R_1-2R_2 \to R_1}$

$\begin{bmatrix} 1 & 0 & -2 & 1 & 1 & -2 & 0 & 0 \\ 0 & 1 & 0 & 1 & 0 & 0 & 1 & 0 \\ 0 & 0 & 1 & 0 & 0 & 1 & -1 & 0 \\ 0 & 0 & 0 & 1 & 1 & 0 & 0 & -1 \end{bmatrix} \xrightarrow[R_2-R_4 \to R_2]{R_1+2R_3 \to R_1}$

$$\begin{bmatrix} 1 & 0 & 0 & 1 & 1 & 0 & -2 & 0 \\ 0 & 1 & 0 & 0 & -1 & 0 & 1 & 1 \\ 0 & 0 & 1 & 0 & 0 & 1 & -1 & 0 \\ 0 & 0 & 0 & 1 & 1 & 0 & 0 & -1 \end{bmatrix} \xrightarrow{R_1 \to R_1 - R_4} \begin{bmatrix} 1 & 0 & 0 & 0 & 0 & 0 & -2 & 1 \\ 0 & 1 & 0 & 0 & -1 & 0 & 1 & 1 \\ 0 & 0 & 1 & 0 & 0 & 1 & -1 & 0 \\ 0 & 0 & 0 & 1 & 1 & 0 & 0 & -1 \end{bmatrix}$$

Therefore, the inverse matrix is $\begin{bmatrix} 0 & 0 & -2 & 1 \\ -1 & 0 & 1 & 1 \\ 0 & 1 & -1 & 0 \\ 1 & 0 & 0 & -1 \end{bmatrix}$.

23. $\begin{cases} 5x + 3y = 4 \\ 3x + 2y = 0 \end{cases}$ is equivalent to the matrix equation $\begin{bmatrix} 5 & 3 \\ 3 & 2 \end{bmatrix}\begin{bmatrix} x \\ y \end{bmatrix} = \begin{bmatrix} 4 \\ 0 \end{bmatrix}$.

Using the inverse from Exercise 3, $\begin{bmatrix} x \\ y \end{bmatrix} = \begin{bmatrix} 2 & -3 \\ -3 & 5 \end{bmatrix}\begin{bmatrix} 4 \\ 0 \end{bmatrix} = \begin{bmatrix} 8 \\ -12 \end{bmatrix}$.

Therefore, $x = 8$ and $y = -12$.

25. $\begin{cases} 2x + 5y = 2 \\ -5x - 13y = 20 \end{cases}$ is equivalent to the matrix equation $\begin{bmatrix} 2 & 5 \\ -5 & -13 \end{bmatrix}\begin{bmatrix} x \\ y \end{bmatrix} = \begin{bmatrix} 2 \\ 20 \end{bmatrix}$.

Using the inverse from Exercise 5, $\begin{bmatrix} x \\ y \end{bmatrix} = \begin{bmatrix} 13 & 5 \\ -5 & -2 \end{bmatrix}\begin{bmatrix} 2 \\ 20 \end{bmatrix} = \begin{bmatrix} 126 \\ -50 \end{bmatrix}$.

Therefore, $x = 126$ and $y = -50$.

27. $\begin{cases} 2x + 4y + z = 7 \\ -x + y - z = 0 \\ x + 4y = -2 \end{cases}$ is equivalent to the matrix equation $\begin{bmatrix} 2 & 4 & 1 \\ -1 & 1 & -1 \\ 1 & 4 & 0 \end{bmatrix}\begin{bmatrix} x \\ y \\ z \end{bmatrix} = \begin{bmatrix} 7 \\ 0 \\ -2 \end{bmatrix}$.

Using the inverse from Exercise 11, $\begin{bmatrix} x \\ y \\ z \end{bmatrix} = \begin{bmatrix} -4 & -4 & 5 \\ 1 & 1 & -1 \\ 5 & 4 & -6 \end{bmatrix}\begin{bmatrix} 7 \\ 0 \\ -2 \end{bmatrix} = \begin{bmatrix} -38 \\ 9 \\ 47 \end{bmatrix}$.

Therefore, $x = -38$, $y = 9$, and $z = 47$.

29. $\begin{cases} -2y + 2z = 12 \\ 3x + y + 3z = -2 \\ x - 2y + 3z = 8 \end{cases}$ is equivalent to the matrix equation $\begin{bmatrix} 0 & -2 & 2 \\ 3 & 1 & 3 \\ 1 & -2 & 3 \end{bmatrix}\begin{bmatrix} x \\ y \\ z \end{bmatrix} = \begin{bmatrix} 12 \\ -2 \\ 8 \end{bmatrix}$.

Using the inverse from Exercise 15, $\begin{bmatrix} x \\ y \\ z \end{bmatrix} = \begin{bmatrix} -\frac{9}{2} & -1 & 4 \\ 3 & 1 & -3 \\ \frac{7}{2} & 1 & -3 \end{bmatrix}\begin{bmatrix} 12 \\ -2 \\ 8 \end{bmatrix} = \begin{bmatrix} -20 \\ 10 \\ 16 \end{bmatrix}$.

Therefore, $x = -20$, $y = 10$, and $z = 16$.

31. Using the calculator you should get the result $(3, 2, 1)$.

33. Using the calculator you should get the result $(3, -2, 2)$.

35. Using the calculator you should get the result $(8, 1, 0, 3)$.

37. This has the form $MX = C$. Since $M^{-1}(MX) = M^{-1}C$ and $M^{-1}(MX) = (M^{-1}M)X = X$.

Now $M^{-1} = \begin{bmatrix} 3 & -2 \\ -4 & 3 \end{bmatrix}^{-1} = \frac{1}{9-8}\begin{bmatrix} 3 & 2 \\ 4 & 3 \end{bmatrix} = \begin{bmatrix} 3 & 2 \\ 4 & 3 \end{bmatrix}$. Since $X = M^{-1}C$ we get

$$\begin{bmatrix} x & y & z \\ u & v & w \end{bmatrix} = \begin{bmatrix} 3 & 2 \\ 4 & 3 \end{bmatrix}\begin{bmatrix} 1 & 0 & -1 \\ 2 & 1 & 3 \end{bmatrix} = \begin{bmatrix} 7 & 2 & 3 \\ 10 & 3 & 5 \end{bmatrix}.$$

39. $$\begin{bmatrix} a & -a \\ a & a \end{bmatrix}^{-1} = \frac{1}{a^2 - (-a^2)}\begin{bmatrix} a & a \\ -a & a \end{bmatrix} = \frac{1}{2a^2}\begin{bmatrix} a & a \\ -a & a \end{bmatrix} = \frac{1}{2a}\begin{bmatrix} 1 & 1 \\ -1 & 1 \end{bmatrix}$$

41. $$\begin{bmatrix} 2 & x \\ x & x^2 \end{bmatrix}^{-1} = \frac{1}{2x^2 - x^2}\begin{bmatrix} x^2 & -x \\ -x & 2 \end{bmatrix} = \frac{1}{x^2}\begin{bmatrix} x^2 & -x \\ -x & 2 \end{bmatrix} = \begin{bmatrix} 1 & -\frac{1}{x} \\ -\frac{1}{x} & \frac{2}{x^2} \end{bmatrix}.$$ Inverse does not exist

when $x = 0$.

43. $$\begin{bmatrix} 1 & e^x & 0 & 1 & 0 & 0 \\ e^x & -e^{2x} & 0 & 0 & 1 & 0 \\ 0 & 0 & 2 & 0 & 0 & 1 \end{bmatrix} \xrightarrow{R_2 - e^x R_1 \to R_2} \begin{bmatrix} 1 & e^x & 0 & 1 & 0 & 0 \\ 0 & -2e^{2x} & 0 & -e^x & 1 & 0 \\ 0 & 0 & 2 & 0 & 0 & 1 \end{bmatrix} \xrightarrow[\frac{1}{2}R_3]{-\frac{1}{2}e^{-2x}R_2}$$

$$\begin{bmatrix} 1 & e^x & 0 & 1 & 0 & 0 \\ 0 & 1 & 0 & \frac{1}{2}e^{-x} & -\frac{1}{2}e^{-2x} & 0 \\ 0 & 0 & 1 & 0 & 0 & \frac{1}{2} \end{bmatrix} \xrightarrow{R_1 - e^x R_2 \to R_1} \begin{bmatrix} 1 & 0 & 0 & \frac{1}{2} & \frac{1}{2}e^{-x} & 0 \\ 0 & 1 & 0 & \frac{1}{2}e^{-x} & -\frac{1}{2}e^{-2x} & 0 \\ 0 & 0 & 1 & 0 & 0 & \frac{1}{2} \end{bmatrix}$$

Therefore, the inverse matrix is $\begin{bmatrix} \frac{1}{2} & \frac{1}{2}e^{-x} & 0 \\ \frac{1}{2}e^{-x} & -\frac{1}{2}e^{-2x} & 0 \\ 0 & 0 & \frac{1}{2} \end{bmatrix}$. Inverse exists for all x.

45. (a) $$\begin{bmatrix} 3 & 1 & 3 & 1 & 0 & 0 \\ 4 & 2 & 4 & 0 & 1 & 0 \\ 3 & 2 & 4 & 0 & 0 & 1 \end{bmatrix} \xrightarrow[R_1 \leftrightarrow R_2]{R_3 - R_1 \to R_3} \begin{bmatrix} 4 & 2 & 4 & 0 & 1 & 0 \\ 3 & 1 & 3 & 1 & 0 & 0 \\ 0 & 1 & 1 & -1 & 0 & 1 \end{bmatrix} \xrightarrow{R_1 - R_2 \to R_1}$$

$$\begin{bmatrix} 1 & 1 & 1 & -1 & 1 & 0 \\ 3 & 1 & 3 & 1 & 0 & 0 \\ 0 & 1 & 1 & -1 & 0 & 1 \end{bmatrix} \xrightarrow{R_2 - 3R_1 \to R_2} \begin{bmatrix} 1 & 1 & 1 & -1 & 1 & 0 \\ 0 & -2 & 0 & 4 & -3 & 0 \\ 0 & 1 & 1 & -1 & 0 & 1 \end{bmatrix} \xrightarrow[R_3 - R_2 \to R_3]{R_1 - R_2 \to R_1, -\frac{1}{2}R_2}$$

$$\begin{bmatrix} 1 & 0 & 1 & 1 & -\frac{1}{2} & 0 \\ 0 & 1 & 0 & -2 & \frac{3}{2} & 0 \\ 0 & 0 & 1 & 1 & -\frac{3}{2} & 1 \end{bmatrix} \xrightarrow{R_1 - R_3 \to R_1} \begin{bmatrix} 1 & 0 & 0 & 0 & 1 & -1 \\ 0 & 1 & 0 & -2 & \frac{3}{2} & 0 \\ 0 & 0 & 1 & 1 & -\frac{3}{2} & 1 \end{bmatrix}$$

Therefore, the inverse of the matrix is $\begin{bmatrix} 0 & 1 & -1 \\ -2 & \frac{3}{2} & 0 \\ 1 & -\frac{3}{2} & 1 \end{bmatrix}$.

(b) $$\begin{bmatrix} A \\ B \\ C \end{bmatrix} = \begin{bmatrix} 0 & 1 & -1 \\ -2 & \frac{3}{2} & 0 \\ 1 & -\frac{3}{2} & 1 \end{bmatrix}\begin{bmatrix} 10 \\ 14 \\ 13 \end{bmatrix} = \begin{bmatrix} 1 \\ 1 \\ 2 \end{bmatrix}$$

Therefore, he should feed the rats 1 oz of A, 1 oz of B, and 2 oz of C.

(c) $\begin{bmatrix} A \\ B \\ C \end{bmatrix} = \begin{bmatrix} 0 & 1 & -1 \\ -2 & \frac{3}{2} & 0 \\ 1 & -\frac{3}{2} & 1 \end{bmatrix} \begin{bmatrix} 9 \\ 12 \\ 10 \end{bmatrix} = \begin{bmatrix} 2 \\ 0 \\ 1 \end{bmatrix}$

Therefore, he should feed the rats 2 oz of A, 0 oz of B, and 1 oz of C.

(d) $\begin{bmatrix} A \\ B \\ C \end{bmatrix} = \begin{bmatrix} 0 & 1 & -1 \\ -2 & \frac{3}{2} & 0 \\ 1 & -\frac{3}{2} & 1 \end{bmatrix} \begin{bmatrix} 2 \\ 4 \\ 11 \end{bmatrix} = \begin{bmatrix} -7 \\ 2 \\ 7 \end{bmatrix}$

Since $A < 0$, there is no combination of foods giving the required supply.

47. (a) $\begin{cases} x + y + 2z = 675 \\ 2x + y + z = 600 \\ x + 2y + z = 625 \end{cases}$

(b) $\begin{bmatrix} 1 & 1 & 2 \\ 2 & 1 & 1 \\ 1 & 2 & 1 \end{bmatrix} \begin{bmatrix} x \\ y \\ z \end{bmatrix} = \begin{bmatrix} 675 \\ 600 \\ 625 \end{bmatrix}$

(c) $\begin{bmatrix} 1 & 1 & 2 & 1 & 0 & 0 \\ 2 & 1 & 1 & 0 & 1 & 0 \\ 1 & 2 & 1 & 0 & 0 & 1 \end{bmatrix} \xrightarrow[R_3 - R_1 \to R_3]{R_2 - 2R_1 \to R_2} \begin{bmatrix} 1 & 1 & 2 & 1 & 0 & 0 \\ 0 & -1 & -3 & -2 & 1 & 0 \\ 0 & 1 & -1 & -1 & 0 & 1 \end{bmatrix}$

$\xrightarrow[R_3 + R_2 \to R_3]{R_1 + R_2 \to R_1} \begin{bmatrix} 1 & 0 & -1 & -1 & 1 & 0 \\ 0 & -1 & -3 & -2 & 1 & 0 \\ 0 & 0 & -4 & -3 & 1 & 1 \end{bmatrix} \xrightarrow[-\frac{1}{4}R_3]{-R_2}$

$\begin{bmatrix} 1 & 0 & -1 & -1 & 1 & 0 \\ 0 & 1 & 3 & 2 & -1 & 0 \\ 0 & 0 & 1 & \frac{3}{4} & -\frac{1}{4} & -\frac{1}{4} \end{bmatrix} \xrightarrow[R_2 - 3R_3 \to R_2]{R_1 + R_3 \to R_1} \begin{bmatrix} 1 & 0 & 0 & -\frac{1}{4} & \frac{3}{4} & -\frac{1}{4} \\ 0 & 1 & 0 & -\frac{1}{4} & -\frac{1}{4} & \frac{3}{4} \\ 0 & 0 & 1 & \frac{3}{4} & -\frac{1}{4} & -\frac{1}{4} \end{bmatrix}$

Therefore, the inverse of the matrix is $\begin{bmatrix} -\frac{1}{4} & \frac{3}{4} & -\frac{1}{4} \\ -\frac{1}{4} & -\frac{1}{4} & \frac{3}{4} \\ \frac{3}{4} & -\frac{1}{4} & -\frac{1}{4} \end{bmatrix}$ and

$\begin{bmatrix} x \\ y \\ z \end{bmatrix} = \begin{bmatrix} -\frac{1}{4} & \frac{3}{4} & -\frac{1}{4} \\ -\frac{1}{4} & -\frac{1}{4} & \frac{3}{4} \\ \frac{3}{4} & -\frac{1}{4} & -\frac{1}{4} \end{bmatrix} \begin{bmatrix} 675 \\ 600 \\ 625 \end{bmatrix} = \begin{bmatrix} 125 \\ 150 \\ 200 \end{bmatrix}.$

Thus he earns $125 on a standard set, $150 on a deluxe set, and $200 on a leather-bound set.

Exercises 7.4

1. The matrix $\begin{bmatrix} 2 & 0 \\ 0 & 3 \end{bmatrix}$ has determinant $|D| = (2)(3) - (0)(0) = 6$.

3. The matrix $\begin{bmatrix} 4 & 5 \\ 0 & -1 \end{bmatrix}$ has determinant $|D| = (4)(-1) - (5)(0) = -4$.

5. The matrix $\begin{bmatrix} 2 & 5 \end{bmatrix}$ does not have a determinant because the matrix is not square.

7. The matrix $\begin{bmatrix} \frac{1}{2} & \frac{1}{8} \\ 1 & \frac{1}{2} \end{bmatrix}$ has determinant $|D| = \frac{1}{2} \cdot \frac{1}{2} - 1 \cdot \frac{1}{8} = \frac{1}{4} - \frac{1}{8} = \frac{1}{8}$.

In Exercises 9 – 13, the matrix is $A = \begin{bmatrix} 1 & 0 & \frac{1}{2} \\ -3 & 5 & 2 \\ 0 & 0 & 4 \end{bmatrix}$.

9. $M_{11} = 5 \cdot 4 - 0 \cdot 2 = 20$, $A_{11} = (-1)^2 \, M_{11} = 20$

11. $M_{12} = -3 \cdot 4 - 0 \cdot 2 = -12$, $A_{12} = (-1)^3 \, M_{12} = 12$

13. $M_{23} = 1 \cdot 0 - 0 \cdot 0 = 0$, $A_{23} = (-1)^5 \, M_{23} = 0$

15. $M = \begin{bmatrix} 2 & 1 & 0 \\ 0 & -2 & 4 \\ 0 & 1 & -3 \end{bmatrix}$. Therefore, expanding by the first column, $|M| = 2 \begin{vmatrix} -2 & 4 \\ 1 & -3 \end{vmatrix}$
$= 2(6-4) = 4$. Since $|M| \neq 0$, the matrix has an inverse.

17. $M = \begin{bmatrix} 1 & 3 & 7 \\ 2 & 0 & -1 \\ 0 & 2 & 6 \end{bmatrix}$. Therefore, expanding by the third row, $|M| = -2 \begin{vmatrix} 1 & 7 \\ 2 & -1 \end{vmatrix} + 6 \begin{vmatrix} 1 & 3 \\ 2 & 0 \end{vmatrix}$
$= -2(-1-14) + 6(0-6) = 30 - 36 = -6$. Since $|M| \neq 0$, the matrix has an inverse.

19. $M = \begin{bmatrix} 30 & 0 & 20 \\ 0 & -10 & -20 \\ 40 & 0 & 10 \end{bmatrix}$. Therefore, expanding by the first row,

$|M| = 30 \begin{vmatrix} -10 & -20 \\ 0 & 10 \end{vmatrix} + 20 \begin{vmatrix} 0 & -10 \\ 40 & 0 \end{vmatrix} = 30(-100+0) + 20(0+400)$
$= -3000 + 8000 = 5000$, and so M^{-1} exists.

21. $M = \begin{bmatrix} 1 & 3 & 3 & 0 \\ 0 & 2 & 0 & 1 \\ -1 & 0 & 0 & 2 \\ 1 & 6 & 4 & 1 \end{bmatrix}$. Therefore, expanding by the third row,

$$|M| = -1 \begin{vmatrix} 3 & 3 & 0 \\ 2 & 0 & 1 \\ 6 & 4 & 1 \end{vmatrix} - 2 \begin{vmatrix} 1 & 3 & 3 \\ 0 & 2 & 0 \\ 1 & 6 & 4 \end{vmatrix} = 1 \begin{vmatrix} 3 & 3 \\ 6 & 4 \end{vmatrix} - 1 \begin{vmatrix} 3 & 3 \\ 2 & 0 \end{vmatrix} - 4 \begin{vmatrix} 1 & 3 \\ 1 & 4 \end{vmatrix} = -6 + 6 - 4 = -4,$$

and so M^{-1} exists.

23. $|M| = \begin{vmatrix} 0 & 0 & 4 & 6 \\ 2 & 1 & 1 & 3 \\ 2 & 1 & 2 & 3 \\ 3 & 0 & 1 & 7 \end{vmatrix} = \begin{vmatrix} 0 & 0 & 4 & 6 \\ 2 & 1 & 1 & 3 \\ 0 & 0 & 1 & 0 \\ 3 & 0 & 1 & 7 \end{vmatrix}$, by replacing R_3 with $R_3 - R_2$. Then, expanding by the

third row, $|M| = 1 \begin{vmatrix} 0 & 0 & 6 \\ 2 & 1 & 3 \\ 3 & 0 & 7 \end{vmatrix} = 6 \begin{vmatrix} 2 & 1 \\ 3 & 0 \end{vmatrix} = 6(2 \cdot 0 - 3 \cdot 1) = -18.$

25. $M = \begin{bmatrix} 1 & 2 & 3 & 4 & 5 \\ 0 & 2 & 4 & 6 & 8 \\ 0 & 0 & 3 & 6 & 9 \\ 0 & 0 & 0 & 4 & 8 \\ 0 & 0 & 0 & 0 & 5 \end{bmatrix}$. Then, expanding by the fifth row, $|M| = 5 \begin{vmatrix} 1 & 2 & 3 & 4 \\ 0 & 2 & 4 & 6 \\ 0 & 0 & 3 & 6 \\ 0 & 0 & 0 & 4 \end{vmatrix}$

$$= 5 \cdot 4 \begin{vmatrix} 1 & 2 & 3 \\ 0 & 2 & 4 \\ 0 & 0 & 3 \end{vmatrix} = 20 \cdot 3 \begin{vmatrix} 1 & 2 \\ 0 & 2 \end{vmatrix} = 60 \cdot 2 = 120.$$

27. $B = \begin{bmatrix} 4 & 1 & 0 \\ -2 & -1 & 1 \\ 4 & 0 & 3 \end{bmatrix}$

(a) $|B| = 2 \begin{vmatrix} 1 & 0 \\ 0 & 3 \end{vmatrix} - 1 \begin{vmatrix} 4 & 0 \\ 4 & 3 \end{vmatrix} - 1 \begin{vmatrix} 4 & 1 \\ 4 & 0 \end{vmatrix} = 6 - 12 + 4 = -2$

(b) $|B| = -1 \begin{vmatrix} 4 & 1 \\ 4 & 0 \end{vmatrix} + 3 \begin{vmatrix} 4 & 1 \\ -2 & -1 \end{vmatrix} = 4 - 6 = -2$

(c) Yes, as expected, the results agree.

29. $\begin{cases} 2x - y = -9 \\ x + 2y = 8 \end{cases}$ Then, $|D| = \begin{vmatrix} 2 & -1 \\ 1 & 2 \end{vmatrix} = 5, |D_x| = \begin{vmatrix} -9 & -1 \\ 8 & 2 \end{vmatrix} = -10,$ and

$|D_y| = \begin{vmatrix} 2 & -9 \\ 1 & 8 \end{vmatrix} = 25.$ Hence, $x = \dfrac{|D_x|}{|D|} = \dfrac{-10}{5} = -2, y = \dfrac{|D_y|}{|D|} = \dfrac{25}{5} = 5,$ and so the solution

is $(-2, 5)$.

31. $\begin{cases} x - 6y = 3 \\ 3x + 2y = 1 \end{cases}$ Then, $|D| = \begin{vmatrix} 1 & -6 \\ 3 & 2 \end{vmatrix} = 20, |D_x| = \begin{vmatrix} 3 & -6 \\ 1 & 2 \end{vmatrix} = 12,$ and

$|D_y| = \begin{vmatrix} 1 & 3 \\ 3 & 1 \end{vmatrix} = -8.$ Hence, $x = \dfrac{|D_x|}{|D|} = \dfrac{12}{20} = 0.6, y = \dfrac{|D_y|}{|D|} = \dfrac{-8}{20} = -0.4,$ and so the solution

is $(0.6, -0.4)$.

33. $\begin{cases} 0.4x + 1.2y = 0.4 \\ 1.2x + 1.6y = 3.2 \end{cases}$ Then, $|D| = \begin{vmatrix} 0.4 & 1.2 \\ 1.2 & 1.6 \end{vmatrix} = -0.8$, $|D_x| = \begin{vmatrix} 0.4 & 1.2 \\ 3.2 & 1.6 \end{vmatrix} = -3.2$, and

$|D_y| = \begin{vmatrix} 0.4 & 0.4 \\ 1.2 & 3.2 \end{vmatrix} = 0.8$. Hence, $x = \dfrac{|D_x|}{|D|} = \dfrac{-3.2}{-0.8} = 4$, $y = \dfrac{|D_y|}{|D|} = \dfrac{0.8}{-0.8} = -1$, and so the

solution is $(4, -1)$.

35. $\begin{cases} x - y + 2z = 0 \\ \quad 3x + z = 11 \\ -x + 2y = 0 \end{cases}$ Then, expanding by the second row,

$|D| = \begin{vmatrix} 1 & -1 & 2 \\ 3 & 0 & 1 \\ -1 & 2 & 0 \end{vmatrix} = -3 \begin{vmatrix} -1 & 2 \\ 2 & 0 \end{vmatrix} - 1 \begin{vmatrix} 1 & -1 \\ -1 & 2 \end{vmatrix} = 12 - 1 = 11,$

$|D_x| = \begin{vmatrix} 0 & -1 & 2 \\ 11 & 0 & 1 \\ 0 & 2 & 0 \end{vmatrix} = -11 \begin{vmatrix} -1 & 2 \\ 2 & 0 \end{vmatrix} = 44$, $|D_y| = \begin{vmatrix} 1 & 0 & 2 \\ 3 & 11 & 1 \\ -1 & 0 & 0 \end{vmatrix} = 11 \begin{vmatrix} 1 & 2 \\ -1 & 0 \end{vmatrix} = 22,$

and $|D_z| = \begin{vmatrix} 1 & -1 & 0 \\ 3 & 0 & 11 \\ -1 & 2 & 0 \end{vmatrix} = -11 \begin{vmatrix} 1 & -1 \\ -1 & 2 \end{vmatrix} = -11$. Therefore, $x = \frac{44}{11} = 4$, $y = \frac{22}{11} = 2$,

$z = \frac{-11}{11} = -1$, and so the solution is $(4, 2, -1)$.

37. $\begin{cases} 2x_1 + 3x_2 - 5x_3 = 1 \\ \quad x_1 + x_2 - x_3 = 2 \\ \quad 2x_2 + x_3 = 8 \end{cases}$ Then, expanding by the third row,

$|D| = \begin{vmatrix} 2 & 3 & -5 \\ 1 & 1 & -1 \\ 0 & 2 & 1 \end{vmatrix} = -2 \begin{vmatrix} 2 & -5 \\ 1 & -1 \end{vmatrix} + \begin{vmatrix} 2 & 3 \\ 1 & 1 \end{vmatrix} = -6 - 1 = -7,$

$|D_{x_1}| = \begin{vmatrix} 1 & 3 & -5 \\ 2 & 1 & -1 \\ 8 & 2 & 1 \end{vmatrix} = \begin{vmatrix} 1 & -1 \\ 2 & 1 \end{vmatrix} - 3 \begin{vmatrix} 2 & -1 \\ 8 & 1 \end{vmatrix} - 5 \begin{vmatrix} 2 & 1 \\ 8 & 2 \end{vmatrix} = 3 - 30 + 20 = -7,$

$|D_{x_2}| = \begin{vmatrix} 2 & 1 & -5 \\ 1 & 2 & -1 \\ 0 & 8 & 1 \end{vmatrix} = -8 \begin{vmatrix} 2 & -5 \\ 1 & -1 \end{vmatrix} + \begin{vmatrix} 2 & 1 \\ 1 & 2 \end{vmatrix} = -24 + 3 = -21$, and

$|D_{x_3}| = \begin{vmatrix} 2 & 3 & 1 \\ 1 & 1 & 2 \\ 0 & 2 & 8 \end{vmatrix} = -2 \begin{vmatrix} 2 & 1 \\ 1 & 2 \end{vmatrix} + 8 \begin{vmatrix} 2 & 3 \\ 1 & 1 \end{vmatrix} = -6 - 8 = -14$. Thus, $x_1 = \frac{-7}{-7} = 1$,

$x_2 = \frac{-21}{-7} = 3$, $x_3 = \frac{-14}{-7} = 2$, and so the solution is $(1, 3, 2)$.

39. $\begin{cases} \frac{1}{3}x - \frac{1}{5}y + \frac{1}{2}z = \frac{7}{10} \\ -\frac{2}{3}x + \frac{2}{5}y + \frac{3}{2}z = \frac{11}{10} \\ \quad x - \frac{4}{5}y + z = \frac{9}{5} \end{cases}$ $\Leftrightarrow$ $\begin{cases} 10x - 6y + 15z = 21 \\ -20x + 12y + 45z = 33 \\ \quad 5x - 4y + 5z = 9 \end{cases}$

Then, $|D| = \begin{vmatrix} 10 & -6 & 15 \\ -20 & 12 & 45 \\ 5 & -4 & 5 \end{vmatrix} = 10\begin{vmatrix} 12 & 45 \\ -4 & 5 \end{vmatrix} + 6\begin{vmatrix} -20 & 45 \\ 5 & 5 \end{vmatrix} + 15\begin{vmatrix} -20 & 12 \\ 5 & -4 \end{vmatrix}$

$= 2400 - 1950 + 300 = 750,$

$|D_x| = \begin{vmatrix} 21 & -6 & 15 \\ 33 & 12 & 45 \\ 9 & -4 & 5 \end{vmatrix} = 21\begin{vmatrix} 12 & 45 \\ -4 & 5 \end{vmatrix} + 6\begin{vmatrix} 33 & 45 \\ 9 & 5 \end{vmatrix} + 15\begin{vmatrix} 33 & 12 \\ 9 & -4 \end{vmatrix}$

$= 5040 - 1440 - 3600 = 0,$

$|D_y| = \begin{vmatrix} 10 & 21 & 15 \\ -20 & 33 & 45 \\ 5 & 9 & 5 \end{vmatrix} = 10\begin{vmatrix} 33 & 45 \\ 9 & 5 \end{vmatrix} - 21\begin{vmatrix} -20 & 45 \\ 5 & 5 \end{vmatrix} + 15\begin{vmatrix} -20 & 33 \\ 5 & 9 \end{vmatrix}$

$= -2400 + 6825 - 5175 = -750,$ and

$|D_z| = \begin{vmatrix} 10 & -6 & 21 \\ -20 & 12 & 33 \\ 5 & -4 & 9 \end{vmatrix} = 10\begin{vmatrix} 12 & 33 \\ -4 & 9 \end{vmatrix} + 6\begin{vmatrix} -20 & 33 \\ 5 & 9 \end{vmatrix} + 21\begin{vmatrix} -20 & 12 \\ 5 & -4 \end{vmatrix}$

$= 2400 - 2070 + 420 = 750.$

Therefore, $x = 0,\ y = -1,\ z = 1,$ and so the solution is $(0, -1, 1)$.

41. $\begin{cases} 3y + 5z = 4 \\ 2x - z = 10 \\ 4x + 7y = 0 \end{cases}$ Then, $|D| = \begin{vmatrix} 0 & 3 & 5 \\ 2 & 0 & -1 \\ 4 & 7 & 0 \end{vmatrix} = -3\begin{vmatrix} 2 & -1 \\ 4 & 0 \end{vmatrix} + 5\begin{vmatrix} 2 & 0 \\ 4 & 7 \end{vmatrix} = -12 + 70 = 58,$

$|D_x| = \begin{vmatrix} 4 & 3 & 5 \\ 10 & 0 & -1 \\ 0 & 7 & 0 \end{vmatrix} = -7\begin{vmatrix} 4 & 5 \\ 10 & -1 \end{vmatrix} = 378,$

$|D_y| = \begin{vmatrix} 0 & 4 & 5 \\ 2 & 10 & -1 \\ 4 & 0 & 0 \end{vmatrix} = 4\begin{vmatrix} 4 & 5 \\ 10 & -1 \end{vmatrix} = -216,$ and

$|D_z| = \begin{vmatrix} 0 & 3 & 4 \\ 2 & 0 & 10 \\ 4 & 7 & 0 \end{vmatrix} = 4\begin{vmatrix} 3 & 4 \\ 0 & 10 \end{vmatrix} - 7\begin{vmatrix} 0 & 4 \\ 2 & 10 \end{vmatrix} = 120 + 56 = 176.$

Thus, $x = \frac{189}{29},\ y = -\frac{108}{29},$ and $z = \frac{88}{29},$ and so the solution is $\left(\frac{189}{29}, -\frac{108}{29}, \frac{88}{29}\right).$

43. $\begin{cases} x + y + z + w = 0 \\ 2z \qquad\quad + w = 0 \\ \quad y - z = 0 \\ x \qquad + 2z = 1 \end{cases}$ Then $|D| = \begin{vmatrix} 1 & 1 & 1 & 1 \\ 2 & 0 & 0 & 1 \\ 0 & 1 & -1 & 0 \\ 1 & 0 & 2 & 0 \end{vmatrix} = -1\begin{vmatrix} 2 & 0 & 1 \\ 0 & -1 & 0 \\ 1 & 2 & 0 \end{vmatrix} - 1\begin{vmatrix} 1 & 1 & 1 \\ 2 & 0 & 1 \\ 1 & 2 & 0 \end{vmatrix}$

$= -\left(2\begin{vmatrix} -1 & 0 \\ 2 & 0 \end{vmatrix} + 1\begin{vmatrix} 0 & 1 \\ -1 & 0 \end{vmatrix}\right) - \left(-1\begin{vmatrix} 2 & 1 \\ 1 & 0 \end{vmatrix} - 2\begin{vmatrix} 1 & 1 \\ 2 & 1 \end{vmatrix}\right)$

$= -2(0) - 1(1) + 1(-1) + 2(-1) = -4,$

$|D_x| = \begin{vmatrix} 0 & 1 & 1 & 1 \\ 0 & 0 & 0 & 1 \\ 0 & 1 & -1 & 0 \\ 1 & 0 & 2 & 0 \end{vmatrix} = -1\begin{vmatrix} 1 & 1 & 1 \\ 0 & 0 & 1 \\ 1 & -1 & 0 \end{vmatrix} = -1(-1)\begin{vmatrix} 1 & 1 \\ 1 & -1 \end{vmatrix} = -2,$

$$|D_y| = \begin{vmatrix} 1 & 0 & 1 & 1 \\ 2 & 0 & 0 & 1 \\ 0 & 0 & -1 & 0 \\ 1 & 1 & 2 & 0 \end{vmatrix} = 1 \begin{vmatrix} 1 & 1 & 1 \\ 2 & 0 & 1 \\ 0 & -1 & 0 \end{vmatrix} = 1 \begin{vmatrix} 0 & 1 \\ -1 & 0 \end{vmatrix} - 2 \begin{vmatrix} 1 & 1 \\ -1 & 0 \end{vmatrix} = 1 - 2(1) = -1,$$

$$|D_z| = \begin{vmatrix} 1 & 1 & 0 & 1 \\ 2 & 0 & 0 & 1 \\ 0 & 1 & 0 & 0 \\ 1 & 0 & 1 & 0 \end{vmatrix} = -1 \begin{vmatrix} 1 & 1 & 1 \\ 2 & 0 & 1 \\ 0 & 1 & 0 \end{vmatrix} = -1 \begin{vmatrix} 0 & 1 \\ 1 & 0 \end{vmatrix} + 2 \begin{vmatrix} 1 & 1 \\ 1 & 0 \end{vmatrix} = -1(-1) + 2(-1) = -1, \text{ and}$$

$$|D_w| = \begin{vmatrix} 1 & 1 & 1 & 0 \\ 2 & 0 & 0 & 0 \\ 0 & 1 & -1 & 0 \\ 1 & 0 & 2 & 1 \end{vmatrix} = 1 \begin{vmatrix} 1 & 1 & 1 \\ 2 & 0 & 0 \\ 0 & 1 & -1 \end{vmatrix} = -2 \begin{vmatrix} 1 & 1 \\ 1 & -1 \end{vmatrix} = -2(-2) = 4. \text{ Hence, we have}$$

$$x = \frac{|D_x|}{|D|} = \frac{-2}{-4} = \tfrac{1}{2}; \; y = \frac{|D_y|}{|D|} = \frac{-1}{-4} = \tfrac{1}{4}; \; z = \frac{|D_z|}{|D|} = \frac{-1}{-4} = \tfrac{1}{4}; \; w = \frac{|D_w|}{|D|} = \frac{4}{-4} = -1, \text{ and the}$$

solution is $\left(\tfrac{1}{2}, \tfrac{1}{4}, \tfrac{1}{4}, -1\right)$.

45. $\begin{vmatrix} a & 0 & 0 & 0 & 0 \\ 0 & b & 0 & 0 & 0 \\ 0 & 0 & c & 0 & 0 \\ 0 & 0 & 0 & d & 0 \\ 0 & 0 & 0 & 0 & e \end{vmatrix} = a \begin{vmatrix} b & 0 & 0 & 0 \\ 0 & c & 0 & 0 \\ 0 & 0 & d & 0 \\ 0 & 0 & 0 & e \end{vmatrix} = ab \begin{vmatrix} c & 0 & 0 \\ 0 & d & 0 \\ 0 & 0 & e \end{vmatrix} = abc \begin{vmatrix} d & 0 \\ 0 & e \end{vmatrix} = abcde.$

47. $\begin{vmatrix} x & 12 & 13 \\ 0 & x-1 & 23 \\ 0 & 0 & x-2 \end{vmatrix} = 0 \quad \Leftrightarrow \quad (x-2) \begin{vmatrix} x & 12 \\ 0 & x-1 \end{vmatrix} = 0 \quad \Leftrightarrow \quad (x-2) \cdot x(x-1) = 0 \quad \Leftrightarrow$

$x = 0, 1$ or 2.

49. $\begin{vmatrix} 1 & 0 & x \\ x^2 & 1 & 0 \\ x & 0 & 1 \end{vmatrix} = 0 \quad \Leftrightarrow \quad 1 \begin{vmatrix} 1 & 0 \\ 0 & 1 \end{vmatrix} + x \begin{vmatrix} x^2 & 1 \\ x & 0 \end{vmatrix} = 0 \quad \Leftrightarrow \quad 1 - x^2 = 0 \quad \Leftrightarrow \quad x^2 = 1 \quad \Leftrightarrow$

$x = \pm 1$.

51. Area $= \pm\tfrac{1}{2} \begin{vmatrix} 0 & 0 & 1 \\ 6 & 2 & 1 \\ 3 & 8 & 1 \end{vmatrix} = \pm\tfrac{1}{2} \begin{vmatrix} 6 & 2 \\ 3 & 8 \end{vmatrix}$

$\quad = \pm\tfrac{1}{2}(48 - 6) = \tfrac{1}{2}(42) = 21.$

53. $\text{Area} = \pm\frac{1}{2}\begin{vmatrix} -1 & 3 & 1 \\ 2 & 9 & 1 \\ 5 & -6 & 1 \end{vmatrix}$

$$= \pm\frac{1}{2}\left[-1\begin{vmatrix} 9 & 1 \\ -6 & 1 \end{vmatrix} - 3\begin{vmatrix} 2 & 1 \\ 5 & 1 \end{vmatrix} + 1\begin{vmatrix} 2 & 9 \\ 5 & -6 \end{vmatrix}\right]$$

$$= \pm\frac{1}{2}[-1(9+6) - 3(2-5) + 1(-12-45)]$$

$$= \pm\frac{1}{2}[-15 - 3(-3) + (-57)] = \pm\frac{1}{2}[-63]$$

$$= \frac{63}{2}.$$

55. $\begin{vmatrix} 1 & x & x^2 \\ 1 & y & y^2 \\ 1 & z & z^2 \end{vmatrix} = 1\begin{vmatrix} y & y^2 \\ z & z^2 \end{vmatrix} - 1\begin{vmatrix} x & x^2 \\ z & z^2 \end{vmatrix} + 1\begin{vmatrix} x & x^2 \\ y & y^2 \end{vmatrix}$

$$= yz^2 - y^2z - (xz^2 - x^2z) + (xy^2 - xy^2)$$

$$= yz^2 - y^2z - xz^2 - x^2z + xy^2 - xy^2 + xyz - xyz$$

$$= xyz - xz^2 - y^2z + yz^2 - x^2y + x^2z + zy^2 - xyz$$

$$= z(xy - xz - y^2 + yz) - x(xy - xz - y^2 + yz)$$

$$= (z - x)(xy - xz - y^2 + yz)$$

$$= (z - x)[x(y - z) - y(y - z)]$$

$$= (z - x)(x - y)(y - z)$$

57. (a) Using the points $(10, 25)$, $(15, 33.75)$, and $(40, 40)$ we substitute for x and y and get the system

$$\begin{cases} 100a + 10b + c = 25 \\ 225a + 15b + c = 33.75 \\ 1600a + 40b + c = 40 \end{cases}$$

(b) $|D| = \begin{vmatrix} 100 & 10 & 1 \\ 225 & 15 & 1 \\ 1600 & 40 & 1 \end{vmatrix} = 1 \cdot \begin{vmatrix} 225 & 15 \\ 1600 & 40 \end{vmatrix} - 1 \cdot \begin{vmatrix} 100 & 10 \\ 1600 & 40 \end{vmatrix} + 1 \cdot \begin{vmatrix} 100 & 10 \\ 225 & 15 \end{vmatrix}$

$$= (9,000 - 24,000) - (4,000 - 16,000) + (1,500 - 2,250) = -15,000 + 12,000 - 750$$

$$= -3,750$$

$|D_a| = \begin{vmatrix} 25 & 10 & 1 \\ 33.75 & 15 & 1 \\ 40 & 40 & 1 \end{vmatrix} = 1 \cdot \begin{vmatrix} 33.75 & 15 \\ 40 & 40 \end{vmatrix} - 1 \cdot \begin{vmatrix} 25 & 10 \\ 40 & 40 \end{vmatrix} + 1 \cdot \begin{vmatrix} 25 & 10 \\ 33.75 & 15 \end{vmatrix}$

$$= (1350 - 600) - (1,000 - 400) + (375 - 337.5) = 750 - 600 + 37.5$$

$$= 187.5$$

$|D_b| = \begin{vmatrix} 100 & 25 & 1 \\ 225 & 33.75 & 1 \\ 1600 & 40 & 1 \end{vmatrix} = 1 \cdot \begin{vmatrix} 225 & 33.75 \\ 1600 & 40 \end{vmatrix} - 1 \cdot \begin{vmatrix} 100 & 25 \\ 1600 & 40 \end{vmatrix} + 1 \cdot \begin{vmatrix} 100 & 25 \\ 225 & 33.75 \end{vmatrix}$

$$= (9,000 - 54,000) - (4,000 - 40,000) + (3375 - 5625) = -45,000 + 36,000 - 2,250$$

$$= -11.250$$

$$|D_c| = \begin{vmatrix} 100 & 10 & 25 \\ 225 & 15 & 33.75 \\ 1600 & 40 & 40 \end{vmatrix}$$

$$= 25 \cdot \begin{vmatrix} 225 & 15 \\ 1600 & 40 \end{vmatrix} - 33.75 \cdot \begin{vmatrix} 100 & 10 \\ 1600 & 40 \end{vmatrix} + 40 \cdot \begin{vmatrix} 100 & 10 \\ 225 & 15 \end{vmatrix}$$

$$= 25 \cdot (9{,}000 - 24{,}000) - 33.75 \cdot (4{,}000 - 16{,}000) + 40 \cdot (1{,}500 - 2{,}250)$$

$$= 25 \cdot (-15{,}000) + 33.75 \cdot 12{,}000 + 40 \cdot (-750) = -375{,}000 + 405{,}000 - 30{,}000$$

$$= 0$$

$$a = \frac{|D_a|}{|D|} = \frac{187.5}{-3750} = 0.05; \; b = \frac{|D_b|}{|D|} = \frac{-11{,}250}{-3{,}750} = 3; \text{ and } c = \frac{|D_c|}{|D|} = \frac{0}{-3{,}750} = 0.$$

Thus model is $y = 0.05x^2 + 3x$.

59. (a) The coordinates of the vertices of the surrounding rectangle are (a_1, b_1), (a_2, b_1), (a_2, b_3), and (a_1, b_3). The area of the surrounding rectangle is given by
$$(a_2 - a_1) \cdot (b_3 - b_1) = a_2 b_3 + a_1 b_1 - a_2 b_1 - a_1 b_3 = a_1 b_1 + a_2 b_3 - a_1 b_3 - a_2 b_1.$$

(b) The area of the three blue triangles are:

Area of $\triangle (a_1, b_1)$, (a_2, b_1), (a_2, b_2) is $\frac{1}{2}(a_2 - a_1) \cdot (b_2 - b_1) = \frac{1}{2}(a_2 b_2 + a_1 b_1 - a_2 b_1 - a_1 b_2)$

Area of $\triangle (a_2, b_2)$, (a_2, b_3), (a_3, b_3) is $\frac{1}{2}(a_2 - a_3) \cdot (b_3 - b_2) = \frac{1}{2}(a_2 b_3 + a_3 b_2 - a_2 b_2 - a_3 b_3)$

Area of $\triangle (a_1, b)$, (a_1, b_3), (a_3, b_3) is $\frac{1}{2}(a_3 - a_1) \cdot (b_3 - b_1) = \frac{1}{2}(a_3 b_3 + a_1 b_1 - a_3 b_1 - a_1 b_3)$.

Thus the sum of the areas of the blue triangles, B, is

$$B = \frac{1}{2}(a_2 b_2 + a_1 b_1 - a_2 b_1 - a_1 b_2) + \frac{1}{2}(a_2 b_3 + a_3 b_2 - a_2 b_2 - a_3 b_3)$$
$$+ \frac{1}{2}(a_3 b_3 + a_1 b_1 - a_3 b_1 - a_1 b_3)$$

$$= \frac{1}{2}(a_1 b_1 + a_1 b_1 + a_2 b_2 + a_2 b_3 + a_3 b_2 + a_3 b_3)$$
$$- \frac{1}{2}(a_1 b_2 + a_1 b_3 + a_2 b_1 + a_2 b_2 + a_3 b_1 + a_3 b_3)$$

$$= a_1 b_1 + \frac{1}{2}(a_2 b_3 + a_3 b_2) - \frac{1}{2}(a_1 b_2 + a_1 b_3 + a_2 b_1 + a_3 b_1)$$

So the area of the red triangle, A, is given by $A = \begin{pmatrix} \text{Area of} \\ \text{rectangle} \end{pmatrix} - \begin{pmatrix} \text{Sum of the areas of} \\ \text{the blue triangles} \end{pmatrix}$ so

$$A = [a_1 b_1 + a_2 b_3 - a_1 b_3 - a_2 b_1]$$
$$- \left[a_1 b_1 + \frac{1}{2}(a_2 b_3 + a_3 b_2) - \frac{1}{2}(a_1 b_2 + a_1 b_3 + a_2 b_1 + a_3 b_1) \right]$$

$$= a_1 b_1 + a_2 b_3 - a_1 b_3 - a_2 b_1 - a_1 b_1 - \frac{1}{2}(a_2 b_3 + a_3 b_2)$$
$$+ \frac{1}{2}(a_1 b_2 + a_1 b_3 + a_2 b_1 + a_3 b_1)$$

$$= \frac{1}{2}(a_1 b_2 + a_2 b_3 + a_3 b_1) - \frac{1}{2}(a_1 b_3 + a_2 b_1 + a_3 b_2)$$

(c) We first find $\begin{vmatrix} a_1 & b_1 & 1 \\ a_2 & b_2 & 1 \\ a_3 & b_3 & 1 \end{vmatrix}$ by expanding about the third column.

$$\begin{vmatrix} a_1 & b_1 & 1 \\ a_2 & b_2 & 1 \\ a_3 & b_3 & 1 \end{vmatrix} = 1 \begin{vmatrix} a_2 & b_2 \\ a_3 & b_3 \end{vmatrix} - 1 \begin{vmatrix} a_1 & b_1 \\ a_3 & b_3 \end{vmatrix} + 1 \begin{vmatrix} a_1 & b_1 \\ a_2 & b_2 \end{vmatrix}$$

$$= a_2 b_3 - a_3 b_2 - (a_1 b_3 - a_3 b_1) + a_1 b_2 - a_2 b_1$$

$$= a_1 b_2 + a_2 b_3 + a_3 b_1 - a_1 b_3 - a_2 b_1 - a_3 b_2$$

So $\dfrac{1}{2}\begin{vmatrix} a_1 & b_1 & 1 \\ a_2 & b_2 & 1 \\ a_3 & b_3 & 1 \end{vmatrix} = \frac{1}{2}(a_1b_2 + a_2b_3 + a_3b_1) - \frac{1}{2}(a_1b_3 - a_2b_1 - a_3b_2) = $ Area of red triangle.

Since $\dfrac{1}{2}\begin{vmatrix} a_1 & b_1 & 1 \\ a_2 & b_2 & 1 \\ a_3 & b_3 & 1 \end{vmatrix}$ is not always positive, we have Area $= \pm\dfrac{1}{2}\begin{vmatrix} a_1 & b_1 & 1 \\ a_2 & b_2 & 1 \\ a_3 & b_3 & 1 \end{vmatrix}$.

61. (a) Let $|M| = \begin{vmatrix} x & y & 1 \\ x_1 & y_1 & 1 \\ x_2 & y_2 & 1 \end{vmatrix}$. Then, expanding by the third column,

$|M| = \begin{vmatrix} x_1 & y_1 \\ x_2 & y_2 \end{vmatrix} - \begin{vmatrix} x & y \\ x_2 & y_2 \end{vmatrix} + \begin{vmatrix} x & y \\ x_1 & y_1 \end{vmatrix} = (x_1y_2 - x_2y_1) - (xy_2 - x_2y) + (xy_1 - x_1y)$

$\quad = x_1y_2 - x_2y_1 - xy_2 + x_2y + xy_1 - x_1y = x_2y - x_1y - xy_2 + xy_1 + x_1y_2 - x_2y_1$

$\quad = (x_2 - x_1)y - (y_2 - y_1)x + x_1y_2 - x_2y_1$.

So $|M| = 0 \iff (x_2 - x_1)y - (y_2 - y_1)x + x_1y_2 - x_2y_1 = 0 \iff$

$(x_2 - x_1)y = (y_2 - y_1)x - x_1y_2 + x_2y_1 \iff$

$(x_2 - x_1)y = (y_2 - y_1)x - x_1y_2 + x_1y_1 - x_1y_1 + x_2y_1 \iff$

$y = \dfrac{y_2 - y_1}{x_2 - x_1}x - \dfrac{x_1(y_2 - y_1)}{x_2 - x_1} + \dfrac{y_1(x_2 - x_1)}{x_2 - x_1} \iff y = \dfrac{y_2 - y_1}{x_2 - x_1}(x - x_1) + y_1 \iff$

$y - y_1 = \dfrac{y_2 - y_1}{x_2 - x_1}(x - x_1)$, which is the "two-point" form of the equation for the line passing

through the points (x_1, y_1) and (x_2, y_2).

(b) Using the result of part (a), the line has the equation

$\begin{vmatrix} x & y & 1 \\ 20 & 50 & 1 \\ -10 & 25 & 1 \end{vmatrix} = 0 \implies \begin{vmatrix} 20 & 50 \\ -10 & 25 \end{vmatrix} - \begin{vmatrix} x & y \\ -10 & 25 \end{vmatrix} + \begin{vmatrix} x & y \\ 20 & 50 \end{vmatrix} = 0 \iff$

$(500 + 500) - (25x + 10y) + (50x - 20y) = 0 \iff 25x - 30y + 1000 = 0 \iff$

$5x - 6y + 200 = 0.$

63. The student should prefer Gaussian elimination, since six 5×5 determinants are much harder to evaluate than one 5-equation Gaussian elimination.

Review Exercises for Chapter 7

1. (a) 2×3

   (b) Yes, this matrix is in echelon form.

   (c) No, this matrix is not in reduced echelon form, since the leading "one" in the second row does not have a zero above it.

   (d) $\begin{cases} x + 2y = -5 \\ y = 3 \end{cases}$

3. (a) 3×4

   (b) Yes, this matrix is in echelon form.

   (c) Yes, this matrix is in reduced echelon form.

   (d) $\begin{cases} x + 8z = 0 \\ y + 5z = -1 \\ 0 = 0 \end{cases}$

5. (a) 3×4

   (b) No, this matrix is not in echelon form, since the leading "one" in the second row is not to the left of the one above it.

   (c) Since this matrix is not in echelon form, it is not in reduced echelon form

   (d) $\begin{cases} y - 3z = 4 \\ x + y = 7 \\ x + 2y + z = 2 \end{cases}$

7. $\begin{bmatrix} 1 & 2 & 2 & 6 \\ 1 & -1 & 0 & -1 \\ 2 & 1 & 3 & 7 \end{bmatrix} \xrightarrow{R_2 \leftrightarrow R_1} \begin{bmatrix} 1 & -1 & 0 & -1 \\ 1 & 2 & 2 & 6 \\ 2 & 1 & 3 & 7 \end{bmatrix} \xrightarrow[R_3 - 2R_1 \to R_3]{R_2 - R_1 \to R_2} \begin{bmatrix} 1 & -1 & 0 & -1 \\ 0 & 3 & 2 & 7 \\ 0 & 3 & 3 & 9 \end{bmatrix}$

   $\xrightarrow{R_3 - R_2 \to R_3} \begin{bmatrix} 1 & -1 & 0 & -1 \\ 0 & 3 & 2 & 7 \\ 0 & 0 & 1 & 2 \end{bmatrix}.$

   Thus, $z = 2$; $3y + 2(2) = 7 \iff 3y = 3 \iff y = 1$; and $x - (1) = -1 \iff x = 0$, and so the solution is $(0, 1, 2)$.

9. $\begin{bmatrix} 1 & -2 & 3 & -2 \\ 2 & -1 & 1 & 2 \\ 2 & -7 & 11 & -9 \end{bmatrix} \xrightarrow[R_3 - 2R_1 \to R_3]{R_2 - 2R_1 \to R_2} \begin{bmatrix} 1 & -2 & 3 & -2 \\ 0 & 3 & -5 & 6 \\ 0 & -3 & 5 & -5 \end{bmatrix} \xrightarrow{R_3 + R_2 \to R_3} \begin{bmatrix} 1 & -2 & 3 & -2 \\ 0 & 3 & -5 & 6 \\ 0 & 0 & 0 & 1 \end{bmatrix}$

   Since the last row corresponds to the equation $0 = 1$ which is always false. Thus there is no solution.

11. $\begin{bmatrix} 1 & 1 & 1 & 1 & 0 \\ 1 & -1 & -4 & -1 & -1 \\ 1 & -2 & 0 & 4 & -7 \\ 2 & 2 & 3 & 4 & -3 \end{bmatrix}$ $\xrightarrow[\substack{R_2-R_1 \to R_2,\ R_3-R_1 \to R_3 \\ R_4-2R_1 \to R_4}]{}$ $\begin{bmatrix} 1 & 1 & 1 & 1 & 0 \\ 0 & -2 & -5 & -2 & -1 \\ 0 & -3 & -1 & 3 & -7 \\ 0 & 0 & 1 & 2 & -3 \end{bmatrix}$ $\xrightarrow[]{-R_3+R_2 \to R_3}$

$\begin{bmatrix} 1 & 1 & 1 & 1 & 0 \\ 0 & 1 & -4 & -5 & 6 \\ 0 & -3 & -1 & 3 & -7 \\ 0 & 0 & 1 & 2 & -3 \end{bmatrix}$ $\xrightarrow[]{R_3+3R_2 \to R_3}$ $\begin{bmatrix} 1 & 1 & 1 & 1 & 0 \\ 0 & 1 & -4 & -5 & 6 \\ 0 & 0 & -13 & -12 & 11 \\ 0 & 0 & 1 & 2 & -3 \end{bmatrix}$ $\xrightarrow[]{R_3 \leftrightarrow R_4}$

$\begin{bmatrix} 1 & 1 & 1 & 1 & 0 \\ 0 & 1 & -4 & -5 & 6 \\ 0 & 0 & 1 & 2 & -3 \\ 0 & 0 & -13 & -12 & 11 \end{bmatrix}$ $\xrightarrow[]{R_4+13R_3 \to R_4}$ $\begin{bmatrix} 1 & 1 & 1 & 1 & 0 \\ 0 & 1 & -4 & -5 & 6 \\ 0 & 0 & 1 & 2 & -3 \\ 0 & 0 & 0 & 14 & -28 \end{bmatrix}$

Therefore, $14w = -28$ $\Leftrightarrow$ $w = -2$; $z + 2(-2) = -3$ $\Leftrightarrow$ $z = 1$; $y - 4(1) - 5(-2) = 6$ $\Leftrightarrow$ $y = 0$; and $x + 0 + 1 + (-2) = 0$ $\Leftrightarrow$ $x = 1$. So, the solution is $(1, 0, 1, -2)$.

13. $\begin{bmatrix} 1 & -1 & 3 & 2 \\ 2 & 1 & 1 & 2 \\ 3 & 0 & 4 & 4 \end{bmatrix}$ $\xrightarrow[\substack{R_2-2R_1 \to R_2, \\ R_3-3R_1 \to R_3}]{}$ $\begin{bmatrix} 1 & -1 & 3 & 2 \\ 0 & 3 & -5 & -2 \\ 0 & 3 & -5 & -2 \end{bmatrix}$ $\xrightarrow[]{R_3-R_2 \to R_3}$

$\begin{bmatrix} 1 & -1 & 3 & 2 \\ 0 & 3 & -5 & -2 \\ 0 & 0 & 0 & 0 \end{bmatrix}$ $\xrightarrow[]{\frac{1}{3}R_2}$ $\begin{bmatrix} 1 & -1 & 3 & 2 \\ 0 & 1 & -\frac{5}{3} & -\frac{2}{3} \\ 0 & 0 & 0 & 0 \end{bmatrix}$ $\xrightarrow[]{R_1+R_2 \to R_1}$ $\begin{bmatrix} 1 & 0 & \frac{4}{3} & \frac{4}{3} \\ 0 & 1 & -\frac{5}{3} & -\frac{2}{3} \\ 0 & 0 & 0 & 0 \end{bmatrix}$

Since the system is dependent, $z = t$; $y - \frac{5}{3}t = -\frac{2}{3}$ $\Leftrightarrow$ $y = \frac{5}{3}t - \frac{2}{3}$; $x + \frac{4}{3}t = \frac{4}{3}$ $\Leftrightarrow$ $x = -\frac{4}{3}t + \frac{4}{3}$. So, the solution is $\left(-\frac{4}{3}t + \frac{4}{3}, \frac{5}{3}t - \frac{2}{3}, t\right)$ where t is any real number.

15. $\begin{bmatrix} 1 & -1 & 1 & -1 & 0 \\ 3 & -1 & -1 & -1 & 2 \end{bmatrix}$ $\xrightarrow[]{R_2-3R_1 \to R_2}$ $\begin{bmatrix} 1 & -1 & 1 & -1 & 0 \\ 0 & 2 & -4 & 2 & 2 \end{bmatrix}$ $\xrightarrow[]{\frac{1}{2}R_4}$

$\begin{bmatrix} 1 & -1 & 1 & -1 & 0 \\ 0 & 1 & -2 & 1 & 1 \end{bmatrix}$ $\xrightarrow[]{R_1+R_2 \to R_1}$ $\begin{bmatrix} 1 & 0 & -1 & 0 & 1 \\ 0 & 1 & -2 & 1 & 1 \end{bmatrix}$

Since the system is dependent, $w = t$ and $z = s$; $y - 2s + t = 1$ $\Leftrightarrow$ $y = 2s - t + 1$; $x - s = 1$ $\Leftrightarrow$ $x = s + 1$. So, the solution is $(s + 1, 2s - t + 1, s, t)$ where t and s are any real number.

17. $\begin{bmatrix} 1 & -1 & 1 & 0 \\ 3 & 2 & -1 & 6 \\ 1 & 4 & -3 & 3 \end{bmatrix}$ $\xrightarrow[\substack{R_2-3R_1 \to R_2 \\ R_3-R_1 \to R_3}]{}$ $\begin{bmatrix} 1 & -1 & 1 & 0 \\ 0 & 5 & -4 & 6 \\ 0 & 5 & -4 & 3 \end{bmatrix}$ $\xrightarrow[]{R_3-R_2 \to R_3}$

$\begin{bmatrix} 1 & -1 & 1 & 0 \\ 0 & 5 & -4 & 6 \\ 0 & 0 & 0 & 3 \end{bmatrix}$

The last row of this matrix corresponds to the equation $0 = 3$, which is always false. Hence there is no solution.

19.
$$\begin{bmatrix} 1 & 1 & -1 & -1 & 2 \\ 1 & -1 & 1 & -1 & 0 \\ 2 & 0 & 0 & 2 & 2 \\ 2 & 4 & -4 & -2 & 6 \end{bmatrix} \xrightarrow{R_1 \leftrightarrow \frac{1}{2}R_3} \begin{bmatrix} 1 & 0 & 0 & 1 & 1 \\ 1 & -1 & 1 & -1 & 0 \\ 1 & 1 & -1 & -1 & 2 \\ 2 & 4 & -4 & -2 & 6 \end{bmatrix} \xrightarrow[R_4-2R_1 \to R_4]{R_2-R_1 \to R_2,\, R_3-R_1 \to R_3}$$

$$\begin{bmatrix} 1 & 0 & 0 & 1 & 1 \\ 0 & -1 & 1 & -2 & -1 \\ 0 & 1 & -1 & -2 & 1 \\ 0 & 4 & -4 & -4 & 4 \end{bmatrix} \xrightarrow[R_4+4R_2 \to R_4]{R_3+R_2 \to R_3} \begin{bmatrix} 1 & 0 & 0 & 1 & 1 \\ 0 & -1 & 1 & -2 & -1 \\ 0 & 0 & 0 & -4 & 0 \\ 0 & 0 & 0 & -12 & 0 \end{bmatrix} \xrightarrow[-\frac{1}{12}R_4]{-\frac{1}{4}R_3}$$

$$\begin{bmatrix} 1 & 0 & 0 & 1 & 1 \\ 0 & -1 & 1 & -2 & -1 \\ 0 & 0 & 0 & 0 & 0 \\ 0 & 0 & 0 & 1 & 0 \end{bmatrix} \xrightarrow[R_4-R_3 \to R_4]{R_1-R_3 \to R_1,\, R_2+2R_3 \to R_2} \begin{bmatrix} 1 & 0 & 0 & 0 & 1 \\ 0 & -1 & 1 & 0 & -1 \\ 0 & 0 & 0 & 1 & 0 \\ 0 & 0 & 0 & 0 & 0 \end{bmatrix}$$

This system is dependent, let $z = t$; $-y + t = -1 \iff y = t+1$; $x = 1 \iff x = 1$. So, the solution is $(1, t+1, t, 0)$ where t is any real number.

In Exercises 21–31,

$A = \begin{bmatrix} 2 & 0 & -1 \end{bmatrix}$

$B = \begin{bmatrix} 1 & 2 & 4 \\ -2 & 1 & 0 \end{bmatrix} \qquad D = \begin{bmatrix} 1 & 4 \\ 0 & -1 \\ 2 & 0 \end{bmatrix} \qquad F = \begin{bmatrix} 4 & 0 & 2 \\ -1 & 1 & 0 \\ 7 & 5 & 0 \end{bmatrix}$

$C = \begin{bmatrix} \frac{1}{2} & 3 \\ 2 & \frac{3}{2} \\ -2 & 1 \end{bmatrix} \qquad E = \begin{bmatrix} 2 & -1 \\ -\frac{1}{2} & 1 \end{bmatrix} \qquad G = \begin{bmatrix} 5 \end{bmatrix}$

21. $A + B$ cannot be performed because the matrix dimensions (1×3 and 2×3) are not compatible.

23. $2C + 3D = 2 \begin{bmatrix} \frac{1}{2} & 3 \\ 2 & \frac{3}{2} \\ -2 & 1 \end{bmatrix} + 3 \begin{bmatrix} 1 & 4 \\ 0 & -1 \\ 2 & 0 \end{bmatrix} = \begin{bmatrix} 1 & 6 \\ 4 & 3 \\ -4 & 2 \end{bmatrix} + \begin{bmatrix} 3 & 12 \\ 0 & -3 \\ 6 & 0 \end{bmatrix} = \begin{bmatrix} 4 & 18 \\ 4 & 0 \\ 2 & 2 \end{bmatrix}$

25. $GA = \begin{bmatrix} 5 \end{bmatrix} \begin{bmatrix} 2 & 0 & -1 \end{bmatrix} = \begin{bmatrix} 10 & 0 & -5 \end{bmatrix}$

27. $BC = \begin{bmatrix} 1 & 2 & 4 \\ -2 & 1 & 0 \end{bmatrix} \begin{bmatrix} \frac{1}{2} & 3 \\ 2 & \frac{3}{2} \\ -2 & 1 \end{bmatrix} = \begin{bmatrix} -\frac{7}{2} & 10 \\ 1 & -\frac{9}{2} \end{bmatrix}$

29. $BF = \begin{bmatrix} 1 & 2 & 4 \\ -2 & 1 & 0 \end{bmatrix} \begin{bmatrix} 4 & 0 & 2 \\ -1 & 1 & 0 \\ 7 & 5 & 0 \end{bmatrix} = \begin{bmatrix} 30 & 22 & 2 \\ -9 & 1 & -4 \end{bmatrix}$

31. $(C + D)E = \left(\begin{bmatrix} \frac{1}{2} & 3 \\ 2 & \frac{3}{2} \\ -2 & 1 \end{bmatrix} + \begin{bmatrix} 1 & 4 \\ 0 & -1 \\ 2 & 0 \end{bmatrix} \right) \begin{bmatrix} 2 & -1 \\ -\frac{1}{2} & 1 \end{bmatrix}$

$$= \begin{bmatrix} \frac{3}{2} & 7 \\ 2 & \frac{1}{2} \\ 0 & 1 \end{bmatrix} \begin{bmatrix} 2 & -1 \\ -\frac{1}{2} & 1 \end{bmatrix} = \begin{bmatrix} -\frac{1}{2} & \frac{11}{2} \\ \frac{15}{4} & -\frac{3}{2} \\ -\frac{1}{2} & 1 \end{bmatrix}$$

33. $AB = \begin{bmatrix} 2 & -5 \\ -2 & 6 \end{bmatrix} \begin{bmatrix} 3 & \frac{5}{2} \\ 1 & 1 \end{bmatrix} = \begin{bmatrix} 1 & 0 \\ 0 & 1 \end{bmatrix}$ and $BA = \begin{bmatrix} 3 & \frac{5}{2} \\ 1 & 1 \end{bmatrix} \begin{bmatrix} 2 & -5 \\ -2 & 6 \end{bmatrix} = \begin{bmatrix} 1 & 0 \\ 0 & 1 \end{bmatrix}$.

In exercises 35-39, $A = \begin{bmatrix} 2 & 1 \\ 3 & 2 \end{bmatrix}$, $B = \begin{bmatrix} 1 & -2 \\ -2 & 4 \end{bmatrix}$, and $C = \begin{bmatrix} 0 & 1 & 3 \\ -2 & 4 & 0 \end{bmatrix}$.

35. $A + 3X = B \quad \Leftrightarrow \quad 3X = B - A \quad \Leftrightarrow \quad X = \frac{1}{3}(B - A)$. Thus

$$X = \frac{1}{3}\left(\begin{bmatrix} 1 & -2 \\ -2 & 4 \end{bmatrix} - \begin{bmatrix} 2 & 1 \\ 3 & 2 \end{bmatrix} \right) = \frac{1}{3}\begin{bmatrix} -1 & -3 \\ -5 & 2 \end{bmatrix}$$

37. $2(X - A) = 3B \quad \Leftrightarrow \quad X - A = \frac{3}{2}B \quad \Leftrightarrow \quad X = A + \frac{3}{2}B$. Thus

$$X = \begin{bmatrix} 2 & 1 \\ 3 & 2 \end{bmatrix} + \frac{3}{2}\begin{bmatrix} 1 & -2 \\ -2 & 4 \end{bmatrix} = \begin{bmatrix} 2 & 1 \\ 3 & 2 \end{bmatrix} + \begin{bmatrix} \frac{3}{2} & -3 \\ -3 & 6 \end{bmatrix} = \begin{bmatrix} \frac{7}{2} & -2 \\ 0 & 8 \end{bmatrix}$$

39. $AX = C \quad \Leftrightarrow \quad A^{-1}AX = X = A^{-1}C$. Since $A = \begin{bmatrix} 2 & 1 \\ 3 & 2 \end{bmatrix}$ we have

$$A^{-1} = \frac{1}{4-3}\begin{bmatrix} 2 & -1 \\ -3 & 2 \end{bmatrix} = \begin{bmatrix} 2 & -1 \\ -3 & 2 \end{bmatrix}.$$

Thus $X = A^{-1}C = \begin{bmatrix} 2 & -1 \\ -3 & 2 \end{bmatrix} \begin{bmatrix} 0 & 1 & 3 \\ -2 & 4 & 0 \end{bmatrix} = \begin{bmatrix} 2 & -2 & 6 \\ -4 & 5 & -9 \end{bmatrix}$

41. $D = \begin{bmatrix} 1 & 4 \\ 2 & 9 \end{bmatrix}$. Then, $|D| = 1(9) - 2(4) = 1$, and so $D^{-1} = \begin{bmatrix} 9 & -4 \\ -2 & 1 \end{bmatrix}$.

43. $D = \begin{bmatrix} 4 & -12 \\ -2 & 6 \end{bmatrix}$. Then, $|D| = 4(6) - 2(12) = 0$, and so D has no inverse.

45. $D = \begin{bmatrix} 3 & 0 & 1 \\ 2 & -3 & 0 \\ 4 & -2 & 1 \end{bmatrix}$. Then, $|D| = 1\begin{vmatrix} 2 & -3 \\ 4 & -2 \end{vmatrix} + 1\begin{vmatrix} 3 & 0 \\ 2 & -3 \end{vmatrix} = -4 + 12 - 9 = -1$. So D^{-1} exists.

$$\begin{bmatrix} 3 & 0 & 1 & 1 & 0 & 0 \\ 2 & -3 & 0 & 0 & 1 & 0 \\ 4 & -2 & 1 & 0 & 0 & 1 \end{bmatrix} \xrightarrow{R_1 - R_2 \to R_1} \begin{bmatrix} 1 & 3 & 1 & 1 & -1 & 0 \\ 2 & -3 & 0 & 0 & 1 & 0 \\ 4 & -2 & 1 & 0 & 0 & 1 \end{bmatrix} \xrightarrow[R_3 - 4R_1 \to R_3]{R_2 - 2R_1 \to R_2}$$

$$\begin{bmatrix} 1 & 3 & 1 & 1 & -1 & 0 \\ 0 & -9 & -2 & -2 & 3 & 0 \\ 0 & -14 & -3 & -4 & 4 & 1 \end{bmatrix} \xrightarrow[-2R_3]{-3R_2} \begin{bmatrix} 1 & 3 & 1 & 1 & -1 & 0 \\ 0 & 27 & 6 & 6 & -9 & 0 \\ 0 & 28 & 6 & 8 & -8 & -2 \end{bmatrix} \xrightarrow{R_3 - R_2 \to R_3}$$

$$\begin{bmatrix} 1 & 3 & 1 & 1 & -1 & 0 \\ 0 & 27 & 6 & 6 & -9 & 0 \\ 0 & 1 & 0 & 2 & 1 & -2 \end{bmatrix} \xrightarrow[\frac{1}{3}R_3]{R_3 \leftrightarrow R_2} \begin{bmatrix} 1 & 3 & 1 & 1 & -1 & 0 \\ 0 & 1 & 0 & 2 & 1 & -2 \\ 0 & 9 & 2 & 2 & -3 & 0 \end{bmatrix} \xrightarrow[R_1 - 3R_2 \to R_1]{R_3 - 9R_2 \to R_3}$$

$$\begin{bmatrix} 1 & 0 & 1 & -5 & -4 & 6 \\ 0 & 1 & 0 & 2 & 1 & -2 \\ 0 & 0 & 2 & -16 & -12 & 18 \end{bmatrix} \xrightarrow[R_1 - R_3 \to R_1]{\frac{1}{2}R_3} \begin{bmatrix} 1 & 0 & 0 & 3 & 2 & -3 \\ 0 & 1 & 0 & 2 & 1 & -2 \\ 0 & 0 & 1 & -8 & -6 & 9 \end{bmatrix}.$$

Thus, $D^{-1} = \begin{bmatrix} 3 & 2 & -3 \\ 2 & 1 & -2 \\ -8 & -6 & 9 \end{bmatrix}.$

47. $D = \begin{bmatrix} 1 & 0 & 0 & 1 \\ 0 & 2 & 0 & 2 \\ 0 & 0 & 3 & 3 \\ 0 & 0 & 0 & 4 \end{bmatrix}.$ Thus, $|D| = \begin{vmatrix} 2 & 0 & 2 \\ 0 & 3 & 3 \\ 0 & 0 & 4 \end{vmatrix} = 2 \begin{vmatrix} 3 & 3 \\ 0 & 4 \end{vmatrix} = 24$ and D^{-1} exists.

$$\begin{bmatrix} 1 & 0 & 0 & 1 & 1 & 0 & 0 & 0 \\ 0 & 2 & 0 & 2 & 0 & 1 & 0 & 0 \\ 0 & 0 & 3 & 3 & 0 & 0 & 1 & 0 \\ 0 & 0 & 0 & 4 & 0 & 0 & 0 & 1 \end{bmatrix} \xrightarrow[\frac{1}{4}R_4]{\frac{1}{2}R_2, \frac{1}{3}R_3} \begin{bmatrix} 1 & 0 & 0 & 1 & 1 & 0 & 0 & 0 \\ 0 & 1 & 0 & 1 & 0 & \frac{1}{2} & 0 & 0 \\ 0 & 0 & 1 & 1 & 0 & 0 & \frac{1}{3} & 0 \\ 0 & 0 & 0 & 1 & 0 & 0 & 0 & \frac{1}{4} \end{bmatrix}$$

$$\xrightarrow[R_3 - R_4 \to R_3]{R_1 - R_4 \to R_1,\ R_2 - R_4 \to R_2} \begin{bmatrix} 1 & 0 & 0 & 0 & 1 & 0 & 0 & -\frac{1}{4} \\ 0 & 1 & 0 & 0 & 0 & \frac{1}{2} & 0 & -\frac{1}{4} \\ 0 & 0 & 1 & 0 & 0 & 0 & \frac{1}{3} & -\frac{1}{4} \\ 0 & 0 & 0 & 1 & 0 & 0 & 0 & \frac{1}{4} \end{bmatrix}.$$

Therefore, $D^{-1} = \begin{bmatrix} 1 & 0 & 0 & -\frac{1}{4} \\ 0 & \frac{1}{2} & 0 & -\frac{1}{4} \\ 0 & 0 & \frac{1}{3} & -\frac{1}{4} \\ 0 & 0 & 0 & \frac{1}{4} \end{bmatrix}.$

49. $\begin{bmatrix} 12 & -5 \\ 5 & -2 \end{bmatrix} \begin{bmatrix} x \\ y \end{bmatrix} = \begin{bmatrix} 10 \\ 17 \end{bmatrix}.$ If we let $A = \begin{bmatrix} 12 & -5 \\ 5 & -2 \end{bmatrix}$, then $A^{-1} = \frac{1}{-24+25} \begin{bmatrix} -2 & 5 \\ -5 & 12 \end{bmatrix}$

$= \begin{bmatrix} -2 & 5 \\ -5 & 12 \end{bmatrix}$, and so $\begin{bmatrix} x \\ y \end{bmatrix} = \begin{bmatrix} -2 & 5 \\ -5 & 12 \end{bmatrix} \begin{bmatrix} 10 \\ 17 \end{bmatrix} = \begin{bmatrix} 65 \\ 154 \end{bmatrix}.$ Therefore, the solution is $(65, 154)$.

51. $\begin{bmatrix} 2 & 1 & 5 \\ 1 & 2 & 2 \\ 1 & 0 & 3 \end{bmatrix} \begin{bmatrix} x \\ y \\ z \end{bmatrix} = \begin{bmatrix} \frac{1}{3} \\ \frac{1}{4} \\ \frac{1}{6} \end{bmatrix}.$ Let $A = \begin{bmatrix} 2 & 1 & 5 \\ 1 & 2 & 2 \\ 1 & 0 & 3 \end{bmatrix}.$ Then,

$$\begin{bmatrix} 2 & 1 & 5 & 1 & 0 & 0 \\ 1 & 2 & 2 & 0 & 1 & 0 \\ 1 & 0 & 3 & 0 & 0 & 1 \end{bmatrix} \xrightarrow{R_1 \leftrightarrow R_2} \begin{bmatrix} 1 & 2 & 2 & 0 & 1 & 0 \\ 2 & 1 & 5 & 1 & 0 & 0 \\ 1 & 0 & 3 & 0 & 0 & 1 \end{bmatrix} \xrightarrow[R_3 - R_1 \to R_3]{R_2 - 2R_1 \to R_2}$$

$$\begin{bmatrix} 1 & 2 & 2 & 0 & 1 & 0 \\ 0 & -3 & 1 & 1 & -2 & 0 \\ 0 & -2 & 1 & 0 & -1 & 1 \end{bmatrix} \xrightarrow{R_2 - 2R_3 \to R_2} \begin{bmatrix} 1 & 2 & 2 & 0 & 1 & 0 \\ 0 & 1 & -1 & 1 & 0 & -2 \\ 0 & -2 & 1 & 0 & -1 & 1 \end{bmatrix} \xrightarrow[R_3 \to R_3 + 2R_2]{R_1 - 2R_2 \to R_1}$$

$$\begin{bmatrix} 1 & 0 & 4 & -2 & 1 & 4 \\ 0 & 1 & -1 & 1 & 0 & -2 \\ 0 & 0 & -1 & 2 & -1 & -3 \end{bmatrix} \xrightarrow{-R_3} \begin{bmatrix} 1 & 0 & 4 & -2 & 1 & 4 \\ 0 & 1 & -1 & 1 & 0 & -2 \\ 0 & 0 & 1 & -2 & 1 & 3 \end{bmatrix} \xrightarrow[R_2 + R_3 \to R_2]{R_1 - 4R_3 \to R_1}$$

$$\begin{bmatrix} 1 & 0 & 0 & 6 & -3 & -8 \\ 0 & 1 & 0 & -1 & 1 & 1 \\ 0 & 0 & 1 & -2 & 1 & 3 \end{bmatrix}.$$

Hence, $A^{-1} = \begin{bmatrix} 6 & -3 & -8 \\ -1 & 1 & 1 \\ -2 & 1 & 3 \end{bmatrix}$ and $\begin{bmatrix} x \\ y \\ z \end{bmatrix} = \begin{bmatrix} 6 & -3 & -8 \\ -1 & 1 & 1 \\ -2 & 1 & 3 \end{bmatrix} \begin{bmatrix} \frac{1}{3} \\ \frac{1}{4} \\ \frac{1}{6} \end{bmatrix} = \begin{bmatrix} -\frac{1}{12} \\ \frac{1}{12} \\ \frac{1}{12} \end{bmatrix}$, and so the

solution is $\left(-\frac{1}{12}, \frac{1}{12}, \frac{1}{12}\right)$.

53. $|D| = \begin{vmatrix} 2 & 7 \\ 6 & 16 \end{vmatrix} = 32 - 42 = -10; \quad |D_x| = \begin{vmatrix} 13 & 7 \\ 30 & 16 \end{vmatrix} = 208 - 210 = -2; \quad$ and

$|D_y| = \begin{vmatrix} 2 & 13 \\ 6 & 30 \end{vmatrix} = 60 - 78 = -18.$ Therefore, $x = \frac{-2}{-10} = \frac{1}{5}$ and $y = \frac{-18}{-10} = \frac{9}{5}$, and so the solution

is $\left(\frac{1}{5}, \frac{9}{5}\right)$.

55. $|D| = \begin{vmatrix} 2 & -1 & 5 \\ -1 & 7 & 0 \\ 5 & 4 & 3 \end{vmatrix} = 5\begin{vmatrix} -1 & 7 \\ 5 & 4 \end{vmatrix} + 3\begin{vmatrix} 2 & -1 \\ -1 & 7 \end{vmatrix} = -195 + 39 = -156;$

$|D_x| = \begin{vmatrix} 0 & -1 & 5 \\ 9 & 7 & 0 \\ -9 & 4 & 3 \end{vmatrix} = 5\begin{vmatrix} 9 & 7 \\ -9 & 4 \end{vmatrix} + 3\begin{vmatrix} 0 & -1 \\ 9 & 7 \end{vmatrix} = 495 + 27 = 522;$

$|D_y| = \begin{vmatrix} 2 & 0 & 5 \\ -1 & 9 & 0 \\ 5 & -9 & 3 \end{vmatrix} = 5\begin{vmatrix} -1 & 9 \\ 5 & -9 \end{vmatrix} + 3\begin{vmatrix} 2 & 0 \\ -1 & 9 \end{vmatrix} = -180 + 54 = -126;$ and

$|D_z| = \begin{vmatrix} 2 & -1 & 0 \\ -1 & 7 & 9 \\ 5 & 4 & -9 \end{vmatrix} = -9\begin{vmatrix} 2 & -1 \\ 5 & 4 \end{vmatrix} - 9\begin{vmatrix} 2 & -1 \\ -1 & 7 \end{vmatrix} = -117 - 117 = -234.$

Therefore, $x = \frac{522}{-156} = -\frac{87}{26}$, $y = \frac{-126}{-156} = \frac{21}{26}$, and $z = \frac{-234}{-156} = \frac{3}{2}$, and so the solution is $\left(-\frac{87}{26}, \frac{21}{26}, \frac{3}{2}\right)$.

57. Area $= \pm\frac{1}{2}\begin{vmatrix} -1 & 3 & 1 \\ 3 & 1 & 1 \\ -2 & -2 & 1 \end{vmatrix} = \pm\frac{1}{2}\left(\begin{vmatrix} 3 & 1 \\ -2 & -2 \end{vmatrix} - \begin{vmatrix} -1 & 3 \\ -2 & -2 \end{vmatrix} + \begin{vmatrix} -1 & 3 \\ 3 & 1 \end{vmatrix}\right) = \pm\frac{1}{2}(-4 - 8 - 10)$

$= 11.$

59. Let $x =$ dollars invested in Bank A, $y =$ dollars invested in Bank B, and $z =$ dollars invested in Bank C. We get the following system:

$$\begin{cases} x + y + z = 60{,}000 \\ 0.02x + 0.025y + 0.03z = 1575 \\ 2x + 2z = y \end{cases} \Leftrightarrow \begin{cases} x + y + z = 60{,}000 \\ 2x + 2.5y + 3z = 157{,}500 \\ 2x - y + 2z = 0 \end{cases}$$

$$\begin{bmatrix} 1 & 1 & 1 & 60{,}000 \\ 2 & 2.5 & 3 & 157{,}500 \\ 2 & -1 & 2 & 0 \end{bmatrix} \xrightarrow[R_3 - 2R_1 \to R_3]{R_2 - 2R_1 \to R_2} \begin{bmatrix} 1 & 1 & 1 & 60{,}000 \\ 0 & 0.5 & 1 & 37{,}500 \\ 0 & -3 & 0 & -120{,}000 \end{bmatrix} \xrightarrow{R_2 \leftrightarrow -\frac{1}{3}R_3}$$

$$\begin{bmatrix} 1 & 1 & 1 & 60{,}000 \\ 0 & 1 & 0 & 40{,}000 \\ 0 & 0.5 & 1 & 37{,}500 \end{bmatrix} \xrightarrow[R_3 - 0.5R_2 \to R_3]{R_1 - R_2 \to R_1} \begin{bmatrix} 1 & 0 & 1 & 20{,}000 \\ 0 & 1 & 0 & 40{,}000 \\ 0 & 0 & 1 & 17{,}500 \end{bmatrix} \xrightarrow{R_1 - R_3 \to R_1}$$

$$\begin{bmatrix} 1 & 0 & 1 & 2{,}500 \\ 0 & 1 & 0 & 40{,}000 \\ 0 & 0 & 1 & 17{,}500 \end{bmatrix}.$$

Thus she invests \$2,500 in Bank A, \$40,000 in Bank B, and \$17,500 in Bank C. .

Chapter 7 Test

1. This matrix is in row-echelon form, but not in reduced row-echelon form, since the leading "one" in the second row does not have a zero above it.

2. This matrix is in row-reduced echelon form.

3. This matrix is neither in row-echelon form nor in row-reduced echelon form.

4. $\begin{cases} x - y + 2z = 0 \\ 2x - 4y + 5z = -5 \\ 2y - 3z = 5 \end{cases}$ This system has the following matrix representation.

$$\begin{bmatrix} 1 & -1 & 2 & 0 \\ 2 & -4 & 5 & -5 \\ 0 & 2 & -3 & 5 \end{bmatrix} \xrightarrow{R_2 - R_1 \to R_2} \begin{bmatrix} 1 & -1 & 2 & 0 \\ 0 & -2 & 1 & -5 \\ 0 & 2 & -3 & 5 \end{bmatrix} \xrightarrow{R_3 + R_2 \to R_3}$$

$$\begin{bmatrix} 1 & -1 & 2 & 0 \\ 0 & -2 & 1 & -5 \\ 0 & 0 & -2 & 0 \end{bmatrix} \xrightarrow{-\frac{1}{2}R_3} \begin{bmatrix} 1 & -1 & 2 & 0 \\ 0 & -2 & 1 & -5 \\ 0 & 0 & 1 & 0 \end{bmatrix}.$$

Thus $z = 0$; $-2y + 0 = -5$ $\Leftrightarrow$ $y = \frac{5}{2}$; and $x - \frac{5}{2} + 2(0) = 0$ $\Leftrightarrow$ $x = \frac{5}{2}$. Thus the solution is $(\frac{5}{2}, \frac{5}{2}, 0)$.

5. $\begin{cases} 2x - 3y + z = 3 \\ x + 2y + 2z = -1 \\ 4x + y + 5z = 4 \end{cases}$ This system has the following matrix representation.

$$\begin{bmatrix} 2 & -3 & 1 & 3 \\ 1 & 2 & 2 & -1 \\ 4 & 1 & 5 & 4 \end{bmatrix} \xrightarrow{R_1 \leftrightarrow R_2} \begin{bmatrix} 1 & 2 & 2 & -1 \\ 2 & -3 & 1 & 3 \\ 4 & 1 & 5 & 4 \end{bmatrix} \xrightarrow[R_3 - 4R_1 \to R_3]{R_2 - 2R_1 \to R_2} \begin{bmatrix} 1 & 2 & 2 & -1 \\ 0 & -7 & -3 & 5 \\ 0 & -7 & -3 & 8 \end{bmatrix}$$

$$\xrightarrow{R_3 - R_2 \to R_3} \begin{bmatrix} 1 & 2 & 2 & -1 \\ 0 & -7 & -3 & 5 \\ 0 & 0 & 0 & 3 \end{bmatrix}$$

Since the row corresponds to the equation $0 = 3$, this system has no solution.

6. $\begin{cases} x + 3y - z = 0 \\ 3x + 4y - 2z = -1 \\ -x + 2y = 1 \end{cases}$ This system has the following matrix representation.

$$\begin{bmatrix} 1 & 3 & -1 & 0 \\ 3 & 4 & -2 & -1 \\ -1 & 2 & 0 & 1 \end{bmatrix} \xrightarrow[R_3 + R_1 \to R_3]{R_2 - 3R_1 \to R_2} \begin{bmatrix} 1 & 3 & -1 & 0 \\ 0 & -5 & 1 & -1 \\ 0 & 5 & -1 & 1 \end{bmatrix} \xrightarrow{R_3 - R_2 \to R_3} \begin{bmatrix} 1 & 3 & -1 & 0 \\ 0 & -5 & 1 & -1 \\ 0 & 0 & 0 & 0 \end{bmatrix}$$

$$\xrightarrow{-\frac{1}{5}R_2} \begin{bmatrix} 1 & 3 & -1 & 0 \\ 0 & 1 & -\frac{1}{5} & \frac{1}{5} \\ 0 & 0 & 0 & 0 \end{bmatrix} \xrightarrow{R_1 - 3R_2 \to R_1} \begin{bmatrix} 1 & 0 & -\frac{2}{5} & -\frac{3}{5} \\ 0 & 1 & -\frac{1}{5} & \frac{1}{5} \\ 0 & 0 & 0 & 0 \end{bmatrix}$$

Since this system is dependent let $z = t$. Then $y - \frac{1}{5}t = \frac{1}{5}$ $\Leftrightarrow$ $y = \frac{1}{5}t + \frac{1}{5}$ and $x - \frac{2}{5}t = -\frac{3}{5}$ $\Leftrightarrow$ $x = \frac{2}{5}t - \frac{3}{5}$. Thus the solution is $(\frac{2}{5}t - \frac{3}{5}, \frac{1}{5}t + \frac{1}{5}, t)$.

In problems 7 – 14, the matrices A, B, and C are defined as follows:

$$A = \begin{bmatrix} 2 & 3 \\ 2 & 4 \end{bmatrix} \qquad B = \begin{bmatrix} 2 & 4 \\ -1 & 1 \\ 3 & 0 \end{bmatrix} \qquad C = \begin{bmatrix} 1 & 0 & 4 \\ -1 & 1 & 2 \\ 0 & 1 & 3 \end{bmatrix}$$

7. $A + B$ is undefined because A is 2×2 and B is 3×2, so they have incompatible dimensions.

8. AB is undefined because A is 2×2 and B is 3×2, so they have incompatible dimensions.

9. $BA - 3B = \begin{bmatrix} 2 & 4 \\ -1 & 1 \\ 3 & 0 \end{bmatrix} \begin{bmatrix} 2 & 3 \\ 2 & 4 \end{bmatrix} - 3 \begin{bmatrix} 2 & 4 \\ -1 & 1 \\ 3 & 0 \end{bmatrix} = \begin{bmatrix} 12 & 22 \\ 0 & 1 \\ 6 & 9 \end{bmatrix} - \begin{bmatrix} 6 & 12 \\ -3 & 3 \\ 9 & 0 \end{bmatrix} = \begin{bmatrix} 6 & 10 \\ 3 & -2 \\ -3 & 9 \end{bmatrix}$

10. $CBA = \begin{bmatrix} 1 & 0 & 4 \\ -1 & 1 & 2 \\ 0 & 1 & 3 \end{bmatrix} \begin{bmatrix} 2 & 4 \\ -1 & 1 \\ 3 & 0 \end{bmatrix} \begin{bmatrix} 2 & 3 \\ 2 & 4 \end{bmatrix} = \begin{bmatrix} 14 & 4 \\ 3 & -3 \\ 8 & 1 \end{bmatrix} \begin{bmatrix} 2 & 3 \\ 2 & 4 \end{bmatrix} = \begin{bmatrix} 36 & 58 \\ 0 & -3 \\ 18 & 28 \end{bmatrix}$

11. $A = \begin{bmatrix} 2 & 3 \\ 2 & 4 \end{bmatrix} \Leftrightarrow A^{-1} = \frac{1}{8-6} \begin{bmatrix} 4 & -3 \\ -2 & 2 \end{bmatrix} = \begin{bmatrix} 2 & -\frac{3}{2} \\ -1 & 1 \end{bmatrix}$

12. B^{-1} does not exist since B is not a square matrix.

13. $\det(B)$ is not defined since B is not a square matrix.

14. $\det(C) = \begin{vmatrix} 1 & 0 & 4 \\ -1 & 1 & 2 \\ 0 & 1 & 3 \end{vmatrix} = 1 \begin{vmatrix} 1 & 2 \\ 1 & 3 \end{vmatrix} + 4 \begin{vmatrix} -1 & 1 \\ 0 & 1 \end{vmatrix} = 1 - 4 = -3$

15. (a) The system $\begin{cases} 4x - 3y = 10 \\ 3x - 2y = 30 \end{cases}$ is equivalent to the matrix equation $\begin{bmatrix} 4 & -3 \\ 3 & -2 \end{bmatrix} \begin{bmatrix} x \\ y \end{bmatrix} = \begin{bmatrix} 10 \\ 30 \end{bmatrix}$.

(b) We have $|D| = \begin{vmatrix} 4 & -3 \\ 3 & -2 \end{vmatrix} = 4(-2) - 3(-3) = 1$. So $D^{-1} = \begin{bmatrix} -2 & 3 \\ -3 & 4 \end{bmatrix}$, and

$\begin{bmatrix} x \\ y \end{bmatrix} = \begin{bmatrix} -2 & 3 \\ -3 & 4 \end{bmatrix} \begin{bmatrix} 10 \\ 30 \end{bmatrix} = \begin{bmatrix} 70 \\ 90 \end{bmatrix}$. Therefore, $x = 70$ and $y = 90$.

16. $|A| = \begin{vmatrix} 1 & 4 & 1 \\ 0 & 2 & 0 \\ 1 & 0 & 1 \end{vmatrix} = 2 \begin{vmatrix} 1 & 1 \\ 1 & 1 \end{vmatrix} = 0; \quad |B| = \begin{vmatrix} 1 & 4 & 0 \\ 0 & 2 & 0 \\ -3 & 0 & 1 \end{vmatrix} = 2 \begin{vmatrix} 1 & 0 \\ -3 & 1 \end{vmatrix} = 2.$

Since $|A| = 0$, A does not have an inverse, and since $|B| \neq 0$, B does have an inverse.

$\begin{bmatrix} 1 & 4 & 0 & 1 & 0 & 0 \\ 0 & 2 & 0 & 0 & 1 & 0 \\ -3 & 0 & 1 & 0 & 0 & 1 \end{bmatrix} \xrightarrow{R_3 + 3R_1 \to R_3} \begin{bmatrix} 1 & 4 & 0 & 1 & 0 & 0 \\ 0 & 2 & 0 & 0 & 1 & 0 \\ 0 & 12 & 1 & 3 & 0 & 1 \end{bmatrix} \xrightarrow[R_3 - 6R_2 \to R_3]{R_1 - 2R_2 \to R_1}$

$\begin{bmatrix} 1 & 0 & 0 & 1 & -2 & 0 \\ 0 & 2 & 0 & 0 & 1 & 0 \\ 0 & 0 & 1 & 3 & -6 & 1 \end{bmatrix} \xrightarrow{\frac{1}{2}R_2} \begin{bmatrix} 1 & 0 & 0 & 1 & -2 & 0 \\ 0 & 1 & 0 & 0 & \frac{1}{2} & 0 \\ 0 & 0 & 1 & 3 & -6 & 1 \end{bmatrix}.$

Therefore, $B^{-1} = \begin{bmatrix} 1 & -2 & 0 \\ 0 & \frac{1}{2} & 0 \\ 3 & -6 & 1 \end{bmatrix}$.

17. $\begin{cases} 2x - z = 14 \\ 3x - y + 5z = 0 \\ 4x + 2y + 3z = -2 \end{cases}$

Then, $\det(D) = \begin{vmatrix} 2 & 0 & -1 \\ 3 & -1 & 5 \\ 4 & 2 & 3 \end{vmatrix} = 2 \begin{vmatrix} -1 & 5 \\ 2 & 3 \end{vmatrix} - 1 \begin{vmatrix} 3 & -1 \\ 4 & 2 \end{vmatrix} = -26 - 10 = -36;$

$\det(D_x) = \begin{vmatrix} 14 & 0 & -1 \\ 0 & -1 & 5 \\ -2 & 2 & 3 \end{vmatrix} = 14 \begin{vmatrix} -1 & 5 \\ 2 & 3 \end{vmatrix} - 1 \begin{vmatrix} 0 & -1 \\ 4 & 2 \end{vmatrix} = -182 + 2 = -180;$

$\det(D_y) = \begin{vmatrix} 2 & 14 & -1 \\ 3 & 0 & 5 \\ 4 & -2 & 3 \end{vmatrix} = -3 \begin{vmatrix} 14 & -1 \\ -2 & 3 \end{vmatrix} - 5 \begin{vmatrix} 2 & 14 \\ 4 & -2 \end{vmatrix} = -120 + 300 = 180;$ and

$\det(D_z) = \begin{vmatrix} 2 & 0 & 14 \\ 3 & -1 & 0 \\ 4 & 2 & -2 \end{vmatrix} = 2 \begin{vmatrix} -1 & 0 \\ 2 & -2 \end{vmatrix} + 14 \begin{vmatrix} 3 & -1 \\ 2 & 2 \end{vmatrix} = 4 + 140 = 144.$

Therefore, $x = \frac{-180}{-36} = 5$, $y = \frac{180}{-36} = -5$, $z = \frac{144}{-36} = -4$, and so the solution is $(5, -5, -4)$.

18. Let x and y represent the pounds of almonds and walnuts respectively. Then the problem is modeled by the system of equations $\begin{cases} x + y = 3 \\ 4.75x + 3.45y = 11.91 \end{cases}$. Then $D = \begin{bmatrix} 1 & 1 \\ 4.75 & 3.45 \end{bmatrix}$ so

$\det(D) = \begin{vmatrix} 1 & 1 \\ 4.75 & 3.45 \end{vmatrix} = 3.45 - 4.75 = -1.3;$

$\det(D_x) = \begin{vmatrix} 3 & 1 \\ 11.91 & 3.45 \end{vmatrix} = 10.35 - 11.91 = -1.56;$

$\det(D_y) = \begin{vmatrix} 1 & 3 \\ 4.75 & 11.91 \end{vmatrix} = 11.91 - 14.25 = -2.34.$

Then $x = \frac{|D_x|}{|D|} = \frac{-1.56}{-1.3} = 1.2$, $y = \frac{|D_y|}{|D|} = \frac{-2.34}{-1.3} = 1.8$. So she bought 1.2 pounds of almonds and 1.8 pounds of walnuts.

Focus on Modeling

1.   The data matrix $D = \begin{bmatrix} 0 & 1 & 1 & 0 \\ 0 & 0 & 1 & 1 \end{bmatrix}$

     represents the square

Reflection: $T = \begin{bmatrix} 1 & 0 \\ 0 & -1 \end{bmatrix}$

$TD = \begin{bmatrix} 1 & 0 \\ 0 & -1 \end{bmatrix}\begin{bmatrix} 0 & 1 & 1 & 0 \\ 0 & 0 & 1 & 1 \end{bmatrix}$

$\quad = \begin{bmatrix} 0 & 1 & 1 & 0 \\ 0 & 0 & -1 & -1 \end{bmatrix}$

Expansion with $c = 2$: $T = \begin{bmatrix} 2 & 0 \\ 0 & 1 \end{bmatrix}$

$TD = \begin{bmatrix} 2 & 0 \\ 0 & 1 \end{bmatrix}\begin{bmatrix} 0 & 1 & 1 & 0 \\ 0 & 0 & 1 & 1 \end{bmatrix}$

$\quad = \begin{bmatrix} 0 & 2 & 2 & 0 \\ 0 & 0 & 1 & 1 \end{bmatrix}$

Shear with $c = 1$: $T = \begin{bmatrix} 1 & 1 \\ 0 & 1 \end{bmatrix}$

$TD = \begin{bmatrix} 1 & 1 \\ 0 & 1 \end{bmatrix}\begin{bmatrix} 0 & 1 & 1 & 0 \\ 0 & 0 & 1 & 1 \end{bmatrix}$

$\quad = \begin{bmatrix} 0 & 1 & 2 & 1 \\ 0 & 0 & 1 & 1 \end{bmatrix}$

3.   (a) $T = \begin{bmatrix} 1 & 1.5 \\ 0 & 1 \end{bmatrix}$ is a shear in the x-direction.

(b) $T^{-1} = \dfrac{1}{1} \cdot \begin{bmatrix} 1 & -1.5 \\ 0 & 1 \end{bmatrix} = \begin{bmatrix} 1 & -1.5 \\ 0 & 1 \end{bmatrix}$

(c) T^{-1} is a shear in the $-x$-direction (left).

(d)   The result is the original matrix. Algebraically, $T^{-1}(TD) = (T^{-1}T))D = ID = D$ where I is
       the 2×2 identity matrix.

$$\begin{bmatrix} 1 & -1.5 \\ 0 & 1 \end{bmatrix}\left(\begin{bmatrix} 1 & 1.5 \\ 0 & 1 \end{bmatrix}\begin{bmatrix} 0 & 1 & 1 & 0 \\ 0 & 0 & 1 & 1 \end{bmatrix}\right) = \begin{bmatrix} 1 & -1.5 \\ 0 & 1 \end{bmatrix}\begin{bmatrix} 0 & 1 & 2.5 & 1.5 \\ 0 & 0 & 1 & 1 \end{bmatrix}$$

$$= \begin{bmatrix} 0 & 1 & 1 & 0 \\ 0 & 0 & 1 & 1 \end{bmatrix}$$

5. (a) $D = \begin{bmatrix} 0 & 1 & 1 & 4 & 4 & 1 & 1 & 6 & 6 & 0 & 0 \\ 0 & 0 & 4 & 4 & 5 & 5 & 7 & 7 & 8 & 8 & 0 \end{bmatrix}.$

   (b) $T = \begin{bmatrix} 0.75 & 0 \\ 0 & 1 \end{bmatrix},$

   $TD = \begin{bmatrix} 0.75 & 0 \\ 0 & 1 \end{bmatrix} \begin{bmatrix} 0 & 1 & 1 & 4 & 4 & 1 & 1 & 6 & 6 & 0 & 0 \\ 0 & 0 & 4 & 4 & 5 & 5 & 7 & 7 & 8 & 8 & 0 \end{bmatrix}$

   $= \begin{bmatrix} 0 & 0.75 & 0.75 & 3 & 3 & 0.75 & 0.75 & 4.5 & 4.5 & 0 & 0 \\ 0 & 0 & 4 & 4 & 5 & 5 & 7 & 7 & 8 & 8 & 0 \end{bmatrix}$

   (c) $S = \begin{bmatrix} 1 & 0.25 \\ 0 & 1 \end{bmatrix},$

   $SD = \begin{bmatrix} 1 & 0.25 \\ 0 & 1 \end{bmatrix} \begin{bmatrix} 0 & 1 & 1 & 4 & 4 & 1 & 1 & 6 & 6 & 0 & 0 \\ 0 & 0 & 4 & 4 & 5 & 5 & 7 & 7 & 8 & 8 & 0 \end{bmatrix}$

   $= \begin{bmatrix} 0 & 1 & 2 & 5 & 5.25 & 2.25 & 2.75 & 7.75 & 8 & 2 & 0 \\ 0 & 0 & 4 & 4 & 5 & 5 & 7 & 7 & 8 & 8 & 0 \end{bmatrix}$

Chapter Eight
Exercises 8.1

1. III. Graph that opens to the right and is not as wide as the graph for Exercise 5.

3. II. Graph opens downward and is not wider than the graph represent in Exercise 6.

5. VI. Graph opens up to the right and is wider than the graph represent in Exercise 1.

7. $y^2 = 4x$. Then $4p = 4 \quad \Leftrightarrow \quad p = 1$.
    Focus: $(1, 0)$
    Directrix: $x = -1$
    Focal diameter: 4

9. $x^2 = 9y$. Then $4p = 9 \quad \Leftrightarrow \quad p = \frac{9}{4}$.
    Focus: $\left(0, \frac{9}{4}\right)$
    Directrix: $y = -\frac{9}{4}$
    Focal diameter: 9

11. $y = 5x^2 \quad \Leftrightarrow \quad x^2 = \frac{1}{5}y$. Then $4p = \frac{1}{5} \quad \Leftrightarrow \quad p = \frac{1}{20}$.
    Focus: $\left(0, \frac{1}{20}\right)$
    Directrix: $y = -\frac{1}{20}$
    Focal diameter: $\frac{1}{5}$

13. $x = -8y^2 \quad \Leftrightarrow \quad y^2 = -\frac{1}{8}x$. Then $4p = -\frac{1}{8} \quad \Leftrightarrow \quad p = -\frac{1}{32}$.
    Focus: $\left(-\frac{1}{32}, 0\right)$
    Directrix: $x = \frac{1}{32}$
    Focal diameter: $\frac{1}{8}$

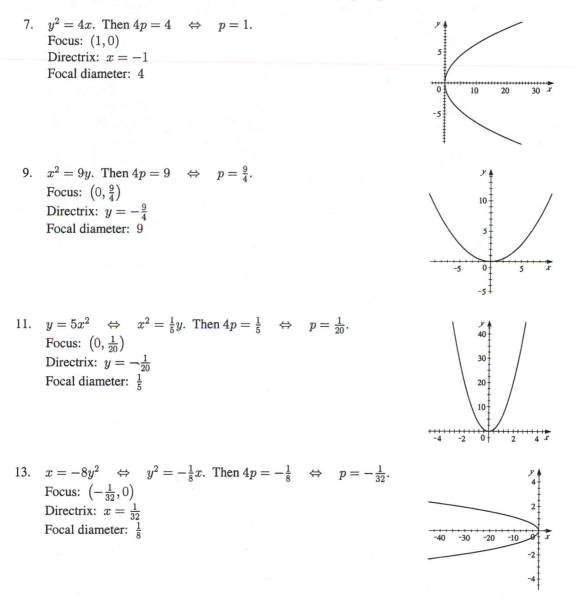

15. $x^2 + 6y = 0$ $\Leftrightarrow$ $x^2 = -6y$. Then $4p = -6$ $\Leftrightarrow$ $p = -\frac{3}{2}$.

Focus: $\left(0, -\frac{3}{2}\right)$

Directrix: $y = \frac{3}{2}$

Focal diameter: 6

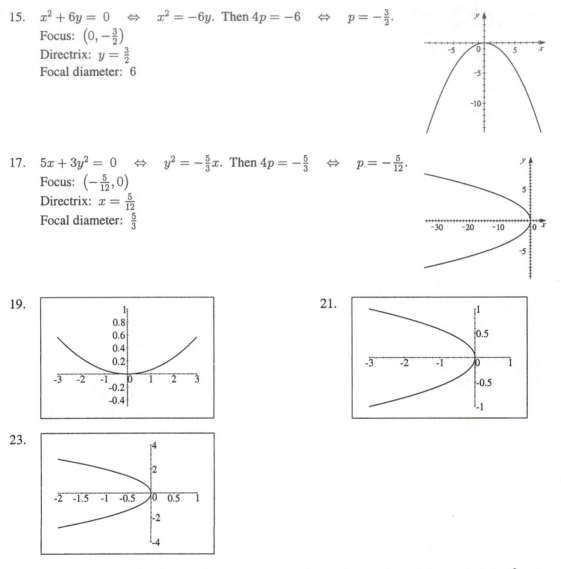

17. $5x + 3y^2 = 0$ $\Leftrightarrow$ $y^2 = -\frac{5}{3}x$. Then $4p = -\frac{5}{3}$ $\Leftrightarrow$ $p = -\frac{5}{12}$.

Focus: $\left(-\frac{5}{12}, 0\right)$

Directrix: $x = \frac{5}{12}$

Focal diameter: $\frac{5}{3}$

19.

21.

23.

25. Since the focus is $(0, 2)$, $p = 2$ $\Leftrightarrow$ $4p = 8$. Hence, the equation of the parabola is $x^2 = 8y$.

27. Since the focus is $(-8, 0)$, $p = -8$ $\Leftrightarrow$ $4p = -32$. Hence, the equation of the parabola is $y^2 = -32x$.

29. Since the directrix is $x = 2$, $p = -2$ $\Leftrightarrow$ $4p = -8$. Hence, the equation of the parabola is $y^2 = -8x$.

31. Since the directrix is $y = -10$, $p = 10$ $\Leftrightarrow$ $4p = 40$. Hence, the equation of the parabola is $x^2 = 40y$.

33. The focus is on the positive x-axis, so the parabola opens horizontally with $2p = 2$ $\Leftrightarrow$ $4p = 4$. So the equation of the parabola is $y^2 = 4x$.

35. Since the parabola opens upward with focus 5 units from the vertex, the focus is $(5, 0)$. So $p = 5$
     $\Leftrightarrow$ $4p = 20$. Thus the equation of the parabola is $x^2 = 20y$.

37. $p = 2$ $\Leftrightarrow$ $4p = 8$. Since the parabola opens upward, its equation is $x^2 = 8y$.

39. $p = 4$ $\Leftrightarrow$ $4p = 16$. Since the parabola opens to the left, its equation is $y^2 = -16x$.

41. The focal diameter is $4p = \frac{3}{2} + \frac{3}{2} = 3$. Since the parabola opens to the left, its equation is
     $y^2 = -3x$.

43. The equation of the parabola has the form $y^2 = 4px$. Since the parabola passes through the point
     $(4, -2)$, $(-2)^2 = 4p(4)$ $\Leftrightarrow$ $4p = 1$, and so the equation is $y^2 = x$.

45. The area of the shaded region is $width \times height = 4p \times p = 8$, and so $p^2 = 2$ $\Leftrightarrow$ $p = -\sqrt{2}$
     (because the parabola opens downward). Therefore, the equation is $x^2 = 4py = -4\sqrt{2}\,y$ $\Leftrightarrow$
     $x^2 = -4\sqrt{2}\,y$.

47. (a) Since a parabola with directrix $y = -p$ has equation $x^2 = 4py$, we have the following.
          The directrix is $y = \frac{1}{2}$ $\Rightarrow$ $p = -\frac{1}{2}$, so the equation is $x^2 = 4\left(-\frac{1}{2}\right)y$ $\Leftrightarrow$ $x^2 = -2y$.
          The directrix is $y = 1$ $\Rightarrow$ $p = -1$, so the equation is $x^2 = 4(-1)y$ $\Leftrightarrow$ $x^2 = -4y$.
          The directrix is $y = 4$ $\Rightarrow$ $p = -4$, so the equation is $x^2 = 4(-4)y$ $\Leftrightarrow$ $x^2 = -16y$.
          The directrix is $y = 8$ $\Rightarrow$ $p = -8$, so the equation is $x^2 = 4(-8)y$ $\Leftrightarrow$ $x^2 = -32y$.

     (b)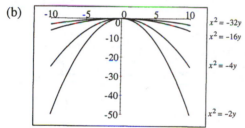

          As the directrix moves further from the vertex,
          the parabola get flatter.

49. (a) Since the focal diameter is 12 cm, $4p = 12$. Hence, the parabola has equation $y^2 = 12x$.

     (b) At a point 20 cm horizontally from the vertex, the parabola passes through the point $(20, y)$,
          and hence from part (a), $y^2 = 12(20)$ $\Leftrightarrow$ $y^2 = 240$ $\Leftrightarrow$ $y = \pm 4\sqrt{15}$. Thus,
          $|CD| = 8\sqrt{15} \approx 31$ cm.

51. With the vertex at the origin, the top of one tower will be at the point $(300, 150)$. Inserting this point
     into the equation $x^2 = 4py$ gives $(300)^2 = 4p(150)$ $\Leftrightarrow$ $90000 = 600p$ $\Leftrightarrow$ $p = 150$. So the
     equation of the parabolic part of the cables is $x^2 = 4(150)y$ $\Leftrightarrow$ $x^2 = 600y$.

53. Many answers are possible: satellite dish TV antennas, sound surveillance equipment, solar
     collectors for hot water heating or electricity generation, bridge pillars, etc.

Exercises 8.2

1. II. Major axis is horizontal, vertices at $(\pm 4, 0)$.

3. I. Major axis is the vertical, vertices at $(0, \pm 2)$.

5. $\dfrac{x^2}{25} + \dfrac{y^2}{9} = 1$. This ellipse has $a = 5$, $b = 3$, and so
    $c^2 = a^2 - b^2 = 16 \quad \Leftrightarrow \quad c = 4$.
    Vertices: $(\pm 5, 0)$; foci: $(\pm 4, 0)$; eccentricity: $e = \frac{c}{a} = \frac{4}{5} = 0.8$;
    length of the major axis: $2a = 10$; length of the minor axis: $2b = 6$.

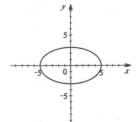

7. $9x^2 + 4y^2 = 36 \quad \Leftrightarrow \quad \dfrac{x^2}{4} + \dfrac{y^2}{9} = 1$. This ellipse has $a = 3$,
    $b = 2$, and so $c^2 = 9 - 4 = 5 \quad \Leftrightarrow \quad c = \sqrt{5}$.
    Vertices: $(0, \pm 3)$; foci: $(0, \pm\sqrt{5})$; eccentricity: $e = \frac{c}{a} = \frac{\sqrt{5}}{3}$;
    length of the major axis: $2a = 6$; length of the minor axis: $2b = 4$.

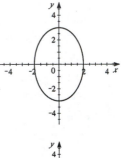

9. $x^2 + 4y^2 = 16 \quad \Leftrightarrow \quad \dfrac{x^2}{16} + \dfrac{y^2}{4} = 1$. This ellipse has $a = 4$,
    $b = 2$, and so $c^2 = 16 - 4 = 12 \quad \Leftrightarrow \quad c = 2\sqrt{3}$.
    Vertices: $(\pm 4, 0)$; foci: $(\pm 2\sqrt{3}, 0)$; eccentricity:
    $e = \frac{c}{a} = \frac{2\sqrt{3}}{4} = \frac{\sqrt{3}}{2}$; length of the major axis: $2a = 8$; length of
    the minor axis: $2b = 4$.

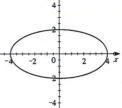

11. $2x^2 + y^2 = 3 \quad \Leftrightarrow \quad \dfrac{x^2}{\frac{3}{2}} + \dfrac{y^2}{3} = 1$. This ellipse has $a = \sqrt{3}$,
    $b = \sqrt{\frac{3}{2}}$, and so $c^2 = 3 - \frac{3}{2} = \frac{3}{2} \quad \Leftrightarrow \quad c = \sqrt{\frac{3}{2}}$.
    Vertices: $\left(0, \pm\sqrt{3}\right)$; foci: $\left(0, \pm\sqrt{\frac{3}{2}}\right)$; eccentricity:

    $e = \frac{c}{a} = \dfrac{\sqrt{\frac{3}{2}}}{\sqrt{3}} = \dfrac{1}{\sqrt{2}} = \dfrac{\sqrt{2}}{2}$; length of the major axis: $2a = 2\sqrt{3}$

    length of the minor axis: $2b = 2\sqrt{\frac{3}{2}} = \sqrt{6}$.

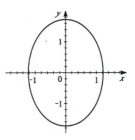

13. $x^2 + 4y^2 = 1 \quad \Leftrightarrow \quad \dfrac{x^2}{1} + \dfrac{y^2}{\frac{1}{4}} = 1.$ This ellipse has $a = 1, b = \frac{1}{2}$,

and so $c^2 = 1 - \frac{1}{4} = \frac{3}{4} \quad \Leftrightarrow \quad c = \frac{\sqrt{3}}{2}.$

Vertices: $(\pm 1, 0)$; foci: $\left(\pm \frac{\sqrt{3}}{2}, 0\right)$; eccentricity:

$e = \frac{c}{a} = \frac{\sqrt{3}/2}{1} = \frac{\sqrt{3}}{2}$; length of the major axis: $2a = 2$; length

of the minor axis: $2b = 1.$

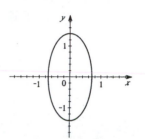

15. $\frac{1}{2}x^2 + \frac{1}{8}y^2 = \frac{1}{4} \quad \Leftrightarrow \quad 2x^2 + \frac{1}{2}y^2 = 1 \quad \Leftrightarrow \quad \dfrac{x^2}{\frac{1}{2}} + \dfrac{y^2}{2} = 1.$

This ellipse has $a = \sqrt{2}, b = \frac{1}{\sqrt{2}}$, and so $c^2 = 2 - \frac{1}{2} = \frac{3}{2}$

$\Leftrightarrow \quad c = \sqrt{\frac{3}{2}}.$

Vertices: $\left(0, \pm\sqrt{2}\right)$; foci: $\left(0, \pm\sqrt{\frac{3}{2}}\right)$; eccentricity:

$= \frac{c}{a} = \dfrac{\sqrt{\frac{3}{2}}}{\sqrt{2}} = \frac{1}{2}\sqrt{3}$; length of the major axis: $2a = 2\sqrt{2}$;

length of the minor axis: $2b = \sqrt{2}.$

17. $y^2 = 1 - 2x^2 \quad \Leftrightarrow \quad 2x^2 + y^2 = 1 \quad \Leftrightarrow \quad \dfrac{x^2}{\frac{1}{2}} + \dfrac{y^2}{1} = 1.$ This

ellipse has $a = 1, b = \frac{1}{\sqrt{2}}$, and so $c^2 = 1 - \frac{1}{2} = \frac{1}{2} \quad \Leftrightarrow \quad c = \frac{1}{\sqrt{2}}.$

Vertices: $(0, \pm 1)$; foci: $\left(0, \pm\frac{1}{\sqrt{2}}\right)$; eccentricity:

$e = \frac{c}{a} = \dfrac{\frac{1}{\sqrt{2}}}{1} = \frac{1}{\sqrt{2}}$; length of the major axis: $2a = 2$; length of

the minor axis: $2b = \sqrt{2}.$

19.   This ellipse has a horizontal major axis with $a = 5$ and $b = 4$, so the equation is $\dfrac{x^2}{(5)^2} + \dfrac{y^2}{(4)^2} = 1$

$\Leftrightarrow \quad \dfrac{x^2}{25} + \dfrac{y^2}{16} = 1.$

21.   This ellipse has a vertical major axis with $c = 2$ and $b = 2$. So $a^2 = c^2 + b^2 = 2^2 + 2^2 = 8 \quad \Leftrightarrow$

$a = 2\sqrt{2}.$ So the equation is $\dfrac{x^2}{(2)^2} + \dfrac{y^2}{(2\sqrt{2})^2} = 1 \quad \Leftrightarrow \quad \dfrac{x^2}{4} + \dfrac{y^2}{8} = 1.$

23.   This ellipse has a horizontal major axis with $a = 16$, so the equation of the ellipse is of the form

$\dfrac{x^2}{16^2} + \dfrac{y^2}{b^2} = 1.$ Substituting the point $(8, 6)$ into the equation, we get $\dfrac{64}{256} + \dfrac{36}{b^2} = 1 \quad \Leftrightarrow$

$\dfrac{36}{b^2} = 1 - \dfrac{1}{4} \quad \Leftrightarrow \quad \dfrac{36}{b^2} = \dfrac{3}{4} \quad \Leftrightarrow \quad b^2 = \dfrac{4(36)}{3} = 48.$ Thus, the equation of the ellipse is

$\dfrac{x^2}{256} + \dfrac{y^2}{48} = 1.$

25. $\dfrac{x^2}{25} + \dfrac{y^2}{20} = 1 \quad \Leftrightarrow \quad \dfrac{y^2}{20} = 1 - \dfrac{x^2}{25} \quad \Leftrightarrow$

$y^2 = 20 - \dfrac{4x^2}{5} \quad \Rightarrow \quad y = \pm\sqrt{20 - \dfrac{4x^2}{5}}.$

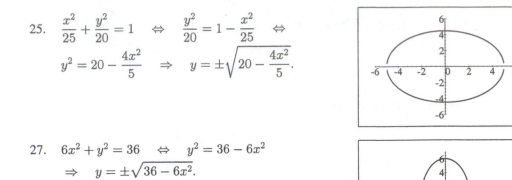

27. $6x^2 + y^2 = 36 \quad \Leftrightarrow \quad y^2 = 36 - 6x^2$

$\Rightarrow \quad y = \pm\sqrt{36 - 6x^2}.$

29. The foci are $(\pm 4, 0)$, and the vertices are $(\pm 5, 0)$. Thus, $c = 4$ and $a = 5$, and so $b^2 = 25 - 16 = 9$. Therefore, the equation of the ellipse is $\dfrac{x^2}{25} + \dfrac{y^2}{9} = 1$.

31. The length of the major axis is $2a = 4 \quad \Leftrightarrow \quad a = 2$, the length of the minor axis is $2b = 2 \quad \Leftrightarrow$ $b = 1$, and the foci are on the y-axis. Therefore, the equation of the ellipse is $x^2 + \dfrac{y^2}{4} = 1$.

33. The foci are $(0, \pm 2)$, and the length of the minor axis is $2b = 6 \quad \Leftrightarrow \quad b = 3$. Thus, $a^2 = 4 + 9 = 13$. Since the foci are on the y-axis, the equation is $\dfrac{x^2}{9} + \dfrac{y^2}{13} = 1$.

35. The endpoints of the major axis are $(\pm 10, 0) \quad \Leftrightarrow \quad a = 10$, and the distance between the foci is $2c = 6 \quad \Leftrightarrow \quad c = 3$. Therefore, $b^2 = 100 - 9 = 91$, and so the equation of the ellipse is $\dfrac{x^2}{100} + \dfrac{y^2}{91} = 1$.

37. The length of the major axis is 10, so $2a = 10 \quad \Leftrightarrow \quad a = 5$, and the foci are on the x-axis, so the form of the equation is $\dfrac{x^2}{25} + \dfrac{y^2}{b^2} = 1$. Since the ellipse passes through $\left(\sqrt{5}, 2\right)$, we know that

$\dfrac{\left(\sqrt{5}\right)^2}{25} + \dfrac{(2)^2}{b^2} = 1 \quad \Leftrightarrow \quad \dfrac{5}{25} + \dfrac{4}{b^2} = 1 \quad \Leftrightarrow \quad \dfrac{4}{b^2} = \dfrac{4}{5} \quad \Leftrightarrow \quad b^2 = 5$, and so the equation is $\dfrac{x^2}{25} + \dfrac{y^2}{5} = 1$.

39. Since the foci are $(\pm 1.5, 0)$, we have $c = \frac{3}{2}$. Since the eccentricity is $0.8 = \frac{c}{a}$, we have $a = \dfrac{\frac{3}{2}}{\frac{4}{5}} = \dfrac{15}{8}$,

and so $b^2 = \frac{225}{64} - \frac{9}{4} = \frac{225 - 16\cdot 9}{64} = \frac{81}{64}$. Therefore, the equation of the ellipse is $\dfrac{x^2}{\left(^{15}/_8\right)^2} + \dfrac{y^2}{^{81}/_{64}} = 1$

$\Leftrightarrow \quad \dfrac{64x^2}{225} + \dfrac{64y^2}{81} = 1.$

41. $\begin{cases} 4x^2 + y^2 = 4 \\ 4x^2 + 9y^2 = 36 \end{cases}$ Subtracting the first equation from the second

equation gives $8y^2 = 32$ $\Leftrightarrow$ $y^2 = 4$ $\Leftrightarrow$ $y = \pm 2$.

Substituting $y = \pm 2$ in the first equation gives $4x^2 + (\pm 2)^2 = 4$

$\Leftrightarrow$ $x = 0$, and so the points of intersection are $(0, \pm 2)$.

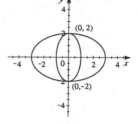

43. $\begin{cases} 100x^2 + 25y^2 = 100 \\ x^2 + \dfrac{y^2}{9} = 1 \end{cases}$ Dividing the first equation by 100 gives

$x^2 + \dfrac{y^2}{4} = 1$. Subtracting the first equation from the second

equation gives $\dfrac{y^2}{9} - \dfrac{y^2}{4} = 0$ $\Leftrightarrow$ $(\frac{1}{9} - \frac{1}{4})y^2 = 0$ $\Leftrightarrow$ $y = 0$.

Substituting $y = 0$ in the second equation gives $x^2 + (0)^2 = 1$

$\Leftrightarrow$ $x = \pm 1$, and so the points of intersection are $(\pm 1, 0)$.

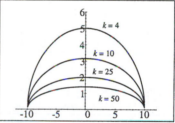

45. (a) $x^2 + ky^2 = 100$ $\Leftrightarrow$ $ky^2 = 100 - x^2$

$\Rightarrow$ $y = \pm\frac{1}{k}\sqrt{100 - x^2}$. For the top

half, we graph $y = \frac{1}{k}\sqrt{100 - x^2}$ for

$k = 4, 10, 25,$ and 50.

(b) This family of ellipses have common major axes and vertices, and the eccentricity increases as k increases.

47. Using the perihelion, $a - c = 147{,}000{,}000$, while using the aphelion, $a + c = 153{,}000{,}000$.
Adding, we have $2a = 300{,}000{,}000$ $\Leftrightarrow$ $a = 150{,}000{,}000$. So $b^2 = a^2 - c^2$
$= (150 \times 10^6)^2 - (3 \times 10^6)^2$ $= 22{,}491 \times 10^{12} = 2.2491 \times 10^{16}$. Thus the equation of the orbit is
$\dfrac{x^2}{2.2500 \times 10^{16}} + \dfrac{y^2}{2.2491 \times 10^{16}} = 1$.

49. Using the perilune, $a - c = 1075 + 68 = 1143$, and using the apolune,
$a + c = 1075 + 195 = 1270$. Adding, we get $2a = 2413$ $\Leftrightarrow$ $a = 1206.5$. So
$c = 1270 - 1206.5$ $\Leftrightarrow$ $c = 63.5$. Therefore, $b^2 = (1206.5)^2 - (63.5)^2 = 1{,}451{,}610$. Since
$a^2 \approx 1{,}455{,}642$, the equation of Apollo 11's orbit is $\dfrac{x^2}{1{,}455{,}642} + \dfrac{y^2}{1{,}451{,}610} = 1$.

51. From the diagram, $a = 40$ and $b = 20$, and so the equation of the ellipse whose top half is the
window is $\dfrac{x^2}{1600} + \dfrac{y^2}{400} = 1$. Since the ellipse passes through the point $(25, h)$, by substituting, we
have $\dfrac{25^2}{1600} + \dfrac{h^2}{400} = 1$ $\Leftrightarrow$ $625 + 4y^2 = 1600$ $\Leftrightarrow$ $y = \frac{\sqrt{975}}{2} = \frac{5\sqrt{39}}{2} \approx 15.61$ in.
Therefore, the window is approximately 15.6 inches high at the specified point.

53.   We start with the flashlight perpendicular to the wall; this shape is a circle. As the angle of elevation increases, the shape of the light changes to an ellipse. When the flashlight is angled so that the outer edge of the light cone is parallel to the wall, the shape of the light is a parabola. Finally, as the angle of elevation increases further, the shape of the light is hyperbolic.

54.   The shape drawn on the paper is *almost*, but not quite, an ellipse. For example, when the bottle has radius 1 unit and the compass legs are set 1 unit apart, then it can be shown that the equation of the resulting curve is $1 + y^2 = 2\cos x$. The graph of this curve differs very slightly from the ellipse with the same major and minor axis. This example shows that in mathematics, things are not always what they appear to be.

Exercises 8.3

1. III. Opens horizontally, vertices at $(\pm 2, 0)$.

3. II. Opens vertically, vertices at $(0, \pm 3)$.

5. The hyperbola $\dfrac{x^2}{4} - \dfrac{y^2}{16} = 1$ has $a = 2$, $b = 4$, and $c^2 = 16 + 4$
   $\Rightarrow \quad c = 2\sqrt{5}$.
   Vertices: $(\pm 2, 0)$; foci: $(\pm 2\sqrt{5}, 0)$; asymptotes: $y = \pm \frac{4}{2}x$
   $\Leftrightarrow \quad y = \pm 2x$.

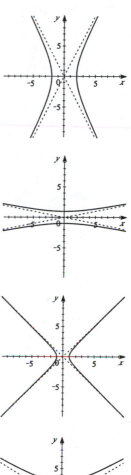

7. The hyperbola $\dfrac{y^2}{1} - \dfrac{x^2}{25} = 1$ has $a = 1$, $b = 5$, and
   $c^2 = 1 + 25 = 26 \quad \Rightarrow \quad c = \sqrt{26}$.
   Vertices: $(0, \pm 1)$; foci: $(0, \pm\sqrt{26})$; asymptotes: $y = \pm \frac{1}{5}x$.

9. The hyperbola $x^2 - y^2 = 1$ has $a = 1$, $b = 1$, and
   $c^2 = 1 + 1 = 2 \quad \Rightarrow \quad c = \sqrt{2}$.
   Vertices: $(\pm 1, 0)$; foci: $(\pm\sqrt{2}, 0)$; asymptotes: $y = \pm x$.

11. The hyperbola $25y^2 - 9x^2 = 225 \quad \Leftrightarrow \quad \dfrac{y^2}{9} - \dfrac{x^2}{25} = 1$ has $a = 3$,
    $b = 5$, and $c^2 = 25 + 9 = 34 \quad \Rightarrow \quad c = \sqrt{34}$.
    Vertices: $(0, \pm 3)$; foci: $(0, \pm\sqrt{34})$; asymptotes: $y = \pm \frac{3}{5}x$.

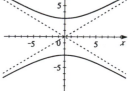

13. The hyperbola $x^2 - 4y^2 - 8 = 0 \iff \dfrac{x^2}{8} - \dfrac{y^2}{2} = 1$ has

$a = \sqrt{8}$, $b = \sqrt{2}$, and $c^2 = 8 + 2 = 10 \implies c = \sqrt{10}$.

Vertices: $(\pm 2\sqrt{2}, 0)$; foci: $(\pm\sqrt{10}, 0)$; asymptotes: $y = \pm\dfrac{\sqrt{2}}{\sqrt{8}} x$

$\iff y = \pm\frac{1}{2}x$.

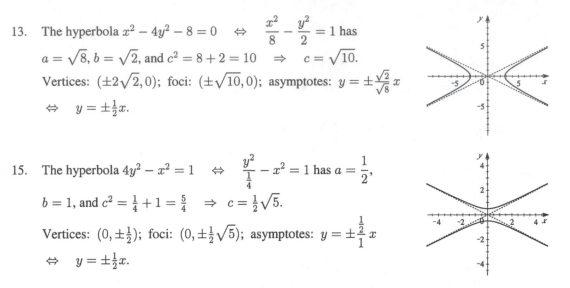

15. The hyperbola $4y^2 - x^2 = 1 \iff \dfrac{y^2}{\frac{1}{4}} - x^2 = 1$ has $a = \dfrac{1}{2}$,

$b = 1$, and $c^2 = \frac{1}{4} + 1 = \frac{5}{4} \implies c = \frac{1}{2}\sqrt{5}$.

Vertices: $(0, \pm\frac{1}{2})$; foci: $(0, \pm\frac{1}{2}\sqrt{5})$; asymptotes: $y = \pm\dfrac{\frac{1}{2}}{1} x$

$\iff y = \pm\frac{1}{2}x$.

17. From the graph, the foci are $(\pm 4, 0)$, and the vertices are $(\pm 2, 0)$, so $c = 4$ and $a = 2$. Thus, $b^2 = 16 - 4 = 12$, and since the vertices are on the x-axis, the equation of the hyperbola is $\dfrac{x^2}{4} - \dfrac{y^2}{12} = 1$.

19. From the graph, the vertices are $(0, \pm 4)$, the foci are on the y-axis, and the hyperbola passes through the point $(3, -5)$. So the equation is of the form $\dfrac{y^2}{16} - \dfrac{x^2}{b^2} = 1$. Substituting the point $(3, -5)$, we have $\dfrac{(-5)^2}{16} - \dfrac{(3)^2}{b^2} = 1 \iff \dfrac{25}{16} - 1 = \dfrac{9}{b^2} \iff \dfrac{9}{16} = \dfrac{9}{b^2} \iff b^2 = 16$. Thus, the equation of the hyperbola is $\dfrac{y^2}{16} - \dfrac{x^2}{16} = 1$.

21. The vertices are $(\pm 3, 0)$, so $a = 3$. Since the asymptotes are $y = \pm\frac{1}{2}x = \pm\frac{b}{a}x$, we have $\frac{b}{3} = \frac{1}{2}$ $\iff b = \frac{3}{2}$. Since the vertices are on the x-axis, the equation is $\dfrac{x^2}{3^2} - \dfrac{y^2}{(3/2)^2} = 1 \iff \dfrac{x^2}{9} - \dfrac{4y^2}{9} = 1$.

23. $x^2 - 2y^2 = 8 \iff 2y^2 = x^2 - 8 \iff$

$y^2 = \dfrac{x^2}{2} - 4 \implies y = \pm\sqrt{\dfrac{x^2}{2} - 4}$.

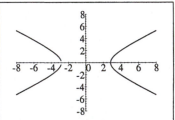

25. $\dfrac{y^2}{2} - \dfrac{x^2}{6} = 1$ $\quad\Leftrightarrow\quad$ $\dfrac{y^2}{2} = \dfrac{x^2}{6} + 1$ $\quad\Leftrightarrow\quad$

$y^2 = \dfrac{x^2}{3} + 2$ $\quad\Rightarrow\quad$ $y = \pm\sqrt{\dfrac{x^2}{3} + 2}$.

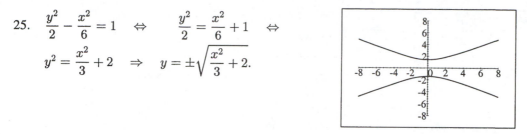

27. The foci are $(\pm5, 0)$, and the vertices are $(\pm3, 0)$, so $c = 5$ and $a = 3$. Then $b^2 = 25 - 9 = 16$, and since the vertices are on the x-axis, the equation of the hyperbola is $\dfrac{x^2}{9} - \dfrac{y^2}{16} = 1$.

29. The foci are $(0, \pm2)$, and the vertices are $(0, \pm1)$, so $c = 2$ and $a = 1$. Then $b^2 = 4 - 1 = 3$, and since the vertices are on the y-axis, the equation is $y^2 - \dfrac{x^2}{3} = 1$.

31. The vertices are $(\pm1, 0)$, and the asymptotes are $y = \pm5x$, so $a = 1$. The asymptotes are $y = \pm\dfrac{b}{a}x$, so $\dfrac{b}{1} = 5$ $\quad\Leftrightarrow\quad$ $b = 5$. Therefore, the equation of the hyperbola is $x^2 - \dfrac{y^2}{25} = 1$.

33. The foci are $(0, \pm8)$, and the asymptotes are $y = \pm\frac{1}{2}x$, so $c = 8$. The asymptotes are $y = \pm\dfrac{a}{b}x$, so $\dfrac{a}{b} = \frac{1}{2}$ and $b = 2a$. Since $a^2 + b^2 = c^2 = 64$, we have $a^2 + 4a^2 = 64$ $\quad\Leftrightarrow\quad$ $a^2 = \frac{64}{5}$ and $b^2 = 4a^2 = \frac{256}{5}$. Thus, the equation of the hyperbola is $\dfrac{y^2}{64/5} - \dfrac{x^2}{256/5} = 1$ $\quad\Leftrightarrow\quad$ $\dfrac{5y^2}{64} - \dfrac{5x^2}{256} = 1$.

35. The asymptotes of the hyperbola are $y = \pm x$ so $b = a$. Since the hyperbola passes through the point $(5, 3)$, its foci are on the x-axis, and the equation is of the form, $\dfrac{x^2}{a^2} - \dfrac{y^2}{a^2} = 1$, so it follows that $\dfrac{25}{a^2} - \dfrac{9}{a^2} = 1$ $\quad\Leftrightarrow\quad$ $a^2 = 16 = b^2$. Therefore, the equation of the hyperbola is $\dfrac{x^2}{16} - \dfrac{y^2}{16} = 1$.

37. The foci are $(\pm5, 0)$, and the length of the transverse axis is 6, so $c = 5$ and $2a = 6$ $\quad\Leftrightarrow\quad$ $a = 3$. Thus, $b^2 = 25 - 9 = 16$, and the equation is $\dfrac{x^2}{9} - \dfrac{y^2}{16} = 1$.

39. (a) The hyperbola $x^2 - y^2 = 5$ $\quad\Leftrightarrow\quad$ $\dfrac{x^2}{5} - \dfrac{y^2}{5} = 1$ has $a = \sqrt{5}$ and $b = \sqrt{5}$. Thus, the asymptotes are $y = \pm x$, and their slopes are $m_1 = 1$ and $m_2 = -1$. Since $m_1 \times m_2 = -1$, the asymptotes are perpendicular.

(b) Since the asymptotes are perpendicular to each other, they must be of slope ±1, so $a = b$. Therefore, $c^2 = 2a^2$ $\quad\Leftrightarrow\quad$ $a^2 = \dfrac{c^2}{2}$, and since the vertices are on the x-axis, the equation is $\dfrac{x^2}{\frac{1}{2}c^2} - \dfrac{y^2}{\frac{1}{2}c^2} = 1$ $\quad\Leftrightarrow\quad$ $x^2 - y^2 = \dfrac{c^2}{2}$.

41. $\sqrt{(x+c)^2+y^2} - \sqrt{(x-c)^2+y^2} = \pm 2a$. Let us consider the positive case only. Then
$\sqrt{(x+c)^2+y^2} = 2a + \sqrt{(x-c)^2+y^2}$, and squaring both sides gives
$x^2 + 2cx + c^2 + y^2 = 4a^2 + 4a\sqrt{(x-c)^2+y^2} + x^2 - 2cx + c^2 + y^2$ $\Leftrightarrow$
$4a\sqrt{(x-c)^2+y^2} = 4cx - 4a^2$. Dividing by 4 and squaring both sides gives
$a^2(x^2 - 2cx + c^2 + y^2) = c^2x^2 - 2a^2cx + a^4$ $\Leftrightarrow$
$a^2x^2 - 2a^2cx + a^2c^2 + a^2y^2 = c^2x^2 - 2a^2cx + a^4$ $\Leftrightarrow$ $a^2x^2 + a^2c^2 + a^2y^2 = c^2x^2 + a^4$.
Rearranging the order, we have $c^2x^2 - a^2x^2 - a^2y^2 = a^2c^2 - a^4$ $\Leftrightarrow$
$(c^2 - a^2)x^2 - a^2y^2 = a^2(c^2 - a^2)$. The negative case gives the same result.

43. (a) From the equation, we have $a^2 = k$ and $b^2 = 16 - k$. Thus, $c^2 = a^2 + b^2 = k + 16 - k = 16$
$\Rightarrow$ $c = \pm 4$. Thus the foci of the family of hyperbolas are $(0, \pm 4)$.

   (b) $\dfrac{y^2}{k} - \dfrac{x^2}{16-k} = 1$ $\Leftrightarrow$ $y^2 = k\left(1 + \dfrac{x^2}{16-k}\right)$ $\Rightarrow$

   $y = \pm\sqrt{k + \dfrac{kx^2}{16-k}}$. For the top branch, we graph

   $y = \sqrt{k + \dfrac{kx^2}{16-k}}$, $k = 1, 4, 8, 12$.

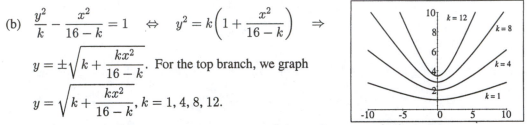

   As k increases the asymptotes get steeper, and the vertices move further apart.

45. Since the asymptotes are perpendicular, $a = b$. Also, since the sun is a focus and the closest distance
is 2×10^9, it follows that $c - a = 2 \times 10^9$. Now $c^2 = a^2 + b^2 = 2a^2$, and so $c = \sqrt{2}a$. Thus,
$\sqrt{2}a - a = 2 \times 10^9$ $\Rightarrow$ $a = \dfrac{2 \times 10^9}{\sqrt{2}-1}$ and $a^2 = b^2 = \dfrac{4 \times 10^{18}}{3 - 2\sqrt{2}} \approx 2.3 \times 10^{19}$. Therefore, the
equation of the hyperbola is $\dfrac{x^2}{2.3 \times 10^{19}} - \dfrac{y^2}{2.3 \times 10^{19}} = 1$ $\Leftrightarrow$ $x^2 - y^2 = 2.3 \times 10^{19}$.

47. Some possible answer are: as cross-sections of nuclear power plant cooling towers, or as reflectors
for camouflaging the location of secret installations (see Figure 7 on page 560 of the text).

Exercises 8.4

1. The ellipse $\dfrac{(x-2)^2}{9} + \dfrac{(y-1)^2}{4} = 1$ is obtained from the ellipse

   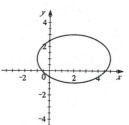

   $\dfrac{x^2}{9} + \dfrac{y^2}{4} = 1$ by shifting it 2 units to the right and 1 unit upward. So

   $a = 3$, $b = 2$, and $c = \sqrt{9-4} = \sqrt{5}$.

   Center: $(2,1)$; foci: $(2 \pm \sqrt{5}, 1)$; vertices: $(2 \pm 3, 1)$, so the vertices
   are $(-1,1)$ and $(5,1)$; length of the major axis: $2a = 6$; length of
   the minor axis: $2b = 4$.

3. The ellipse $\dfrac{x^2}{9} + \dfrac{(y+5)^2}{25} = 1$ is obtained from the ellipse

   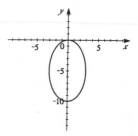

   $\dfrac{x^2}{9} + \dfrac{y^2}{4} = 1$ by shifting it 5 units downward. So $a = 5$, $b = 3$, and

   $c = \sqrt{25-9} = 4$.

   Center: $(0,-5)$; foci: $(0, -5 \pm 4)$ or $(0,-9)$ and $(0,-1)$;
   vertices: $(0, -5 \pm 5)$, so the vertices are $(0,-10)$ and $(0,0)$; length
   of the major axis: $2a = 10$; length of the minor axis: $2b = 6$.

5. The parabola $(x-3)^2 = 8(y+1)$ is obtained from the parabola

   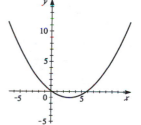

   $x^2 = 8y$ by shifting it 3 units to the right and 1 unit down. So $4p = 8$
   $\Leftrightarrow \quad p = 2$.

   Vertex: $(3,-1)$; focus: $(3, -1+2) = (3,1)$; directrix:
   $y = -1-2 = -3$.

7. The parabola $-4\left(x + \tfrac{1}{2}\right)^2 = y \quad \Leftrightarrow \quad \left(x + \tfrac{1}{2}\right)^2 = -\tfrac{1}{4}y$ is

   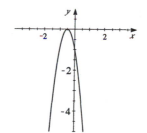

   obtained from the parabola $x^2 = -\tfrac{1}{4}y$ by shifting it $\tfrac{1}{2}$ unit to the
   left. So $4p = -\tfrac{1}{4} \quad \Leftrightarrow \quad p = -\tfrac{1}{16}$.

   Vertex: $\left(-\tfrac{1}{2}, 0\right)$; focus: $\left(-\tfrac{1}{2}, 0 - \tfrac{1}{16}\right) = \left(-\tfrac{1}{2}, -\tfrac{1}{16}\right)$; directrix:
   $y = 0 + \tfrac{1}{16} = \tfrac{1}{16}$.

9. The hyperbola $\dfrac{(x+1)^2}{9} - \dfrac{(y-3)^2}{16} = 1$ is obtained from the

   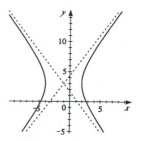

   hyperbola $\dfrac{x^2}{9} - \dfrac{y^2}{16} = 1$ by shifting it 1 unit to the left and 3 units

   up. So $a = 3$, $b = 4$, and $c = \sqrt{9+16} = 5$.

   Center: $(-1,3)$; foci: $(-1 \pm 5, 3)$, so the foci are $(-6,3)$ and $(4,3)$,
   vertices: $(-1 \pm 3, 3)$, so the vertices are $(-4,3)$ and $(2,3)$;
   asymptotes: $(y-3) = \pm \tfrac{4}{3}(x+1) \quad \Leftrightarrow \quad y = \pm \tfrac{4}{3}(x+1) + 3$

⇔ $3y = 4x + 13$ and $3y = -4x + 5$.

11.   The hyperbola $y^2 - \dfrac{(x+1)^2}{4} = 1$ is obtained from hyperbola

$y^2 - \frac{x^2}{4} = 1$ by shifting it 1 to the unit left. So $a = 1$, $b = 2$, and

$c = \sqrt{1 + 4} = \sqrt{5}$.

Center: $(-1, 0)$; foci: $(-1, \pm\sqrt{5})$, so the foci are $(-1, -\sqrt{5})$ and

$(-1, \sqrt{5})$; vertices: $(-1, \pm 1)$, so the vertices are $(-1, -1)$ and

$(-1, 1)$; asymptotes: $y = \pm\frac{1}{2}(x + 1)$ ⇔ $y = \frac{1}{2}(x + 1)$ and $y = -\frac{1}{2}(x + 1)$.

13.   This is a parabola that opens down with its vertex at $(0, 4)$, so its equation is of the form
$x^2 = a(y - 4)$. Since $(1, 0)$ is a point on this parabola, we have $(1)^2 = a(0 - 4)$ ⇔ $1 = -4a$
⇔ $a = -\frac{1}{4}$. Thus, the equation is $x^2 = -\frac{1}{4}(y - 4)$.

15.   This is an ellipse with the major axis parallel to the x-axis, with one vertex at $(0, 0)$, the other vertex
at $(10, 0)$, and one focus at $(8, 0)$. The center is at $\left(\frac{0+10}{2}, 0\right) = (5, 0)$, $a = 5$, and $c = 3$ (the distance
from one focus to the center). So $b^2 = a^2 - c^2 = 25 - 9 = 16$. Thus, the equation is
$\dfrac{(x - 5)^2}{25} + \dfrac{y^2}{16} = 1$.

17.   This is a hyperbola with center $(0, 1)$ and vertices $(0, 0)$ and $(0, 2)$. Since a is the distance form the
center to a vertex, we have $a = 1$. The slope of the given asymptote is 1, so $\frac{a}{b} = 1$ ⇔ $b = 1$.
Thus, the equation of the hyperbola is $(y - 1)^2 - x^2 = 1$.

19. $9x^2 - 36x + 4y^2 = 0$ ⇔ $9(x^2 - 4x + 4) - 36 + 4y^2 = 0$ ⇔

$9(x - 2)^2 + 4y^2 = 36$ ⇔ $\dfrac{(x - 2)^2}{4} + \dfrac{y^2}{9} = 1$. This is an

ellipse that has $a = 3$, $b = 2$, and $c = \sqrt{9 - 4} = \sqrt{5}$.

Center: $(2, 0)$; foci: $(2, \pm\sqrt{5})$; vertices: $(2, \pm 3)$; length of major
axis: $2a = 6$; length of minor axis: $2b = 4$.

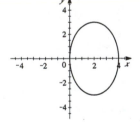

21. $x^2 - 4y^2 - 2x + 16y = 20$ ⇔

$(x^2 - 2x + 1) - 4(y^2 - 4y + 4) = 20 + 1 - 16$ ⇔

$(x - 1)^2 - 4(y - 2)^2 = 5$ ⇔ $\dfrac{(x - 1)^2}{5} - \dfrac{(y - 2)^2}{\frac{5}{4}} = 1$. This

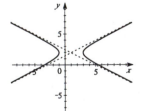

is a hyperbola that has $a = \sqrt{5}$, $b = \frac{1}{2}\sqrt{5}$, and $c = \sqrt{5 + \frac{5}{4}} = \frac{5}{2}$.

Center: $(1, 2)$; foci: $\left(1 \pm \frac{5}{2}, 2\right)$, so the foci are $\left(-\frac{3}{2}, 2\right)$ and $\left(\frac{7}{2}, 2\right)$; vertices: $(1 \pm \sqrt{5}, 2)$; asymptotes:
$y - 2 = \pm\frac{1}{2}(x - 1)$ ⇔ $y = \pm\frac{1}{2}(x - 1) + 2$ ⇔ $y = \frac{1}{2}x + \frac{3}{2}$ and $y = -\frac{1}{2}x + \frac{5}{2}$.

23. $4x^2 + 25y^2 - 24x + 250y + 561 = 0 \quad \Leftrightarrow$
$4(x^2 - 6x + 9) + 25(y^2 + 10y + 25) = -561 + 36 + 625$

$\Leftrightarrow \quad 4(x-3)^2 + 25(y+5)^2 = 100 \quad \Leftrightarrow \quad \dfrac{(x-3)^2}{25} + \dfrac{(y+5)^2}{4} = 1.$

This is an ellipse that has $a = 5$, $b = 2$, and $c = \sqrt{25-4} = \sqrt{21}$.
Center: $(3,-5)$; foci: $(3 \pm \sqrt{21}, -5)$; vertices: $(3 \pm 5, -5)$,
so the vertices are $(-2,-5)$ and $(8,-5)$;
length of the major axis: $2a = 10$; length of the minor axis: $2b = 4$.

25. $16x^2 - 9y^2 - 96x + 288 = 0 \quad \Leftrightarrow \quad 16(x^2 - 6x) - 9y^2 + 288 = 0$
$\Leftrightarrow \quad 16(x^2 - 6x + 9) - 9y^2 = 144 - 288 \quad \Leftrightarrow$

$16(x-3)^2 - 9y^2 = -144 \quad \Leftrightarrow \quad \dfrac{y^2}{16} - \dfrac{(x-3)^2}{9} = 1.$ This is a

hyperbola that has $a = 4$, $b = 3$, and $c = \sqrt{16+9} = 5$.
Center: $(3,0)$; foci: $(3, \pm 5)$; vertices: $(3, \pm 4)$; asymptotes:
$y = \pm \frac{4}{3}(x-3) \quad \Leftrightarrow \quad y = \frac{4}{3}x - 4$ and $y = 4 - \frac{4}{3}x$.

27. $x^2 + 16 = 4(y^2 + 2x) \quad \Leftrightarrow \quad x^2 - 8x - 4y^2 + 16 = 0 \quad \Leftrightarrow$
$(x^2 - 8x + 16) - 4y^2 = -16 + 16 \quad \Leftrightarrow \quad 4y^2 = (x-4)^2 \quad \Leftrightarrow$
$y = \pm \frac{1}{2}(x-4)$. Thus, the conic is degenerate, and its graph is a
pair of lines, $y = \frac{1}{2}(x-4)$ and $y = -\frac{1}{2}(x-4)$.

29. $3x^2 + 4y^2 - 6x - 24y + 39 = 0 \quad \Leftrightarrow$
$3(x^2 - 2x) + 4(y^2 - 6y) = -39 \quad \Leftrightarrow$
$3(x^2 - 2x + 1) + 4(y^2 - 6y + 9) = -39 + 3 + 36 \quad \Leftrightarrow$
$3(x-1)^2 + 4(y-3)^2 = 0 \quad \Leftrightarrow \quad x = 1$ and $y = 3$. This is a
degenerate conic whose graph is the point $(1,3)$.

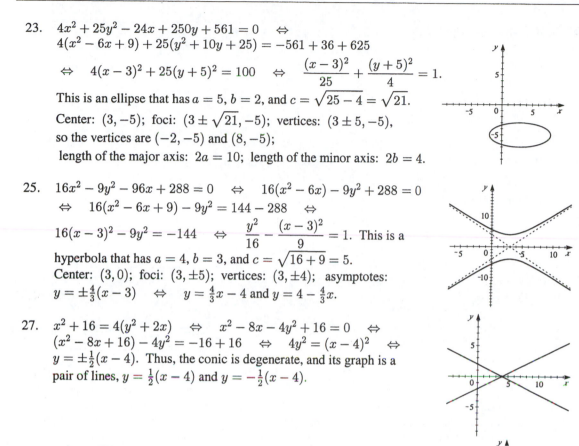

31. $2x^2 - 4x + y + 5 = 0 \quad \Leftrightarrow \quad y = -2x^2 + 4x - 5.$

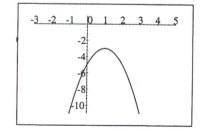

33. $9x^2 + 36 = y^2 + 36x + 6y$ $\Leftrightarrow$ $x^2 - 36x + 36 = y^2 + 6y$

    $\Leftrightarrow$ $9x^2 - 36x + 45 = y^2 + 6y + 9$ $\Leftrightarrow$

    $9(x^2 - 4x + 5) = (y + 3)^2$ $\Leftrightarrow$

    $y + 3 = \pm\sqrt{9(x^2 - 4x + 5)}$ $\Leftrightarrow$

    $y = -3 \pm 3\sqrt{x^2 - 4x + 5}.$

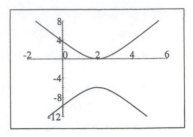

35. $4x^2 + y^2 + 4(x - 2y) + F = 0$ $\Leftrightarrow$ $4(x^2 + x) + (y^2 - 8y) = -F$ $\Leftrightarrow$

    $4(x^2 + x + \frac{1}{4}) + (y^2 - 8y + 16) = 16 + 1 - F$ $\Leftrightarrow$ $4(x + \frac{1}{2})^2 + (y - 1)^2 = 17 - F.$

    (a)   For an ellipse, $17 - F > 0$ $\Leftrightarrow$ $F < 17.$

    (b)   For a single point, $17 - F = 0$ $\Leftrightarrow$ $F = 17.$

    (c)   For the empty set, $17 - F < 0$ $\Leftrightarrow$ $F > 17.$

37.   (a) $x^2 = 4p(y + p)$, for $p = -2, -\frac{3}{2}, -1, -\frac{1}{2}, \frac{1}{2}, 1, \frac{3}{2}, 2.$

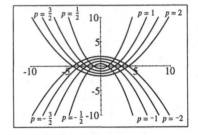

    (b)   The graph of $x^2 = 4p(y + p)$ is obtained by shifting the graph of $x^2 = 4py$ vertically $-p$ units so that the vertex is at $(0, -p)$. The focus of $x^2 = 4py$ is at $(0, p)$, so this point is also shifted $-p$ units vertically to the point $(0, p - p) = (0, 0)$. Thus, the focus is located at the origin.

    (c)   The parabolas become narrower as the vertex moves toward the origin.

39.   Since the height of the satellite *above the earth* varies between 140 and 440, the length of the major axis is $2a = 140 + 2(3960) + 440 = 8500$ $\Leftrightarrow$ $a = 4250$. Since the center of the earth is at one focus, we have $a - c = (earth\ radius) + 140 = 3960 + 140 = 4100$ $\Leftrightarrow$ $c = a - 4100 = 4250 - 4100 = 150$. Thus the center of the ellipse is $(-150, 0)$. So $b^2 = a^2 - c^2 = 4250^2 - 150^2$

    $= 18{,}062{,}500 - 22500 = 18{,}040{,}000.$

Hence the equation is $\dfrac{(x + 150)^2}{18{,}062{,}500} + \dfrac{y^2}{18{,}040{,}000} = 1.$

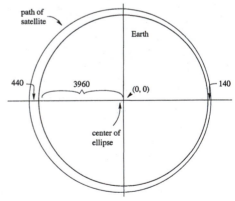

Review Exercises for Chapter 8

1. $y^2 = 4x$. This is a parabola that has $4p = 4 \iff p = 1$.
    Vertex: $(0,0)$; focus: $(1,0)$; directrix: $x = -1$.

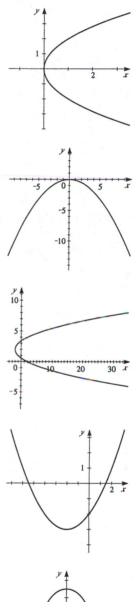

3. $x^2 + 8y = 0 \iff x^2 = -8y$. This is a parabola that has
    $4p = -8 \iff p = -2$.
    Vertex: $(0,0)$; focus: $(0,-2)$; directrix: $y = 2$.

5. $x - y^2 + 4y - 2 = 0 \iff x - (y^2 - 4y + 4) - 2 = -4 \iff$
    $x - (y-2)^2 = -2 \iff (y-2)^2 = x + 2$. This is a parabola
    that has $4p = 1 \iff p = \frac{1}{4}$.
    Vertex: $(-2,2)$; focus: $\left(-2+\frac{1}{4}, 2\right) = \left(-\frac{7}{4}, 2\right)$;
    directrix: $x = -2 - \frac{1}{4} = -\frac{9}{4}$.

7. $\frac{1}{2}x^2 + 2x = 2y + 4 \iff x^2 + 4x = 4y + 8 \iff$
    $x^2 + 4x + 4 = 4y + 8 \iff (x+2)^2 = 4(y+3)$. This is a
    parabola that has $4p = 4 \iff p = 1$.
    Vertex: $(-2,-3)$; focus: $(-2, -3+1) = (-2,-2)$;
    directrix: $y = -3 - 1 = -4$.

9. $\dfrac{x^2}{9} + \dfrac{y^2}{25} = 1$. This is an ellipse with $a = 5, b = 3$, and
    $c = \sqrt{25 - 9} = 4$.
    Center: $(0,0)$; vertices: $(0, \pm 5)$; foci: $(0, \pm 4)$;
    length of the major axis: $2a = 10$;
    length of the minor axis: $2b = 6$.

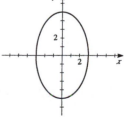

11. $x^2 + 4y^2 = 16$ $\Leftrightarrow$ $\dfrac{x^2}{16} + \dfrac{y^2}{4} = 1$. This is an ellipse with

$a = 4$, $b = 2$, and $c = \sqrt{16 - 4} = 2\sqrt{3}$.

Center: $(0,0)$; vertices: $(\pm 4, 0)$; foci: $(\pm 2\sqrt{3}, 0)$; length of the

major axis: $2a = 8$; length of the minor axis: $2b = 4$.

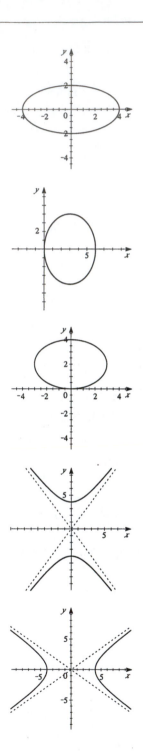

13. $\dfrac{(x - 3)^2}{9} + \dfrac{y^2}{16} = 1$. This is an ellipse with $a = 4$,

$b = 3$, and $c = \sqrt{16 - 9} = \sqrt{7}$.

Center: $(3, 0)$; vertices: $(3, \pm 4)$; foci: $\left(3, \pm\sqrt{7}\right)$;

length of the major axis: $2a = 8$;

length of the minor axis: $2b = 6$.

15. $4x^2 + 9y^2 = 36y$ $\Leftrightarrow$ $4x^2 + 9(y^2 - 4y + 4) = 36$ $\Leftrightarrow$

$4x^2 + 9(y - 2)^2 = 36$ $\Leftrightarrow$ $\dfrac{x^2}{9} + \dfrac{(y - 2)^2}{4} = 1$. This is an

ellipse with $a = 3$, $b = 2$, and $c = \sqrt{9 - 4} = \sqrt{5}$.

Center: $(0, 2)$; vertices: $(\pm 3, 2)$; foci: $(\pm\sqrt{5}, 2)$; length of the

major axis: $2a = 6$; length of the minor axis: $2b = 4$.

17. $-\dfrac{x^2}{9} + \dfrac{y^2}{16} = 1$ $\Leftrightarrow$ $\dfrac{y^2}{16} - \dfrac{x^2}{9} = 0$. This is a hyperbola that

has $a = 4$, $b = 3$, and $c = \sqrt{16 + 9} = \sqrt{25} = 5$.

Center: $(0,0)$; vertices: $(0, \pm 4)$; foci: $(0, \pm 5)$;

asymptotes: $y = \pm\frac{4}{3}x$.

19. $x^2 - 2y^2 = 16$ $\Leftrightarrow$ $\dfrac{x^2}{16} - \dfrac{y^2}{8} = 1$. This is a hyperbola that has

$a = 4$, $b = 2\sqrt{2}$, and $c = \sqrt{16 + 8} = \sqrt{24} = 2\sqrt{6}$.

Center: $(0,0)$; vertices: $(\pm 4, 0)$; foci: $(\pm 2\sqrt{6}, 0)$; asymptotes:

$y = \pm\frac{2\sqrt{2}}{4}x$ $\Leftrightarrow$ $y = \pm\frac{1}{\sqrt{2}}x$.

21. $\dfrac{(x+4)^2}{16} - \dfrac{y^2}{16} = 1$. This is a hyperbola that has $a = 4$,
$b = 4$ and $c = \sqrt{16+16} = 4\sqrt{2}$.
Center: $(-4, 0)$; vertices: $(-4 \pm 4, 0)$ which are $(-8, 0)$ and
$(0, 0)$; foci: $(-4 \pm 4\sqrt{2}, 0)$; asymptotes: $y = \pm(x+4)$.

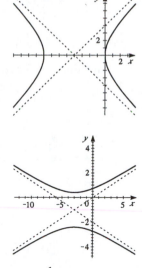

23. $9y^2 + 18y = x^2 + 6x + 18 \quad\Leftrightarrow$
$9(y^2 + 2y + 1) = (x^2 + 6x + 9) + 9 - 9 + 18 \quad\Leftrightarrow$
$9(y+1)^2 - (x+3)^2 = 18 \quad\Leftrightarrow \quad \dfrac{(y+1)^2}{2} - \dfrac{(x+3)^2}{18} = 1$.
This is a hyperbola that has $a = \sqrt{2}$, $b = 3\sqrt{2}$, and
$c = \sqrt{2+18} = 2\sqrt{5}$.
Center: $(-3, -1)$; vertices: $(-3, -1 \pm \sqrt{2})$; foci:
$(-3, -1 \pm 2\sqrt{5})$; asymptotes: $y + 1 = \pm\frac{1}{3}(x+3) \quad\Leftrightarrow \quad y = \frac{1}{3}x$ and $y = -\frac{1}{3}x - 2$.

25. This is a parabola that opens to the right with its vertex at $(0, 0)$ and the focus at $(2, 0)$. So $p = 2$, and the equation is $y^2 = 4(2)x \quad\Leftrightarrow \quad y^2 = 8x$.

27. From the graph, the center is $(0, 0)$, and the vertices are $(0, -4)$ and $(0, 4)$. Since a is the distance from the center to a vertex, we have $a = 4$. Because one focus is $(0, 5)$, we have $c = 5$, and since $c^2 = a^2 + b^2$, we have $25 = 16 + b^2 \quad\Leftrightarrow \quad b^2 = 9$. Thus the equation of the hyperbola is $\dfrac{y^2}{16} - \dfrac{x^2}{9} = 1$.

29. From the graph, the center of the ellipse is $(4, 2)$, and so $a = 4$ and $b = 2$. The equation is $\dfrac{(x-4)^2}{4^2} + \dfrac{(y-2)^2}{2^2} = 1 \quad\Leftrightarrow \quad \dfrac{(x-4)^2}{16} + \dfrac{(y-2)^2}{4} = 1$.

31. $\dfrac{x^2}{12} + y = 1 \quad\Leftrightarrow \quad \dfrac{x^2}{12} = -(y-1) \quad\Leftrightarrow \quad x^2 = -12(y-1)$.
This is a parabola that has $4p = -12 \quad\Leftrightarrow \quad p = -3$.
Vertex: $(0, 1)$; focus: $(0, 1-3) = (0, -2)$.

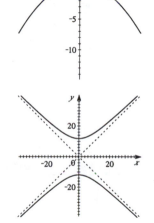

33. $x^2 - y^2 + 144 = 0 \quad\Leftrightarrow \quad \dfrac{y^2}{144} - \dfrac{x^2}{144} = 1$. This is a
hyperbola that has $a = 12$, $b = 12$, and
$c = \sqrt{144 + 144} = 12\sqrt{2}$.
Vertices: $(0, \pm 12)$; foci: $(0, \pm 12\sqrt{2})$.

35. $4x^2 + y^2 = 8(x + y)$ $\Leftrightarrow$ $4(x^2 - 2x) + (y^2 - 8y) = 0$ $\Leftrightarrow$
$4(x^2 - 2x + 1) + (y^2 - 8y + 16) = 4 + 16$ $\Leftrightarrow$
$4(x - 1)^2 + (y - 4)^2 = 20$ $\Leftrightarrow$ $\dfrac{(x - 1)^2}{5} + \dfrac{(y - 4)^2}{20} = 1$. This

is an ellipse that has $a = 2\sqrt{5}$, $b = \sqrt{5}$, and $c = \sqrt{20 - 5} = \sqrt{15}$.

Vertices: $(1, 4 \pm 2\sqrt{5})$; foci: $(1, 4 \pm \sqrt{15})$.

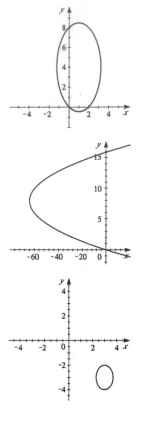

37. $x = y^2 - 16y$ $\Leftrightarrow$ $x + 64 = y^2 - 16y + 64$ $\Leftrightarrow$
$(y - 8)^2 = x + 64$. This is a parabola that has $4p = 1$ $\Leftrightarrow$ $p = \frac{1}{4}$.

Vertex: $(-64, 8)$; focus: $\left(-64 + \frac{1}{4}, 8\right) = \left(-\frac{255}{4}, 8\right)$.

39. $2x^2 - 12x + y^2 + 6y + 26 = 0$ $\Leftrightarrow$
$2(x^2 - 6x) + (y^2 + 6y) = -26$ $\Leftrightarrow$
$2(x^2 - 6x + 9) + (y^2 + 6y + 9) = -26 + 18 + 9$ $\Leftrightarrow$
$2(x - 3)^2 + (y + 3)^2 = 1$ $\Leftrightarrow$ $\dfrac{(x - 3)^2}{\frac{1}{2}} + (y + 3)^2 = 1$. This is

an ellipse that has $a = 1$, $b = \frac{1}{\sqrt{2}}$, and $c = \sqrt{1 - \frac{1}{2}} = \frac{1}{\sqrt{2}}$.

Vertices: $(3, -3 \pm 1)$, so the vertices are $(3, -4)$ and $(3, -2)$; foci:
$\left(3, -3 \pm \frac{1}{\sqrt{2}}\right)$.

41. $9x^2 + 8y^2 - 15x + 8y + 27 = 0$ $\Leftrightarrow$ $9\left(x^2 - \frac{5}{3}x + \frac{25}{36}\right) + 8\left(y^2 + y + \frac{1}{4}\right) = -27 + \frac{25}{4} + 2$ $\Leftrightarrow$
$9\left(x - \frac{5}{6}\right)^2 + 8\left(y + \frac{1}{2}\right)^2 = -\frac{75}{4}$. However, since the left-hand side of the equation is greater than or
equal to 0, there is no point that satisfies this equation. The graph is empty.

43. The parabola has focus $(0, 1)$ and directrix $y = -1$. Therefore, $p = 1$ and so $4p = 4$. Since the
focus is on the y-axis and the vertex is $(0, 0)$, the equation of the parabola is $x^2 = 4y$.

45. The hyperbola has vertices $(0, \pm 2)$ and asymptotes $y = \pm\frac{1}{2}x$. Therefore, $a = 2$, and the foci are on
the y-axis. Since the slopes of the asymptotes are $\pm\frac{1}{2} = \pm\dfrac{a}{b}$ $\Leftrightarrow$ $b = 2a = 4$, the equation of
the hyperbola is $\dfrac{y^2}{4} - \dfrac{x^2}{16} = 1$.

47. The ellipse has foci $F_1(1, 1)$ and $F_2(1, 3)$, and one vertex is on the x-axis. Thus, $2c = 3 - 1 = 2$
$\Leftrightarrow$ $c = 1$, and so the center of the ellipse is $C(1, 2)$. Also, since one vertex is on the x-axis,
$a = 2 - 0 = 2$, and thus $b^2 = 4 - 1 = 3$. So the equation of the ellipse is
$\dfrac{(x - 1)^2}{3} + \dfrac{(y - 2)^2}{4} = 1$.

49. The ellipse has vertices $V_1(7, 12)$ and $V_2(7, -8)$ and passes through the point $P(1, 8)$. Thus, $2a = 12 - (-8) = 20 \quad \Leftrightarrow \quad a = 10$, and the center is $\left(7, \frac{-8+12}{2}\right) = (7, 2)$. Thus the equation of the ellipse has the form $\dfrac{(x-7)^2}{b^2} + \dfrac{(y-2)^2}{100} = 1$. Since the point $P(1, 8)$ is on the ellipse,

$\dfrac{(1-7)^2}{b^2} + \dfrac{(8-2)^2}{100} = 1 \quad \Leftrightarrow \quad 3600 + 36b^2 = 100b^2 \quad \Leftrightarrow \quad 64b^2 = 3600 \quad \Leftrightarrow \quad b^2 = \dfrac{225}{4}$.

Therefore, the equation of the ellipse is $\dfrac{(x-7)^2}{225/4} + \dfrac{(y-2)^2}{100} = 1 \quad \Leftrightarrow$

$\dfrac{4(x-7)^2}{225} + \dfrac{(y-2)^2}{100} = 1$.

51. The length of the major axis is $2a = 186{,}000{,}000 \quad \Leftrightarrow \quad a = 93{,}000{,}000$. The eccentricity is $e = \frac{c}{a} = 0.017$, and so $c = 0.017(93{,}000{,}000) = 1{,}581{,}000$.

(a) The earth is closest to the sun when the distance is $a - c = 93{,}000{,}000 - 1{,}581{,}000$
$= 91{,}419{,}000$.

(b) The earth is furthest from the sun when the distance is $a + c = 93{,}000{,}000 + 1{,}581{,}000$
$= 94{,}581{,}000$.

53. (a) The graphs of $\dfrac{x^2}{16+k^2} + \dfrac{y^2}{k^2} = 1$ for
$k = 1, 2, 4$, and 8 are shown in the figure.

(b) $c^2 = (16 + k^2) - k^2 = 16 \quad \Rightarrow \quad c = \pm 4$.
Since the center is $(0, 0)$, the foci of each of
the ellipses are $(\pm 4, 0)$.

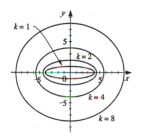

Chapter 8 Test

1. $x^2 + 8y = 0 \Leftrightarrow x^2 = -8y$. This is a parabola that has
   $4p = -8 \Leftrightarrow p = -2$.

   Focus: $(0, -2)$; directrix: $y = 2$.

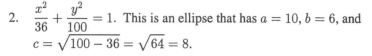

2. $\dfrac{x^2}{36} + \dfrac{y^2}{100} = 1$. This is an ellipse that has $a = 10$, $b = 6$, and
   $c = \sqrt{100 - 36} = \sqrt{64} = 8$.

   Vertices: $(0, \pm 10)$; foci: $(0, \pm 8)$; length of the major axis: $2a = 20$;
   length of the minor axis: $2b = 12$.

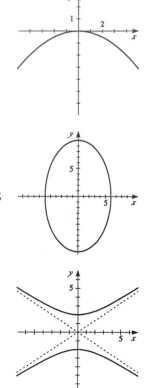

3. $-\dfrac{x^2}{9} + \dfrac{y^2}{4} = 1 \Leftrightarrow \dfrac{y^2}{4} - \dfrac{x^2}{9} = 1$. This is a hyperbola that
   has $a = 2$, $b = 3$, and $c = \sqrt{9 + 4} = \sqrt{13}$.

   Vertices: $(0, \pm 2)$; foci: $(0, \pm\sqrt{13})$; asymptotes: $y = \pm\frac{2}{3}x$.

4. This is a parabola that opens to the left with its vertex at $(0, 0)$. So its equation is of the form
   $y^2 = 4px$ with $p < 0$. Substituting the point $(-4, 2)$, we have $2^2 = 4p(-4) \Leftrightarrow 4 = -16p$
   $\Leftrightarrow p = -\frac{1}{4}$. So the equation is $y^2 = 4\left(-\frac{1}{4}\right)x \Leftrightarrow y^2 = -x$.

5. This is an ellipse tangent to the x-axis at $(0, 0)$ and with one vertex at the point $(4, 3)$. The center is
   $(0, 3)$, and $a = 4$ and $b = 3$. Thus the equation is $\dfrac{x^2}{16} + \dfrac{(y - 3)^2}{9} = 1$.

6. This a hyperbola with a horizontal transverse axis, vertices at $(1, 0)$ and $(3, 0)$, and foci at $(0, 0)$ and
   $(4, 0)$. Thus the center is $(2, 0)$, and $a = 3 - 2 = 1$ and $c = 4 - 2 = 2$. Thus $b^2 = 2^2 - 1^2 = 3$. So
   the equation is $\dfrac{(x - 2)^2}{1^2} - \dfrac{y^2}{3} = 1 \Leftrightarrow (x - 2)^2 - \dfrac{y^2}{3} = 1$.

7. $16x^2 + 36y^2 - 96x + 36y + 9 = 0$ ⇔
$16(x^2 - 6x) + 36(y^2 + y) = -9$ ⇔
$16(x^2 - 6x + 9) + 36\left(y^2 + y + \frac{1}{4}\right) = -9 + 144 + 9$ ⇔

$16(x - 3)^2 + 36\left(y + \frac{1}{2}\right)^2 = 144$ ⇔ $\dfrac{(x-3)^2}{9} + \dfrac{\left(y + \frac{1}{2}\right)^2}{4} = 1$.

This is an ellipse that has $a = 3$, $b = 2$, and $c = \sqrt{9 - 4} = \sqrt{5}$.

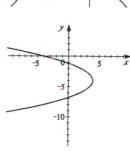

Center: $\left(3, -\frac{1}{2}\right)$; vertices: $\left(3 \pm 3, -\frac{1}{2}\right)$, so vertices are $\left(0, -\frac{1}{2}\right)$ and $\left(6, -\frac{1}{2}\right)$; foci: $(h \pm c, k)$ which are $\left(3 \pm \sqrt{5}, -\frac{1}{2}\right)$, or $\left(3 + \sqrt{5}, -\frac{1}{2}\right)$ and $\left(3 - \sqrt{5}, -\frac{1}{2}\right)$.

8. $9x^2 - 8y^2 + 36x + 64y = 92$ ⇔ $9(x^2 + 4x) - 8(y^2 - 8y) = 92$
⇔ $9(x^2 + 4x + 4) - 8(y^2 - 8y + 16) = 92 + 36 - 128$
⇔ $9(x + 2)^2 - 8(y - 4)^2 = 0$.
This is a degenerate conic.

9. $2x + y^2 + 8y + 8 = 0$ ⇔ $y^2 + 8y + 16 = -2x - 8 + 16$ ⇔
$(y + 4)^2 = -2(x - 4)$. This is a parabola that has $4p = -2$ ⇔
$p = -\frac{1}{2}$.
Vertex: $(4, -4)$; focus: $\left(4 - \frac{1}{2}, -4\right) = \left(\frac{7}{2}, -4\right)$. No asymptotes.

10. The hyperbola has foci $(\pm\sqrt{2}, 0)$ and asymptotes $y = \pm x$. Since the foci are $(\pm\sqrt{2}, 0)$, $c = \sqrt{2}$, the foci are on the x-axis, and the center is $(0, 0)$. Also, since $y = \pm\frac{1}{1}x = \pm\frac{a}{b}x$, it follows that $a = b$. Then $c^2 = 2 = a^2 + b^2 = 2b^2$ ⇔ $b^2 = 1$ ⇒ $b = 1$. Hence $a = 1$. Therefore, the equation of the hyperbola is $x^2 - y^2 = 1$.

11. The parabola has focus $(2, 4)$ and directrix the x-axis ($y = 0$). Therefore, $2p = 4 - 0 = 4$ ⇔ $p = 2$ ⇔ $4p = 8$, and the vertex is $(2, 4 - p) = (2, 2)$. Hence, the equation of the parabola is $(x - 2)^2 = 8(y - 2)$ ⇔ $x^2 - 4x + 4 = 8y - 16$ ⇔ $x^2 - 4x - 8y + 20 = 0$.

12. We place the vertex of the parabola at the origin, so the parabola contains the points $(3, \pm 3)$, and the equation is of the form $y^2 = 4px$. Substituting the point $(3, 3)$, we get $3^2 = 4p(3)$ ⇔ $9 = 12p$ ⇔ $p = \frac{3}{4}$. So the focus is $\left(\frac{3}{4}, 0\right)$, and we should place the light bulb $\frac{3}{4}$ inches from the vertex.

Focus on Modeling

1. Many answers possible.

3. Many answers possible.

5. Many answers possible.

7. (a) Using the point-slope equation with slope m and point (a, a^2) we get the equation of the tangent line is $y - a^2 = m(x - a)$.

    (b) First we note that the point (a, a^2) satisfies both equation, so it is a solution the system. Since the tangent line to $y = x^2$ at (a, a^2) must have an equation of the form $y - a^2 = m(x - a)$ and the tangent line intersects the parabola at exactly one point, we conclude that (a, a^2) is the only solution.

    (c) Substituting for y we get $x^2 - a^2 = m(x - a)$ $\Leftrightarrow$ $x^2 - mx + (am - a^2) = 0$. The discriminant of this quadratic equation is
    $\sqrt{(-m)^2 - 4(1)(am - a^2)} = \sqrt{m^2 - 4am + 4a^2} = \sqrt{(m - 2a)^2}$. Since the system has exactly one solution $\sqrt{(m - 2a)^2} = 0$ so $m - 2a = 0$ $\Leftrightarrow$ $m = 2a$.

    (d) Substituting for m, we get $y - a^2 = 2a(x - a)$ $\Leftrightarrow$ $y = 2ax - a^2$.

Chapter Nine
Exercises 9.1

1. $a_n = n + 1$. Then $a_1 = 1 + 1 = 2$; $a_2 = 2 + 1 = 3$; $a_3 = 3 + 1 = 4$; $a_4 = 4 + 1 = 5$; and $a_{100} = 100 + 1 = 101$.

3. $a_n = \dfrac{1}{n + 1}$. Then $a_1 = \dfrac{1}{1 + 1} = \dfrac{1}{2}$; $a_2 = \dfrac{1}{2 + 1} = \dfrac{1}{3}$; $a_3 = \dfrac{1}{3 + 1} = \dfrac{1}{4}$; $a_4 = \dfrac{1}{4 + 1} = \dfrac{1}{5}$; and $a_{100} = \dfrac{1}{100 + 1} = \dfrac{1}{101}$.

5. $a_n = \dfrac{(-1)^n}{n^2}$. Then $a_1 = \dfrac{(-1)^1}{1^2} = -1$; $a_2 = \dfrac{(-1)^2}{2^2} = \dfrac{1}{4}$; $a_3 = \dfrac{(-1)^3}{3^2} = -\dfrac{1}{9}$; $a_4 = \dfrac{(-1)^4}{4^2} = \dfrac{1}{16}$; and $a_{100} = \dfrac{(-1)^{100}}{100^2} = \dfrac{1}{10,000}$.

7. $a_n = 1 + (-1)^n$. Then $a_1 = 1 + (-1)^1 = 0$; $a_2 = 1 + (-1)^2 = 2$; $a_3 = 1 + (-1)^3 = 0$; $a_4 = 1 + (-1)^4 = 2$; and $a_{100} = 1 + (-1)^{100} = 2$.

9. $a_n = n^n$. Then $a_1 = 1^1 = 1$; $a_2 = 2^2 = 4$; $a_3 = 3^3 = 27$; $a_4 = 4^4 = 256$; and $a_{100} = 100^{100} = 10^{200}$.

11. $a_n = 2(a_{n-1} - 2)$ and $a_1 = 3$. Then $a_2 = 2[(3) - 2] = 2$; $a_3 = 2[(2) - 2] = 0$; $a_4 = 2[(0) - 2] = -4$; and $a_5 = 2[(-4) - 2] = -12$.

13. $a_n = 2a_{n-1} + 1$ and $a_1 = 1$. Then $a_2 = 2(1) + 1 = 3$; $a_3 = 2(3) + 1 = 7$; $a_4 = 2(7) + 1 = 15$; and $a_5 = 2(15) + 1 = 31$.

15. $a_n = a_{n-1} + a_{n-2}$; $a_1 = 1$; and $a_2 = 2$. Then $a_3 = 2 + 1 = 3$; $a_4 = 3 + 2 = 5$; and $a_5 = 5 + 3 = 8$.

17. (a)

$a_1 = 7$	$a_2 = 11$
$a_3 = 15$	$a_4 = 19$
$a_5 = 23$	$a_6 = 27$
$a_7 = 31$	$a_8 = 35$
$a_9 = 39$	$a_{10} = 43$

(b)

19. (a) $a_1 = \frac{12}{1} = 12$ $a_2 = \frac{12}{2} = 6$ (b)

$a_3 = \frac{12}{3} = 4$ $a_4 = \frac{12}{4} = 3$

$a_5 = \frac{12}{5}$ $a_6 = \frac{12}{6} = 2$

$a_7 = \frac{12}{7}$ $a_8 = \frac{12}{8} = \frac{3}{2}$

$a_9 = \frac{12}{9} = \frac{4}{3}$ $a_{10} = \frac{12}{10} = \frac{6}{5}$

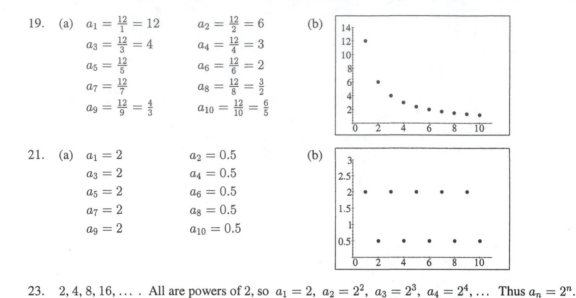

21. (a) $a_1 = 2$ $a_2 = 0.5$ (b)

$a_3 = 2$ $a_4 = 0.5$

$a_5 = 2$ $a_6 = 0.5$

$a_7 = 2$ $a_8 = 0.5$

$a_9 = 2$ $a_{10} = 0.5$

23. $2, 4, 8, 16, \ldots$. All are powers of 2, so $a_1 = 2$, $a_2 = 2^2$, $a_3 = 2^3$, $a_4 = 2^4, \ldots$ Thus $a_n = 2^n$.

25. $1, 4, 7, 10, \ldots$. The difference between any two consecutive terms is 3, so $a_1 = 3(1) - 2$, $a_2 = 3(2) - 2$, $a_3 = 3(3) - 2$, $a_4 = 3(4) - 2, \ldots$ Thus $a_n = 3n - 2$.

27. $1, \frac{3}{4}, \frac{5}{9}, \frac{7}{16}, \frac{9}{25}, \ldots$. We consider the numerator separately from the denominator. The numerators of the terms differ by 2, and the denominators are perfect squares. So $a_1 = \dfrac{2(1) - 1}{1^2}$, $a_2 = \frac{2(2)-1}{2^2}$, $a_3 = \dfrac{2(3) - 1}{3^2}$, $a_4 = \dfrac{2(4) - 1}{4^2}$, $a_5 = \dfrac{2(5) - 1}{5^2}, \ldots$. Thus $a_n = \dfrac{2n - 1}{n^2}$.

29. $0, 2, 0, 2, 0, 2, \ldots$. These terms alternate between 0 and 2. So $a_1 = 1 - 1$, $a_2 = 1 + 1$, $a_3 = 1 - 1$, $a_4 = 1 + 1$, $a_5 = 1 - 1$, $a_6 = 1 + 1, \ldots$ Thus $a_n = 1 + (-1)^n$.

31. $a_1 = 1$, $a_2 = 3$, $a_3 = 5$, $a_4 = 7, \ldots$. Therefore, $a_n = 2n - 1$. So $S_1 = 1$; $S_2 = 1 + 3 = 4$; $S_3 = 1 + 3 + 5 + \ = 9$; $S_4 = 1 + 3 + 5 + 7 = 16$; $S_5 = 1 + 3 + 5 + 7 + 9 = 25$; and $S_6 = 1 + 3 + 5 + 7 + 9 + 11 = 36$.

33. $a_1 = \frac{1}{3}$, $a_2 = \frac{1}{3^2}$, $a_3 = \frac{1}{3^3}$, $a_4 = \frac{1}{3^4}, \ldots$. Therefore, $a_n = \frac{1}{3^n}$. So $S_1 = \frac{1}{3}$; $S_2 = \frac{1}{3} + \frac{1}{3^2} = \frac{4}{9}$; $S_3 = \frac{1}{3} + \frac{1}{3^2} + \frac{1}{3^3} = \frac{13}{27}$; $S_4 = \frac{1}{3} + \frac{1}{3^2} + \frac{1}{3^3} + \frac{1}{3^4} = \frac{40}{81}$; and $S_5 = \frac{1}{3} + \frac{1}{3^2} + \frac{1}{3^3} + \frac{1}{3^4} + \frac{1}{3^5} = \frac{121}{243}$; $S_6 = \frac{1}{3} + \frac{1}{3^2} + \frac{1}{3^3} + \frac{1}{3^4} + \frac{1}{3^5} + \frac{1}{3^6} = \frac{364}{729}$.

35. $a_n = \frac{2}{3^n}$. So $S_1 = \frac{2}{3}$; $S_2 = \frac{2}{3} + \frac{2}{3^2} = \frac{8}{9}$; $S_3 = \frac{2}{3} + \frac{2}{3^2} + \frac{2}{3^3} = \frac{26}{27}$; and $S_4 = \frac{2}{3} + \frac{2}{3^2} + \frac{2}{3^3} + \frac{2}{3^4} = \frac{80}{81}$. Therefore, $S_n = \frac{3^n - 1}{3^n}$.

37. $a_n = \sqrt{n} - \sqrt{n+1}$. So $S_1 = \sqrt{1} - \sqrt{2} = \underline{1 - \sqrt{2}}$;

$S_2 = \left(\sqrt{1} - \sqrt{2}\right) + \left(\sqrt{2} - \sqrt{3}\right) = 1 + \left(-\sqrt{2} + \sqrt{2}\right) - \sqrt{3} = \underline{1 - \sqrt{3}}$;

$S_3 = \left(\sqrt{1} - \sqrt{2}\right) + \left(\sqrt{2} - \sqrt{3}\right) + \left(\sqrt{3} - \sqrt{4}\right)$

$= 1 + \left(-\sqrt{2} + \sqrt{2}\right) + \left(-\sqrt{3} + \sqrt{3}\right) - \sqrt{4} = \underline{1 - \sqrt{4}}$;

$$S_4 = \left(\sqrt{1} - \sqrt{2}\right) + \left(\sqrt{2} - \sqrt{3}\right) + \left(\sqrt{3} - \sqrt{4}\right) + \left(\sqrt{4} - \sqrt{5}\right)$$

$$= 1 + \left(-\sqrt{2} + \sqrt{2}\right) + \left(-\sqrt{3} + \sqrt{3}\right) + \left(-\sqrt{4} + \sqrt{4}\right) - \sqrt{5} = \underline{1 - \sqrt{5}}. \text{ Therefore,}$$

$$S_n = \left(\sqrt{1} - \sqrt{2}\right) + \left(\sqrt{2} - \sqrt{3}\right) + \cdots + \left(\sqrt{n} - \sqrt{n+1}\right)$$

$$= 1 + \left(-\sqrt{2} + \sqrt{2}\right) + \left(-\sqrt{3} + \sqrt{3}\right) + \cdots + \left(-\sqrt{n} + \sqrt{n}\right) - \sqrt{n+1} = \underline{1 - \sqrt{n+1}}.$$

39. $\displaystyle\sum_{k=1}^{4} k = 1 + 2 + 3 + 4 = 10$

41. $\displaystyle\sum_{k=1}^{3} \frac{1}{k} = 1 + \frac{1}{2} + \frac{1}{3} = \frac{6}{6} + \frac{3}{6} + \frac{2}{6} = \frac{11}{6}$

43. $\displaystyle\sum_{i=1}^{8} [1 + (-1)^i] = 0 + 2 + 0 + 2 + 0 + 2 + 0 + 2 = 8$

45. $\displaystyle\sum_{k=1}^{5} 2^{k-1} = 2^0 + 2^1 + 2^2 + 2^3 + 2^4 = 1 + 2 + 4 + 8 + 16 = 31$

47. 385   49.   46,438   51.   22

53. $\displaystyle\sum_{k=1}^{5} \sqrt{k} = \sqrt{1} + \sqrt{2} + \sqrt{3} + \sqrt{4} + \sqrt{5}$

55. $\displaystyle\sum_{k=0}^{6} \sqrt{k+4} = \sqrt{4} + \sqrt{5} + \sqrt{6} + \sqrt{7} + \sqrt{8} + \sqrt{9} + \sqrt{10}$

57. $\displaystyle\sum_{k=3}^{100} x^k = x^3 + x^4 + x^5 + \cdots + x^{100}$

59. $1 + 2 + 3 + 4 + \cdots + 100 = \displaystyle\sum_{k=1}^{100} k$

61. $1^2 + 2^2 + 3^2 + \cdots + 10^2 = \displaystyle\sum_{k=1}^{10} k^2$

63. $\dfrac{1}{1 \cdot 2} + \dfrac{1}{2 \cdot 3} + \dfrac{1}{3 \cdot 4} + \cdots + \dfrac{1}{999 \cdot 1000} = \displaystyle\sum_{k=1}^{999} \dfrac{1}{k(k+1)}$

65. $1 + x + x^2 + x^3 + \cdots + x^{100} = \displaystyle\sum_{k=0}^{100} x^k$

67. $\sqrt{2}, \sqrt{2\sqrt{2}}, \sqrt{2\sqrt{2\sqrt{2}}}, \sqrt{2\sqrt{2\sqrt{2\sqrt{2}}}}, \ldots$. We simplify each term in an attempt to determine a

formula for a_n. So $a_1 = 2^{1/2}$; $a_2 = \sqrt{2 \cdot 2^{1/2}} = \sqrt{2^{3/2}} = 2^{3/4}$; $a_3 = \sqrt{2 \cdot 2^{3/4}} = \sqrt{2^{7/4}} = 2^{7/8}$;

$a_4 = \sqrt{2 \cdot 2^{7/8}} = \sqrt{2^{15/8}} = 2^{15/16}; \ldots$. Thus $a_n = 2^{(2^n - 1)/2^n}$.

69. (a) $A_1 = \$2004$; $A_2 = \$2008.01$; $A_3 = \$2012.02$; $A_4 = \$2016.05$; $A_5 = \$2020.08$; $A_6 = \$2024.12$.

(b) Since 3 years is 36 months, we get $A_{36} = \$2149.16$.

71. (a) $P_1 = 35,700$; $P_2 = 36,414$; $P_3 = 37,142$; $P_4 = 37,885$; $P_5 = 38,643$.

(b) Since 2014 is 10 years after 2004, $P_{10} = 42,665$.

73. (a) The number of catfish at the end of the month, P_n, is the population at the start of the month, P_{n-1}, plus the increase in population, $0.08P_{n-1}$, minus the 300 catfish harvested. Thus $P_n = P_{n-1} + 0.08P_{n-1} - 300 \quad \Leftrightarrow \quad P_n = 1.08P_{n-1} - 300$.

(b) $P_1 = 5100$; $P_2 = 5208$; $P_3 = 5325$; $P_4 = 5451$; $P_5 = 5587$; $P_6 = 5734$; $P_7 = 5892$; $P_8 = 6064$; $P_9 = 6249$; $P_{10} = 6449$; $P_{11} = 6665$; $P_{12} = 6898$. Thus there should be 6898 catfish in the pond at the end of 12 months.

75. (a) Let S_n be his salary in the nth year. Then $S_1 = \$30,000$. Since his salary increase by 2000 each year, $S_n = S_{n-1} + 2000$. Thus $S_1 = \$30,000$ and $S_n = S_{n-1} + 2000$.

(b) $S_5 = S_4 + 2000 = (S_3 + 2000) + 2000 = (S_2 + 2000) + 4000 = (S_1 + 2000) + 6000$ $= \$38,000$.

77. Let F_n be the number of pairs of rabbits in the nth month. Clearly $F_1 = F_2 = 1$. In the nth month each pair that is two or more months old (that is, F_{n-2} pairs) will add a pair of offspring to the F_{n-1} pairs already present. Thus $F_n = F_{n-1} + F_{n-2}$. So F_n is the Fibonacci sequence.

79. $a_{n+1} = \begin{cases} \dfrac{a_n}{2} & \text{if } a_n \text{ is an even number} \\ 3a_n + 1 & \text{if } a_n \text{ is an odd number} \end{cases}$

With $a_1 = 11$, we have $a_2 = 34$; $a_3 = 17$; $a_4 = 52$; $a_5 = 26$; $a_6 = 13$; $a_7 = 40$; $a_8 = 20$; $a_9 = 10$; $a_{10} = 5$; $a_{11} = 16$; $a_{12} = 8$; $a_{13} = 4$; $a_{14} = 2$; $a_{15} = 1$; $a_{16} = 4$; $a_{17} = 2$; $a_{18} = 1$; ... (with 4, 2, 1 repeating). So $a_{3n+1} = 4$; $a_{3n+2} = 2$; and $a_{3n} = 1$, for $n \geq 5$.

With $a_1 = 25$, we have $a_2 = 76$; $a_3 = 38$; $a_4 = 19$; $a_5 = 58$; $a_6 = 29$; $a_7 = 88$; $a_8 = 44$; $a_9 = 22$; $a_{10} = 11$; $a_{11} = 34$; $a_{12} = 17$; $a_{13} = 52$; $a_{14} = 26$; $a_{15} = 13$; $a_{16} = 40$; $a_{17} = 20$; $a_{18} = 10$; $a_{19} = 5$; $a_{20} = 16$; $a_{21} = 8$; $a_{22} = 4$; $a_{23} = 2$; $a_{24} = 1$; $a_{25} = 4$; $a_{26} = 2$; $a_{27} = 1$; ... (with 4, 2, 1 repeating). So $a_{3n+1} = 4$; $a_{3n+2} = 2$; and $a_{3n} = 1$, for $n \geq 8$.

Conjecture: The sequence will always return to the numbers 4, 2, 1 repeating.

Exercises 9.2

1. (a) $a_1 = 5 + 2(1 - 1) = 5$
   $a_2 = 5 + 2(2 - 1) = 5 + 2 = 7$
   $a_3 = 5 + 2(3 - 1) = 5 + 4 = 9$
   $a_4 = 5 + 2(4 - 1) = 5 + 6 = 11$
   $a_5 = 5 + 2(5 - 1) = 5 + 8 = 13$

   (b) The common difference is 2.

   (c)

3. (a) $a_1 = \frac{5}{2} - (1 - 1) = \frac{5}{2}$
   $a_2 = \frac{5}{2} - (2 - 1) = \frac{3}{2}$
   $a_3 = \frac{5}{2} - (3 - 1) = \frac{1}{2}$
   $a_4 = \frac{5}{2} - (4 - 1) = -\frac{1}{2}$
   $a_5 = \frac{5}{2} - (5 - 1) = -\frac{3}{2}$

   (b) The common difference is -1.

   (c)

5. $a = 3$, $d = 5$, $a_n = a + d(n - 1) = 3 + 5(n - 1)$. So $a_{10} = 3 + 5(10 - 1) = 48$.

7. $a = \frac{5}{2}$, $d = -\frac{1}{2}$, $a_n = a + d(n - 1) = \frac{5}{2} - \frac{1}{2}(n - 1)$. So $a_{10} = \frac{5}{2} - \frac{1}{2}(10 - 1) = \frac{-4}{2} = -2$.

9. $a_4 - a_3 = 14 - 11 = 3$; $a_3 - a_2 = 11 - 8 = 3$; $a_2 - a_1 = 8 - 5 = 3$. This sequence is arithmetic with the common difference 3.

11. Since $a_2 - a_1 = 4 - 2 = 2$ and $a_4 - a_3 = 16 - 8 = 8$, the terms of the sequence do not have a common difference. This sequence is not arithmetic.

13. $a_4 - a_3 = -\frac{3}{2} - 0 = -\frac{3}{2}$; $a_3 - a_2 = 0 - \frac{3}{2} = -\frac{3}{2}$; $a_2 - a_1 = \frac{3}{2} - 3 = -\frac{3}{2}$. This sequence is arithmetic with the common difference $-\frac{3}{2}$.

15. $a_4 - a_3 = 7.7 - 6.0 = 1.7$; $a_3 - a_2 = 6.0 - 4.3 = 1.7$; $4. - a_1 = 4.3 - 2.6 = 1.7$. This sequence is arithmetic with the common difference 1.7.

17. $a_1 = 4 + 7(1) = 11$; $a_2 = 4 + 7(2) = 18$; $a_3 = 4 + 7(3) = 25$; $a_4 = 4 + 7(4) = 32$; $a_5 = 4 + 7(5) = 39$. This sequence is arithmetic, the common difference is $d = 7$ and $a_n = 4 + 7n = 4 + 7n - 7 + 7 = 11 + 7(n - 1)$.

19. $a_1 = \frac{1}{1+2(1)} = \frac{1}{3}$; $a_2 = \frac{1}{1+2(2)} = \frac{1}{5}$; $a_3 = \frac{1}{1+2(3)} = \frac{1}{7}$; $a_4 = \frac{1}{1+2(4)} = \frac{1}{9}$; $a_5 = \frac{1}{1+2(5)} = \frac{1}{11}$. Since $a_4 - a_3 = \frac{1}{9} - \frac{1}{7} = -\frac{2}{63}$ and $a_3 - a_2 = \frac{1}{7} - \frac{1}{3} = -\frac{2}{21}$, the terms of the sequence do not have a common difference. This sequence is not arithmetic.

21. $a_1 = 6(1) - 10 = -4$; $a_2 = 6(2) - 10 = 2$; $a_3 = 6(3) - 10 = 8$; $a_4 = 6(4) - 10 = 14$; $a_5 = 6(5) - 10 = 20$. This sequence is arithmetic, the common difference is $d = 6$ and $a_n = 6n - 10 = 6n - 6 + 6 - 10 = -4 + 6(n - 1)$.

23. $2, 5, 8, 11, \ldots$. Then $d = a_2 - a_1 = 5 - 2 = 3$; $a_5 = a_4 + 3 = 11 + 3 = 14$; $a_n = 2 + 3(n-1)$; and $a_{100} = 2 + 3(99) = 299$.

25. $4, 9, 14, 19, \ldots$. Then $d = a_2 - a_1 = 9 - 4 = 5$; $a_5 = a_4 + 5 = 19 + 5 = 24$; $a_n = 4 + 5(n-1)$; and $a_{100} = 4 + 5(99) = 499$.

27. $-12, -8, -4, 0, \ldots$. Then $d = a_2 - a_1 = -8 - (-12) = 4$; $a_5 = a_4 + 4 = 0 + 4 = 4$; $a_n = -12 + 4(n-1)$; and $a_{100} = -12 + 4(99) = 384$.

29. $25, 26.5, 28, 29.5, \ldots$. Then $d = a_2 - a_1 = 26.5 - 25 = 1.5$; $a_5 = a_4 + 1.5 = 29.5 + 1.5 = 31$; $a_n = 25 + 1.5(n-1)$; $a_{100} = 25 + 1.5(99) = 173.5$.

31. $2, 2 + s, 2 + 2s, 2 + 3s, \ldots$. Then $d = a_2 - a_1 = 2 + s - 2 = s$; $a_5 = a_4 + s = 2 + 3s + s = 2 + 4s$; $a_n = 2 + (n-1)s$; and $a_{100} = 2 + 99s$.

33. $a_{10} = \frac{55}{2}$, $a_2 = \frac{7}{2}$, and $a_n = a + d(n-1)$. Then $a_2 = a + d = \frac{7}{2}$ $\Leftrightarrow$ $d = \frac{7}{2} - a$. Substituting into $a_{10} = a + 9d = \frac{55}{2}$ gives $a + 9\left(\frac{7}{2} - a\right) = \frac{55}{2}$ $\Leftrightarrow$ $a = \frac{1}{2}$. Thus, the first term is $a_1 = \frac{1}{2}$.

35. $a_{100} = 98$ and $d = 2$. Note that $a_{100} = a + 99d = a + 99(2) = a + 198$. Since $a_{100} = 98$, we have $a + 198 = a_{100} = 98$ $\Leftrightarrow$ $a = -100$. Hence, $a_1 = -100$, $a_2 = -100 + 2 = -98$, and $a_3 = -100 + 4 = -96$.

37. The arithmetic sequence is $1, 4, 7, \ldots$. So $d = 4 - 1 = 3$ and $a_n = 1 + 3(n-1)$. Then $a_n = 88$ $\Leftrightarrow$ $1 + 3(n-1) = 88$ $\Leftrightarrow$ $3(n-1) = 87$ $\Leftrightarrow$ $n - 1 = 29$ $\Leftrightarrow$ $n = 30$. So, 88 is the 30th term.

39. $a = 1$, $d = 2$, $n = 10$. Then $S_{10} = \frac{10}{2}[2a + (10-1)d] = \frac{10}{2}[2 \cdot 1 + 9 \cdot 2] = 100$.

41. $a = 4$, $d = 2$, $n = 20$. Then $S_{20} = \frac{20}{2}[2a + (20-1)d] = \frac{20}{2}[2 \cdot 4 + 19 \cdot 2] = 460$.

43. $a_1 = 55$, $d = 12$, $n = 10$. Then $S_{10} = \frac{10}{2}[2a + (10-1)d] = \frac{10}{2}[2 \cdot 55 + 9 \cdot 12] = 1090$.

45. $1 + 5 + 9 + \cdots + 401$ is a partial sum of an arithmetic series, where $a = 1$ and $d = 5 - 1 = 4$. The last term is $401 = a_n = 1 + 4(n-1)$, so $n - 1 = 100$ $\Leftrightarrow$ $n = 101$. So, the partial sum is $S_{101} = \frac{101}{2}(1 + 401) = 101 \cdot 201 = 20{,}301$.

47. $0.7 + 2.7 + 4.7 + \cdots + 56.7$ is a partial sum of an arithmetic series, where $a = 0.7$ and $d = 2.7 - 0.7 = 2$. The last term is $56.7 = a_n = 0.7 + 2(n-1)$ $\Leftrightarrow$ $28 = n - 1$ $\Leftrightarrow$ $n = 29$. So, the partial sum is $S_{29} = \frac{29}{2}(0.7 + 56.7) = 832.3$.

49. $\sum\limits_{k=0}^{10}(3 + 0.25k)$ is a partial sum of an arithmetic series, where $a = 3 + 0.25 \cdot 0 = 3$ and $d = 0.25$. The last term is $a_{11} = 3 + 0.25 \cdot 10 = 5.5$. So the partial sum is $S_{11} = \frac{11}{2}(3 + 5.5) = 46.75$.

51. Let x denote the length of the side between the length of the other two sides. Then the lengths of the three sides of the triangle are $x - a$, x, and $x + a$, for some $a > 0$. Since $x + a$ is the longest side, it is the hypotenuse, and by the Pythagorean Theorem, we know that $(x - a)^2 + x^2 = (x + a)^2$ $\Leftrightarrow$ $x^2 - 2ax + a^2 + x^2 = x^2 + 2ax + a^2$ $\Leftrightarrow$ $x^2 - 4ax = 0$ $\Leftrightarrow$ $x(x - 4a) = 0$ $\Rightarrow$

$x = 4a$ ($x = 0$ is not a possible solution). Thus, the lengths of the three sides are $x - a = 4a - a = 3a$, $x = 4a$, and $x + a = 4a + a = 5a$. The lengths $3a$, $4a$, $5a$ are proportional to 3, 4, 5, and so the triangle is similar to a 3 - 4 - 5 triangle.

53. The sequence $1, \frac{3}{5}, \frac{3}{7}, \frac{1}{3}, \ldots$ is harmonic if $1, \frac{5}{3}, \frac{7}{3}, 3, \ldots$ forms an arithmetic sequence. Since $\frac{5}{3} - 1 = \frac{7}{3} - \frac{5}{3} = 3 - \frac{7}{3} = \frac{2}{3}$, the sequence of reciprocals is arithmetic and thus the original sequence is harmonic.

55. We have an arithmetic sequence with $a = 5$ and $d = 2$. We seek n such that $2700 = S_n = \frac{n}{2}[2a + (n-1)d]$. Solving for n, we have $2700 = \frac{n}{2}[10 + 2(n-1)]$ $\Leftrightarrow$ $5400 = 10n + 2n^2 - 2n$ $\Leftrightarrow$ $n^2 + 4n - 2700 = 0$ $\Leftrightarrow$ $(n - 50)(n + 54) = 0$ $\Leftrightarrow$ $n = 50$ or $n = -54$. Since n is a positive integer, 50 terms of the sequence must be added to get 2700.

57. The diminishing values of the computer form an arithmetic sequence with $a_1 = 12500$ and common difference $d = -1875$. Thus the value of the computer after 6 years is $a_7 = 12500 + (7 - 1)(-1875) = \$1250$.

59. The increasing values of the man's salary form an arithmetic sequence with $a_1 = 30000$ and common difference $d = 2300$. Then his total earnings for a ten-year period are $S_{10} = \frac{10}{2}[2(30000) + 9(2300)] = 403500$. Thus his total earnings for the 10 year period are $\$403,500$.

61. The number of seats in the nth row is given by the nth term of an arithmetic sequence with $a_1 = 15$ and common difference $d = 3$. We need to find n such that $S_n = 870$. So we solve $870 = S_n = \frac{n}{2}[2(15) + (n-1)3]$ for n. We have $870 = \frac{n}{2}(27 + 3n)$ $\Leftrightarrow$ $1740 = 3n^2 + 27n$ $\Leftrightarrow$ $3n^2 + 27n - 1740 = 0$ $\Leftrightarrow$ $n^2 + 9n - 580 = 0$ $\Leftrightarrow$ $(x - 20)(x + 29) = 0$ $\Rightarrow$ $n = 20$ or $n = -29$. Since the number of rows is positive, the theater must have 20 rows.

63. The number of gifts on the 12th day is $1 + 2 + 3 + 4 + \cdots + 12$. Since $a_2 - a_1 = a_3 - a_2$ $= a_4 - a_3 = \cdots = 1$, the number of gifts on the 12th day is the partial sum of an arithmetic sequence with $a = 1$ and $d = 1$. So the sum is $S_{12} = 12\left(\frac{1+12}{2}\right) = 6 \cdot 13 = 78$.

Exercises 9.3

1. (a) $a_1 = 5(2)^0 = 5$
    $a_2 = 5(2)^1 = 10$
    $a_3 = 5(2)^2 = 20$
    $a_4 = 5(2)^3 = 40$
    $a_5 = 5(2)^4 = 80$

    (b) The common ratio is 2.

    (c)

3. (a) $a_1 = \frac{5}{2}(-\frac{1}{2})^0 = \frac{5}{2}$
    $a_2 = \frac{5}{2}(-\frac{1}{2})^1 = -\frac{5}{4}$
    $a_3 = \frac{5}{2}(-\frac{1}{2})^2 = \frac{5}{8}$
    $a_4 = \frac{5}{2}(-\frac{1}{2})^3 = -\frac{5}{16}$
    $a_5 = \frac{5}{2}(-\frac{1}{2})^4 = \frac{5}{32}$

    (b) The common ratio is $-\frac{1}{2}$.

    (c)

5. $a = 3, r = 5$. So $a_n = ar^{n-1} = 3(5)^{n-1}$ and $a_4 = 3 \cdot 5^3 = 375$.

7. $a = \frac{5}{2}, r = -\frac{1}{2}$. So $a_n = ar^{n-1} = \frac{5}{2}(-\frac{1}{2})^{n-1}$ and $a_4 = \frac{5}{2} \cdot (-\frac{1}{2})^3 = -\frac{5}{16}$.

9. $\frac{a_2}{a_1} = \frac{4}{2} = 2; \frac{a_3}{a_2} = \frac{8}{4} = 2; \frac{a_4}{a_3} = \frac{16}{8} = 2$. Since these ratios are the same, the sequence is geometric with the common ratio 2.

11. $\frac{a_2}{a_1} = \frac{3/2}{3} = \frac{1}{2}; \frac{a_3}{a_2} = \frac{3/4}{3/2} = \frac{1}{2}; \frac{a_4}{a_3} = \frac{3/8}{3/4} = \frac{1}{2}$. Since these ratios are the same, the sequence is geometric with the common ratio $\frac{1}{2}$.

13. $\frac{a_2}{a_1} = \frac{1/3}{1/2} = \frac{2}{3}; \frac{a_4}{a_3} = \frac{1/5}{1/4} = \frac{4}{5}$. Since these ratios are not the same, this is not a geometric sequence.

15. $\frac{a_2}{a_1} = \frac{1.1}{1.0} = 1.1; \frac{a_3}{a_2} = \frac{1.21}{1.1} = 1.1; \frac{a_4}{a_3} = \frac{1.331}{1.21} = 1.1$. Since these ratios are the same, the sequence is geometric with the common ratio 1.1.

17. $a_1 = 2(3)^1 = 6; a_2 = 2(3)^2 = 18; a_3 = 2(3)^3 = 54; a_4 = 2(3)^4 = 162; a_5 = 2(3)^5 = 486$. This sequence is geometric, the common ratio is $r = 3$ and $a_n = a_1 r^{n-1} = 6(3)^{n-1}$.

19. $a_1 = \frac{1}{4}; a_2 = \frac{1}{4^2} = \frac{1}{16}; a_3 = \frac{1}{4^3} = \frac{1}{64}; a_4 = \frac{1}{4^4} = \frac{1}{256}; a_5 = \frac{1}{4^5} = \frac{1}{1024}$. This sequence is geometric , the common ratio is $r = \frac{1}{4}$ and $a_n = a_1 r^{n-1} = \frac{1}{4}(\frac{1}{4})^{n-1}$.

21. Since $\ln a^b = b \ln a$ we have: $a_1 = \ln(5^0) = \ln 1 = 0; a_2 = \ln(5^1) = \ln 5; a_3 = \ln(5^2) = 2 \ln 5;$ $a_4 = \ln(5^3) = 3 \ln 5; a_5 = \ln(5^4) = 4 \ln 5$. Since $a_1 = 0$ and $a_2 \neq 0$, this sequence is not geometric.

23. $2, 6, 18, 54, \ldots$. Then $r = \dfrac{a_2}{a_1} = \dfrac{6}{2} = 3$; $a_5 = a_4 \cdot 3 = 54(3) = 162$; and $a_n = 2 \cdot 3^{n-1}$.

25. $0.3, -0.09, 0.027, -0.0081, \ldots$. Then $r = \dfrac{a_2}{a_1} = \dfrac{-0.09}{0.3} = -0.3$; $a_5 = a_4 \cdot (-0.3) =$ $-0.0081(-0.3) = 0.00243$; and $a_n = 0.3(-0.3)^{n-1}$.

27. $144, -12, 1, -\frac{1}{12}, \ldots$. Then $r = \dfrac{a_2}{a_1} = \dfrac{-12}{144} = -\frac{1}{12}$; $a_5 = a_4 \cdot \left(-\frac{1}{12}\right) = -\frac{1}{12}\left(-\frac{1}{12}\right) = \frac{1}{144}$; $a_n = 144\left(-\frac{1}{12}\right)^{n-1}$.

29. $3, 3^{5/3}, 3^{7/3}, 27, \ldots$. Then $r = \dfrac{a_2}{a_1} = \dfrac{3^{5/3}}{3} = 3^{2/3}$; $a_5 = a_4 \cdot \left(3^{2/3}\right) = 27 \cdot 3^{2/3} = 3^{11/3}$; and $a_n = 3\left(3^{2/3}\right)^{n-1} = 3 \cdot 3^{(2n-2)/3} = 3^{(2n+1)/3}$.

31. $1, s^{2/7}, s^{4/7}, s^{6/7}, \ldots$. Then $r = \dfrac{a_2}{a_1} = \dfrac{s^{2/7}}{1} = s^{2/7}$; $a_5 = a_4 \cdot s^{2/7} = s^{6/7} \cdot s^{2/7} = s^{8/7}$; and $a_n = \left(s^{2/7}\right)^{n-1} = s^{(2n-2)/7}$.

33. $a_1 = 8$, $a_2 = 4$. Thus $r = \dfrac{a_2}{a_1} = \dfrac{4}{8} = \dfrac{1}{2}$ and $a_5 = a_1 r^{5-1} = 8\left(\dfrac{1}{2}\right)^4 = \dfrac{8}{16} = \dfrac{1}{2}$.

35. $r = \frac{2}{5}$, $a_4 = \frac{5}{2}$. Since $r = \dfrac{a_4}{a_3}$, we have $a_3 = \dfrac{a_4}{r} = \dfrac{\frac{5}{2}}{\frac{2}{5}} = \dfrac{25}{4}$.

37. The geometric sequence is $2, 6, 18, \ldots$. Thus $r = \dfrac{a_2}{a_1} = \dfrac{6}{2} = 3$. We need to find n so that $a_n = 2 \cdot 3^{n-1} = 118{,}098 \quad \Leftrightarrow \quad 3^{n-1} = 59{,}049 \quad \Leftrightarrow \quad n - 1 = \log_3 59{,}049 = 10 \quad \Leftrightarrow \quad n = 11$. Therefore, $118{,}098$ is the 11th term of the geometric sequence.

39. $a = 5$, $r = 2$, $n = 6$. Then $S_6 = 5\,\dfrac{1 - 2^6}{1 - 2} = (-5)(-63) = 315$.

41. $a_3 = 28$, $a_6 = 224$, $n = 6$. So $\dfrac{a_6}{a_3} = \dfrac{ar^5}{ar^2} = r^3$. So we have $r^3 = \dfrac{a_6}{a_3} = \dfrac{224}{28} = 8$, and hence $r = 2$. Since $a_3 = a \cdot r^2$, we get $a = \dfrac{a_3}{r^2} = \dfrac{28}{2^2} = 7$. So $S_6 = 7\dfrac{1 - 2^6}{1 - 2} = (-7)(-63) = 441$.

43. $1 + 3 + 9 + \cdots + 2187$ is a partial sum of a geometric sequence, where $a = 1$ and $r = \dfrac{a_2}{a_1} = \dfrac{3}{1} = 3$. Then the last term is $2187 = a_n = 1 \cdot 3^{n-1} \quad \Leftrightarrow \quad n - 1 = \log_3 2187 = 7 \quad \Leftrightarrow \quad n = 8$. So the partial sum is $S_8 = (1)\dfrac{1 - 3^8}{1 - 3} = 3280$.

45. $\displaystyle\sum_{k=0}^{10} 3\left(\frac{1}{2}\right)^k$ is partial sum of a geometric sequence, where $a = 3$, $r = \frac{1}{2}$, and $n = 11$. So the partial sum is $S_{11} = (3)\dfrac{1 - \left(\frac{1}{2}\right)^{11}}{1 - \left(\frac{1}{2}\right)} = 6\left[1 - \left(\frac{1}{2}\right)^{11}\right] = 5.997070313$.

47. $1 + \frac{1}{3} + \frac{1}{9} + \frac{1}{27} + \cdots$ is an infinite geometric series with $a = 1$ and $r = \frac{1}{3}$. Therefore, the sum of the series is $S = \frac{a}{1-r} = \frac{1}{1 - \left(\frac{1}{3}\right)} = \frac{3}{2}$.

49. $1 - \frac{1}{3} + \frac{1}{9} - \frac{1}{27} + \cdots$ is an infinite geometric series with $a = 1$ and $r = -\frac{1}{3}$. Therefore, the sum of the series is $S = \frac{a}{1-r} = \frac{1}{1 - \left(-\frac{1}{3}\right)} = \frac{3}{4}$.

51. $\frac{1}{3^6} + \frac{1}{3^8} + \frac{1}{3^{10}} + \frac{1}{3^{12}} + \cdots$ is an infinite geometric series with $a = \frac{1}{3^6}$ and $r = \frac{1}{3^2} = \frac{1}{9}$. Therefore, the sum of the series is $S = \frac{a}{1-r} = \frac{\frac{1}{3^6}}{1 - \left(\frac{1}{9}\right)} = \frac{1}{3^6} \cdot \frac{9}{8} = \frac{1}{648}$.

53. $-\frac{100}{9} + \frac{10}{3} - 1 + \frac{3}{10} - \cdots$ is an infinite geometric series with $a = -\frac{100}{9}$ and $r = \frac{\frac{10}{3}}{-\frac{100}{9}} = \frac{3}{10}$.

    Therefore, the sum of the series is $S = \frac{a}{1-r} = \frac{-\frac{100}{9}}{1 - \left(-\frac{3}{10}\right)} = \frac{-\frac{100}{9}}{\frac{13}{10}} = -\frac{100}{9} \cdot \frac{10}{13} = -\frac{1000}{117}$.

55. $0.777\ldots = \frac{7}{10} + \frac{7}{100} + \frac{7}{1000} + \cdots$ is an infinite geometric series with $a = \frac{7}{10}$ and $r = \frac{1}{10}$. Thus $0.777\ldots = \frac{a}{1-r} = \frac{\frac{7}{10}}{1 - \frac{1}{10}} = \frac{7}{9}$.

57. $0.030303\ldots = \frac{3}{100} + \frac{3}{10,000} + \frac{3}{1,000,000} + \cdots$ is an infinite geometric series with $a = \frac{3}{100}$ and $r = \frac{1}{100}$. Thus $0.030303\ldots = \frac{a}{1-r} = \frac{\frac{3}{100}}{1 - \frac{1}{100}} = \frac{3}{99} = \frac{1}{33}$.

59. $0.\overline{112} = 0.112112112\ldots = \frac{112}{1000} + \frac{112}{1,000,000} + \frac{112}{1,000,000,000} + \cdots$ is an infinite geometric series with $a = \frac{112}{1000}$ and $r = \frac{1}{1000}$. Thus $0.112112112\ldots = \frac{a}{1-r} = \frac{\frac{112}{1000}}{1 - \frac{1}{1000}} = \frac{112}{999}$.

61. Since we have 5 terms, let us denote $a_1 = 5$ and $a_5 = 80$. Also, $\frac{a_5}{a_1} = r^4$ because the sequence is geometric, and so $r^4 = \frac{80}{5} = 16 \Leftrightarrow r = \pm 2$. If $r = 2$, the three geometric means are $a_2 = 10$, $a_3 = 20$, and $a_4 = 40$. (If $r = -2$, the three geometric means are $a_2 = -10$, $a_3 = 20$, and $a_4 = -40$, but these are not between 5 and 80.)

63. (a) *Value at the end of the year = value at beginning − depreciation*, so
    $V_n = V_{n-1} - 0.2V_{n-1} = 0.8V_{n-1}$ with $V_0 = 160,000$. Thus $V_n = 160,000 \cdot 0.8^n$.

    (b) $V_n < 100,000 \Leftrightarrow 0.8^n \cdot 160,000 < 100,000 \Leftrightarrow 0.8^n < 0.625 \Leftrightarrow$
    $n \log 0.8 < \log 0.625 \Leftrightarrow n > \frac{\log 0.625}{\log 0.8} = 2.11$. Thus it will depreciate to below $100,000 during the third year.

65. Since the ball is dropped from a height of 80 feet, $a = 80$. Also since the ball rebounds three-fourths of the distance fallen, $r = \frac{3}{4}$. So on the nth bounce, the ball attains a height of $a_n = 80\left(\frac{3}{4}\right)^n$. Hence, on the 5th bounce, the ball goes $a_5 = 80\left(\frac{3}{4}\right)^5 = \frac{80 \cdot 243}{1024} \approx 19$ ft high.

67. Let a_n be the amount of water remaining at the nth stage. We start with 5 gallons, so $a = 5$. When 1 gallon, that is $\frac{1}{5}$ of the mixture, is removed, $\frac{4}{5}$ of the mixture (and hence $\frac{4}{5}$ of the water in the mixture) remains. Thus, $a_1 = 5 \cdot \frac{4}{5}$, $a_2 = 5 \cdot \frac{4}{5} \cdot \frac{4}{5}, \dots$, and in general, $a_n = 5\left(\frac{4}{5}\right)^n$. The amount of water remaining after 3 repetitions is $a_3 = 5\left(\frac{4}{5}\right)^3 = \frac{64}{25}$, and after 5 repetitions is $a_5 = 5\left(\frac{4}{5}\right)^5 = \frac{1024}{625}$.

69. Let a_n be the height the ball reaches on the nth bounce. From the given information, a_n is the geometric sequence $a_n = 9 \cdot \left(\frac{1}{3}\right)^n$. (Notice that the ball hits the ground for the fifth time after the 4th bounce.)

(a) $a_0 = 9$, $a_1 = 9 \cdot \frac{1}{3} = 3$, $a_2 = 9 \cdot \left(\frac{1}{3}\right)^2 = 1$, $a_3 = 9 \cdot \left(\frac{1}{3}\right)^3 = \frac{1}{3}$, and $a_4 = 9 \cdot \left(\frac{1}{3}\right)^4 = \frac{1}{9}$. The total distance traveled is $a_0 + 2a_1 + 2a_2 + 2a_3 + 2a_4 =$
$9 + 2 \cdot 3 + 2 \cdot 1 + 2 \cdot \frac{1}{3} + 2 \cdot \frac{1}{9} = \frac{161}{9} = 17\frac{8}{9}$ ft.

(b) The total distance traveled at the instant the ball hits the ground for the nth time is
$D_n = 9 + 2 \cdot 9 \cdot \frac{1}{3} + 2 \cdot 9 \cdot \left(\frac{1}{3}\right)^2 + 2 \cdot 9 \cdot \left(\frac{1}{3}\right)^3 + 2 \cdot 9 \cdot \left(\frac{1}{3}\right)^4 + \cdots + 2 \cdot 9 \cdot \left(\frac{1}{3}\right)^{n-1}$
$= 2\left[9 + 9 \cdot \frac{1}{3} + 9 \cdot \left(\frac{1}{3}\right)^2 + 9 \cdot \left(\frac{1}{3}\right)^3 + 9 \cdot \left(\frac{1}{3}\right)^4 + \cdots + 9 \cdot \left(\frac{1}{3}\right)^{n-1}\right] - 9$
$= 2\left[9 \cdot \frac{1 - \left(\frac{1}{3}\right)^n}{1 - \frac{1}{3}}\right] - 9 = 27\left[1 - \left(\frac{1}{3}\right)^n\right] - 9 = 18 - \left(\frac{1}{3}\right)^{n-3}$.

71. Let $a_1 = 1$ be the man with 7 wives. Also, let $a_2 = 7$ (the wives), $a_3 = 7a_2 = 7^2$ (the sacks), $a_4 = 7a_3 = 7^3$ (the cats), and $a_5 = 7a_4 = 7^4$ (the kits). The total is $a_1 + a_2 + a_3 + a_4 + a_5 = 1 + 7 + 7^2 + 7^3 + 7^4$, which is a partial sum of a geometric sequence with $a = 1$ and $r = 7$. Thus, the number in the party is $S_5 = 1 \cdot \frac{1 - 7^5}{1 - 7} = 2801$.

73. Let a_n be the height the ball reaches on the nth bounce. We have $a_0 = 1$ and $a_n = \frac{1}{2}a_{n-1}$. Since the total distance d traveled includes the bounce up as well and the distance down, we have
$d = a_0 + 2 \cdot a_1 + 2 \cdot a_2 + \dots = 1 + 2\left(\frac{1}{2}\right) + 2\left(\frac{1}{2}\right)^2 + 2\left(\frac{1}{2}\right)^3 + 2\left(\frac{1}{2}\right)^4 + \cdots$
$= 1 + 1 + \frac{1}{2} + \left(\frac{1}{2}\right)^2 + \left(\frac{1}{2}\right)^3 + \cdots = 1 + \sum_{i=0}^{\infty}\left(\frac{1}{2}\right)^i = 1 + \frac{1}{1 - \frac{1}{2}} = 3$. Thus the total distance traveled is about 3 m.

75. (a) If a square has side x, then the side of the square formed by joining the midpoints is (by the Pythagorean Theorem) $\sqrt{\left(\frac{x}{2}\right)^2 + \left(\frac{x}{2}\right)^2} = \sqrt{\frac{x^2}{4} + \frac{x^2}{4}} = \frac{x}{\sqrt{2}}$. In our case, $x = 1$ and the side of the first inscribed square is $\frac{1}{\sqrt{2}}$, the side of the second inscribed square is
$\frac{1}{\sqrt{2}} \cdot \frac{1}{\sqrt{2}} = \left(\frac{1}{\sqrt{2}}\right)^2$, the side of the third inscribed square is $\left(\frac{1}{\sqrt{2}}\right)^3, \dots$. Since this pattern continues, the total area of all the squares is $A = 1^2 + \left(\frac{1}{\sqrt{2}}\right)^2 + \left(\frac{1}{\sqrt{2}}\right)^4 + \left(\frac{1}{\sqrt{2}}\right)^6 + \cdots$
$= 1 + \frac{1}{2} + \left(\frac{1}{2}\right)^2 + \left(\frac{1}{2}\right)^3 + \cdots = \frac{1}{1 - \frac{1}{2}} = 2$.

(b)   As in part (a), the sides of the squares are $1, \frac{1}{\sqrt{2}}, \left(\frac{1}{\sqrt{2}}\right)^2, \left(\frac{1}{\sqrt{2}}\right)^3, \ldots$. Thus the sum of the

perimeters is $S = 4 \cdot 1 + 4 \cdot \frac{1}{\sqrt{2}} + 4 \cdot \left(\frac{1}{\sqrt{2}}\right)^2 + 4 \cdot \left(\frac{1}{\sqrt{2}}\right)^3 + \cdots$, which is an infinite

geometric series with $a = 4$ and $r = \frac{1}{\sqrt{2}}$. Thus the sum of the perimeters is

$$S = \frac{4}{1 - \frac{1}{\sqrt{2}}} = \frac{4\sqrt{2}}{\sqrt{2}-1} = \frac{4\sqrt{2}}{\sqrt{2}-1} \cdot \frac{\sqrt{2}+1}{\sqrt{2}+1} = \frac{4 \cdot 2 + 4\sqrt{2}}{2-1} = 8 + 4\sqrt{2}.$$

77.   Let a_n denote the area colored blue at nth stage. Since only the middle squares are colored blue, $a_n = \frac{1}{9} \times$ (area remaining yellow at the $(n-1)$st stage). Also, the area remaining yellow at the nth stage is $\frac{8}{9}$ of the area remaining yellow at the preceding stage. So $a_1 = \frac{1}{9}, a_2 = \frac{1}{9}\left(\frac{8}{9}\right), a_3 = \frac{1}{9}\left(\frac{8}{9}\right)^2$, $a_4 = \frac{1}{9}\left(\frac{8}{9}\right)^3, \ldots$. Thus the total area colored blue $A = \frac{1}{9} + \frac{1}{9}\left(\frac{8}{9}\right) + \frac{1}{9}\left(\frac{8}{9}\right)^2 + \frac{1}{9}\left(\frac{8}{9}\right)^3 + \ldots$ is an infinite geometric series with $a = \frac{1}{9}$ and $r = \frac{8}{9}$. So the total area is $A = \frac{\frac{1}{9}}{1 - \frac{8}{9}} = 1$.

79.   Let $a_1, a_2, a_3, \ldots$ be a geometric sequence with common ratio r. Thus $a_2 = a_1 r$, $a_3 = a_1 \cdot r^2, \ldots$, $a_n = a_1 \cdot r^{n-1}$. Hence, $\frac{1}{a_2} = \frac{1}{a_1 \cdot r} = \frac{1}{a_1} \cdot \frac{1}{r}, \frac{1}{a_3} = \frac{1}{a_1 \cdot r^2} = \frac{1}{a_1} \cdot \frac{1}{r^2} = \frac{1}{a_1} \cdot \left(\frac{1}{r}\right)^2, \ldots$
$\frac{1}{a_n} = \frac{1}{a_1 \cdot r^{n-1}} = \frac{1}{a_1} \cdot \frac{1}{r^{n-1}} = \frac{1}{a_1}\left(\frac{1}{r}\right)^{n-1}$, and so $\frac{1}{a_1}, \frac{1}{a_2}, \frac{1}{a_3}, \ldots$ is a geometric sequence with common ratio $\frac{1}{r}$.

81.   Since $a_1, a_2, a_3, \ldots$ is an arithmetic sequence with common difference d, the terms can be expressed as $a_2 = a_1 + d, a_3 = a_1 + 2d, \ldots, a_n = a_1 + (n-1)d$. So $10^{a_2} = 10^{a_1+d} = 10^{a_1} \cdot 10^d$, $10^{a_3} = 10^{a_1+2d} = 10^{a_1} \cdot (10^d)^2, \ldots, 10^{a_n} = 10^{a_1+(n-1)d} = 10^{a_1} \cdot (10^d)^{n-1}$, and so $10^{a_1}, 10^{a_2}, 10^{a_3}$, $\ldots$ is a geometric sequence with common ratio $r = 10^d$.

Exercises 9.4

1. $n = 10, R = \$1000, i = 0.06.$ So, $A_f = R\dfrac{(1+i)^n - 1}{i} = 1000\dfrac{(1+0.06)^{10} - 1}{0.06} = \$13,180.79.$

3. $n = 20, R = \$5000, i = 0.12.$ So, $A_f = R\dfrac{(1+i)^n - 1}{i} = 5000\dfrac{(1+0.12)^{20} - 1}{0.12} = \$360,262.21.$

5. $n = 16, R = \$300, i = \dfrac{0.08}{4} = 0.02.$ So, $A_f = R\dfrac{(1+i)^n - 1}{i} = 300\dfrac{(1+0.02)^{16} - 1}{0.02}$
    $= \$5,591.79.$

7. $A_f = \$2000, i = \dfrac{0.06}{12} = 0.005, n = 8.$ Then $R = \dfrac{iA_f}{(1+i)^n - 1} = \dfrac{(0.005)(2000)}{(1+0.005)^8 - 1} = \$245.66.$

9. $R = \$200, n = 20, i = \dfrac{0.09}{2} = 0.045.$ So, $A_p = R\dfrac{1 - (1+i)^{-n}}{i} = (200)\dfrac{1 - (1+0.045)^{-20}}{0.045}$
    $= \$2601.59.$

11. $A_p = \$12,000, i = \dfrac{0.105}{12} = 0.00875, n = 48.$ Then $R = \dfrac{iA_p}{1 - (1+i)^{-n}} = \dfrac{(0.00875)(12000)}{1 - (1+0.00875)^{-48}}$
    $= \$307.24.$

13. $A_p = \$100,000, i = \dfrac{0.08}{12} \approx 0.006667, n = 360.$ Then $R = \dfrac{iA_p}{1 - (1+i)^{-n}}$
    $= \dfrac{(0.006667)(100,000)}{1 - (1+0.006667)^{-360}} = \$733.76.$ Therefore, the total amount paid on this loan over the 30
    year period is $(360)(733.76) = \$264,153.60.$

15. $A_p = 100,000, n = 360, i = \dfrac{0.0975}{12} = 0.008125.$

    (a) $R = \dfrac{i A_p}{1 - (1+i)^{-n}} = \dfrac{(0.008125)(100,000)}{1 - (1+0.008125)^{-360}} = \$859.15.$

    (b) The total amount that will be paid over the 30 year period is $(360)(859.15) = \$309,294.00.$

    (c) $R = \$859.15, i = \dfrac{0.0975}{12} = 0.008125, n = 360.$ So, $A_f = 859.15\dfrac{(1+0.008125)^{360} - 1}{0.008125}$
    $= \$1,841,519.29.$

17. $R = \$30, i = \dfrac{0.10}{12} \approx 0.008333, n = 12.$ Then $A_p = R\dfrac{1 - (1+i)^{-n}}{i}$
    $= 30\dfrac{1 - \left(1 + 0.008333\right)^{-12}}{0.008333} = \$341.24.$

19. $A_p = \$640, R = \$32, n = 24.$ We want to solve the equation $R = \dfrac{iA_p}{1 - (1+i)^n}$ for the interest
    rate i. Let x be the interest rate, then $i = \frac{x}{12}.$ So we can express R as a function of x by

$R(x) = \dfrac{\frac{x}{12} \cdot 640}{1 - \left(1 + \frac{x}{12}\right)^{-24}}$. We graph $R(x)$ and $y = 32$ in the rectangle $[0.12, 0.22] \times [30, 34]$. The x-coordinate of the intersection is about 0.1816, which corresponds to an interest rate of 18.16%.

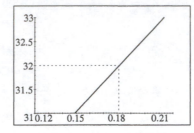

21. $A_p = \$189.99$, $R = \$10.50$, $n = 20$. We want to solve the equation $R = \dfrac{iA_p}{1 - (1+i)^n}$ for the interest rate i. Let x be the interest rate, then $i = \frac{x}{12}$. So we can express R as a function of x by $R(x) = \dfrac{\frac{x}{12} \cdot 189.99}{1 - \left(1 + \frac{x}{12}\right)^{-20}}$. We graph $R(x)$ and $y = 10.50$ in the rectangle $[0.10, 0.18] \times [10, 11]$. The x-coordinate of the intersection is about 0.1168, which corresponds to an interest rate of 11.68%.

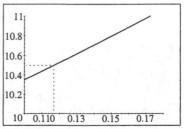

23. (a) The present value of the kth payment is $PV = \dfrac{R}{(1+i)^k}$. The amount of money to be invested now (A_p) to ensure an annuity in perpetuity is the (infinite) sum of the present values of each of the payments, as the following time line shows.

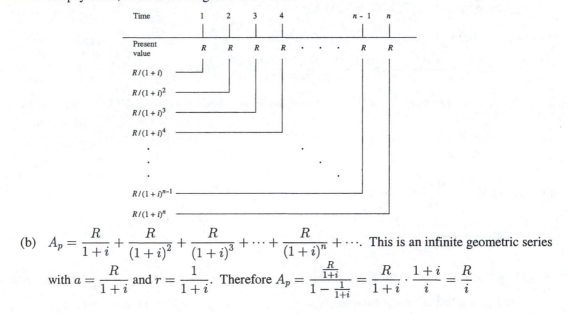

(b) $A_p = \dfrac{R}{1+i} + \dfrac{R}{(1+i)^2} + \dfrac{R}{(1+i)^3} + \cdots + \dfrac{R}{(1+i)^n} + \cdots$. This is an infinite geometric series with $a = \dfrac{R}{1+i}$ and $r = \dfrac{1}{1+i}$. Therefore $A_p = \dfrac{\frac{R}{1+i}}{1 - \frac{1}{1+i}} = \dfrac{R}{1+i} \cdot \dfrac{1+i}{i} = \dfrac{R}{i}$

(c) Using the result from Part (b), we have $R = 5000$ and $i = 0.10$. Then $A_p = \dfrac{R}{i} = \dfrac{5000}{0.10}$
$= \$50,000.$

(d) We are given two different time periods: the interest is compounded quarterly, while the annuity is paid yearly. In order to use the formula in Part (b), we need to find the effective annual interest rate produced by 8% interest compounded quarterly, which is $\left(1 + \frac{i}{n}\right)^n - 1$. Now $i = 8\%$ and it is compounded quarterly, $n = 4$, so the effective annual yield is $(1 + 0.02)^4 - 1 \approx 0.08243216$. Thus by the formula in Part (b), the amount that must be invested is $A_p = \dfrac{3000}{0.08243216} = \$36,393.56.$

Exercises 9.5

1. Let $P(n)$ denote the statement $2 + 4 + 6 + \cdots + 2n = n(n+1)$.

    Step 1 $P(1)$ is the statement that $2 = 1(1+1)$, which is true.

    Step 2 Assume that $P(k)$ is true; that is, $2 + 4 + 6 + \cdots + 2k = k(k+1)$. We want to use this to show that $P(k+1)$ is true. Now,

    $$2 + 4 + 6 + \cdots + 2k + 2(k+1) =$$
    $$k(k+1) + 2(k+1) = \qquad \text{induction hypothesis}$$
    $$(k+1)(k+2) = (k+1)[(k+1)+1].$$

    Thus, $P(k+1)$ follows from $P(k)$. So by the Principle of Mathematical Induction, $P(n)$ is true for all n.

3. Let $P(n)$ denote the statement $5 + 8 + 11 + \cdots + (3n+2) = \dfrac{n(3n+7)}{2}$.

    Step 1 We need to show that $P(1)$ is true. But $P(1)$ says that $5 = \frac{1 \cdot (3 \cdot 1 + 7)}{2}$, which is true.

    Step 2 Assume that $P(k)$ is true; that is, $5 + 8 + 11 + \cdots + (3k+2) = \dfrac{k(3k+7)}{2}$. We want to use this to show that $P(k+1)$ is true. Now,

    $$5 + 8 + 11 + \cdots + (3k+2) + [3(k+1)+2] =$$
    $$\frac{k(3k+7)}{2} + (3k+5) = \qquad \text{induction hypothesis}$$
    $$\frac{3k^2 + 7k}{2} + \frac{6k+10}{2} =$$
    $$\frac{3k^2 + 13k + 10}{2} = \frac{(3k+10)(k+1)}{2} = \frac{(k+1)[3(k+1)+7]}{2}.$$

    Thus, $P(k+1)$ follows from $P(k)$. So by the Principle of Mathematical Induction, $P(n)$ is true for all n

5. Let $P(n)$ denote the statement $1 \cdot 2 + 2 \cdot 3 + 3 \cdot 4 + \cdots + n(n+1) = \dfrac{n(n+1)(n+2)}{3}$.

    Step 1 $P(1)$ is the statement that $1 \cdot 2 = \frac{1 \cdot (1+1) \cdot (1+2)}{3}$, which is true.

    Step 2 Assume that $P(k)$ is true; that is, $1 \cdot 2 + 2 \cdot 3 + 3 \cdot 4 + \cdots + k(k+1) = \dfrac{k(k+1)(k+2)}{3}$. We want to use this to show that $P(k+1)$ is true. Now,

    $$1 \cdot 2 + 2 \cdot 3 + 3 \cdot 4 + \cdots + k(k+1) + (k+1)[(k+1)+1] =$$
    $$\frac{k(k+1)(k+2)}{3} + (k+1)(k+2) = \qquad \text{induction hypothesis}$$
    $$\frac{k(k+1)(k+2)}{3} + \frac{3(k+1)(k+2)}{3} = \frac{(k+1)(k+2)(k+3)}{3}.$$

    Thus, $P(k+1)$ follows from $P(k)$. So by the Principle of Mathematical Induction, $P(n)$ is true for all n.

7. Let $P(n)$ denote the statement $1^3 + 2^3 + 3^3 + \cdots + n^3 = \dfrac{n^2(n+1)^2}{4}$.

Step 1 $P(1)$ is the statement that $1^3 = \frac{1^2 \cdot (1+1)^2}{4}$, which is clearly true.

Step 2 Assume that $P(k)$ is true; that is, $1^3 + 2^3 + 3^3 + \cdots + k^3 = \dfrac{k^2(k+1)^2}{4}$. We want to use this to show that $P(k+1)$ is true. Now,

$1^3 + 2^3 + 3^3 + \cdots + k^3 + (k+1)^3 =$

$\dfrac{k^2(k+1)^2}{4} + (k+1)^3 =$ \qquad induction hypothesis

$\dfrac{(k+1)^2[k^2 + 4(k+1)]}{4} =$

$\dfrac{(k+1)^2[k^2 + 4k + 4]}{4} = \dfrac{(k+1)^2(k+2)^2}{4} = \dfrac{(k+1)^2[(k+1)+1]^2}{4}$.

Thus, $P(k+1)$ follows from $P(k)$. So by the Principle of Mathematical Induction, $P(n)$ is true for all n.

9. Let $P(n)$ denote the statement $2^3 + 4^3 + 6^3 + \cdots + (2n)^3 = 2n^2(n+1)^2$.

Step 1 $P(1)$ is true since $2^3 = 2(1)^2(1+1)^2 = 2 \cdot 4 = 8$.

Step 2 Assume that $P(k)$ is true; that is, $2^3 + 4^3 + 6^3 + \cdots + (2k)^3 = 2k^2(k+1)^2$. We want to use this to show that $P(k+1)$ is true. Now,

$2^3 + 4^3 + 6^3 + \cdots + (2k)^3 + [2(k+1)]^3 =$

$2k^2(k+1)^2 + [2(k+1)]^3 =$ \qquad induction hypothesis

$2k^2(k+1)^2 + 8(k+1)(k+1)^2 =$

$(k+1)^2(2k^2 + 8k + 8) = 2(k+1)^2(k+2)^2 = 2(k+1)^2[(k+1)+1]^2$.

Thus, $P(k+1)$ follows from $P(k)$. So by the Principle of Mathematical Induction, $P(n)$ is true for all n.

11. Let $P(n)$ denote the statement $1 \cdot 2 + 2 \cdot 2^2 + 3 \cdot 2^3 + 4 \cdot 2^4 + \cdots + n \cdot 2^n = 2[1 + (n-1)2^n]$.

Step 1 $P(1)$ is the statement that $1 \cdot 2 = 2[1 + 0]$, which is clearly true.

Step 2 Assume that $P(k)$ is true; that is,

$1 \cdot 2 + 2 \cdot 2^2 + 3 \cdot 2^3 + 4 \cdot 2^4 + \cdots + k \cdot 2^k = 2[1 + (k-1)2^k]$. We want to use this to show that $P(k+1)$ is true. Now,

$1 \cdot 2 + 2 \cdot 2^2 + 3 \cdot 2^3 + 4 \cdot 2^4 + \cdots + k \cdot 2^k + (k+1) \cdot 2^{(k+1)} =$

$2[1 + (k-1)2^k] + (k+1) \cdot 2^{(k+1)} =$ \qquad induction hypothesis

$2[1 + (k-1) \cdot 2^k + (k+1) \cdot 2^k] =$

$2[1 + 2k \cdot 2^k] = 2[1 + k \cdot 2^{k+1}] = 2\{1 + [(k+1) - 1]2^{k+1}\}$.

Thus $P(k+1)$ follows from $P(k)$. So by the Principle of Mathematical Induction, $P(n)$ is true for all n.

13. Let $P(n)$ denote the statement $n^2 + n$ is divisible by 2.

Step 1 $P(1)$ is the statement that $1^2 + 1 = 2$ is divisible by 2, which is clearly true.

Step 2 Assume that $P(k)$ is true; that is, $k^2 + k$ is divisible by 2. Now,

$(k+1)^2 + (k+1) = k^2 + 2k + 1 + k + 1 = (k^2 + k) + 2k + 2 = (k^2 + k) + 2(k+1)$.

By the induction hypothesis, $k^2 + k$ is divisible by 2, and clearly $2(k+1)$ is divisible by 2. Thus, the sum is divisible by 2, so $P(k+1)$ true. Therefore, $P(k+1)$ follows from $P(k)$. So by the Principle of Mathematical Induction, $P(n)$ is true for all n.

15. Let $P(n)$ denote the statement that $n^2 - n + 41$ is odd.

Step 1 $P(1)$ is the statement that $1^2 - 1 + 41 = 41$ is odd, which is clearly true.

Step 2 Assume that $P(k)$ is true; that is, $k^2 - k + 41$ is odd. We want to use this to show that $P(k+1)$ is true. Now,
$(k+1)^2 - (k+1) + 41 = k^2 + 2k + 1 - k - 1 + 41 = (k^2 - k + 41) + 2k$, which is also odd because $k^2 - k + 41$ is odd by the induction hypothesis, $2k$ is always even, and an odd number plus an even number is always odd. Therefore, $P(k+1)$ follows from $P(k)$. So by the Principle of Mathematical Induction, $P(n)$ is true for all n.

17. Let $P(n)$ denote the statement that $8^n - 3^n$ is divisible by 5.

Step 1 $P(1)$ is the statement that $8^1 - 3^1 = 5$ is divisible by 5, which is clearly true.

Step 2 Assume that $P(k)$ is true; that is, $8^k - 3^k$ is divisible by 5. We want to use this to show that $P(k+1)$ is true. Now, $8^{k+1} - 3^{k+1} = 8 \cdot 8^k - 3 \cdot 3^k = 8 \cdot 8^k - (8-5) \cdot 3^k$
$= 8 \cdot (8^k - 3^k) + 5 \cdot 3^k$, which is divisible by 5 because $8^k - 3^k$ is divisible by 5 by our induction hypothesis, and $5 \cdot 3^k$ is divisible by 5. Thus $P(k+1)$ follows from $P(k)$. So by the Principle of Mathematical Induction, $P(n)$ is true for all n.

19. Let $P(n)$ denote the statement $n < 2^n$.

Step 1 $P(1)$ is the statement that $1 < 2^1 = 2$, which is clearly true.

Step 2 Assume that $P(k)$ is true; that is, $k < 2^k$. We want to use this to show that $P(k+1)$ is true. Adding 1 to both sides of $P(k)$ we have $k + 1 < 2^k + 1$. Since $1 < 2^k$ for $k \geq 1$, we have $2^k + 1 < 2^k + 2^k = 2 \cdot 2^k = 2^{k+1}$. Thus $k + 1 < 2^{k+1}$, which is exactly $P(k+1)$. Therefore, $P(k+1)$ follows from $P(k)$. So by the Principle of Mathematical Induction, $P(n)$ is true for all n.

21. Let $P(n)$ denote the statement $(1 + x)^n \geq 1 + nx$, if $x > -1$.

Step 1 $P(1)$ is the statement that $(1 + x)^1 \geq 1 + 1x$, which is clearly true.

Step 2 Assume that $P(k)$ is true; that is, $(1 + x)^k \geq 1 + kx$. Now,
$(1 + x)^{k+1} = (1 + x)(1 + x)^k \geq (1 + x)(1 + kx)$, by the induction hypothesis. Since $(1 + x)(1 + kx) = 1 + (k+1)x + kx^2 \geq 1 + (k+1)x$ (since $kx^2 \geq 0$), we have $(1 + x)^{k+1} \geq 1 + (k+1)x$, which is $P(k+1)$. Thus $P(k+1)$ follows from $P(k)$. So the Principle of Mathematical Induction, $P(n)$ is true for all n.

23. Let $P(n)$ be the statement that $a_n = 5 \cdot 3^{n-1}$.

Step 1 $P(1)$ is the statement that $a_1 = 5 \cdot 3^0 = 5$, which is true.

Step 2 Assume that $P(k)$ is true; that is, $a_k = 5 \cdot 3^{k-1}$. We want to use this to show that $P(k+1)$ is true. Now, $a_{k+1} = 3a_k = 3 \cdot (5 \cdot 3^{k-1})$, by the induction hypothesis. Therefore, $a_{k+1} = 3 \cdot (5 \cdot 3^{k-1}) = 5 \cdot 3^k$, which is exactly $P(k+1)$. Thus, $P(k+1)$ follows from $P(k)$. So by the Principle of Mathematical Induction, $P(n)$ is true for all n.

25. Let $P(n)$ be the statement that $x - y$ is a factor of $x^n - y^n$ for all natural numbers n.

    <u>Step 1</u> $P(1)$ is the statement that $x - y$ is a factor of $x^1 - y^1$, which is clearly true.

    <u>Step 2</u> Assume that $P(k)$ is true; that is, $x - y$ is a factor of $x^k - y^k$. We want to use this to show that $P(k + 1)$ is true. Now, $x^{k+1} - y^{k+1} = x^{k+1} - x^k y + x^k y - y^{k+1}$ $= x^k(x - y) + (x^k - y^k)y$, for which $x - y$ is a factor because $x - y$ is a factor of $x^k(x - y)$, and $x - y$ is a factor of $(x^k - y^k)y$, by the induction hypothesis. Thus $P(k + 1)$ follows from $P(k)$. So by the Principle of Mathematical Induction, $P(n)$ is true for all n.

27. Let $P(n)$ denote the statement that F_{3n} is even for all natural numbers n.

    <u>Step 1</u> $P(1)$ is the statement that F_3 is even. Since $F_3 = F_2 + F_1 = 1 + 1 = 2$, this statement is true.

    <u>Step 2</u> Assume that $P(k)$ is true; that is, F_{3k} is even. We want to use this to show that $P(k + 1)$ is true. Now, $F_{3(k+1)} = F_{3k+3} = F_{3k+2} + F_{3k+1} = F_{3k+1} + F_{3k} + F_{3k+1} = F_{3k} + 2 \cdot F_{3k+1}$, which is even because F_{3k} is even by the induction hypothesis, and $2 \cdot F_{3k+1}$ is even. Thus $P(k + 1)$ follows from $P(k)$. So by the Principle of Mathematical Induction, $P(n)$ is true for all n.

29. Let $P(n)$ denote the statement that $F_1^2 + F_2^2 + F_3^2 + \cdots + F_n^2 = F_n \cdot F_{n+1}$.

    <u>Step 1</u> $P(1)$ is the statement that $F_1^2 = F_1 \cdot F_2$ or $1^2 = 1 \cdot 1$, which is true.

    <u>Step 2</u> Assume that $P(k)$ is true, that is, $F_1^2 + F_2^2 + F_3^2 + \cdots + F_k^2 = F_k \cdot F_{k+1}$. We want to use this to show that $P(k + 1)$ is true. Now,

    $$F_1^2 + F_2^2 + F_3^2 + \cdots + F_k^2 + F_{k+1}^2 =$$
    $$F_k \cdot F_{k+1} + F_{k+1}^2 = \qquad \text{induction hypothesis}$$
    $$F_{k+1}(F_k + F_{k+1}) = F_{k+1} \cdot F_{k+2}. \qquad \text{by definition of the Fibonacci sequence}$$

    Thus $P(k + 1)$ follows from $P(k)$. So by the Principle of Mathematical Induction, $P(n)$ is true for all n.

31. Let $P(n)$ denote the statement $\begin{bmatrix} 1 & 1 \\ 1 & 0 \end{bmatrix}^n = \begin{bmatrix} F_{n+1} & F_n \\ F_n & F_{n-1} \end{bmatrix}$.

    <u>Step 1</u> Since $\begin{bmatrix} 1 & 1 \\ 1 & 0 \end{bmatrix}^2 = \begin{bmatrix} 1 & 1 \\ 1 & 0 \end{bmatrix}\begin{bmatrix} 1 & 1 \\ 1 & 0 \end{bmatrix} = \begin{bmatrix} 2 & 1 \\ 1 & 1 \end{bmatrix} = \begin{bmatrix} F_3 & F_2 \\ F_2 & F_1 \end{bmatrix}$, it follows that $P(2)$ is true.

    <u>Step 2</u> Assume that $P(k)$ is true; that is, $\begin{bmatrix} 1 & 1 \\ 1 & 0 \end{bmatrix}^k = \begin{bmatrix} F_{k+1} & F_k \\ F_k & F_{k-1} \end{bmatrix}$. We show that $P(k + 1)$ follows from this. Now,

    $$\begin{bmatrix} 1 & 1 \\ 1 & 0 \end{bmatrix}^{k+1} = \begin{bmatrix} 1 & 1 \\ 1 & 0 \end{bmatrix}^k \begin{bmatrix} 1 & 1 \\ 1 & 0 \end{bmatrix} = \begin{bmatrix} F_{k+1} & F_k \\ F_k & F_{k-1} \end{bmatrix}\begin{bmatrix} 1 & 1 \\ 1 & 0 \end{bmatrix} \quad \text{(by the induction hypothesis)}$$
    $$= \begin{bmatrix} F_{k+1} + F_k & F_{k+1} \\ F_k + F_{k-1} & F_k \end{bmatrix} = \begin{bmatrix} F_{k+2} & F_{k+1} \\ F_{k+1} & F_k \end{bmatrix}. \quad \text{(by definition of the Fibonacci sequence)}$$

    Thus $P(k + 1)$ follows from $P(k)$. So by the Principle of Mathematical Induction, $P(n)$ is true for all $n \geq 2$.

33. Since $F_1 = 1$, $F_2 = 1$, $F_3 = 2$, $F_4 = 3$, $F_5 = 5$, $F_6 = 8$, $F_7 = 13, \ldots$ our conjecture is that $F_n \geq n$, for all $n \geq 5$. Let $P(n)$ denote the statement that $F_n \geq n$.

    <u>Step 1</u> $P(5)$ is the statement that $F_5 = 5 \geq 5$, which is clearly true.

Step 2 Assume that $P(k)$ is true; that is, $F_k \geq k$, for some $k \geq 5$. We want to use this to show that $P(k+1)$ is true. Now, $F_{k+1} = F_k + F_{k-1} \geq k + F_{k-1}$ (by the induction hypothesis) $\geq k + 1$ (because $F_{k-1} \geq 1$). Thus $P(k+1)$ follows from $P(k)$. So by the Principle of Mathematical Induction, $P(n)$ is true for all $n \geq 5$.

35. (a) $P(n) = n^2 - n + 11$ is prime for all n. This is false as the case for $n = 11$ demonstrates: $P(11) = 11^2 - 11 + 11 = 121$, which is not prime since $11^2 = 121$.

(b) $n^2 > n$, for all $n \geq 2$. This is true. Let $P(n)$ denote the statement that $n^2 > n$.

Step 1 $P(2)$ is the statement that $2^2 = 4 > 2$, which is clearly true.

Step 2 Assume that $P(k)$ is true; that is, $k^2 > k$. We want to use this to show that $P(k+1)$ is true. Now $(k+1)^2 = k^2 + 2k + 1$. Using the induction hypothesis (to replace k^2), we have $k^2 + 2k + 1 > k + 2k + 1 = 3k + 1 > k + 1$, since $k \geq 2$. Therefore, $(k+1)^2 > k + 1$, which is exactly $P(k+1)$. Thus $P(k+1)$ follows from $P(k)$. So by the Principle of Mathematical Induction, $P(n)$ is true for all n.

(c) $2^{2n+1} + 1$ is divisible by 3, for all $n \geq 1$. This is true. Let $P(n)$ denote the statement that $2^{2n+1} + 1$ is divisible by 3.

Step 1 $P(1)$ is the statement that $2^3 + 1 = 9$ is divisible by 3, which is clearly true.

Step 2 Assume that $P(k)$ is true; that is, $2^{2k+1} + 1$ is divisible by 3. We want to use this to show that $P(k+1)$ is true. Now, $2^{2(k+1)+1} + 1 = 2^{2k+3} + 1 = 4 \cdot 2^{2k+1} + 1$ $= (3+1)2^{2k+1} + 1 = 3 \cdot 2^{2k+1} + (2^{2k+1} + 1)$, which is divisible by 3 since $2^{2k+1} + 1$ is divisible by 3 by the induction hypothesis, and $3 \cdot 2^{2k+1}$ is clearly divisible by 3. Thus $P(k+1)$ follows from $P(k)$. So by the Principle of Mathematical Induction, $P(n)$ is true for all n.

(d) The statement $n^3 \geq (n+1)^2$ for all $n \geq 2$ is false. The statement fails when $n = 2$: $2^3 = 8 < (2+1)^2 = 9$.

(e) $n^3 - n$ is divisible by 3, for all $n \geq 2$. This is true. Let $P(n)$ denote the statement that $n^3 - n$ is divisible by 3.

Step 1 $P(2)$ is the statement that $2^3 - 2 = 6$ is divisible by 3, which is clearly true.

Step 2 Assume that $P(k)$ is true; that is, $k^3 - k$ is divisible by 3. We want to use this to show that $P(k+1)$ is true. Now, $(k+1)^3 - (k+1) = k^3 + 3k^2 + 3k + 1 - (k+1) = k^3 + 3k^2 + 2k = k^3 - k + 3k^2 + 2k + k = (k^3 - k) + 3(k^2 + k)$. The term $k^3 - k$ is divisible by 3 by our induction hypothesis, and the term $3(k^2 + k)$ is clearly divisible by 3. Thus $(k+1)^3 - (k+1)$ is divisible by 3, which is exactly $P(k+1)$. So by the Principle of Mathematical Induction, $P(n)$ is true for all n.

(f) $n^3 - 6n^2 + 11n$ is divisible by 6, for all $n \geq 1$. This is true. Let $P(n)$ denote the statement that $n^3 - 6n^2 + 11n$ is divisible by 6.

Step 1 $P(1)$ is the statement that $(1)^3 - 6(1)^2 + 11(1) = 6$ is divisible by 6, which is clearly true.

Step 2 Assume that $P(k)$ is true; that is, $k^3 - 6k^2 + 11k$ is divisible by 6. We show that $P(k+1)$ is then also true. Now, $(k+1)^3 - 6(k+1)^2 + 11(k+1)$ $= k^3 + 3k^2 + 3k + 1 - 6k^2 - 12k - 6 + 11k + 11 = k^3 - 3k^2 + 2k + 6$ $= k^3 - 6k^2 + 11k + (3k^2 - 9k + 6) = (k^3 - 6k^2 + 11k) + 3(k^2 - 3k + 2)$

$= (k^3 - 6k^2 + 11k) + 3(k-1)(k-2)$. In this last expression, the first term is divisible by 6 by our induction hypothesis. The second term is also divisible by 6. To see this, notice that $(k-1)$ and $(k-2)$ are consecutive natural numbers, and so one of them must be even (divisible by 2). Since 3 also appears in this second term, it follows that this term is divisible by 2 and 3 and so is divisible by 6. Thus $P(k+1)$ follows from $P(k)$. So by the Principle of Mathematical Induction, $P(n)$ is true for all n.

Exercises 9.6

1. $(x + y)^6 = x^6 + 6x^5y + 15x^4y^2 + 20x^3y^3 + 15x^2y^4 + 6xy^5 + y^6$

3. $\left(x + \dfrac{1}{x}\right)^4 = x^4 + 4x^3 \cdot \dfrac{1}{x} + 6x^2\left(\dfrac{1}{x}\right)^2 + 4x\left(\dfrac{1}{x}\right)^3 + \left(\dfrac{1}{x}\right)^4 = x^4 + 4x^2 + 6 + \dfrac{4}{x^2} + \dfrac{1}{x^4}$

5. $(x - 1)^5 = x^5 - 5x^4 + 10x^3 - 10x^2 + 5x - 1$

7. $(x^2y - 1)^5 = (x^2y)^5 - 5(x^2y)^4 + 10(x^2y)^3 - 10(x^2y)^2 + 5x^2y - 1$
$= x^{10}y^5 - 5x^8y^4 + 10x^6y^3 - 10x^4y^2 + 5x^2y - 1$

9. $(2x - 3y)^3 = (2x)^3 - 3(2x)^2 3y + 3 \cdot 2x(3y)^2 - (3y)^3 = 8x^3 - 36x^2y + 54xy^2 - 27y^3$

11. $\left(\dfrac{1}{x} - \sqrt{x}\right)^5 = \left(\dfrac{1}{x}\right)^5 - 5\left(\dfrac{1}{x}\right)^4\sqrt{x} + 10\left(\dfrac{1}{x}\right)^3 x - 10\left(\dfrac{1}{x}\right)^2 x\sqrt{x} + 5\left(\dfrac{1}{x}\right)x^2 - x^2\sqrt{x}$
$= \dfrac{1}{x^5} - \dfrac{5}{x^{7/2}} + \dfrac{10}{x^2} - \dfrac{10}{x^{1/2}} + 5x - x^{5/2}$

13. $\dbinom{6}{4} = \dfrac{6!}{4!\,2!} = \dfrac{6 \cdot 5 \cdot 4!}{2 \cdot 1 \cdot 4!} = 15$

15. $\dbinom{100}{98} = \dfrac{100!}{98!\,2!} = \dfrac{100 \cdot 99 \cdot 98!}{98! \cdot 2 \cdot 1} = 4950$

17. $\dbinom{3}{1}\dbinom{4}{2} = \dfrac{3!}{1!\,2!}\dfrac{4!}{2!\,2!} = \dfrac{3 \cdot 2! \cdot 4 \cdot 3 \cdot 2!}{1 \cdot 2! \cdot 2 \cdot 1 \cdot 2!} = 18$

19. $\dbinom{5}{0} + \dbinom{5}{1} + \dbinom{5}{2} + \dbinom{5}{3} + \dbinom{5}{4} + \dbinom{5}{5} = (1 + 1)^5 = 2^5 = 32$

21. $(x + 2y)^4 = \dbinom{4}{0}x^4 + \dbinom{4}{1}x^3 \cdot 2y + \dbinom{4}{2}x^2 \cdot 4y^2 + \dbinom{4}{3}x \cdot 8y^3 + \dbinom{4}{4}16y^4$
$= x^4 + 8x^3y + 24x^2y^2 + 32xy^3 + 16y^4$

23. $\left(1 + \dfrac{1}{x}\right)^6 = \dbinom{6}{0}1^6 + \dbinom{6}{1}1^5\left(\dfrac{1}{x}\right) + \dbinom{6}{2}1^4\left(\dfrac{1}{x}\right)^2 + \dbinom{6}{3}1^3\left(\dfrac{1}{x}\right)^3 + \dbinom{6}{4}1^2\left(\dfrac{1}{x}\right)^4$
$+ \dbinom{6}{5}1\left(\dfrac{1}{x}\right)^5 + \dbinom{6}{6}\left(\dfrac{1}{x}\right)^6 = 1 + \dfrac{6}{x} + \dfrac{15}{x^2} + \dfrac{20}{x^3} + \dfrac{15}{x^4} + \dfrac{6}{x^5} + \dfrac{1}{x^6}$

25. The first three terms in the expansion of $(x + 2y)^{20}$ are $\dbinom{20}{0}x^{20} = x^{20}$, $\dbinom{20}{1}x^{19} \cdot 2y = 40x^{19}y$,
and $\dbinom{20}{2}x^{18} \cdot (2y)^2 = 760x^{18}y^2$.

27. The last two terms in the expansion of $(a^{2/3} + a^{1/3})^{25}$ are $\binom{25}{24} a^{2/3} \cdot \left(a^{1/3}\right)^{24} = 25a^{26/3}$, and $\binom{25}{25} a^{25/3} = a^{25/3}$.

29. The middle term in the expansion of $(x^2 + 1)^{18}$ occurs when both terms are raised to the 9th power. So this term is $\binom{18}{9} \left(x^2\right)^9 1^9 = 48{,}620x^{18}$.

31. The 24th term in the expansion of $(a + b)^{25}$ is $\binom{25}{23} a^2 b^{23} = 300a^2 b^{23}$.

33. The 100th term in the expansion of $(1 + y)^{100}$ is $\binom{100}{99} 1^1 \cdot y^{99} = 100y^{99}$.

35. The term that contains x^4 in the expansion of $(x + 2y)^{10}$ has exponent $r = 4$. So this term is $\binom{10}{4} x^4 \cdot (2y)^{10-4} = 13{,}440x^4 y^6$.

37. The rth term is $\binom{12}{r} a^r (b^2)^{12-r} = \binom{12}{r} a^r b^{24-2r}$. Thus the term that contains b^8 occurs where $24 - 2r = 8 \quad \Leftrightarrow \quad r = 8$. So the term is $\binom{12}{8} a^8 b^8 = 495a^8 b^8$.

39. $x^4 + 4x^3 y + 6x^2 y^2 + 4xy^3 + y^4 = (x + y)^4$.

41. $8a^3 + 12a^2 b + 6ab^2 + b^3 = \binom{3}{0}(2a)^3 + \binom{3}{1}(2a)^2 b + \binom{3}{2} 2ab^2 + \binom{3}{3} b^3 = (2a + b)^3$.

43. $\dfrac{(x + h)^3 - x^3}{h} = \dfrac{x^3 + 3x^2 h + 3xh^2 + h^3 - x^3}{h} = \dfrac{3x^2 h + 3xh^2 + h^3}{h} = \dfrac{h(3x^2 + 3xh + h^2)}{h}$
$= 3x^2 + 3xh + h^2$.

45. $(1.01)^{100} = (1 + 0.01)^{100}$. Now the 1st term in the expansion is $\binom{100}{0} 1^{100} = 1$, and the 2nd term is $\binom{100}{1} 1^{99}(0.01) = 1$, and the 3rd term is $\binom{100}{2} 1^{98}(0.01)^2 = 0.495$. Now each term is non-negative, so $(1.01)^{100} = (1 + 0.01)^{100} > 1 + 1 + .0.495 > 2$. Thus $(1.01)^{100} > 2$.

47. $\binom{n}{1} = \dfrac{n!}{1!\,(n-1)!} = \dfrac{n(n-1)!}{1(n-1)!} = \dfrac{n}{1} = n$. $\binom{n}{n-1} = \dfrac{n!}{(n-1)!\,1!} = \dfrac{n(n-1)!}{(n-1)!\,1} = n$.
Therefore, $\binom{n}{1} = \binom{n}{n-1} = n$.

49. (a) $\binom{n}{r-1} + \binom{n}{r} = \dfrac{n!}{(r-1)!\,[n - (r-1)]!} + \dfrac{n!}{r!\,(n-r)!}$.

(b) $\dfrac{n!}{(r-1)!\,[n-(r-1)]!} + \dfrac{n!}{r!\,(n-r)!} = \dfrac{r\cdot n!}{r\cdot(r-1)!\,(n-r+1)!} + \dfrac{(n-r+1)\,n!}{r!\,(n-r+1)(n-r)!}$

$= \dfrac{r\cdot n!}{r!\,(n-r+1)!} + \dfrac{(n-r+1)\,n!}{r!\,(n-r+1)!}$. Thus a common denominator is $r!\,(n-r+1)!$.

(c) Therefore, using the results of parts (a) and (b),

$\dbinom{n}{r-1} + \dbinom{n}{r} = \dfrac{n!}{(r-1)!\,[n-(r-1)]!} + \dfrac{n!}{r!\,(n-r)!} = \dfrac{r\cdot n!}{r!\,(n-r+1)!} + \dfrac{(n-r+1)\,n!}{r!\,(n-r+1)!}$

$= \dfrac{r\cdot n! + (n-r+1)n!}{r!\,(n-r+1)!} = \dfrac{n!\cdot(r+n-r+1)}{r!\,(n-r+1)!} = \dfrac{n!\cdot(n+1)}{r!\,(n+1-r)!}$

$= \dfrac{(n+1)!}{r\,!(n+1-r)!} = \dbinom{n+1}{r}$.

51. Notice that $(100!)^{101} = (100!)^{100}\cdot 100!$ and $(101!)^{100} = (101\cdot 100!)^{100} = 101^{100}\cdot(100!)^{100}$.
Now, $100! = 1\cdot 2\cdot 3\cdot 4\cdots 99\cdot 100$ and $101^{100} = 101\cdot 101\cdot 101\cdots 101$. Thus each of
these last two expressions consists of 100 factors multiplied together, and since each factor in the
product for 101^{100} is larger than each factor in the product for $100!$, it follows that $100! < 101^{100}$.
Thus $(100!)^{100}\cdot 100! < (100!)^{100}\cdot 101^{100}$. So, $(100!)^{101} < (101!)^{100}$.

53. $0 = 0^n = (-1+1)^n$

$= \dbinom{n}{0}(-1)^0(1)^n + \dbinom{n}{1}(-1)^1(1)^{n-1} + \dbinom{n}{2}(-1)^2(1)^{n-2} + \cdots + \dbinom{n}{n}(-1)^n(1)^0$

$= \dbinom{n}{0} - \dbinom{n}{1} + \dbinom{n}{2} - \cdots + (-1)^k\dbinom{n}{k} + \cdots + (-1)^n\dbinom{n}{n}$.

Review Exercises for Chapter 9

1. $a_n = \dfrac{n^2}{n+1}$. Then $a_1 = \dfrac{1^2}{1+1} = \dfrac{1}{2}$; $a_2 = \dfrac{2^2}{2+1} = \dfrac{4}{3}$; $a_3 = \dfrac{3^2}{3+1} = \dfrac{9}{4}$; $a_4 = \dfrac{4^2}{4+1} = \dfrac{16}{5}$; and

   $a_{10} = \dfrac{10^2}{10+1} = \dfrac{100}{11}$.

3. $a_n = \dfrac{(-1)^n + 1}{n^3}$. Then $a_1 = \dfrac{(-1)^1 + 1}{1^3} = 0$; $a_2 = \dfrac{(-1)^2 + 1}{2^3} = \dfrac{2}{8} = \dfrac{1}{4}$; $a_3 = \dfrac{(-1)^3 + 1}{3^3} = 0$;

   $a_4 = \dfrac{(-1)^4 + 1}{4^3} = \dfrac{2}{64} = \dfrac{1}{32}$; and $a_{10} = \dfrac{(-1)^{10} + 1}{10^3} = \dfrac{1}{500}$.

5. $a_n = \dfrac{(2n)!}{2^n n!}$. Then $a_1 = \dfrac{(2 \cdot 1)!}{2^1 \cdot 1!} = 1$; $a_2 = \dfrac{(2 \cdot 2)!}{2^2 \cdot 2!} = 3$; $a_3 = \dfrac{(2 \cdot 3)!}{2^3 \cdot 3!} = \dfrac{6 \cdot 5 \cdot 4}{8} = 15$;

   $a_4 = \dfrac{(2 \cdot 4)!}{2^4 \cdot 4!} = \dfrac{8 \cdot 7 \cdot 6 \cdot 5}{16} = 105$; and $a_{10} = \dfrac{(2 \cdot 10)!}{2^{10} \cdot 10!} = 654{,}729{,}075$.

7. $a_n = a_{n-1} + 2n - 1$ and $a_1 = 1$. Then $a_2 = a_1 + 4 - 1 = 4$; $a_3 = a_2 + 6 - 1 = 9$;
   $a_4 = a_3 + 8 - 1 = 16$; $a_5 = a_4 + 10 - 1 = 25$; $a_6 = a_5 + 12 - 1 = 36$; and
   $a_7 = a_6 + 14 - 1 = 49$.

9. $a_n = a_{n-1} + 2a_{n-2}$, $a_1 = 1$ and $a_2 = 3$. Then $a_3 = a_2 + 2a_1 = 5$; $a_4 = a_3 + 2a_2 = 11$;
   $a_5 = a_4 + 2a_3 = 21$; $a_6 = a_5 + 2a_4 = 43$; and $a_7 = a_6 + 2a_5 = 85$.

11. (a) $a_1 = 2(1) + 5 = 7$
         $a_2 = 2(2) + 5 = 9$
         $a_3 = 2(3) + 5 = 11$
         $a_4 = 2(4) + 5 = 13$
         $a_5 = 2(5) + 5 = 15$

    (b)

    (c)   This sequence is arithmetic, the common difference is 2.

13. (a) $a_1 = \dfrac{3^1}{2^2} = \dfrac{3}{4}$

         $a_2 = \dfrac{3^2}{2^3} = \dfrac{9}{8}$

         $a_3 = \dfrac{3^3}{2^4} = \dfrac{27}{16}$

         $a_4 = \dfrac{3^4}{2^5} = \dfrac{81}{32}$

         $a_5 = \dfrac{3^5}{2^6} = \dfrac{243}{64}$

    (b)

    (c)   This sequence is geometric, the common ratio is $\dfrac{3}{2}$.

15. $5, 5.5, 6, 6.5, \ldots$. Since $5.5 - 5 = 6 - 5.5 = 6.5 - 6 = 0.5$, this is an arithmetic sequence with
    $a_1 = 5$ and $d = 0.5$. Then $a_5 = a_4 + 0.5 = 7$.

17. $\sqrt{2}, 2\sqrt{2}, 3\sqrt{2}, 4\sqrt{2}, \ldots$. Since $2\sqrt{2} - \sqrt{2} = 3\sqrt{2} - 2\sqrt{2} = 4\sqrt{2} - 3\sqrt{2} = \sqrt{2}$, this is an arithmetic sequence with $a_1 = \sqrt{2}$ and $d = \sqrt{2}$. Then $a_5 = a_4 + \sqrt{2} = 4\sqrt{2} + \sqrt{2} = 5\sqrt{2}$.

19. $t - 3, t - 2, t - 1, t, \ldots$. Since $(t-2) - (t-3) = (t-1) - (t-2) = t - (t-1) = 1$, this is an arithmetic sequence with $a_1 = t - 3$ and $d = 1$. Then $a_5 = a_4 + 1 = t + 1$.

21. $\frac{3}{4}, \frac{1}{2}, \frac{1}{3}, \frac{2}{9}, \ldots$. Since $\frac{\frac{1}{2}}{\frac{3}{4}} = \frac{\frac{1}{3}}{\frac{1}{2}} = \frac{\frac{2}{9}}{\frac{1}{3}} = \frac{2}{3}$, this is a geometric sequence with $a_1 = \frac{3}{4}$ and $r = \frac{2}{3}$.

    Then $a_5 = a_4 \cdot r = \frac{2}{9} \cdot \frac{2}{3} = \frac{4}{27}$.

23. $3, 6i, -12, -24i, \ldots$. Since $\frac{6i}{3} = 2i$, $\frac{-12}{6i} = \frac{-2}{i} = \frac{-2i}{i^2} = 2i$, $\frac{-24i}{-12} = 2i$, this is a geometric sequence with common ratio $r = 2i$.

25. $a_6 = 17 = a + 5d$ and $a_4 = 11 = a + 3d$. Then, $a_6 - a_4 = 17 - 11 \Leftrightarrow (a+5d) - (a+3d) = 6 \Leftrightarrow 6 = 2d \Leftrightarrow d = 3$. Substituting into $11 = a + 3d$ gives $11 = a + 3 \cdot 3$, and so $a = 2$. Thus $a_2 = a + (2-1)d = 2 + 3 = 5$.

27. $a_3 = 9$ and $r = \frac{3}{2}$. Then $a_5 = a_3 \cdot r^2 = 9 \cdot \left(\frac{3}{2}\right)^2 = \frac{81}{4}$.

29. (a) $A_n = 32{,}000 \cdot 1.05^{n-1}$.

    (b) $A_1 = \$32{,}000$; $A_5 = 32{,}000 \cdot 1.05^4 = \$38{,}896.20$;
    $A_2 = 32{,}000 \cdot 1.05^1 = \$33{,}600$; $A_6 = 32{,}000 \cdot 1.05^5 = \$40{,}841.01$;
    $A_3 = 32{,}000 \cdot 1.05^2 = \$35{,}280$; $A_7 = 32{,}000 \cdot 1.05^6 = \$42{,}883.06$;
    $A_4 = 32{,}000 \cdot 1.05^3 = \$37{,}044$; $A_8 = 32{,}000 \cdot 1.05^7 = \$45{,}027.21$.

31. Let a_n be the number of bacteria in the dish at the end of $5n$ seconds. So, $a_0 = 3$, $a_1 = 3 \cdot 2$, $a_2 = 3 \cdot 2^2$, $a_3 = 3 \cdot 2^3, \ldots$. Then, clearly, a_n is a geometric sequence with $r = 2$ and $a = 3$. Thus at the end of $60 = 5(12)$ seconds, the number of bacteria is $a_{12} = 3 \cdot 2^{12} = 12{,}288$.

33. Suppose that the common ratio in the sequence $a_1, a_2, a_3, \ldots$ is r. Also, suppose that the common ratio in the sequence $b_1, b_2, b_3, \ldots$ is s. Then $a_n = a_1 r^{n-1}$ and $b_n = b_1 s^{n-1}$, $n = 1, 2, 3, \ldots$. Thus $a_n b_n = a_1 r^{n-1} \cdot b_1 s^{n-1} = (a_1 b_1)(rs)^{n-1}$. So the sequence $a_1 b_1, a_2 b_2, a_3 b_3, \ldots$ is geometric with first term $a_1 b_1$ and common ratio rs.

35. (a) $6, x, 12, \ldots$ is arithmetic if $x - 6 = 12 - x \Leftrightarrow 2x = 18 \Leftrightarrow x = 9$.

    (b) $6, x, 12, \ldots$ is geometric if $\frac{x}{6} = \frac{12}{x} \Leftrightarrow x^2 = 72 \Leftrightarrow x = \pm 6\sqrt{2}$.

37. $\sum_{k=3}^{6} (k+1)^2 = (3+1)^2 + (4+1)^2 + (5+1)^2 + (6+1)^2 = 16 + 25 + 36 + 49 = 126$

39. $\sum_{k=1}^{6} (k+1)2^{k-1} = 2 \cdot 2^0 + 3 \cdot 2^1 + 4 \cdot 2^2 + 5 \cdot 2^3 + 6 \cdot 2^4 + 7 \cdot 2^5 =$
    $2 + 6 + 16 + 40 + 96 + 224 = 384$

41. $\displaystyle\sum_{k=1}^{10}(k-1)^2 = 0^2 + 1^2 + 2^2 + 3^2 + 4^2 + 5^2 + 6^2 + 7^2 + 8^2 + 9^2$

43. $\displaystyle\sum_{k=1}^{50}\frac{3^k}{2^{k+1}} = \frac{3}{2^2} + \frac{3^2}{2^3} + \frac{3^3}{2^4} + \frac{3^4}{2^5} + \cdots + \frac{3^{49}}{2^{50}} + \frac{3^{50}}{2^{51}}$

45. $3 + 6 + 9 + 12 + \cdots + 99 = 3(1) + 3(2) + 3(3) + \cdots + 3(33) = \displaystyle\sum_{k=1}^{33} 3k$

47. $1 \cdot 2^3 + 2 \cdot 2^4 + 3 \cdot 2^5 + 4 \cdot 2^6 + \cdots + 100 \cdot 2^{102} = (1)2^{(1)+2} + (2)2^{(2)+2} + (3)2^{(3)+2} + (4)2^{(4)+2}$

$+ \cdots + (100)2^{(100)+2} = \displaystyle\sum_{k=1}^{100} k \cdot 2^{k+2}$

49. $1 + 0.9 + (0.9)^2 + \cdots + (0.9)^5$ is a geometric series with $a = 1$ and $r = \frac{0.9}{1} = 0.9$. Thus, the sum of

the series is $S_6 = \dfrac{1 - (0.9)^6}{1 - 0.9} = \dfrac{1 - 0.531441}{0.1} = 4.68559$.

51. $\sqrt{5} + 2\sqrt{5} + 3\sqrt{5} + \cdots + 100\sqrt{5}$ is an arithmetic series with $a = \sqrt{5}$ and $d = \sqrt{5}$. Then

$100\sqrt{5} = a_n = \sqrt{5} + \sqrt{5}(n-1) \quad \Leftrightarrow \quad n = 100$. So the sum is

$S_{100} = \frac{100}{2}\left(\sqrt{5} + 100\sqrt{5}\right) = 50(101\sqrt{5}) = 5050\sqrt{5}$.

53. $\displaystyle\sum_{n=0}^{6} 3 \cdot (-4)^n$ is a geometric series with $a = 3$, $r = -4$, and $n = 7$. Therefore, the sum of the series

is $S_7 = 3 \cdot \dfrac{1 - (-4)^7}{1 - (-4)} = \dfrac{3}{5}(1 + 4^7) = 9831$.

55. We have an arithmetic sequence with $a = 7$ and $d = 3$. Then $S_n = 325 = \dfrac{n}{2}[2a + (n-1)d]$

$= \dfrac{n}{2}[14 + 3(n-1)] = \dfrac{n}{2}(11 + 3n) \quad \Leftrightarrow \quad 650 = 3n^2 + 11n \quad \Leftrightarrow \quad (3n + 50)(n - 13) = 0$

$\Leftrightarrow \quad n = 13$ (because $n = -\frac{50}{3}$ is inadmissible). Thus, 13 terms must be added.

57. This is a geometric sequence with $a = 2$ and $r = 2$. Then

$S_{15} = 2 \cdot \dfrac{1 - 2^{15}}{1 - 2} = 2(2^{15} - 1) = 65{,}534$, and so the total number of ancestors is $65{,}534$.

59. $A = 10{,}000$, $i = 0.03$, and $n = 4$. Thus, $10{,}000 = R\dfrac{(1.03)^4 - 1}{0.03} \quad \Leftrightarrow \quad R = \dfrac{10{,}000 \cdot 0.03}{(1.03)^4 - 1}$

$= \$2390.27$.

61. $1 - \frac{2}{5} + \frac{4}{25} - \frac{8}{125} + \cdots$ is a geometric series with $a = 1$ and $r = -\frac{2}{5}$. Therefore, the sum is

$S = \dfrac{a}{1 - r} = \dfrac{1}{1 - \left(-\frac{2}{5}\right)} = \dfrac{5}{7}$.

63. $1 + \dfrac{1}{3^{1/2}} + \dfrac{1}{3} + \dfrac{1}{3^{3/2}} + \cdots$ is an infinite geometric series with $a = 1$ and $r = \dfrac{1}{\sqrt{3}}$. Thus, the sum is

$$S = \frac{1}{1 - \left(\frac{1}{\sqrt{3}}\right)} = \frac{\sqrt{3}}{\sqrt{3} - 1} = \frac{1}{2}(3 + \sqrt{3}).$$

65. Let $P(n)$ denote the statement that $1 + 4 + 7 + \cdots + (3n - 2) = \dfrac{n(3n - 1)}{2}$.

Step 1 $P(1)$ is the statement that $1 = \dfrac{1[3(1) - 1]}{2} = \dfrac{1 \cdot 2}{2}$, which is true.

Step 2 Assume that $P(k)$ is true; that is, $1 + 4 + 7 + \cdots + (3k - 2) = \dfrac{k(3k - 1)}{2}$. We want to use
this to show that $P(k + 1)$ is true. Now,
$$1 + 4 + 7 + 10 + \cdots + (3k - 2) + [3(k + 1) - 2] =$$
$$\frac{k(3k - 1)}{2} + 3k + 1 = \qquad\qquad \text{induction hypothesis}$$
$$\frac{k(3k - 1)}{2} + \frac{6k + 2}{2} = \frac{3k^2 - k + 6k + 2}{2} =$$
$$\frac{3k^2 + 5k + 2}{2} = \frac{(k + 1)(3k + 2)}{2} = \frac{(k + 1)[3(k + 1) - 1]}{2}.$$

Thus, $P(k + 1)$ follows from $P(k)$. So by the Principle of Mathematical Induction, $P(n)$ is
true for all n.

67. Let $P(n)$ denote the statement that $\left(1 + \dfrac{1}{1}\right)\left(1 + \dfrac{1}{2}\right)\left(1 + \dfrac{1}{3}\right) \cdots \cdot \left(1 + \dfrac{1}{n}\right) = n + 1$.

Step 1 $P(1)$ is the statement that $\left(1 + \frac{1}{1}\right) = 1 + 1$, which is clearly true.

Step 2 Assume that $P(k)$ is true; that is, $\left(1 + \frac{1}{1}\right)\left(1 + \frac{1}{2}\right)\left(1 + \frac{1}{3}\right) \cdots \cdot \left(1 + \frac{1}{k}\right) = k + 1$. We
want to use this to show that $P(k + 1)$ is true. Now,
$$\left(1 + \tfrac{1}{1}\right)\left(1 + \tfrac{1}{2}\right)\left(1 + \tfrac{1}{3}\right) \cdots \cdot \left(1 + \tfrac{1}{k}\right)\left(1 + \tfrac{1}{k+1}\right) =$$
$$\left[\left(1 + \tfrac{1}{1}\right)\left(1 + \tfrac{1}{2}\right)\left(1 + \tfrac{1}{3}\right) \cdots \cdot \left(1 + \tfrac{1}{k}\right)\right]\left(1 + \tfrac{1}{k+1}\right) =$$
$$(k + 1)\left(1 + \tfrac{1}{k+1}\right) = (k + 1) + 1. \quad \text{induction hypothesis}$$

Thus, $P(k + 1)$ follows from $P(k)$. So by the Principle of Mathematical Induction, $P(n)$ is
true for all n.

69. $a_{n+1} = 3a_n + 4$ and $a_1 = 4$. Let $P(n)$ denote the statement that $a_n = 2 \cdot 3^n - 2$.

Step 1 $P(1)$ is the statement that $a_1 = 2 \cdot 3^1 - 2 = 4$, which is clearly true.

Step 2 Assume that $P(k)$ is true; that is, $a_k = 2 \cdot 3^k - 2$. We want to use this to show that $P(k + 1)$
is true. Now,
$$\begin{aligned}
a_{k+1} &= 3a_k + 4 & &\text{definition of } a_{k+1} \\
&= 3(2 \cdot 3^k - 2) + 4 & &\text{induction hypothesis} \\
&= 2 \cdot 3^{k+1} - 6 + 4 \\
&= 2 \cdot 3^{k+1} - 2.
\end{aligned}$$

Thus $P(k+1)$ follows from $P(k)$. So by the Principle of Mathematical Induction, $P(n)$ is true for all n.

71. Let $P(n)$ denote the statement that $n! > 2^n$, for all natural numbers $n \geq 4$.

Step 1 $P(4)$ is the statement that $4! = 24 > 2^4 = 16$, which is true.

Step 2 Assume that $P(k)$ is true; that is, $k! > 2^k$. We want to use this to show that $P(k+1)$ is true. Now,

$$(k+1)! = (k+1)k!$$
$$> (k+1) \cdot 2^k \qquad \text{induction hypothesis}$$
$$> 2 \cdot 2^k \qquad \text{because } k+1 > 2 \text{ (for } k \geq 4)$$
$$= 2^{k+1}.$$

Thus $P(k+1)$ follows from $P(k)$. So by the Principle of Mathematical Induction, $P(n)$ is true for all $n \geq 4$.

73. $\dbinom{10}{2} + \dbinom{10}{6} = \dfrac{10!}{2!\,8!} + \dfrac{10!}{6!\,4!} = \dfrac{10 \cdot 9}{2} + \dfrac{10 \cdot 9 \cdot 8 \cdot 7}{4 \cdot 3 \cdot 2} = 45 + 210 = 255.$

75. $\displaystyle\sum_{k=0}^{8} \dbinom{8}{k}\dbinom{8}{8-k} = 2\dbinom{8}{0}\dbinom{8}{8} + 2\dbinom{8}{1}\dbinom{8}{7} + 2\dbinom{8}{2}\dbinom{8}{6} + 2\dbinom{8}{3}\dbinom{8}{5} + \dbinom{8}{4}\dbinom{8}{4}$

$= 2 + 2 \cdot 8^2 + 2 \cdot (28)^2 + 2 \cdot (56)^2 + (70)^2 = 12{,}870.$

77. $(2x+y)^4 = \dbinom{4}{0}(2x)^4 + \dbinom{4}{1}(2x)^3 y + \dbinom{4}{2}(2x)^2 y^2 + \dbinom{4}{3} \cdot 2xy^3 + \dbinom{4}{4}y^4$

$= 16x^4 + 32x^3 y + 24x^2 y^2 + 8xy^3 + y^4.$

79. The first three terms in the expansion of $(b^{-2/3} + b^{1/3})^{20}$ are $\dbinom{20}{0}(b^{-2/3})^{20} = b^{-40/3}$;

$\dbinom{20}{1}(b^{-2/3})^{19}(b^{1/3}) = 20b^{-37/3}$; and $\dbinom{20}{2}(b^{-2/3})^{18}(b^{1/3})^2 = 190b^{-34/3}$.

Chapter 9 Test

1. $a_1 = 1^2 - 1 = 0$; $a_2 = 2^2 - 1 = 3$; $a_3 = 3^2 - 1 = 8$; $a_4 = 4^2 - 1 = 15$; and $a_{10} = 10^2 - 1 = 99$.

2. $a_{n+2} = (a_n)^2 - a_{n+1}$, $a_1 = 1$ and $a_2 = 1$. Then $a_3 = a_1^2 - a_2 = 1^2 - 1 = 0$,
    $a_4 = a_2^2 - a_3 = 1^2 - 0 = 1$, and $a_5 = a_3^2 - a_4 = 0^2 - 1 = -1$.

3. $d = 76 - 80 = -4$. So $a_4 = a_3 + (-4) = 72 + (-4) = 68$ and
    $a_n = a_1 + (n-1)d = 80 + (n-1)(-4) = 84 - 4n$.

4. Since $a_n = a_1 \, r^{n-1}$, we substitute $a_1 = 25$, $a_4 = \frac{1}{5}$, and $n = 4$ and solve for r. This gives $\frac{1}{5} = 25 \, r^3$
    $\Leftrightarrow \quad r^3 = \frac{1}{125} = \frac{1}{5^3} \quad \Leftrightarrow \quad r = \frac{1}{5}$. So the common ratio is $\frac{1}{5}$ and $a_5 = a_1 r^4 = 25\left(\frac{1}{5}\right)^4 = \frac{1}{25}$.

5. (a) If $a_1, a_2, a_3, \ldots$ is an arithmetic sequence, then the sequence $a_1^2, a_2^2, a_3^2, \ldots$ is also arithmetic.
        This statement is FALSE. For a counterexample, consider $a_1 = 1$, $a_2 = 2$, $a_3 = 3 \ldots$. Then
        $a_1, a_2, a_3, \ldots$ is an arithmetic sequence with $a = 1$ and $d = 1$. However, the sequence a_1^2, a_2^2,
        $a_3^2, \ldots$ becomes $1, 4, 9, \ldots$, which is NOT arithmetic because there is no common difference.

    (b) If $a_1, a_2, a_3, \ldots$ is a geometric sequence, then the sequence $a_1^2, a_2^2, a_3^2, \ldots$ is also geometric.
        This statement is TRUE. Let the common ratio for the geometric series $a_1, a_2, a_3, \ldots$ be r so
        that $a_n = a_1 r^{n-1}$, $n = 1, 2, 3, \ldots$. Then $a_n^2 = (a_1 r^{n-1})^2 = (a_1^2)(r^2)^{n-1}$. Therefore, the
        sequence $a_1^2, a_2^2, a_3^2, \ldots$ is geometric with common ratio r^2.

6. (a) The nth partial sum an arithmetic sequence is $S_n = \dfrac{n}{2}(a + a_n)$ or $S_n = \dfrac{n}{2}[2a + (n-1)d]$.

    (b) $a_1 = 10$ and $a_{10} = 2$. Then $S_{10} = \frac{10}{2}(10 + 2) = 60$.

    (c) Since $S_{10} = \frac{10}{2}(2 \cdot 10 + 9d) = 60$, then $5(20 + 9d) = 60 \quad \Leftrightarrow \quad d = -\frac{8}{9}$, and so the common
        difference is $-\frac{8}{9}$. Then $a_{100} = a + 99d = 10 - \frac{8}{9} \cdot 99 = -78$.

7. (a) The nth partial sum a geometric sequence is $S_n = a\dfrac{1 - r^n}{1 - r}$, where $r \neq 1$.

    (b) The geometric series $\dfrac{1}{3} + \dfrac{2}{3^2} + \dfrac{2^2}{3^3} + \dfrac{2^3}{3^4} + \cdots + \dfrac{2^9}{3^{10}}$ has $a = \frac{1}{3}, r = \frac{2}{3}$, and $n = 10$. So

    $$S_{10} = \frac{1}{3} \cdot \frac{1 - \left(\frac{2}{3}\right)^{10}}{1 - \left(\frac{2}{3}\right)} = \frac{1}{3} \cdot 3\left(1 - \frac{1024}{59,049}\right) = \frac{58,025}{59,049}.$$

8. The infinite geometric series $1 + \dfrac{1}{2^{1/2}} + \dfrac{1}{2} + \dfrac{1}{2^{3/2}} + \cdots$ has $a = 1$ and $r = 2^{-1/2} = \dfrac{1}{\sqrt{2}}$. Thus,
    $S = \dfrac{1}{1 - \left(\frac{1}{\sqrt{2}}\right)} = \dfrac{\sqrt{2}}{\sqrt{2} - 1} = \dfrac{\sqrt{2}}{\sqrt{2} - 1} \cdot \dfrac{\sqrt{2} + 1}{\sqrt{2} + 1} = 2 + \sqrt{2}$.

9. Let $P(n)$ denote the statement that $1^2 + 2^2 + 3^2 + \cdots + n^2 = \dfrac{n(n+1)(2n+1)}{6}$.

    <u>Step 1</u> Show that $P(1)$ is true. But $P(1)$ says that $1^2 = \frac{1 \cdot 2 \cdot 3}{6}$, which is true.

Step 2 Assume that $P(k)$ is true; that is, $1^2 + 2^2 + 3^2 + \cdots + k^2 = \dfrac{k(k+1)(2k+1)}{6}$. We want to use this to show that $P(k+1)$ is true. Now,

$$1^2 + 2^2 + 3^2 + \cdots + k^2 + (k+1)^2 = \frac{k(k+1)(2k+1)}{6} + (k+1)^2 \qquad \text{induction hypothesis}$$

$$= \frac{k(k+1)(2k+1) + 6(k+1)^2}{6}$$

$$= \frac{(k+1)(2k^2 + k) + (6k+6)(k+1)}{6}$$

$$= \frac{(k+1)(2k^2 + k + 6k + 6)}{6}$$

$$= \frac{(k+1)(2k^2 + 7k + 6)}{6}$$

$$= \frac{(k+1)(k+2)(2k+3)}{6}$$

$$= \frac{(k+1)[(k+1)+1][2(k+1)+1]}{6}.$$

Thus $P(k+1)$ follows from $P(k)$. So by the Principle of Mathematical Induction, $P(n)$ is true for all n.

10. (a) $\displaystyle\sum_{n=1}^{5}(1-n^2) = (1-1^2) + (1-2^2) + (1-3^2) + (1-4^2) + (1-5^2)$

$= 0 - 3 - 8 - 15 - 24 = -50$

(b) $\displaystyle\sum_{n=3}^{6}(-1)^n 2^{n-2} = (-1)^3 2^{3-2} + (-1)^4 2^{4-2} + (-1)^5 2^{5-2} + (-1)^6 2^{6-2}$

$= -2 + 4 - 8 + 16 = 10$

11. $(2x + y^2)^5 = \dbinom{5}{0}(2x)^5 + \dbinom{5}{1}(2x)^4 y^2 + \dbinom{5}{2}(2x)^3 (y^2)^2 + \dbinom{5}{3}(2x)^2 (y^2)^3$

$$+ \dbinom{5}{4}(2x)(y^2)^4 + \dbinom{5}{5}(y^2)^5$$

$$= 32x^5 + 80x^4 y^2 + 80x^3 y^4 + 40x^2 y^6 + 10xy^8 + y^{10}.$$

12. $\dbinom{10}{3}(3x)^3 (-2)^7 = 120 \cdot 27x^3(-128) = -414{,}720x^3.$

Focus on Problem Solving

1. (a) Since there are 365 days in a year, the interest earned per day is $\dfrac{0.0365}{365} = 0.0001$. Thus the amount in the account at the end of the nth day is $A_n = 1.0001 A_{n-1}$ with $A_0 = \$275,000$.

   (b) $A_0 = \$275,000$, $\ A_1 = 1.0001 A_0 = 1.0001 \cdot 275,000 = \$275,027.50$,
   $A_2 = 1.0001 A_1 = 1.0001(1.0001 A_0) = 1.0001^2 A_0 = \$275,055.00$,
   $A_3 = 1.0001 A_2 = 1.0001(1.0001^2 A_0) = 1.0001^3 A_0 = \$275,082.51$,
   $A_4 = 1.0001 A_3 = 1.0001(1.0001^3 A_0) = 1.0001^4 A_0 = \$275,110.02$,
   $A_5 = 1.0001 A_4 = 1.0001(1.0001^4 A_0) = 1.0001^5 A_0 = \$275,137.53$,
   $A_6 = 1.0001 A_5 = 1.0001(1.0001^5 A_0) = 1.0001^6 A_0 = \$275,165.04$,
   $A_7 = 1.0001 A_6 = 1.0001(1.0001^6 A_0) = 1.0001^7 A_0 = \$275,192.56$.

   (c) $A_n = 1.0001^n \cdot 275,000$.

3. (a) Since there are 12 months in a year, the interest earned per day is $\dfrac{0.03}{12} = 0.0025$. Thus the amount in the account at the end of the nth month is $A_n = 1.0025 A_{n-1} + 100$ with $A_0 = \$100$.

   (b) $A_0 = \$100$, $\ A_1 = 1.0025 A_0 + 100 = 1.0025 \cdot 100 + 100 = \$200.25$,
   $A_2 = 1.0025 A_1 + 100 = 1.0025(1.0025 \cdot 100 + 100) + 100$
   $\quad = 1.0025^2 \cdot 100 + 1.0025 \cdot 100 + 100 = \$300.75$,
   $A_3 = 1.0025 A_2 + 100 = 1.0025(1.0025^2 \cdot 100 + 1.0025 \cdot 100 + 100) + 100$
   $\quad = 1.0025^3 \cdot 100 + 1.0025^2 \cdot 100 + 1.0025 \cdot 100 + 100= = \$401.50$,
   $A_4 = 1.0025 A_3 + 100 = 1.0025(1.0025^3 \cdot 100 + 1.0025^2 \cdot 100 + 1.0025 \cdot 100 + 100) + 100$
   $\quad = 1.0025^4 \cdot 100 + 1.0025^3 \cdot 100 + 1.0025^2 \cdot 100 + 1.0025 \cdot 100 + 100 = \$502.51$.

   (c) $A_n = 1.0025^n \cdot 100 + \cdots + 1.0025^2 \cdot 100 + 1.0025 \cdot 100 + 100$ the partial sum of a geometric series, so $A_n = 100 \cdot \dfrac{1 - 1.0025^{n+1}}{1 - 1.0025} = 100 \cdot \dfrac{1.0025^{n+1} - 1}{0.0025}$.

   (d) Since 5 years is 60 months we have $A_{60} = 100 \cdot \dfrac{1.0025^{61} - 1}{0.0025} \approx \$6,580.83$.

5. (a) The amount (A_n) of pollutants in the lake in the nth year is 30% of the amount form the preceding year ($0.30 A_{n-1}$) plus the amount discharged that year (2400 tons). Thus $A_n = 0.30 A_{n-1} + 2400$.

   (b) $A_0 = 2400$, $\ A_1 = 0.30(2400) + 2400 = 3120$;
   $A_2 = 0.30[0.30(2400) + 2400] + 2400 = 0.30^2(2400) + 2400(2400) + 2400 = 3336$;
   $A_3 = 0.30[0.30^2(2400) + 2400(2400) + 2400] + 2400$
   $\quad = 0.03^3(2400) + 0.30^2(2400) + 2400(2400) + 2400 = 3400.8$;
   $A_4 = 0.30[0.03^3(2400) + 0.30^2(2400) + 2400(2400) + 2400] + 2400$
   $\quad = 0.03^4(2400) + 0.03^3(2400) + 0.30^2(2400) + 2400(2400) + 2400 = 3420.2$.

(c) A_n is the partial sum of a geometric series, so

$$A_n = 2400 \cdot \frac{1 - 0.30^{n+1}}{1 - 0.30} = 2400 \cdot \frac{1 - 0.30^{n+1}}{0.70} \approx 3428.6(1 - 0.30^{n+1}).$$

(d) $A_6 = 2400 \cdot \dfrac{1 - 0.30^7}{0.70} = 3427.8$ tons. The sum of a geometric series, is

$$A = 2400 \cdot \frac{1}{0.70} = 3428.6 \text{ tons.}$$

(e)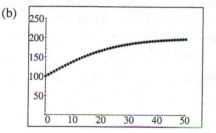

7.   (a)   In the nth year since Victoria's initial deposit the amount (V_n) in her CD is the amount from the preceding year (V_{n-1}), plus the 5% interest earned on that amount ($0.05V_{n-1}$), plus $500 times the number of years since her initial deposit ($500n$). Thus $V_n = 1.05V_{n-1} + 500n$.

(b)   Ursula's savings surpass Victoria savings in the 35th year.

9.   (a) $R_1 = 104$; $R_2 = 108.0$; $R_3 = 112.0$; $R_4 = 115.9$; $R_5 = 119.8$; $R_6 = 123.7$; $R_7 = 127.4$.

(b)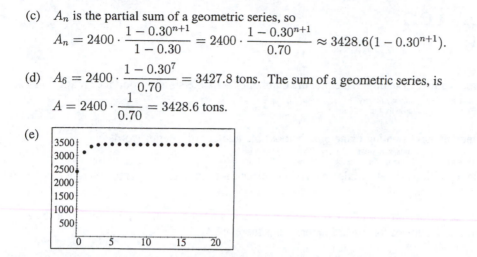

As n becomes large the raccoon population approaches 200.

Chapter Ten
Exercises 10.1

1. By the Fundamental Counting Principle, the number of possible single-scoop ice cream cones is
$$\left(\begin{smallmatrix}\text{number of ways to}\\\text{choose the flavor}\end{smallmatrix}\right) \cdot \left(\begin{smallmatrix}\text{number of ways to}\\\text{choose the type of cone}\end{smallmatrix}\right) = 4 \cdot 3 = 12.$$

3. By the Fundamental Counting Principle, the possible number of 3-letter words is
$$\left(\begin{smallmatrix}\text{number of ways to}\\\text{choose the 1st letter}\end{smallmatrix}\right) \cdot \left(\begin{smallmatrix}\text{number of ways to}\\\text{choose the 2nd letter}\end{smallmatrix}\right) \cdot \left(\begin{smallmatrix}\text{number of ways to}\\\text{choose the 3rd letter}\end{smallmatrix}\right).$$

    (a) Since repetitions are allowed, we have 4 choices for each letter. Thus there are $4 \cdot 4 \cdot 4 = 64$ words.

    (b) Since repetitions are *not* allowed, we have 4 choices for the 1st letter, 3 choices for the 2nd letter, and 2 choices for the 3rd letter. Thus there are $4 \cdot 3 \cdot 2 = 24$ words.

5. Since there are four choices for each of the five questions, by the Fundamental Counting Principle there are $4 \cdot 4 \cdot 4 \cdot 4 \cdot 4 = 1024$ different ways the test can be completed.

7. Since a runner can only finish once, there are no repetitions. And since we are assuming that there is no tie, the number of different finishes is $\left(\begin{smallmatrix}\text{number of ways to}\\\text{choose the 1st runner}\end{smallmatrix}\right) \cdot \left(\begin{smallmatrix}\text{number of ways to}\\\text{choose the 2nd runner}\end{smallmatrix}\right) \cdot \left(\begin{smallmatrix}\text{number of ways to}\\\text{choose the 3rd runner}\end{smallmatrix}\right) \cdot$
$\left(\begin{smallmatrix}\text{number of ways to}\\\text{choose the 4th runner}\end{smallmatrix}\right) \cdot \left(\begin{smallmatrix}\text{number of ways to}\\\text{choose the 5th runner}\end{smallmatrix}\right) = 5 \cdot 4 \cdot 3 \cdot 2 \cdot 1 = 120.$

9. Since there are 4 main courses, there are 6 ways to choose a main course. Likewise, there are 5 drinks and 3 desserts so there are 5 ways to choose a drink and 3 ways to choose a dessert. So the number of different meals consisting of a main course, a drink, and a dessert is
$$\left(\begin{smallmatrix}\text{number of ways to}\\\text{choose the main course}\end{smallmatrix}\right) \cdot \left(\begin{smallmatrix}\text{number of ways to}\\\text{choose a drink}\end{smallmatrix}\right) \cdot \left(\begin{smallmatrix}\text{number of ways to}\\\text{choose a dessert}\end{smallmatrix}\right) = (4)(5)(3) = 60.$$

11. By the Fundamental Counting Principle, the number of different routes from town A to town D via towns B and C is $\left(\begin{smallmatrix}\text{number of routes}\\\text{from A to B}\end{smallmatrix}\right) \cdot \left(\begin{smallmatrix}\text{number of routes}\\\text{from B to C}\end{smallmatrix}\right) \cdot \left(\begin{smallmatrix}\text{number of routes}\\\text{from C to D}\end{smallmatrix}\right) = (4)(5)(6) = 120.$

13. The number of possible sequences of heads and tails when a coin is flipped 5 times is
$\left(\begin{smallmatrix}\text{number of possible}\\\text{outcomes on the 1st flip}\end{smallmatrix}\right) \cdot \left(\begin{smallmatrix}\text{number of possible}\\\text{outcomes on the 2nd flip}\end{smallmatrix}\right) \cdot \left(\begin{smallmatrix}\text{number of possible}\\\text{outcomes on the 3rd flip}\end{smallmatrix}\right) \cdot \left(\begin{smallmatrix}\text{number of possible}\\\text{outcomes on the 4th flip}\end{smallmatrix}\right) \cdot$
$\left(\begin{smallmatrix}\text{number of possible}\\\text{outcomes on the 5th flip}\end{smallmatrix}\right) = (2)(2)(2)(2)(2) = 2^5 = 32.$ (Here there are only two choices, heads or tails, for each flip.)

15. Since there are six different faces on each die, the number of possible outcomes when a red die and a blue die and a white die are rolled is
$\left(\begin{smallmatrix}\text{number of possible}\\\text{outcomes on the red die}\end{smallmatrix}\right) \cdot \left(\begin{smallmatrix}\text{number of possible}\\\text{outcomes on the blue die}\end{smallmatrix}\right) \cdot \left(\begin{smallmatrix}\text{number of possible}\\\text{outcomes on the white die}\end{smallmatrix}\right) = (6)(6)(6) = 6^3 = 216.$

17. The number of possible skirt-blouse-shoe outfits is
$\left(\begin{smallmatrix}\text{number of ways}\\\text{to choose a skirt}\end{smallmatrix}\right) \cdot \left(\begin{smallmatrix}\text{number of ways}\\\text{to choose a blouse}\end{smallmatrix}\right) \cdot \left(\begin{smallmatrix}\text{number of ways}\\\text{to choose shoes}\end{smallmatrix}\right) = (5)(8)(12) = 480.$

19. The number of possible ID numbers of one letter followed by two digits is
$\left(\begin{smallmatrix}\text{number of ways}\\\text{to choose a letter}\end{smallmatrix}\right) \cdot \left(\begin{smallmatrix}\text{number of ways}\\\text{to choose a digit}\end{smallmatrix}\right) \cdot \left(\begin{smallmatrix}\text{number of ways}\\\text{to choose a digit}\end{smallmatrix}\right) = (26)(10)(10) = 2600.$ Since $2600 < 2844$, it is not possible to give each of the company's employees a different ID number using this scheme.

21. The number of different California license plates possible is

$$\left(\begin{smallmatrix}\text{number of ways to}\\\text{choose a nonzero digit}\end{smallmatrix}\right) \cdot \left(\begin{smallmatrix}\text{number of ways}\\\text{to choose 3 letters}\end{smallmatrix}\right) \cdot \left(\begin{smallmatrix}\text{number of ways}\\\text{to choose 3 digits}\end{smallmatrix}\right) = (9)(26^3)(10^3) = 158{,}184{,}000.$$

23. There are two possible ways to answer each question on a true-false test. Thus the possible number of different ways to complete a true-false test containing 10 questions is

$$\left(\begin{smallmatrix}\text{number of ways}\\\text{to answer question 1}\end{smallmatrix}\right) \cdot \left(\begin{smallmatrix}\text{number of ways}\\\text{to answer question 2}\end{smallmatrix}\right) \cdots \left(\begin{smallmatrix}\text{number of ways}\\\text{to answer question 10}\end{smallmatrix}\right) = 2^{10} = 1024.$$

25. The number of different classifications possible is

$$\left(\begin{smallmatrix}\text{number of ways}\\\text{to choose a major}\end{smallmatrix}\right) \cdot \left(\begin{smallmatrix}\text{number of ways to}\\\text{choose a minor}\end{smallmatrix}\right) \cdot \left(\begin{smallmatrix}\text{number of ways}\\\text{to choose a year}\end{smallmatrix}\right) \cdot \left(\begin{smallmatrix}\text{number of ways}\\\text{to choose a sex}\end{smallmatrix}\right) = (32)(32)(4)(2) = 8192.$$ To
see that there are 32 ways to choose a minor, we start with 32 fields, subtract the major (you can not major and minor in the same subject), and then add one for the possibility of *NO MINOR*.

27. The number of possible license plates of two letters followed by three digits is

$$\left(\begin{smallmatrix}\text{number of ways to}\\\text{choose 2 letters}\end{smallmatrix}\right) \cdot \left(\begin{smallmatrix}\text{number of ways}\\\text{to choose 3 digits}\end{smallmatrix}\right) = (26^2)(10^3) = 676{,}000.$$ Since $676{,}000 < 8{,}000{,}000$, there will
not be enough different license plates for the state's 8 million registered cars.

29. Since a student can hold only one office, the number of ways that a president, a vice-president and a secretary can be chosen from a class of 30 students is

$$\left(\begin{smallmatrix}\text{number of ways}\\\text{to choose a president}\end{smallmatrix}\right) \cdot \left(\begin{smallmatrix}\text{number of ways to}\\\text{choose a vice-president}\end{smallmatrix}\right) \cdot \left(\begin{smallmatrix}\text{number of ways}\\\text{to choose a secretary}\end{smallmatrix}\right) = (30)(29)(28) = 24{,}360.$$

31. The number of ways a chairman, vice-chairman, and a secretary can be chosen if the chairman must be a Democrat and the vice-chairman must be a Republican is

$$\left(\begin{smallmatrix}\text{number of ways to}\\\text{choose a Dem. chairman}\\\text{from the 10 Dem.}\end{smallmatrix}\right) \cdot \left(\begin{smallmatrix}\text{number of ways to choose}\\\text{a Rep. vice-chairman}\\\text{from the 7 Rep}\end{smallmatrix}\right) \cdot \left(\begin{smallmatrix}\text{number of ways to choose}\\\text{a secretary from the}\\\text{remaining 15 members}\end{smallmatrix}\right) = (10)(7)(15) = 1050.$$

33. The possible number of 5-letter words formed using the letters A, B, C, D, E, F, and G is calculated as follows:

(a) If repetition of letters is allowed, then there are 7 ways to pick each letter in the word. Thus the number of 5-letter words is

$$\left(\begin{smallmatrix}\text{number of ways to}\\\text{choose the 1st letter}\end{smallmatrix}\right) \cdot \left(\begin{smallmatrix}\text{number of ways to}\\\text{choose the 2nd letter}\end{smallmatrix}\right) \cdots \left(\begin{smallmatrix}\text{number of ways to}\\\text{choose the 5th letter}\end{smallmatrix}\right) = 7^5 = 16{,}807.$$

(b) If no letter can be repeated in a word, then the number of 5-letter words is

$$\left(\begin{smallmatrix}\text{number of ways to}\\\text{choose the 1st letter}\end{smallmatrix}\right) \cdot \left(\begin{smallmatrix}\text{number of ways to}\\\text{choose the 2nd letter}\end{smallmatrix}\right) \cdots \left(\begin{smallmatrix}\text{number of ways to}\\\text{choose the 5th letter}\end{smallmatrix}\right) = (7)(6)(5)(4)(3) = 2520.$$

(c) If each word must begin with the letter A, then the number of 5-letter words (letters can be repeated) is

$$\left(\begin{smallmatrix}\text{number of ways to}\\\text{choose the letter A}\end{smallmatrix}\right) \cdot \left(\begin{smallmatrix}\text{number of ways to}\\\text{choose the 2nd letter}\end{smallmatrix}\right) \cdots \left(\begin{smallmatrix}\text{number of ways to}\\\text{choose the 5th letter}\end{smallmatrix}\right) = (1)(7)(7)(7)(7) = 7^4 = 2401.$$

(d) If the letter C must be in the middle, then the number of 5-letter words (letters can be repeated) is

$$\left(\begin{smallmatrix}\text{number of ways to}\\\text{choose the 1st letter}\end{smallmatrix}\right) \cdot \left(\begin{smallmatrix}\text{number of ways to}\\\text{choose the 2nd letter}\end{smallmatrix}\right) \cdot \left(\begin{smallmatrix}\text{number of ways to}\\\text{choose the letter C}\end{smallmatrix}\right) \cdot \left(\begin{smallmatrix}\text{number of ways to}\\\text{choose the 4th letter}\end{smallmatrix}\right) \cdot \left(\begin{smallmatrix}\text{number of ways to}\\\text{choose the 5th letter}\end{smallmatrix}\right)$$
$$= (7)(7)(1)(7)(7) = 7^4 = 2401.$$

(e) If the middle letter must be a vowel, then the number of 5-letter words is $(7)(7)(2)(7)(7)$
$$= 2 \cdot 7^4 = 4802.$$

35.   The number of possible variable names is

$$\binom{\text{number of ways to}}{\text{choose a letter}} \cdot \binom{\text{number of ways to choose}}{\text{a letter or a digit}} = (26)(36) = 936.$$

37.   The number of ways 4 men and 4 women may be seated in a row of 8 seats is calculated as follows:

   (a)   If the women are to be seated together and the men together, then the number of ways they can
         be seated is $\binom{\text{number of ways to seat the}}{\text{women first and the men second}} + \binom{\text{number of ways to seat the}}{\text{the men first and the women second}}$
         $= (4)(3)(2)(1)(4)(3)(2)(1) + (4)(3)(2)(1)(4)(3)(2)(1) = (4!)(4!) + (4!)(4!) = 576 + 576$
         $= 1152.$

   (b)   If they are to be seated alternately by gender , then the number of ways they can be seated is
         $\binom{\text{number of ways to seat them}}{\text{alternately if a woman is first}} + \binom{\text{number of ways to seat them}}{\text{alternately if a man is first}}$
         $= (4)(4)(3)(3)(2)(2)(1)(1) + (4)(4)(3)(3)(2)(2)(1)(1) = 1152.$

39.   The number of ways that 8 mathematics books and 3 chemistry books may be placed on a shelf if the
      math books are to be next to each other and the chemistry books next to each other is
      $\binom{\text{number of ways to place them}}{\text{if the math books are first}} + \binom{\text{number of ways to place them}}{\text{if the chemistry books are first}} = (8!)(3!) + (3!)(8!) = 483{,}840.$

41.   In order for a number to be odd, the last digit must be odd. In this case, it must be a 1. Thus, the
      other two digits must be chosen from the three digits 2, 4, and 6. So, the number of 3-digit odd
      numbers that can be formed using the digits 1, 2, 4, and 6, if no digit may be used more than once is
      $\binom{\text{number of ways}}{\substack{\text{to choose a digit}\\\text{from the 3 digits}}} \cdot \binom{\text{number of ways}}{\substack{\text{to choose from the}\\\text{2 remaining digits}}} \cdot \binom{\text{number of ways}}{\substack{\text{to choose}\\\text{an odd digit}}} = (3)(2)(1) = 6.$

43.   (a)   From Exercise 6 the number of seven-digit phone numbers is 8,000,000. Under the old rules,
             the possible number of telephone area codes consisting of 3 digits with the given conditions is
             $(8)(2)(10) = 160$. So the possible number of area code + telephone number combinations
             under the old rules is $160 \cdot 8{,}000{,}000 = 1{,}280{,}000{,}000$.

      (b)   Under the new rules, where the second digit can be any number, the number of area codes
             possible is $(8)(10)(10) = 800$. Thus the possible number of area code + telephone number
             combinations under the new rules is $800 \cdot 8{,}000{,}000 = 6{,}400{,}000{,}000$.

      (c)   Although over $1\frac{1}{4}$ billion telephone numbers may seem to be enough for a population of
             300,000,000, so many individuals and businesses now have multiple lines for fax, Internet
             access, and cellular phones that this number has proven insufficient.

      (d)   There are now more "new" area codes than "old" area codes. Numerous answers are possible.

Exercises 10.2

1. $P(8,3) = \dfrac{8!}{(8-3)!} = \dfrac{8!}{5!} = 8 \cdot 7 \cdot 6 = 336$

3. $P(11,4) = \dfrac{11!}{(11-4)!} = \dfrac{11!}{7!} = 11 \cdot 10 \cdot 9 \cdot 8 = 7920$

5. $P(100,1) = \dfrac{100!}{(100-1)!} = \dfrac{100!}{99!} = 100$

7. Here we have 6 letters, of which 3 are A's, 2 are B's, and 1 is a C. Thus the number of distinguishable permutations is $\dfrac{6!}{3!\,2!\,1!} = \dfrac{(6)(5)(4)(3!)}{(2)(3!)} = 60.$

9. Here we have 5 letters, of which 2 are A's, 1 is a B, 1 is a C, and 1 is a D. Thus the number of distinguishable permutations is $\dfrac{5!}{2!\,1!\,1!\,1!} = \dfrac{(5)(4)(3)(2!)}{2!} = 60.$

11. Here we have 9 letters, of which 2 are X's, 3 are Y, and 4 a Z. Thus the number of distinguishable permutations is $\dfrac{9!}{2!\,3!\,4!} = \dfrac{9 \cdot 8 \cdot 7 \cdot 6 \cdot 5 \cdot 4!}{2 \cdot 3 \cdot 2 \cdot 4!} = 1260.$

13. $C(8,3) = \dfrac{8!}{3!\,(8-3)!} = \dfrac{8!}{3!\,5!} = \dfrac{8 \cdot 7 \cdot 6}{3 \cdot 2 \cdot 1} = 56$

15. $C(11,4) = \dfrac{11!}{4!\,7!} = \dfrac{11 \cdot 10 \cdot 9 \cdot 8}{4 \cdot 3 \cdot 2 \cdot 1} = 330$

17. $C(100,1) = \dfrac{100!}{1!\,99!} = \dfrac{100}{1} = 100$

19. We need the number of ways of selecting three students *in order* for the positions of president, vice president, and secretary from the 15 students in the class. Since a student cannot hold more than one position, this number is $P(15,3) = \dfrac{15!}{(15-3)!} = \dfrac{15!}{12!} = 15 \cdot 14 \cdot 13 = 2730.$

21. Since a person cannot occupy more than one chair at a time, the number of ways of seating 6 people from 10 people in a row of 6 chairs is $P(10,6) = 10 \cdot 9 \cdot 8 \cdot 7 \cdot 6 \cdot 5 = 151{,}200.$

23. The number of ways of selecting 3 objects in order (a 3-letter word) from 6 distinct objects (the 6 letters) assuming that the letters cannot be repeated is $P(6,3) = 6 \cdot 5 \cdot 4 = 120.$

25. The number of ways of selecting 3 objects in order (a 3-digit number) from 4 distinct objects (the 4 digits) with no repetition of the digits is $P(4,3) = 4 \cdot 3 \cdot 2 = 24.$

27. The number of ways of ordering 9 distinct objects (the contestants) is $P(9,9) = 9! = 362{,}880.$ Here a runner cannot finish more than once, so no repetitions are allowed, and order is important.

29. The number of ways of ordering 1000 distinct objects (the contestants) taking 3 at a time is $P(1000, 3) = 1000 \cdot 999 \cdot 998 = 997{,}002{,}000$. We are assuming that a person cannot win more than once, that is, there are no repetitions.

31. We first place Jack in the first seat, and then seat the remaining 4 students. Thus the number of these arrangements is $\left(\begin{smallmatrix} \text{number of ways to} \\ \text{seat Jack in the 1st seat} \end{smallmatrix} \right) \cdot \left(\begin{smallmatrix} \text{number of ways to seat} \\ \text{the remaining 4 students} \end{smallmatrix} \right) = P(1,1) \cdot P(4,4) = 1!\,4! = 24$.

33. Here we have 6 objects, of which 2 are blue marbles and 4 are red marbles. Thus the number of distinguishable permutations is $\dfrac{6!}{2!\,4!} = \dfrac{6 \cdot 5 \cdot 4!}{2 \cdot 4!} = 15$.

35. The number of distinguishable permutations of 12 objects (the 12 coins), from like groups of size 4 (the pennies), of size 3 (the nickels), of size 2 (the dimes) and of size 3 (the quarters) is
$$\dfrac{12!}{4!\,3!\,2!\,3!} = 277{,}200.$$

37. The number of distinguishable permutations of 12 objects (the 12 ice cream cones) from like groups of size 3 (the vanilla cones), of size 2 (the chocolate cones), of size 4 (the strawberry cones), and of size 5 (the butterscotch cones) is $\dfrac{14!}{3!\,2!\,4!\,5!} = 2{,}522{,}520$.

39. The number of distinguishable permutations of 8 objects (the 8 cleaning tasks) from like groups of size 5, 2, and 1 workers, respectively is $\dfrac{8!}{5!\,2!\,1!} = 168$.

41. We want the number of ways of choosing a group of three from a group of six. This number is $C(6,3) = \dfrac{6!}{3!\,3!} = 20$.

43. We want the number of ways of choosing a group of six people from a group of ten people. The number of combinations of 10 objects (10 people) taken 6 at a time is $C(10,6) = \dfrac{10!}{6!\,4!} = 210$.

45. We want the number of ways of choosing a group (the 5-card hand) where order of selection is not important. The number of combinations of 52 objects (the 52 cards) taken 5 at a time is $C(52,5) = \dfrac{52!}{5!\,47!} = 2{,}598{,}960$.

47. The order of selection is not important, hence we must calculate the number of combinations of 10 objects (the 10 questions) taken 7 at a time, this gives $C(10,7) = \dfrac{10!}{7!\,3!} = 120$.

49. We assume that the order in which he plays the pieces in the recital is *not* important, so the number of combinations of 12 objects (the 12 pieces) taken 8 at a time is $C(12,8) = \dfrac{12!}{8!\,4!} = 495$.

51. We are just interested in the group of seven students taken from the class of 30 students, not the order in which they are picked. Thus the number is $C(30,7) = \dfrac{30!}{7!\,23!} = 2{,}035{,}800$.

53. We first take Jack out of the class of 30 students and select the 7 students from the remaining 29 students. Thus there are $C(29, 7) = \dfrac{29!}{7!\,22!} = 1{,}560{,}780$ ways to pick the 7 students for the field trip.

55. We must count the number of different ways to choose the group of 6 numbers from the 53 numbers. There are $C(53, 6) = \dfrac{53!}{6!\,47!} = 22{,}957{,}480$ possible tickets, so it would cost \$22,957,480.

57. (a)   The number of ways to select 5 of the 8 objects is $C(8, 5) = \dfrac{8!}{5!\,3!} = 56$.

     (b)   A set with 8 elements has $2^8 = 256$ subsets.

59. Each subset of toppings constitutes a different way a hamburger can be ordered. Since a set with 10 elements has $2^{10} = 1024$ subsets, there are 1024 different ways to order a hamburger.

61. We pick the two men from the group of ten men, and we pick the two women from the group of ten women. So the number of ways 2 men and 2 women can be chosen is
$$\binom{\text{number of ways to pick}}{\text{2 of the 10 men}} \cdot \binom{\text{number of ways to pick}}{\text{2 of the 10 women}} = C(10, 2) \cdot C(10, 2) = (45)(45) = 2025.$$

63. The leading and supporting roles are different (order counts), while the extra roles are not (order doesn't count). Also, the male roles must be filled by the male actors and the female roles filled by the female actresses. Thus, number of ways the actors and actresses can be chosen is
$$\begin{pmatrix}\text{number of ways the}\\ \text{leading and the supporting}\\ \text{actors can be chosen}\end{pmatrix} \cdot \begin{pmatrix}\text{number of ways the}\\ \text{leading and supporting}\\ \text{actresses can be chosen}\end{pmatrix} \cdot \begin{pmatrix}\text{number of ways to choose}\\ \text{5 of 8 male extras}\end{pmatrix} \cdot \begin{pmatrix}\text{number of ways to choose}\\ \text{3 of 10 female extras}\end{pmatrix}$$
$$= P(10, 2) \cdot P(12, 2) \cdot C(8, 5) \cdot C(10, 3) = (90)(132)(56)(120) = 79{,}833{,}600.$$

65. To order a pizza, we must make several choices. First the size (4 choices), the type of crust (2 choices), and then the toppings. Since there are 14 toppings, the number of possible choices is the number of subsets of the 14 toppings, that is 2^{14} choices. So by the Fundamental Counting Principle, the number of possible pizzas is $(4)(2) \cdot 2^{14} = 131{,}072$.

67. We treat John and Jane as one object and Mike and Molly as one object, so we need the number of ways of permuting 8 objects. We then multiply this by the number of ways of arranging John and Jane within their group and arranging Mike and Molly within their group. Thus the number of possible arrangements is $P(8, 8) \cdot P(2, 2) \cdot P(2, 2) = 8! \cdot 2! \cdot 2! = 161{,}280$.

69. (a)   To find the number of ways the men and women can be seated we first select and place a man in the first seat and then arrange the other 7 people. Thus we get
$$\binom{\text{select 1 of}}{\text{the 4 men}} \cdot \binom{\text{arrange the}}{\text{remaining 7 people}} = C(4, 1) \cdot P(7, 7) = 4 \cdot 7! = 20{,}160.$$

     (b)   To find the number of ways the men and women can be seated we first select and place a woman in the first and last seats and then arrange the other 6 people. Thus we get
$$\binom{\text{arrange 2 of}}{\text{the 4 women}} \cdot \binom{\text{arrange the}}{\text{remaining 6 people}} = P(4, 2) \cdot P(6, 6) = 6 \cdot 6! = 8{,}640.$$

71. The number of ways the top finalist can be chosen is
$$\binom{\text{number of ways to choose}}{\text{the 6 semifinalists from the 30}} \cdot \binom{\text{number of ways to choose}}{\text{the 2 finalist from the 6}} \cdot \binom{\text{number of ways to pick}}{\text{the top finalist from the 2}}$$
$$= C(30, 6) \cdot C(6, 2) \cdot C(2, 1) = (593775)(15)(2) = 17{,}813{,}250.$$

73. We find the number of committees that can be formed and subtract those committees that contain the two picky people. There are $C(10, 4)$ ways to select the committees without restrictions. The number of committees that contain the two picky people is $C(8, 2)$ because once we place these two people on the committee, then we need only choose two more members from the remaining eight candidates to complete the committee. Thus the number of committees that do not contain these two people is $C(10, 4) - C(8, 2) = 210 - 28 = 182$.

75. We show two methods to solve this exercise.

    Method 1: We consider the number of 5 member committees that can be formed from the 26 interested people and subtract off those that contain no teacher and those that contain no students. Thus the number of committees is

    $C(26, 5) - C(12, 5) - C(14, 5) = 65{,}780 - 792 - 2002 = 62{,}986$.

    Method 2: In the method we construct all the committees that are possible. Thus the number of

    committees is $\begin{pmatrix} \text{committees} \\ \text{with 1 student} \\ \text{and 4 teachers} \end{pmatrix} + \begin{pmatrix} \text{committees} \\ \text{with 2 students} \\ \text{and 3 teachers} \end{pmatrix} + \begin{pmatrix} \text{committees} \\ \text{with 3 students} \\ \text{and 2 teachers} \end{pmatrix} + \begin{pmatrix} \text{committees} \\ \text{with 4 students} \\ \text{and 1 teacher} \end{pmatrix}$

    $= C(14, 1) \cdot C(12, 4) + C(14, 2) \cdot C(12, 3) + C(14, 3) \cdot C(12, 2) + C(14, 4) \cdot C(12, 1)$

    $= (14)(495) + (91)(220) + (364)(66) + (1001)(12) = 6{,}930 + 20{,}020 + 24{,}024 + 12{,}012$

    $= 62{,}986$.

77. We are only interested in selecting a set of three marbles to give to Luke and a set of two marbles to give to Mark, not the order in which we hand out the marbles. Since both $C(10, 3) \cdot C(7, 2)$ and $C(10, 2) \cdot C(8, 3)$ count the number of ways this can be done, these numbers must be equal. (Calculating these values shows that they are indeed equal.) In general, if we wish to find two distinct sets of k and r objects selected from n objects ($k + r \leq n$), then we can either first select the k objects from the n objects and then select the r objects from the $n - k$ remaining objects, or we can first select the r objects from the n objects and then the k objects from the $n - r$ remaining objects. Thus $\begin{pmatrix} n \\ r \end{pmatrix} \cdot \begin{pmatrix} n - r \\ k \end{pmatrix} = \begin{pmatrix} n \\ k \end{pmatrix} \cdot \begin{pmatrix} n - k \\ r \end{pmatrix}$.

Exercises 10.3

1. Let H stand for head and T for tails.

   (a) The sample space is $S = \{HH, HT, TH, TT\}$.

   (b) Let E be the event of getting exactly two heads, so $E = \{HH\}$. Then $P(E) = \dfrac{n(E)}{n(S)} = \frac{1}{4}$.

   (c) Let F be the event of getting at least one head. Then $F = \{HH, HT, TH\}$, and
   $$P(F) = \frac{n(F)}{n(S)} = \tfrac{3}{4}.$$

   (d) Let G be the event of getting exactly one head, that is, $G = \{HT, TH\}$. Then
   $$P(G) = \frac{n(G)}{n(S)} = \tfrac{2}{4} = \tfrac{1}{2}.$$

3. (a) Let E be the event of rolling a six. Then $P(E) = \dfrac{n(E)}{n(S)} = \frac{1}{6}$.

   (b) Let F be the event of rolling an even number. Then $F = \{2, 4, 6\}$. So
   $$P(F) = \frac{n(F)}{n(S)} = \tfrac{3}{6} = \tfrac{1}{2}.$$

   (c) Let G be the event of rolling a number greater than 5. Since 6 is the only face greater than 5,
   $$P(G) = \frac{n(G)}{n(S)} = \tfrac{1}{6}.$$

5. (a) Let E be the event of choosing a king. Since a deck has 4 kings, $P(E) = \dfrac{n(E)}{n(S)} = \tfrac{4}{52} = \tfrac{1}{13}$.

   (b) Let F be the event of choosing a face card. Since there are 3 face cards per suit and 4 suits,
   $$P(F) = \frac{n(F)}{n(S)} = \tfrac{12}{52} = \tfrac{3}{13}.$$

   (c) Let F be the event of choosing a face card. Then $P(F') = 1 - P(F) = 1 - \tfrac{3}{13} = \tfrac{10}{13}$.

7. (a) Let E be the event of selecting a red ball. Since the jar contains 5 red balls,
   $$P(E) = \frac{n(E)}{n(S)} = \tfrac{5}{8}.$$

   (b) Let F be the event of selecting a yellow ball. Since there is only one yellow ball,
   $$P(F') = 1 - P(F) = 1 - \frac{n(F)}{n(S)} = 1 - \tfrac{1}{8} = \tfrac{7}{8}.$$

   (c) Let G be the event of selecting a black ball. Since there are no black balls in the jar,
   $$P(G) = \frac{n(G)}{n(S)} = \tfrac{0}{8} = 0.$$

9. (a) Let E be the event of drawing a red sock. Since 3 pairs are red, the drawer contains 6 red socks, and so $P(E) = \dfrac{n(E)}{n(S)} = \tfrac{6}{18} = \tfrac{1}{3}$.

(b) Let F be the event of drawing another red sock. Since there are 17 socks left of which 5 are
     red, $P(F) = \dfrac{n(F)}{n(S)} = \frac{5}{17}$.

11. (a) Let E be the event of choosing a "T". Since 3 of the 16 letters are T's, $P(E) = \frac{3}{16}$.

     (b) Let F be the event of choosing a vowel. Since there are 6 vowels, $P(F) = \frac{6}{16} = \frac{3}{8}$.

     (c) Let F be the event of choosing a vowel. Then $P(F') = 1 - \frac{3}{8} = \frac{5}{8}$.

13. Let E be the event of choosing 5 cards of the same suit. Since there are 4 suits and 13 cards in each
     suit, $n(E) = 4 \cdot C(13,5)$. Also by Exercise 43, Section 10.2, $n(S) = C(52,5)$. Therefore,
     $$P(E) = \frac{4 \cdot C(13,5)}{C(52,5)} = \frac{5{,}148}{2{,}598{,}960} \approx 0.00198.$$

15. Let E be the event of dealing a royal flush (ace, king, queen, jack, and 10 of the same suit). Since
     there is only one such sequence for each suit, there are only 4 royal flushes, so
     $$P(E) = \frac{4}{C(52,5)} = \frac{4}{2{,}598{,}960} \approx 1.53908 \times 10^{-6}.$$

17. (a) Let B stand for "boy" and G stand for "girl". Then $S = \{$BBBB, GBBB, BGBB, BBGB,
          BBBG, GGBB, GBGB, GBBG, BGGB, BGBG, BBGG, BGGG, GBGG, GGBG, GGGB,
          GGGG$\}$.

     (b) Let E be the event that the couple has only boys. Then $E = \{$BBBB$\}$ and $P(E) = \frac{1}{16}$.

     (c) Let F be the event that the couple has 2 boys and 2 girls. Then $F = \{$GGBB, GBGB, GBBG,
          BGGB, BGBG, BBGG$\}$, so $P(F) = \frac{6}{16} = \frac{3}{8}$.

     (d) Let G be the event that the couple has 4 children of the same sex. Then $G = \{$BBBB, GGGG$\}$,
          and $P(G) = \frac{2}{16} = \frac{1}{8}$.

     (e) Let H be the event that the couple has at least 2 girls. Then H' is the event that the couple has
          fewer than two girls. Thus, $H' = \{$BBBB, GBBB, BGBB, BBGB, BBBG$\}$, so $n(H') = 5$,
          and $P(H) = 1 - P(H') = 1 - \frac{5}{16} = \frac{11}{16}$.

19. Let E be the event that the ball lands in an odd numbered slot. Since there are 18 odd numbers
     between 1 and 36, $P(E) = \frac{18}{38} = \frac{9}{19}$.

21. Let E be the event of picking the 6 winning numbers. Since there is only one way to pick these,
     $$P(E) = \frac{1}{C(49,6)} = \frac{1}{13{,}983{,}816} \approx 7.15 \times 10^{-8}.$$

23. The sample space consist of all possible True-False combinations, so $n(S) = 2^{10}$.

     (a) Let E be the event that the student answers all 10 questions correctly. Since there is only one
          way to answer all 10 questions correctly, $P(E) = \frac{1}{2^{10}} = \frac{1}{1024}$.

     (b) Let F be the event that the student answers exactly 7 questions correctly. The number of
          ways to answer exactly 7 of the 10 questions correctly is the number of ways to choose 7 of the
          10 questions, so $n(F) = C(10,7)$. Therefore, $P(E) = \dfrac{C(10,7)}{2^{10}} = \frac{120}{1024} = \frac{15}{128}$.

25. (a) Let E be the event that the monkey types "Hamlet" as his first word. Since "Hamlet" contains 6 letters and there are 48 typewriter keys, $P(E) = \frac{1}{48^6} \approx 8.18 \times 10^{-11}$.

(b) Let F be the event that the monkey types "to be or not to be" as his first words. Since this phrase has 18 characters (including the blanks), $P(F) = \frac{1}{48^{18}} \approx 5.47 \times 10^{-31}$.

27. Let E be the event that the monkey will arrange the 11 blocks to spell "PROBABILITY" as his first word. The number of ways of arranging these blocks is the number of distinguishable permutations of 11 blocks. Since there are two blocks labeled `B' and two blocks labeled `I', the number of distinguishable permutations is $\frac{11!}{2!\,2!}$. Only one of these arrangements spells the word "PROBABILITY". Thus $P(E) = \dfrac{1}{\frac{11!}{2!\,2!}} = \frac{2!\,2!}{11!} \approx 1.00 \times 10^{-7}$.

29. (a) Let E be the event that the pea is tall. Since tall is dominant, $E = \{\text{TT, Tt, tT}\}$. So $P(E) = \frac{3}{4}$.

(b) E' is the event that the pea is short. So $P(E') = 1 - P(E) = 1 - \frac{3}{4} = \frac{1}{4}$.

31. (a) YES, the events are mutually exclusive since a person cannot be both male and female.

(b) NO, the events are not mutually exclusive since a person can be both tall and blond.

33. (a) YES, the events are mutually exclusive since the number cannot be both even and odd. So $P(E \cup F) = P(E) + P(F) = \frac{3}{6} + \frac{3}{6} = 1$.

(b) NO, the events are not mutually exclusive since 6 is both even and greater than 4. So $P(E \cup F) = P(E) + P(F) - P(E \cap F) = \frac{3}{6} + \frac{2}{6} - \frac{1}{6} = \frac{2}{3}$.

35. (a) NO, the events E and F are not mutually exclusive since the Jack, Queen, and King of spades are both face cards and spades. So $P(E \cup F) = P(E) + P(F) - P(E \cap F) = \frac{13}{52} + \frac{12}{52} - \frac{3}{52} = \frac{11}{26}$.

(b) YES, the events E and F are mutually exclusive since the card cannot be both a heart and a spade. So $P(E \cup F) = P(E) + P(F) = \frac{13}{52} + \frac{13}{52} = \frac{1}{2}$.

37. (a) Let E be the event that the spinner stops on red. Since 12 of the regions are red, $P(E) = \frac{12}{16} = \frac{3}{4}$.

(b) Let F be the event that the spinner stops on an even number. Since 8 of the regions are even-numbered, $P(F) = \frac{8}{16} = \frac{1}{2}$.

(c) Since 4 of the even-numbered regions are red, $P(E \cup F) = P(E) + P(F) - P(E \cap F)$ $= \frac{3}{4} + \frac{1}{2} - \frac{4}{16} = 1$.

39. Let E be the event that the ball lands in an odd numbered slot and F be the event that it lands in a slot with a number higher than 31. Since there are two odd-numbered slots with numbers greater than 31, $P(E \cup F) = P(E) + P(F) - P(E \cap F) = \frac{18}{38} + \frac{5}{38} - \frac{2}{38} = \frac{21}{38}$.

41. Let E be the event that the committee is all male and F the event it is all female. The sample space is the set of all ways that 5 people can be chosen from the group of 14. These events are mutually

exclusive, so $P(E \cup F) = P(E) + P(F) = \dfrac{C(6,5)}{C(14,5)} + \dfrac{C(8,5)}{C(14,5)} = \frac{6+56}{2002} = \frac{31}{1001}$.

43. Let E be the event that the marble is red and F be the event that the number is odd-numbered. Then E' is the event that the marble is blue, and F' is the event that the marble is even-numbered.

(a) $P(E) = \frac{6}{16} = \frac{3}{8}$

(b) $P(F) = \frac{8}{16} = \frac{1}{2}$

(c) $P(E \cup F) = P(E) + P(F) - P(E \cap F) = \frac{6}{16} + \frac{8}{16} - \frac{3}{16} = \frac{11}{16}$

(d) $P(E' \cup F') = P(E') + P(F') - P(E' \cap F') = \frac{10}{16} + \frac{8}{16} - \frac{5}{16} = \frac{13}{16}$.

45. (a) YES, the first roll does not influence the outcome of the second roll.

(b) The probability of getting a six on both rolls is $P(E \cap F) = P(E) \cdot P(F) = \left(\frac{1}{6}\right)\left(\frac{1}{6}\right) = \frac{1}{36}$.

47. (a) Let E_A and E_B be the event that the respective spinners stop on a purple region. Since these events are independent, $P(E_A \cap E_B) = P(E_A) \cdot P(E_B) = \frac{1}{4} \cdot \frac{2}{8} = \frac{1}{16}$.

(b) Let F_A and F_B be the event that the respective spinners stop on a blue region. Since these events are independent, $P(F_A \cap F_B) = P(F_A) \cdot P(F_B) = \frac{1}{4} \cdot \frac{1}{8} = \frac{1}{32}$.

49. Let E be the event of getting a 1 on the first roll, and let F be the event of getting an even number on the second roll. Since these events are independent, $P(E \cap F) = P(E) \cdot P(F) = \frac{1}{6} \cdot \frac{3}{6} = \frac{1}{12}$.

51. Let E be the event that the player wins on spin 1, and let F be the event that the player wins on spin 2. What happens on the first spin does not influence what happens on the second spin, so the events are independent. Thus, $P(E \cap F) = P(E) \cdot P(F) = \frac{1}{38} \cdot \frac{1}{38} = \frac{1}{1444}$.

53. Let E, F and G denote the events of rolling two ones on the first, second, and third rolls, respectively, of a pair of dice. The events are independent, so
$P(E \cap F \cap G) = P(E) \cdot P(F) \cdot P(G) = \frac{1}{36} \cdot \frac{1}{36} \cdot \frac{1}{36} = \frac{1}{36^3} \approx 2.14 \times 10^{-5}$.

55. The probability of getting 2 red balls by picking from jar B is $\left(\frac{5}{7}\right)\left(\frac{4}{6}\right) = \frac{10}{21}$. The probability of getting 2 red balls by picking one ball from each jar is $\left(\frac{3}{7}\right)\left(\frac{5}{7}\right) = \frac{15}{49}$. The probability of getting 2 red balls after putting all balls in one jar is $\left(\frac{8}{14}\right)\left(\frac{7}{13}\right) = \frac{4}{13}$. Hence, picking both balls from jar B gives the greatest probability.

57. Let E be the event that two of the students have the same birthday. It is easier to consider the complementary event E': no two students have the same birthday. Then
$$P(E') = \frac{\text{number of ways to assign 8 different birthdays}}{\text{number of ways to assign 8 birthdays}} = \frac{P(365,8)}{365^8}$$
$$= \frac{365 \cdot 364 \cdot 363 \cdot 362 \cdot 361 \cdot 360 \cdot 359 \cdot 358}{365 \cdot 365 \cdot 365 \cdot 365 \cdot 365 \cdot 365 \cdot 365 \cdot 365} \approx 0.92566. \text{ So } P(E) = 1 - P(E') \approx 0.07434.$$

59. Let E be the event that she opens the lock within an hour. The number of combinations she can try in one hour is $10 \cdot 60 = 600$. The number of possible combinations is $P(40,3)$ assuming that no number can be repeated. Thus $P(E) = \dfrac{600}{P(40,3)} = \dfrac{600}{59,280} = \dfrac{5}{494} \approx 0.010$.

61. Let E be the event that Paul stands next to Phyllis. To find $n(E)$ we treat Paul and Phyllis as one object and find the number of ways to arrange the 19 objects and then multiply the result by the number of ways to arrange Paul and Phyllis. So $n(E) = 19! \cdot 2!$. The sample space is all the ways that 20 people can be arranged. Thus $P(E) = \dfrac{19! \cdot 2!}{20!} = \dfrac{2}{20} = 0.10$.

63. The sample space for the genders of Mrs. Smith's children is $S_S = \{(b, g), (b, b)\}$, where we list the first born first. The sample space for the genders of Mrs. Jones' children is $S_J = \{(b, g), (g, b), (b, b)\}$. Thus, the probability that Mrs. Smith's other child is a boy is $\frac{1}{2}$, while the probability that Mrs. Jones' other child is a boy is $\frac{1}{3}$. The reason there is a difference is that we *know* that Mrs. Smith's first born is a boy.

Exercises 10.4

1. $P(2 \text{ successes in } 5) = C(5,2) \cdot (0.7^2)(0.3^3) = 0.13230.$

3. $P(0 \text{ successes in } 5) = C(5,0) \cdot (0.7^0)(0.3^5) = 0.00243.$

5. $P(1 \text{ successes in } 5) = C(5,1) \cdot (0.7^1)(0.3^4) = 0.02835.$

7. $P(\text{at least 4 successes}) = P(4 \text{ successes}) + P(5 \text{ successes}) = 0.36015 + 0.16807 = 0.52822.$

9. $P(\text{at most 1 failure}) = P(0 \text{ failure}) + P(1 \text{ failure}) = P(5 \text{ successes}) + P(4 \text{ successes}) = 0.52822.$

11. $P(\text{at least 2 successes}) = P(2 \text{ successes}) + P(3 \text{ successes}) + P(4 \text{ successes}) + P(5 \text{ successes})$
    $= 0.13230 + 0.30870 + 0.36015 + 0.16807 = 0.96922.$

13.   Here "success" is "face is 4" and $P(\text{face is } 4) = \frac{1}{6}$. Then
    $P(2 \text{ successes in } 6) = C(6,2) \cdot \left(\frac{1}{6}\right)^2 \left(\frac{5}{6}\right)^4 = 0.20094.$

15. $P(4 \text{ successes in } 10) = C(10,4) \cdot (0.4^4)(0.6^6) \approx 0.25082.$

17.   (a) $P(5 \text{ in } 10) = C(10,5) \cdot (0.45^5)(0.55^5) \approx 0.23403.$

     (b) $P(\text{at least } 3) = 1 - P(\text{at most } 2) = P(0 \text{ in } 10) + P(1 \text{ in } 10) + P(2 \text{ in } 10)$
        $= 1 - [C(10,0) \cdot (0.45^0)(0.55^{10}) + C(10,1) \cdot (0.45^1)(0.55^9) + C(10,2) \cdot (0.45^2)(0.55^8)]$
        $\approx 0.90044.$

19.   (a)   The complement *at least 1 germinates* is *no seeds germinate*, so
        $P(\text{at least 1 germinates}) = 1 - P(0 \text{ germinates}) = 1 - C(4,0) \cdot (0.75^0)(0.25^4) \approx 0.99609.$

     (b) $P(\text{at least 2 germinates}) = P(2 \text{ germinates}) + P(3 \text{ germinates}) + P(4 \text{ germinates})$
        $= C(4,2) \cdot (0.75^2)(0.25^2) + C(4,3) \cdot (0.75^3)(0.25^1) + C(4,4) \cdot (0.75^4)(0.25^0) \approx 0.94922.$

     (c) $P(4 \text{ germinates}) = C(4,4) \cdot (0.75^4)(0.25^0) \approx 0.31641.$

21.   (a) $P(\text{all 10 are boys}) = C(10,10) \cdot (0.52^{10})(0.48^0) \approx 0.0014456.$

     (b) $P(\text{all 10 are girls}) = C(10,0) \cdot (0.52^0)(0.48^{10}) \approx 0.00064925.$

     (c) $P(5 \text{ in 10 are boys}) = C(10,5) \cdot (0.52^5)(0.48^5) \approx 0.24413.$

23.   (a) $P(3 \text{ in } 3) = C(3,3) \cdot (0.005^3)(0.995^0) \approx 0.000000125.$

     (b)   The complement of "one or more bulbs is defective" is "none of the bulbs are defective." So
        $P(\text{at least 1 defective}) = 1 - P(0 \text{ in 3 are defective}) = 1 - C(3,0) \cdot (0.005^0)(0.995^3)$
        $\approx 0.014925.$

25.   The complement of "2 or more workers call in sick" is "0 or 1 workers call in sick." So
    $P(2 \text{ or more}) = 1 - [P(0 \text{ in } 8) + P(1 \text{ in } 8)]$
    $= 1 - [C(8,0) \cdot (0.04^0)(0.96^8) + C(8,1) \cdot (0.04^1)(0.96^7)] \approx 0.038147.$

27.   (a) $P(6 \text{ in } 6) = C(6,6) \cdot (0.75^6)(0.25^0) \approx 0.17798.$

(b) $P(0 \text{ in } 6) = C(6, 0) \cdot (0.75^0)(0.25^6) \approx 0.00024414.$

(c) $P(3 \text{ in } 6) = C(6, 3) \cdot (0.75^3)(0.25^3) \approx 0.13184.$

(d) $P(\text{at least } 2) = 1 - P(\text{at most } 1) = 1 - [P(0 \text{ in } 6) + P(1 \text{ in } 6)]$
$= 1 - [C(6, 0) \cdot (0.75^0)(0.25^6) + C(6, 1) \cdot (0.75^1)(0.25^5)] \approx 0.99536.$

29. (a) The complement of "at least one gets the disease" is "none gets the disease." Then
$P(\text{at least 1 gets the disease}) = 1 - P(0 \text{ gets the disease}) = 1 - C(4, 0) \cdot (0.25^0)(0.75^4)$
$\approx 0.68359.$

(b) $P(\text{at least 3 gets the disease}) = P(3 \text{ gets the disease}) + P(4 \text{ gets the disease})$
$= C(4, 3) \cdot (0.25^3)(0.75^1) + C(4, 4) \cdot (0.25^4)(0.75^0) \approx 0.05078.$

31. Fred, a nonsmoker is already in the room, so this exercise concerns the remaining 4 participants assigned to the room.

(a) $P(1 \text{ in } 4 \text{ is a smoker}) = C(4, 1) \cdot (0.3^1)(0.7^3) = 0.4116.$

(b) $P(\text{at least 1 smoker}) = 1 - P(0 \text{ in } 4 \text{ are smokers}) = 1 - C(4, 0) \cdot (0.3^0)(0.7^4) = 0.7599.$

33. (a)

Number of heads	Probability
0	0.001953
1	0.017578
2	0.070313
3	0.164063
4	0.246094
5	0.246094
6	0.164063
7	0.070313
8	0.017578
9	0.001953

(b) Between 4 and 5 heads.

(c) If the coin is flipped 101 times, then 50 and 51 heads has the greatest probability of occurring. If the coin is flipped 100 times, then 50 heads has the greatest probability of occurring.

Exercises 10.5

1. Mike gets \$2 with probability $\frac{1}{2}$ and \$1 with probability $\frac{1}{2}$. Thus, $E = (2)\left(\frac{1}{2}\right) + (1)\left(\frac{1}{2}\right) = 1.5$, and so his expected winnings are \$1.50 per game.

3. Since the probability of drawing the ace of spades is $\frac{1}{52}$, the expected value of this game is $E = (100)\left(\frac{1}{52}\right) + (-1)\left(\frac{51}{52}\right) = \frac{49}{52} \approx 0.94$. So your expected winnings are \$0.94 per game.

5. Since the probability that Carol rolls a six is $\frac{1}{6}$, the expected value of this game is $E = (3)\left(\frac{1}{6}\right) + (0.50)\left(\frac{5}{6}\right) = \frac{5.5}{6} \approx 0.9167$. So Carol expects to win \$0.92 per game.

7. Since the probability that the die shows an even number equals the probability that that die shows an odd number, the expected value of this game is $E = (2)\left(\frac{1}{2}\right) + (-2)\left(\frac{1}{2}\right) = 0$. So Tom should expect to break even after playing this game many times.

9. Since it cost 50¢ to play, if you get a silver dollar, you only win $1 - 0.50 = \$.50$. Thus the expected value of this game is $E = (0.50)\left(\frac{2}{10}\right) + (-0.50)\left(\frac{8}{10}\right) = -0.30$. So your expected winnings are $-\$0.30$ per game. In other words, you should expect to lose \$0.30 per game.

11. You can either win \$35 or lose \$1, so the expected value of this game is $E = (35)\left(\frac{1}{38}\right) + (-1)\left(\frac{37}{38}\right) = -\frac{2}{38} = -0.0526$. Thus the expected value is $-\$0.0526$ per game.

13. By the rules of the game, a player can win \$10, \$5, \$0 or lose \$100. Thus the expected value of this game is $E = (10)\left(\frac{10}{100}\right) + (5)\left(\frac{10}{100}\right) + (-100)\left(\frac{2}{100}\right) + (0)\left(\frac{78}{100}\right) = -0.50$. So the expected winnings per game are $-\$0.50$.

15. If the stock goes up to \$20, she expects to make $\$20 - \$5 = \$15$. And if the stock falls to \$1, then she has lost $\$5 - \$1 = \$4$. So the expected value of her profit is $E = (15)(0.1) - (4)(0.9) = -2.1$. Thus, her expected profit per share is $-\$2.10$, that is, she should expect to lose \$2.10 per share. She did not make a wise investment.

17. There are $C(49, 6)$ ways to select a group of six numbers from the group of 49 numbers, of which only one is a winning set. Thus the expected value of this game is
    $$E = (10^6 - 1)\left(\frac{1}{C(49, 6)}\right) + (-1)\left(1 - \frac{1}{C(49, 6)}\right) \approx -\$0.93.$$

19. Let x be the fair price to pay to play this game. Then the game is fair whenever $E = 0 \quad \Leftrightarrow \quad (13 - x)\left(\frac{4}{52}\right) + (-x)\left(\frac{48}{52}\right) = 0 \quad \Leftrightarrow \quad 52 - 52x = 0 \quad \Leftrightarrow \quad x = 1$. Thus, a fair price to pay to play this game is \$1.

Review Exercises for Chapter 10

1. The number of possible outcomes is $\left(\begin{smallmatrix}\text{number of outcomes}\\\text{when a coin is tossed}\end{smallmatrix}\right) \cdot \left(\begin{smallmatrix}\text{number of outcomes}\\\text{a die is rolled}\end{smallmatrix}\right) \cdot \left(\begin{smallmatrix}\text{number of ways}\\\text{to draw a card}\end{smallmatrix}\right)$
    $= (2)(6)(52) = 624.$

3. (a) Order is not important, and there are no repetitions, so the number of different two-element
        subsets is $C(5,2) = \dfrac{5!}{2!\,3!} = \dfrac{5 \cdot 4}{2} = 10.$

    (b) Order is important, and there are no repetitions, so the number of different two-letter words is
        $P(5,2) = \frac{5!}{3!} = 20.$

5. You earn a score of 70% by answering exactly 7 of the 10 questions correctly. The number of
    different ways to answer the questions correctly is $C(10,7) = \dfrac{10!}{7!\,3!} = 120.$

7. You must choose two of the ten questions to omit, and the number of ways of choosing these two
    questions is $C(10,2) = \dfrac{10!}{2!\,8!} = 45.$

9. The maximum number of employees using this security system is
    $\left(\begin{smallmatrix}\text{number of choices}\\\text{for the first letter}\end{smallmatrix}\right) \cdot \left(\begin{smallmatrix}\text{number of choices}\\\text{for the second letter}\end{smallmatrix}\right) \cdot \left(\begin{smallmatrix}\text{number of choices}\\\text{for the third letter}\end{smallmatrix}\right) = (26)(26)(26) = 17{,}576.$

11. We could count the number of ways of choosing 7 of the flips to be HEADS; equivalently we could
    count the number of ways of choosing 3 of the flips to be TAILS. Thus, the number of different
    ways this can occur is $C(10,7) = C(10,3) = \dfrac{10!}{3!\,7!} = 120.$

13. Let x be the number of people in the group. Then $C(x,2) = 10 \quad\Leftrightarrow\quad \dfrac{x!}{2!(x-2)!} = 10 \quad\Leftrightarrow$

    $\dfrac{x!}{(x-2)!} = 20 \quad\Leftrightarrow\quad x(x-1) = 20 \quad\Leftrightarrow\quad x^2 - x - 20 = 0 \quad\Leftrightarrow\quad (x-5)(x+4) = 0 \quad\Leftrightarrow$
    $x = 5$ or $x = -4$. So there are 5 people in this group.

15. A letter can be represented by a sequence of length 1, a sequence of length 2, or a sequence of length
    3. Since each symbol is either a dot or a dash, the possible number of letters is
    $\left(\begin{smallmatrix}\text{number of letters}\\\text{using 3 symbols}\end{smallmatrix}\right) + \left(\begin{smallmatrix}\text{number of letters}\\\text{using 2 symbols}\end{smallmatrix}\right) + \left(\begin{smallmatrix}\text{number of letters}\\\text{using 1 symbol}\end{smallmatrix}\right) = 2^3 + 2^2 + 2 = 14.$

17. (a) Since we cannot choose a major and a minor in the same subject, the number of ways a student
        can select a major and a minor is $P(16,2) = 16 \cdot 15 = 240.$

    (b) Again, since we cannot have repetitions and the order of selection is important, the number of
        ways to select a major, a first minor, and a second minor is $P(16,3) = 16 \cdot 15 \cdot 14 = 3360.$

    (c) When we select a major and 2 minors, the order in which we choose the minors is not
        important. Thus the number of ways to select a major and 2 minors is
        $\left(\begin{smallmatrix}\text{number of ways}\\\text{to select a major}\end{smallmatrix}\right) \cdot \left(\begin{smallmatrix}\text{number of ways to}\\\text{select two minors}\end{smallmatrix}\right) = 16 \cdot C(15,2) = 16 \cdot 105 = 1680.$

19.   Since the letters are distinct, the number of anagrams of the word TRIANGLE is $8! = 40{,}320$.

21.   (a)   The possible number of committees is $C(18, 7) = 31{,}824$.

(b)   Since we must select the 4 men from the group 10 men and the 3 women from the group of 8 women, the possible number of committees is $\left(\begin{smallmatrix}\text{number of ways to}\\\text{choose 4 of 10 men}\end{smallmatrix}\right) \cdot \left(\begin{smallmatrix}\text{number of ways to}\\\text{choose 3 of 8 women}\end{smallmatrix}\right)$
$= C(10, 4) \cdot C(8, 3) = 210 \cdot 56 = 11{,}760$

(c)   We remove Susie from the group of 18, so the possible number of committees is $C(17, 7) = 19{,}448$.

(d)   The possible number of committees is
$\left(\begin{smallmatrix}\text{possible number of}\\\text{committees with 5 women}\end{smallmatrix}\right) + \left(\begin{smallmatrix}\text{possible number of}\\\text{committees with 6 women}\end{smallmatrix}\right) + \left(\begin{smallmatrix}\text{possible number of}\\\text{committees with 7 women}\end{smallmatrix}\right)$
$= C(8, 5) \cdot C(10, 2) + C(8, 6) \cdot C(10, 1) + C(8, 7) \cdot C(10, 0) = 56 \cdot 45 + 28 \cdot 10 + 8 \cdot 1$
$= 2808$.

(e)   Since the committee is to have 7 members, "at most two men" is the same as "at least five women" which we found in part (d). So the number is also 2808.

(f)   We select the specific offices first, then complete the committee from the remaining members of the group. So the number of possible committees is
$\left(\begin{smallmatrix}\text{number of ways to choose}\\\text{a chairman, a vice-chairman, a secretary}\end{smallmatrix}\right) \cdot \left(\begin{smallmatrix}\text{number of ways to}\\\text{choose 4 other members}\end{smallmatrix}\right) = P(18, 3) \cdot C(15, 4)$
$= 4896 \cdot 1365 = 6{,}683{,}040$.

23.   Let R_n denote the event that the nth ball is red and let W_n denote the event that the nth ball is white.

(a) $P(\text{both balls are red}) = P(R_1 \cap R_2) = P(R_1) \cdot P(R_2 \mid R_1) = \frac{10}{15} \cdot \frac{9}{14} = \frac{3}{7}$.

(b)   <u>Solution 1</u>: The probability that one is white and that the other is red is
$$\frac{\text{number of ways to select one white \& one red}}{\text{number of ways to select two balls}} = \frac{C(10, 1) \cdot C(5, 1)}{C(15, 2)} = \frac{10}{21}.$$
<u>Solution 2</u>: $P(\text{one white and one red}) = P(W_1 \cap R_2) + P(R_1 \cap W_2)$
$= P(W_1) \cdot P(R_2 \mid W_1) + P(R_1) \cdot P(W_2 \mid R_1) = \frac{5}{15} \cdot \frac{10}{14} + \frac{10}{15} \cdot \frac{5}{14} = \frac{10}{21}$.

(c)   <u>Solution 1</u>: Let E be the event "at least one is red". Then E' is the event "both are white".
$P(E') = P(W_1 \cap W_2) = P(W_1) \cdot P(W_2 \mid W_1) = \frac{5}{15} \cdot \frac{4}{14} = \frac{2}{21}$. Thus $P(E) = 1 - \frac{2}{21} = \frac{19}{21}$.
<u>Solution 2</u>: $P(\text{at least one is red}) = P(\text{one red and one white}) + P(\text{both red})$
$= \frac{10}{21} + \frac{9}{21} = \frac{19}{21}$ (from (a) and (b)).

(d)   Since 5 of the 15 balls are both red and even-numbered, the probability that both balls are red and even-numbered is $\frac{5}{15} \cdot \frac{4}{14} = \frac{2}{21}$.

(e)   Since 2 of the 15 balls are both white and odd-numbered, the probability that both are white is $\frac{2}{15} \cdot \frac{1}{14} = \frac{1}{105}$.

25.   The probability that you select a mathematics book is $\dfrac{\text{number of ways to select a mathematics book}}{\text{number of ways to select a book}} = \frac{4}{10}$
$= \frac{2}{5} = 0.4$.

27.   (a) $P(\text{ace}) = \frac{4}{52} = \frac{1}{13}$.

(b)   Let E be the event the card chosen is an ace, and let F be the event the card chosen is a jack.
Then $P(E \cup F) = P(E) + P(F) = \frac{4}{52} + \frac{4}{52} = \frac{2}{13}$.

(c)   Let E be the event the card chosen is an ace, and let F be the event the card chosen is a spade.
Then $P(E \cup F) = P(E) + P(F) - P(E \cap F) = \frac{4}{52} + \frac{13}{52} - \frac{1}{52} = \frac{4}{13}$.

(d)   Let E be the event the card chosen is an ace, and let F be the event the card chosen is a red
card. Then $P(E \cap F) = \dfrac{n(E \cap F)}{n(S)} = \frac{2}{52} = \frac{1}{26}$.

29.   (a)   The probability the first die shows some number is 1, and the probability the second die shows
the same number is $\frac{1}{6}$. So the probability each die shows the same number is $1 \cdot \frac{1}{6} = \frac{1}{6}$.

(b)   By part (a), the event of showing the same number has probability of $\frac{1}{6}$, and the complement of
this event is that the dice show different numbers. Thus the probability that the dice show
different numbers is $1 - \frac{1}{6} = \frac{5}{6}$.

31.   In the numbers game lottery, there are 1000 possible "winning" numbers.

(a)   The probability that John wins \$500 is $\frac{1}{1000}$.

(b)   There are $P(3,3) = 6$ ways to arrange the digits "1", "5", "9". However, if John wins only
\$50, it means that his number 159 was not the winning number. Thus the probability is
$\frac{5}{1000} = \frac{1}{200}$.

33.   There are 36 possible outcomes in rolling two dice and 6 ways in which both dice show the same
numbers, namely, $(1,1)$, $(2,2)$, $(3,3)$, $(4,4)$, $(5,5)$, and $(6,6)$. So the expected value of this game
is $E = (5)\left(\frac{6}{36}\right) + (-1)\left(\frac{30}{36}\right) = 0$.

35.   Since Mary makes a guess as to the order of ratification of the 13 original states, the number of such
guesses is $P(13,13) = 13!$, while the probability that she guesses the correct order is $\frac{1}{13!}$. Thus the
expected value is $E = (1{,}000{,}000)\left(\frac{1}{13!}\right) + (0)\left(\frac{13!-1}{13!}\right) = 0.00016$. So Mary's expected winnings are
\$0.00016.

37.   (a)   Since there are only two colors of socks, any 3 socks must contain a matching pair.

(b)   Method 1: If the two socks drawn form a matching pair then they are either both red or both
blue. So $P(\text{choosing a matching pair}) = P(\text{both red or both blue})$
$$= P(\text{both red}) + P(\text{both blue}) = \frac{C(20,2)}{C(50,2)} + \frac{C(30,2)}{C(50,2)} \approx 0.51.$$

Method 2: The complement of choosing a matching pair is choosing one sock of each color.
So $P(\text{choosing a matching pair}) = 1 - P(\text{different colors}) = 1 - \dfrac{C(20,1) \cdot C(30,1)}{C(50,2)}$
$\approx 1 - .49 = 0.51$.

39.   (a) $\begin{pmatrix} \text{number of different} \\ \text{zip codes} \end{pmatrix} = \begin{pmatrix} \text{number of ways to} \\ \text{choose the 1st digit} \end{pmatrix} \cdot \begin{pmatrix} \text{number of ways to} \\ \text{choose the 2nd digit} \end{pmatrix} \cdot \ \cdots \ \cdot \begin{pmatrix} \text{number of ways to} \\ \text{choose the 5th digit} \end{pmatrix}$
$= 10 \cdot 10 \cdot 10 \cdot 10 \cdot 10 = 10^5 = 100{,}000$.

(b) Since there are five numbers (0, 1, 6, 8 and 9) that can be read upside down, we have

$$\begin{pmatrix} \text{number of different} \\ \text{zip codes} \end{pmatrix} = \begin{pmatrix} \text{number of ways to} \\ \text{choose the 1st digit} \end{pmatrix} \cdot \begin{pmatrix} \text{number of ways to} \\ \text{choose the 2nd digit} \end{pmatrix} \cdot \ldots \cdot \begin{pmatrix} \text{number of ways to} \\ \text{choose the 5th digit} \end{pmatrix} = 5^5 = 3125.$$

(c) Let E be the event that a zip code can be read upside down. Then by parts (a) and (b),

$$P(E) = \frac{n(E)}{n(S)} = \frac{5^5}{10^5} = \frac{1}{32}.$$

(d) Suppose a zip code is turned upside down. Then the middle digit remains the middle digit, so it must be a digit that reads the same when turned upside down, that is, a 0, 1 or 8. Also, the last digit becomes the first digit, and the next to last digit becomes the second digit. Thus, once the first two digits are chosen, the last two are determined. Therefore, the number of zip codes that read the same upside down as right side up is

$$\begin{pmatrix} \text{number of ways to} \\ \text{choose the 1st digit} \end{pmatrix} \cdot \begin{pmatrix} \text{number of ways to} \\ \text{choose the 2nd digit} \end{pmatrix} \cdot \begin{pmatrix} \text{number of ways to} \\ \text{choose the 3rd digit} \end{pmatrix} \cdot \begin{pmatrix} \text{number of ways to} \\ \text{choose the 4th digit} \end{pmatrix} \cdot \begin{pmatrix} \text{number of ways to} \\ \text{choose the 5th digit} \end{pmatrix}$$
$$= 5 \cdot 5 \cdot 3 \cdot 1 \cdot 1 = 75.$$

41. (a) Using the rule for the number of distinguishable combinations, the number of divisors of N is $(7+1)(2+1)(5+1) = 144.$

(b) An even divisor of N must contain 2 as a factor. Thus we place a 2 as one of the factors and count the number of distinguishable combinations of $M = 2^6 3^2 5^5$. So using the rule for the number of distinguishable combinations, the number of even divisors of N is $(6+1)(2+1)(5+1) = 126.$

(c) A divisor is a multiple of 6 if 2 is a factor and 3 is a factor. Thus we place a 2 as one of the factors and a 3 as one of the factors. Then we count the number of distinguishable combinations of $K = 2^6 3^1 5^5$. So using the rule for the number of distinguishable combinations, the number of even divisors of N is $(6+1)(1+1)(5+1) = 84.$

(d) Let E be the event that the divisor is even. Then using parts (a) and (b), $P(E) = \dfrac{n(E)}{n(S)}$
$$= \frac{126}{144} = \frac{7}{8}.$$

43. (a) $P(4 \text{ sixes in 8 rolls}) = C(8,4) \cdot (\frac{1}{6})^4 (\frac{5}{6})^4 \approx 0.026048.$

(b) There are three even numbers on a die and three odds numbers, thus $P(\text{even}) = P(\text{odd}) = 0.5.$
Thus $P(2 \text{ or more evens in 8 rolls}) = 1 - P(\text{less than 2 even numbers rolled})$
$$= 1 - [P(0 \text{ evens in 8 rolls}) + P(1 \text{ even in 8 rolls})]$$
$$= 1 - [C(8,0) \cdot (0.5^0)(0.5^8) + C(8,1) \cdot (0.5^1)(0.5^7)] \approx 0.96484$$

Chapter 10 Test

1. (a) If repetition is allowed, then each letter of the word can be chosen in 10 ways since there are 10 letters. Thus the number of possible 5 letter words is $10 \cdot 10 \cdot 10 \cdot 10 \cdot 10 = 10^5 = 100{,}000$.

   (b) If repetition is not allowed, then since order is important, we need the number of permutations of 10 objects (the 10 letters) taken 5 at a time. Therefore, the number of possible 5 letter words is $P(10, 5) = 10 \cdot 9 \cdot 8 \cdot 7 \cdot 6 = 30{,}240$.

2. There are three choices to be made: one choice each of a main course, a dessert, and a drink. Since a main course can be chosen in one of five ways, a dessert in one of three ways, and a drink in one of four ways, there are $5 \cdot 3 \cdot 4 = 60$ possible ways that a customer could order a meal.

3. (a) Order is important in the arrangement, therefore the number of ways to arrange $P(30, 4) = 657{,}720$.

   (b) Here we are interested in the group of books to be taken on vacation so order is not important, therefore the number of ways to choose these books is $C(30, 4) = 27{,}405$.

4. There are two choices to be made: choose a road to travel from Ajax to Barrie, and then choose a different road from Barrie to Ajax. Since there are 4 roads joining the two cities, we need the number of permutations of 4 objects (the roads) taken 2 at a time (the road there and the road back). This number is $P(4, 2) = 4 \cdot 3 = 12$.

5. A customer must choose a size of pizza and must make a choice of toppings. There are 4 sizes of pizza, and each choice of toppings from the 14 available corresponds to a subset of the 14 objects. Since a set with 14 objects has 2^{14} subsets, the number of different pizzas this parlor offers is $4 \cdot 2^{14} = 65{,}536$.

6. (a) We want the number of ways of arranging 4 distinct objects (the letters L, O, V, E). This is the number of permutations of 4 objects taken 4 at a time. Therefore, the number of anagrams of the word LOVE is $P(4, 4) = 4! = 24$.

   (b) We want the number of distinguishable permutations of 6 objects (the letters K, I, S, S, E, S) consisting of three like groups of size 1 and a like group of size 3 (the S's). Therefore, the number of different anagrams of the word KISSES is $\dfrac{6!}{1!\,1!\,1!\,3!} = \dfrac{6!}{3!} = 120$.

7. We choose the officers first. Here order is important, because the officers are different. Thus there are $P(30, 3)$ ways to do this. Next we choose the other 5 members from the remaining 27 members. Here order is not important, so there are $C(27, 5)$ ways to do this. Therefore the number of ways that the board of directors can be chosen is
$P(30, 3) \cdot C(27, 5) = 30 \cdot 29 \cdot 28 \cdot \frac{27!}{5!\,22!} = 1{,}966{,}582{,}800$.

8. One card is drawn from a deck.

   (a) Since there are 26 red cards, the probability that the card is red is $\frac{26}{52} = \frac{1}{2}$.

   (b) Since there are 4 kings, the probability that the card is a king is $\frac{4}{52} = \frac{1}{13}$.

   (c) Since there are 2 red kings, the probability that the card is a red king is $\frac{2}{52} = \frac{1}{26}$.

9. Let R be the event that the ball chosen is red. Let E be the event that the ball chosen is even-numbered.

   (a) Since 5 of the 13 balls are red, $P(R) = \frac{5}{13} \approx 0.3846$.

   (b) Since 6 of the 13 balls are even-numbered, $P(E) = \frac{6}{13} \approx .4615$.

   (c) $P(R \text{ or } E) = P(R) + P(E) - P(R \cap E) = \frac{5}{13} + \frac{6}{13} - \frac{2}{13} = \frac{9}{13} \approx 0.6923$.

10. Let E be the event of choosing 3 men. Then
$$P(E) = \frac{n(E)}{n(S)} = \frac{\text{number of ways to choose 3 men}}{\text{number of ways to choose 3 people}} = \frac{C(5,3)}{C(15,3)} \approx 0.022.$$

11. Two dice are rolled. Let E be the event of getting doubles. Since a double may occur in 6 ways,
$$P(E) = \frac{n(E)}{n(S)} = \frac{6}{36} = \frac{1}{6} \ .$$

12. There are 4 students and 12 astrological signs. Let E be the event that at least 2 have the same astrological sign. Then E' is the event that no 2 have the same astrological sign. It is easier to find E'. So $P(E') = \dfrac{\text{number of ways to assign 4 different astrological signs}}{\text{number of ways to assign 4 astrological signs}} = \dfrac{P(12,4)}{12^4}$
$$= \frac{12 \cdot 11 \cdot 10 \cdot 9}{12 \cdot 12 \cdot 12 \cdot 12} = \frac{55}{96}. \text{ Therefore, } P(E) = 1 - P(E') = 1 - \frac{55}{96} = \frac{41}{96} \approx 0.427.$$

13. (a) $P(6 \text{ heads in 10 tosses}) = C(10,6) \cdot (0.55^6)(0.45^4) = 0.23837.$

    (b) Since " less than 3 heads" is the same as "0, 1, or 2 heads." So
    $P(\text{less than 3 heads}) = P(0 \text{ heads in 10}) + P(1 \text{ heads in 10}) + P(2 \text{ heads in 10})$
    $= C(10,0) \cdot (0.55^0)(0.45^{10}) + C(10,1) \cdot (0.55^1)(0.45^9) + C(10,2) \cdot (0.55^2)(0.45^8)$
    $= 0.02739.$

14. A deck of cards contains 4 aces, 12 face cards, and 36 other cards. So the probability of an ace is $\frac{4}{52} = \frac{1}{13}$, the probability of a face card is $\frac{12}{52} = \frac{3}{13}$, and the probability of a non-ace, non-face card is $\frac{36}{52} = \frac{9}{13}$. Thus the expected value of this game is $E = (10)\left(\frac{1}{13}\right) + (1)\left(\frac{3}{13}\right) + (-.5)\left(\frac{9}{13}\right) = \frac{8.5}{13}$ ≈ 0.654, that is, about $0.65.

Focus on Modeling

1. (a) You should find that with the switching strategy, you win about 90% of the time. The more games you play, the closer to 90% your winning ratio will be.

   (b) The probability that the contestant has selected the winning door to begin with is $\frac{1}{10}$, since there are ten doors and only one is a winner. So the probability that he has selected a losing door is $\frac{9}{10}$. If the contestant switches, he exchanges a losing door for a winning door (and vice versa), so the probability that he loses is now $\frac{1}{10}$, and the probability that he wins is now $\frac{9}{10}$.

3. (a) You should find that A wins about $\frac{7}{8}$ of the time. That is, if you play this game 80 times, A should win approximately 70 times.

   (b) The game will end when either A gets one more "head" or B gets three more "tails". Each toss is independent, and both "heads" and "tails" have probability $\frac{1}{2}$, so we obtain the following probabilities.

Outcome	Probability
H	$\frac{1}{2}$
TH	$\frac{1}{2} \cdot \frac{1}{2} = \frac{1}{4}$
TTH	$\frac{1}{2} \cdot \frac{1}{2} \cdot \frac{1}{2} = \frac{1}{8}$
TTT	$\frac{1}{2} \cdot \frac{1}{2} \cdot \frac{1}{2} = \frac{1}{8}$

   Since A wins for any outcome that ends in "heads", the probability that he wins is $\frac{1}{2} + \frac{1}{4} + \frac{1}{8} = \frac{7}{8}$.

5. With 1000 trials, you should obtain an estimate for π that is reasonably close to 3.1 or 3.2.

7. (a) We can use the following TI-83 program to model this experiment. It is a minor modification of the one given in Problem 5.

```
PROGRAM: PROB6
:0 → P
:For(N,1,1000)
:rand → X:rand → Y
:P+((X+Y)<1) → P
:End
:Disp "PROBABILITY IS APPROX"
:Disp P/1000
```

   You should find that the probability is very close to $\frac{1}{2}$.

   (b) Following the hint, the points in the square for which $x + y < 1$ are the ones that lie below the line $x + y = 1$. This triangle has area $\frac{1}{2}$ (it takes up half the square), so the probability that $x + y < 1$ is $\frac{1}{2}$.